D0936509

TAPEWORMS,
LICE, AND PRIONS

Tapeworms, Lice, and Prions

A compendium of unpleasant infections

DAVID I GROVE

Great Clarendon Street, Oxford, OX2 6DP,
United Kingdom

Oxford University Press is a department of the University of Oxford.
It furthers the University's objective of excellence in research, scholarship,
and education by publishing worldwide. Oxford is a registered trade mark of
Oxford University Press in the UK and in certain other countries

First Edition published in 2014
Impression: 1

Published in the United States of America by Oxford University Press
198 Madison Avenue, New York, NY 10016, United States of America

British Library Cataloguing in Publication Data
Data available

Library of Congress Control Number: 2013942199

ISBN 978–0–19–964102–4

Printed in Great Britain by
Clays Ltd, St Ives plc

CONTENTS

ACKNOWLEDGEMENTS

This book had its genesis in the prodding of my wife, Marilyn, who persuaded me to write it. She had to pay her penance by reading the manuscript and making comments and corrections. My grateful thanks are due to Margaret Wigglesworth who scoured every page looking for errors. That this book saw the light of day is due to the enthusiasm of Latha Menon, Senior Commissioning Editor at *Oxford University Press* who successfully put the case to the Delegates of the *Press*. To them, I am most grateful for agreeing to its publication. My thanks also to Emma Ma, Assistant Commissioning Editor, Jenny Rogers, Production Editor, and to all those unknown to me, who expertly put this book in the form which you see before you. Finally, this book could never have been written without the staff (who were forever helpful) and the facilities of the Barr Smith Library and its branches of the University of Adelaide.

> We need reminding, now more than ever, that the capacity of medicine to deal with infectious diseases was not a lucky fluke, nor was it something that happened simply as the result of the passage of time. It was the direct outcome of many years of hard work, done by imaginative and skilled scientists, none of whom had the faintest idea that penicillin and streptomycin lay somewhere in the decades ahead. It was basic science of a very high order, storing up a great mass of interesting knowledge for its own sake, creating, so to speak, a bank of information, ready for drawing on when the time for intelligent use arrived.

LEWIS THOMAS
The Medusa and the Snail:
More notes of a biology watcher, 1979

PART I

INFECTION

the search for its causes

Infectious diseases have been the major cause of sickness and death throughout recorded history, and no doubt in pre-history as well. We have all heard of the Black Death, tuberculosis, meningococcal disease, and swine flu. But what are infectious diseases? My trusty *Shorter Oxford English Dictionary* says they are diseases communicated by infections. But what is an infection? That same dictionary says in its third definition that an infection is 'the agency, substance, germ or principle by which an infectious disease is communicated or transmitted'. This seems suitably vague and tautological. It reflects humanity's understanding, or rather lack of understanding, of infectious diseases until recent times. The word itself is derived from the Latin word 'infectus', the past participle of 'inficere' meaning 'to dip in', 'put into', 'taint', or 'stain'. Its use long antedated the knowledge of pathogenic micro-organisms and implied tainting with morbid matter, or contamination with noxious effluvia, vapours, and miasmata. In other words, nobody had a clue.

What about contagious diseases? Are they the same thing as infectious diseases? Some people use the two terms as synonyms but strictly speaking contagious diseases are a subset of infectious diseases acquired by touch. The word is derived from the Latin 'tangere' meaning 'to touch'. Contagion has rather gone out of fashion these days so perhaps 'contagious diseases' is an expression best forgotten.

A rather better definition for an infectious disease is that it is an illness caused by a microbe, a word which was coined by the French surgeon Charles Sédillot (1804–1883) in 1878. This has the advantage of allowing us to include diseases caused by microbes which come from other people, from animals, from the environment, and from ourselves (for we all carry micro-organisms on our skin and in our mouth and bowel). But what is a microbe? Again the dictionary defines a microbe, which comes from Greek words *micros* and *bios* meaning 'small' and 'life', as 'an extremely minute living creature whether plant or animal; chiefly applied to the bacteria causing diseases and fermentation'. So, a microbiologist is a person who studies the agents that cause infectious diseases whereas an infectious diseases doctor is someone who tries to cure a patient with an infection.

This is a rather more practical definition but it still has some problems. As you will soon see when we consider worms, these are not microscopic at all. Indeed some are huge; a tapeworm may reach 5 metres in length although its eggs are certainly microscopic. And are all infectious agents living? Until a couple of decades ago, everyone would have agreed that they are. Most infectious organisms carry the building blocks of life, DNA and RNA. Then it was found that viruses are made up of either DNA or RNA. Then, most confusingly of all, prions were discovered. We shall consider these in our penultimate chapter; they have neither DNA nor RNA and are simply proteins. I don't think anyone would consider these as living.

This book is concerned with the discovery of the most important infectious agents throughout history and in our own time, who discovered them, and how they were related to the diseases that they caused. In 1894, Robert Louis Stevenson died in Samoa at the age of 44 from tuberculosis. He once said that it is not a hard thing to know what to write; the hard thing is to know what to leave out. That has been my problem too. It is perforce a matter of my judgement.

You can read any individual chapter with profit if you are particularly interested in a certain disease or particular organism. However, you will gain more benefit if you read the book from beginning to end. This is because it is arranged in a manner which roughly reflects both the size of the organisms and the order in which they were discovered. It therefore somewhat matches the development and maturation of ideas in infectious diseases and the progressive introduction of novel, powerful technologies that have allowed new discoveries. Furthermore, technical terms or words tend to be defined the first time that they are used. If you are uncertain of the meaning of a word, flip over to the glossary near the end of the book and it may well be defined there. There is also a section there telling you how to pronounce unfamiliar words.

We begin with multicellular infectious agents—the worms, arthropods, and some fungi (the moulds). We then turn to single-celled creatures, first yeasts (also fungi) and protozoa. All of these organisms are like us in that they have a well-defined nucleus, so they are classified as eukaryotes. Next we examine the bacteria, a group of organisms now classified as prokaryotes, which although they have DNA and RNA, do not have a well-defined nucleus. Then we review viruses, which have no nucleus at all and must live inside another eukaryotic cell. Finally we reach the enigmatic proteinaceous prions.

Medical scientists have made enormous progress, especially over the last 150 years or so. But it has not been easy. It is a tale at one time or another of dogged determination, perseverance, flashes of insight, luck, serendipity, argument, dispute, and, on occasion, tremendous bravery. We owe an incalculable debt to those who have gone before us.

This book is meant to entertain and inform, not to be dreary and didactic. But like any historical inquiry, it is subject to error. It is dependent upon my assessments and judgements of the significance of events. There may even be one or two errors of fact, although I hope not. Limitations of space mean there are many stories that cannot be told. Perhaps there are important discoveries of which I am simply unaware. Enlightenment and correction from you, dear reader, are always welcome.

PART II

WORMS

Most people don't like worms very much. The Bible tells us that in Old Testament times, Job in his agony cried 'And though after my skin worms destroy this body, yet in my flesh shall I see God' (King James Version) while just after the time of Christ, King Herod 'was eaten of worms, and gave up the ghost'. Actually, they were probably not worms at all but fly maggots. It is not surprising then that worms are popularly thought of as any elongated, creeping, loathsome thing that is not immediately recognizable as something else. The ancient Romans had a word to describe these obnoxious creatures—'vermes'—from which we get our English word 'verminous' meaning 'wormy'. Another word you might come across is 'helminth', which means much the same thing. It is derived from the word *helmins* which was used by the ancient Greeks to denote worms found in the intestines of humans and animals. Nowadays the diseases caused by worms or helminths are called 'helminthiases' (singular: helminthiasis).

In contrast to those pathogens described in this book which are unicellular or subcellular and can only be seen with a microscope or electron microscope, worms are made up of many cells. This means they are usually big enough to be visible to the naked eye; in fact, intestinal tapeworms can reach an extraordinary several metres in length. So it was that cavemen knew and doubtless were terrified by the horrid, motile creatures that occasionally passed from the anus.

We now know that there are thousands of species of worms but it is only in the last two or three centuries that they have all been discovered and sorted out. Worms are soft-bodied invertebrates, that is, animals without backbones. Structurally, there are two forms—roundworms and flatworms. Functionally, they also fall into one of two groups. The first category contains the free-living worms that live happily and freely in favourable niches in the environment, feeding on any detritus they find tasty, and passing on from one generation to the next. They are not our concern. The second group, for at least part of their lives, must live in and feed on another animal, often in the process doing damage to their unfortunate host. A few of these worms will only live in humans, some live in humans and a variety of animals, but most will only live in particular species of fish, amphibians, reptiles, birds, or animals. These worms are parasites, a term which is derived from two Greek words, *para* meaning 'beside' and *sitos* indicating 'food'. You might like to think of a parasite as one who eats at the table of another.

Worms that live in the human intestine are called, naturally enough, 'intestinal worms'. Other worms live in the tissues of the body; you can think of these as 'tissue worms'. But parasitic worms, whether living in the intestines or tissues, have a problem—they have to get from one host to another host in order to allow the next generation of worms to appear. The way in which they do this is called a life cycle. For some intestinal worms, this is very simple; microscopic eggs are passed in the faeces of one person, contaminate some food, and are ingested by another unsuspecting human in whom they develop once more into adult worms. Other worms find it much more difficult and have to be carried by a vector (such as a mosquito) from one person to another. Others again have to spend some time in a non-human animal, which is called an intermediate host. Here they look completely different from the worm found in humans. Indeed, in the case of tapeworms, they become cysts and do not look like worms at all.

We shall consider different sets of worms in the next five chapters. The first two are the giant intestinal roundworm, *Ascaris lumbricoides*, and tapeworms, especially the pork tapeworm. These are all parasites that dwell in the human bowel. They are large and obvious and the challenge has been to determine what illness, if any, they cause. Then we shall look at hookworms, which have a particularly cunning way of getting into you or me. After that we shall examine schistosomes (which cause a disease sometimes called bilharziasis) and which are particularly partial to snails and water. Finally, we shall review the filarial worms that cause elephantiasis but are only able to survive because mosquitoes like to suck our blood.

No matter which worm we are concerned with, the really exciting and sometimes controversial detective work has been to unravel the differing life cycles of these noxious beasts. Such understanding has often led to ways in which transmission can be interrupted and infection prevented.

1

Ascaris—the giant intestinal roundworm

SCIENTIFIC NAME:	*Ascaris lumbricoides*
COMMON NAME:	giant intestinal roundworm
DISEASE NAME:	ascariasis
DISTRIBUTION:	widespread especially in the tropics and subtropics
TRANSMISSION:	ingestion of eggs
LOCATION OF THE WORM IN A PERSON:	small intestine
CLINICAL FEATURES:	frequently without symptoms but may cause abdominal pain or discomfort and rarely intestinal obstruction
DIAGNOSIS:	passage of adult worms from the anus or finding eggs on microscopical examination of the stool
TREATMENT:	various anthelmintics such as mebendazole and pyrantel
PREVENTION:	avoid eating uncooked vegetables or ground-fruits that may have been grown in gardens contaminated with human faeces

Ancient man must have been well aware of, if not terrified by, the large motile creatures that he passed in his faeces from time to time. So impressive were these beasts that they were recorded in the writings of Egyptians, Greeks, Romans, and Chinese in the centuries before Christ. Furthermore, they recognized that there were two types of these worms, some round and some flattened. Thus Aulus Cornelius Celsus (*c.* 25 BC–AD *c.* 50), a Roman nobleman and writer of an encyclopaedia, wrote:

> Worms occasionally take possession of the bowel, and these are discharged at times from the lower bowels, or more nastily from the mouth; and we observe them sometimes to be flattened, which are the worse, at times to be rounded.[68]

The large rounded worms (Figure 1) were about 15–30 cm long and the Romans called them 'lumbricus teres' in view of their fancied resemblance to the common earthworm.

No-one paid much attention to the precise nature of these creatures. It was not until the late seventeenth century when an English doctor, Edward Tyson (p.16), described details of the anatomical structure of these worms and clearly distinguished them from earthworms. What is more, he used a new-fangled magnifying instrument called a microscope and in 1683 described how he had found eggs in these worms. Tyson still called this worm 'lumbricus teres' and it was not until 1758 that the Swedish physician turned naturalist, Carolus Linnaeus (1707–1778), gave it the scientific name by which we still know it—*Ascaris lumbricoides*. The first (genus) name was a Greek word for a worm while the second (species) name means that it was like the 'lumbricus' of the Romans.

But the big question was: 'Where do these worms come from?' For many centuries it was generally believed that intestinal worms arose by a process of 'spontaneous generation'. Belief in the spontaneous generation of animals and plants goes back to ancient times. How else could the sudden appearance of mushrooms after a heavy rain or the plagues of locusts or rodents in certain seasons be explained? Intestinal worms seemed excellent proof of this doctrine. It seemed impossible to account in any other way for the existence of such large organisms in the human intestine as they clearly had not been ingested as such.

These beliefs persisted from the ancients well into the Middle Ages. A famous Persian physician and philosopher named ibn Sina (known as Avicenna in the West, 980–1037) thought that with a proper mixing of the elements and under the influence of the stars, all animals and even man could be produced by spontaneous generation. In the sixteenth century, Paracelsus (1493–1541), a Swiss physician, botanist, alchemist, and astrologer, had similar fantastic ideas writing, 'Many

Figure 1. Adult *Ascaris lumbricoides* about 20 cm long. Female on the bottom, male on the top.

things will be changed in putrefaction so that they give birth to a noble fruit.'[239] Thus Paracelsus thought that worms were produced from putrefaction and a bird could be recreated from its own ashes in horse manure. Even the English doctor William Harvey (1578–1657), who discovered the circulation of the blood, was a prisoner of this idea and thought that worms arose spontaneously as a result of a special principle existing in putrid material.

Of course, all these notions were only theory and were utterly wrong. What was needed was a completely new approach. What was required was an *experiment* to prove or disprove the existence of spontaneous generation. The first to recognize this and do something about it was Francisco Redi (Figure 2), court physician to the Duke of Tuscany. Redi was born in Tuscany in Italy, was educated by the Jesuits, and graduated in medicine and philosophy from the University of Pisa. He was a great man of science and a brilliant investigator. For example, he was the first to experiment with snake venom, finding that it was innocuous by mouth but toxic when injected. He never married, was epileptic for the last nine years of his life, and died during a fit.

Redi thought it was possible that 'worms' were generated by insemination into putrefying matter, the latter merely serving as a suitable nest in which animals could deposit their eggs and in which the resultant offspring could find nourishment and grow.

First of all, Redi killed three snakes and put them in an open box to decay. Soon afterwards, he found that they were covered with maggots (which he called worms). Once all the meat had been consumed, however, the maggots all disappeared. In order to find out what had become of them, he repeated the experiment but this time, once the maggots had appeared, he covered all the exits from the box. He observed that some of the maggots became quiet, appeared to shrink and assumed a shape similar to an egg; we now call these structures pupae. There were pupae of different shapes so he separated them into their types and put them into glass containers covered with paper. After a week or so, he saw the pupae break

Figure 2. Francisco Redi (1626–1697).

open and flies came forth, the same kind of fly appearing from the same kind of pupa. So now he knew what happened to the maggoty worms. He repeated variations of this experiment many times and noticed that the meats became covered with eggs from which the maggots hatched. So he wrote in 1663:

> Having considered these things, I began to believe that all the worms found in meat were derived from the droppings of flies, and not from the putrefaction of meat.[263]

This seemed especially likely because flies of the same kind as those that were bred had hovered over the meat before it became maggoty. But Redi made the crucial step, remarking that 'Belief would be vain without the confirmation of experiment'.[263]

Thereupon Redi put a snake, some fish, some eels and a slice of veal from a milk-fed cow, each into separate, large, wide-mouthed flasks. Each flask was carefully sealed then a duplicate series was set up except that each flask was left open to the atmosphere. You can guess what happened. No maggots developed in the closed flasks whereas Redi saw flies go in and out of the open flasks and in them maggots appeared. So not only had Redi shown that maggots did not breed spontaneously in meat but he had traced the development of eggs through larval (the maggot) and pupal stages to adulthood. Even though he had designed and executed these wonderful series of innovative experiments, Redi had some mental blocks. He still thought that galls in plants and intestinal worms in humans probably arose by spontaneous generation through the agency of some vital force. How could he have been so muddled in his thinking on this important subject? Robert Bigelow, the translator into English of Redi's books, surmised that 'constant friendship with the Jesuits must have had a maleficent effect on (his) mind, as it exacted blind faith and put a limit to his logic'.[126]

Nevertheless, others gradually realized over the next century or two that Redi's findings were generally applicable to all organisms although, as we shall see, the final death knell had to await the experiments of Louis Pasteur with bacteria in the middle of the nineteenth century. So, if worms such as *Ascaris* did not arise spontaneously in people, how did they get there?

Medical scientists began to pay more and more attention to eggs, especially what they looked like under the microscope. In 1849 in Moscow an investigator of natural history named George Gros obtained some eggs from an adult *Ascaris* worm, moistened them and put them in an incubator at a temperature of about 15°C, then examined them periodically under a microscope. He noticed that things began to happen within each egg shell within 24 hours and there slowly assumed the shape of a coiled-up larva, although it took four months to reach a perfect state. These eggs were called embryonated eggs because they had an embryo or larva inside them.

Surely these embryonated eggs were infective? Ten years later in Paris, Casimir Davaine (p.197) took some of these eggs and put them in gastric (stomach) juice in a test tube to see whether the juice would destroy the egg shell and allow the larva to escape. Nothing happened! Not to be put off, Davaine then put some embryonated eggs into one fabric container and fresh, unembryonated eggs into another, then induced a dog to swallow them. Two days later, the fabric containers were recovered from the dog's faeces. On microscopic examination, he discovered that the unembryonated eggs were still there while in the other container, only a few free larvae could be found. He concluded that intestinal juices softened the egg shell so that if a larva was present within, it could pierce the wall and escape.

So it seemed easy; all you had to do was feed embryonated eggs to an animal and they would each hatch a larva in the intestine which would then develop into the big adult worms in the bowel. In 1861 Davaine gave 400 of these eggs to a cow. Four months later, there were absolutely no worms in the intestines. He then tried infecting a rat. This time he found that the eggs hatched larvae but they were expelled in the faeces within a day or two. What Davaine and his contemporaries did not know but which we now know is that this worm will only complete its development in humans. All other animals are resistant to infection with it, a phenomenon called 'natural immunity'. Why this should be, we still have little idea.

Thus everything was a big puzzle. Some investigators postulated that perhaps the eggs had to be taken up first by another animal, an intermediate host, in which they partly developed and then the animals harbouring these intermediate stage worms were eaten by people in whom they developed into the big adult worms. But all experiments were fruitless. This parasite did not seem to develop either partly or completely in the tissues or intestines of any other animal.

There seemed to be only one solution; give embryonated eggs to a human. In 1879, Battista Grassi (p.131), then a young medical practitioner who had been born in Lombardy, Italy, decided to undertake an experiment to settle the matter. On 30 August he ingested about 100 embryonated eggs. Twenty-two days later he triumphantly declared that he had found *Ascaris* eggs in his faeces, thus proving that direct infection had occurred. This perhaps seemed fair enough at the time, but in hindsight we know that something probably went wrong with the experiment because observations over the years were to show that it usually takes at least two to three months after ingestion of eggs for ova released by the newly developed adult worms to appear in the stool. Perhaps Grassi already had a low-level infection which he acquired naturally because it was common in his community. Or perhaps he mixed up his specimen with someone else's in the laboratory.

A few years later, Salvatore Calandruccio (1858–1908), a colleague of Grassi, repeated the experiment and swallowed a large number of embryonated eggs but failed to infect himself. However, he did have more success with a seven-year-old boy who had been infected naturally and had been cured. Calandruccio gave the lad 150 eggs in a pill at the end of September 1886. He searched the stools assiduously for the next 20 days but found nothing and abandoned the search until the end of November, when he found the faeces to be packed with eggs. Following treatment with an anthelmintic (anti-worm drug) at the end of January, four months after infection, the boy expelled 143 worms, each about 20 cm in length. Grassi then published the results of this experiment without giving acknowledgement to Calandruccio. Calandruccio was furious. Scientists do not always behave properly towards each other!

But the point was made and it was becoming clear that *Ascaris* worms were not generated spontaneously but could be transmitted directly from person to person via eggs, provided the eggs had been left long enough in the external environment to become properly embryonated. The natural assumption was that a larva hatched from an egg in the intestine, stayed in the intestine and grew there into adult worms. This seemed to be supported by the earlier finding in 1872 by Arnold Heller (1840–1913) in Germany, during the autopsy of a madman, of 18 small worms between 2.75 and 13 mm in length in the small intestine. But assumptions can be wrong, and so was this one.

A clue was given in 1887 when Adolpho Lutz (1855–1950), a physician in São Paulo, Brazil, repeated Calandruccio's experiment. He gave 96 embryonated eggs to a 32-year-old woman over a period of one month. A few days later she developed severe bronchitis and a fever. When she was later given an anthelmintic, she passed 35 adult *Ascaris* worms. Did you spot the clue? Lutz didn't. It was the bronchitis the patient developed. Why that happened will become clear shortly.

You might also be wondering about the foolhardiness of parasitologists infecting themselves or others. You will find this is a recurrent theme in the study of infectious diseases. Even I have done it. But that is another story.

It was not until 1915 that the next stage in the saga unfolded. Francis Stewart (1879–1951) was a British doctor who had joined the Indian Medical Service. During World War I he was posted to Hong Kong. Perhaps to pass the time because there was no fighting in the region, he began to experiment with *Ascaris lumbricoides* and with *Ascaris suum*, a very similar worm which infects pigs. After a number of false starts he gave *A. lumbricoides* or *A. suum* embryonated eggs to some rats. One rat died and the rest seemed to have pneumonia so they too were killed and the lungs examined; they were teeming with larvae. The larvae had hatched from the eggs in the intestines and had apparently penetrated through the wall of the intestine, entered the blood vessels, and passed through the liver to the lungs, where they left the blood vessels and entered the airways and migrated upwards towards the mouth. Stewart found that these larvae were then swallowed and re-entered the intestinal tract. But instead of developing there into adult worms, the larvae passed right through the gut and were passed in the faeces. Neither *A. lumbricoides* nor *A. suum* was able to complete their development in rats so he could not follow and prove the whole cycle.

Clearly you could not perform similar experiments of feeding eggs to people then cutting them up to see what had happened. But what about *A. suum* and pigs? Stewart infected pigs with *A. suum* eggs but was only able to find intestinal worms on some occasions. With masterly understatement, he wrote in 1919 that these experiments were very puzzling.

But help was at hand from two different Japanese parasitologists. In 1918, Sadao Yoshida (1878–1964) fed *A. lumbricoides* eggs to a guinea pig. He then recovered 50 larvae from the airways of the guinea pig then swallowed them himself; 75 days later he found eggs in his faeces. Thus, larvae recovered from the respiratory tract of another animal could develop into adult worms in the human bowel and there produce their own eggs. But did this mean that *Ascaris* eggs had to pass through an intermediate host before they could infect humans?

Final proof that this was not the case was provided in 1922 by the intrepid Shimesu Koino. On 28 August he ingested 2,000 *A. lumbricoides* eggs. A single larva was found in his sputum three days after infection, five on the next day, and 178 on the fifth day. He was unable to collect any sputum on the succeeding two days because he was seriously ill but larvae were then found on the following four days. Fifty days after ingestion of eggs, he took an anthelmintic and recovered 667 immature worms! Thus Koino had proven that *A. lumbricoides* larvae both migrate through the lungs and develop within the intestine of the same human host.

The life cycle was now clear and is illustrated in Figure 3. Infection occurs in areas where sanitation is poor and faeces are deposited on the ground. Transmission is particularly likely to happen when human faeces containing

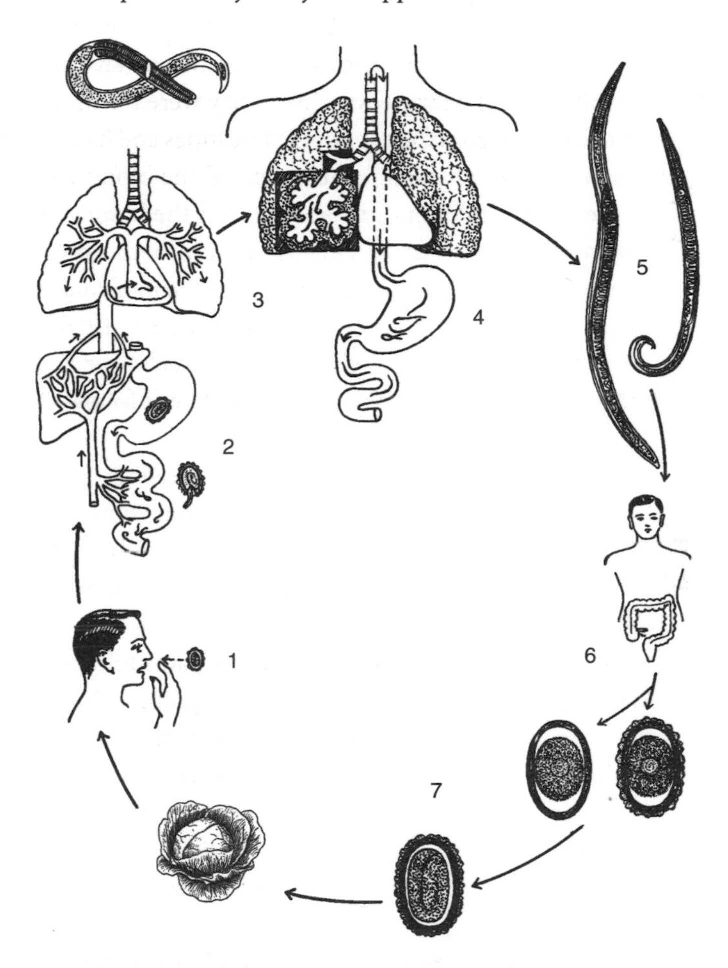

Figure 3. Life cycle of *Ascaris lumbricoides*. Infection begins when embryonated eggs are swallowed (1). The larvae hatch in the small intestine (2), penetrate the bowel wall, enter blood vessels, and are carried to the lungs (3). Here they leave the capillaries (small blood vessels) and enter the air spaces (alveoli), travel up the bronchi and trachea to the mouth where they are swallowed and return to the small intestine (4). There they grow over a couple of months and become adult worms (5) which eventually produce unembryonated eggs that are passed in the stools (6). Eggs deposited on the ground develop an infective larva within each one (7) over the next few weeks or months. Not to scale.

eggs are used as fertilizer in vegetable patches as happens in many poor rural areas. Finally, the warmer and moister the climate, the faster do eggs embryonate so that they are ready to infect the next person.

What do these worms actually do when they infect a person? All sorts of fanciful symptoms and signs were ascribed to this infection until late in the nineteenth century. Friedrich Küchenmeister, who we will meet in the next chapter, was closest to the truth when he wrote in his textbook in 1855 'as a general rule, the host and his guests agree very well together and give one another very little mutual trouble'.[172] Very heavy infections though can be quite serious. As we saw with Dr Koino, if very large numbers of eggs are ingested at much the same time, then there may be very severe inflammation of the lungs resembling pneumonia or bronchitis. We also know that large numbers of these big worms sometimes aggregate into a ball in the small intestine and block the flow of intestinal fluids, a condition known as intestinal obstruction. This can be particularly serious in small children. Finally, a wayward worm has been known to come out of the mouth or nostril, a very frightening experience indeed!

2

Tapeworms

SCIENTIFIC NAMES:	*Taenia solium* and *T. saginata*
COMMON NAMES:	pork and beef tapeworms
DISEASE NAMES:	taeniasis and cysticercosis
DISTRIBUTION:	widespread
INTERMEDIATE HOSTS:	pigs (*T. solium*) and cattle (*T. saginata*)
TRANSMISSION:	ingestion of undercooked infected meat
LOCATION OF THE ADULT WORM IN A PERSON:	small intestine
LOCATION OF CYSTICERCI IN PEOPLE:	soft tissues
CLINICAL FEATURES:	a. intestinal taeniasis: abdominal discomfort, spontaneous passage of proglottids or segments b. cysticercosis: often asymptomatic but can produce extremely variable symptoms depending upon location, most commonly epilepsy
DIAGNOSIS OF INTESTINAL TAENIASIS:	passage of proglottids or finding eggs on microscopical examination of the stool
TREATMENT:	various anthelmintics such as niclosamide and praziquantel
PREVENTION:	avoid eating undercooked beef or pork

Just like *Ascaris lumbricoides* considered in the previous chapter, bits of tapeworms have been known for generations. They too were recorded in the writings of the ancient Egyptians, Greeks, Romans, Indians, and Chinese and, as we have seen, these observers noticed that some worms were round and others flat. An individual roundworm vaguely resembled a common earthworm but bits of tapeworms passed in faeces did not resemble anything in particular.

These individual bits were whitish-grey in colour and about 2 cm long by 5 mm wide and 2–3 mm thick. But they had that incontrovertible evidence of life—independent *movement*. We now call each of these 'bits' a proglottid. Sometimes a string of proglottids stuck to each other, now called a segment, was passed through the anus—a truly horrifying experience.

Some Greeks thought these things were perhaps worms and gave them a name, in fact two names. One was 'helmins plateia' meaning 'flatworm' and the other was 'taenia' meaning 'band' or 'ribbon worm'. The Romans called them 'lumbricus latus', with the first name meaning 'worm' and the second indicating that it was 'broad' or 'wide'. When Carolus Linnaeus (p.7) came to give them our current scientific name in 1755, he called them *Taenia solium*. Why he called them 'solium', which is a Latin word meaning 'seat' or 'throne', has been disputed for years. But we are getting ahead of ourselves. Disputes were to be the stuff of life as far as tapeworms were concerned.

The first dispute was 'what were these things?' Were they really worms? As we have seen, many observers thought they were worms. Others such as the Byzantine Greek physician Paulus Aegineta (*c.* AD 640) thought they represented pieces of intestinal lining. The ninth-century Syrian physician Serapion believed that each individual proglottid was a separate worm, and since they resembled pumpkin seeds (*Cucurbita* species), he called them cucurbitini. But what about the segments or strings of proglottids? It was becoming increasingly recognized that these segments may reach two or more metres in length. Many Arab writers did not consider the whole tapeworm was a worm at all but believed it was a membranous bag formed by the intestine to hold these cucurbitini. Yet again, Ibn Siva (Avicenna, AD 981–1037) thought that pumpkin seed worms (cucurbitini) were completely different creatures to segments which he called gigantic worms. Confusion reigned supreme.

You might have noticed that the people who studied roundworms also tended to investigate tapeworms. So it was that the man who began to shed light into these dark corners was Edward Tyson (1650–1708). He was born near Bristol in England then went to Oxford to study arts and medicine. After graduation he moved to London and was appointed reader in anatomy at Surgeon's Hall and physician to Bethlehem Hospital. He made major contributions to human and comparative anatomy. In particular, he produced a monograph on the orang-utan and was the first to distinguish anthropoid apes as a group separate from both monkeys and man.

Figure 4. Prof. Marshall Lightowlers holding a tapeworm (*T. saginata*) nearly 4 metres long.

As you can see from Figures 4 and 5, tapeworms were broad at one end and narrow at the other. Did a tapeworm have a head? Which end was the tail? No-one could find a head in segments of worms that people passed in their stools and it seemed natural that the narrow end was the tail. Or was it? Tyson summed up the problem:

> The head of the Nile does not seem to be more perplex't and obscure to the Ancients, than that of the Worm, which has caused as many controversies among Anatomists of late, as that has with the Geographers of old.[304]

How could the problem be solved? In 1683 Tyson found a dog that had a tapeworm so he killed the animal and dissected the dog's intestine. This allowed him to orientate the worm properly in the bowel. He recorded what he found:

> It was a dog I opened at our private meeting in the Anatomical Theater of the College of Physicians where I observed the worm alive and in the ilion [the lower part of the small intestine], not lying straight but in many places, winding and doubling.... I traced it up to the smallest Extream where I expected the head to be and which did lye towards the Duodenum whereas the broader end was free and did nothing adhere whereas the small extream did firmly stick... to the inward coat of the intestine.[304]

So far so good. Tyson had worked out the disposition of the tapeworm in the small intestine and had found that the narrow end was firmly stuck to the lining of the bowel. Thereupon he detached the tapeworm and examined the narrow end with a rudimentary microscope and found that the head:

> very plainly appeared . . . beset with two orders of Spikes or Hooks . . . I could not upon my strictest Enquiry and with extraordinary Glasses too, inform myself of any orifice there which we may suppose to be the mouth.[304]

So, one problem had been solved but another one appeared. What sort of an animal was this? The beast did not have a mouth. How did it feed itself? And where did it come from? Just as with *Ascaris*, most people thought tapeworms were another clear-cut case of spontaneous generation. Tyson himself was not so sure about this but he had no better explanation to put forward. Thus he mused 'how to account for those that are bred in Animal bodies. . . . with whom we cannot meet with a parallel or of the same Species out of the body'.[304]

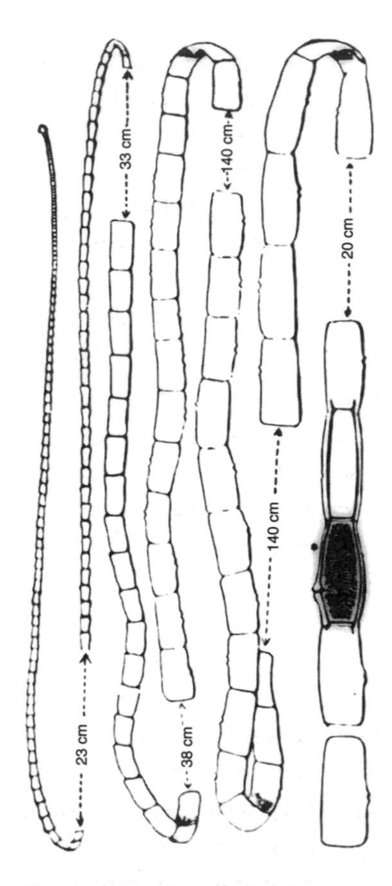

Figure 5. Diagram of a Taenia showing a minute head at the top left and proglottids progressively increasing in size. A string of proglottids is a segment.

A few years later in 1700, Nicholas Andry (1658–1742), later to become dean of medicine at the University of Paris, published the first description of the head of a human tapeworm (it happened to be that of the worm we now call *Taenia saginata*). What is more, he thought that he had found eggs in proglottids and was a firm upholder of the view that worms must develop from eggs. In fact what he thought were eggs were probably globular structures now called calcareous bodies. But tapeworms did have eggs and these were clearly demonstrated in 1782 by the German pastor Johannes Goeze (1731–1793), minister of the protestant church of St Blas in Quedlinberg, Germany. What is more, he showed that the small pore with an opening in it on each side of a proglottid (Figure 5), which had previously been mistaken for a mouth or for an airway, was in fact connected with the reproductive system and eggs came out of it. What particularly puzzled Goeze was that every tapeworm he had seen was packed with eggs. How were they fertilized? He wrote:

> Are there therefore two sexes amongst them? Or is every tapeworm sufficient unto itself and does it fertilise its own eggs? How are the organs for this purpose constituted and where are they?[115]

Goeze was postulating that tapeworms were hermaphroditic, i.e. contained both male and female sex organs. A few decades later, he was shown to be correct.

No-one had any idea how these tiny eggs became great big tapeworms. A long tortuous path had yet to be followed. But first there was another question. Were all tapeworms in humans the same or were there different types (later called species)? To answer this we have to return to 1602 and Felix Platter (1536–1614), chief physician of the city of Basel in Switzerland. When he looked at proglottids, particularly at strings of them, he realized that they took one of two forms. Either they were wider than they were long and he called them *Taenia intestinorum* or they were longer than they were wide (as in Figure 5) in which case he called them *Taenia longissima*. This was confirmed years later when the heads of these worms were examined and found to be different. Platter's *Taenia intestinorum* is now called *Diphyllobothrium latum* and has a complex life cycle involving fish but space does not permit us to pursue this.

But we are concerned with Platter's *Taenia longissima*. Various investigators looked at and drew diagrams of its head. Goeze, for example, described one: 'The head of the worm is like a small box and not round. At the four corners of the head are four suckers.'[115] But the diagram Goeze drew also showed a crown with two rows of hooks. Sometimes people saw these hooks and sometimes they didn't but any significance of this was unnoticed or ignored for 70 years until Friedrich Küchenmeister, of whom much more anon, in 1852 declared that based upon the appearances of the heads and the structures of the proglottids, there were definitely two species of Platter's *Taenia longissima*. Eventually these came to be called *Taenia solium* and *T. saginata* (Figure 6). Thus there were now known to be three species of big human tapeworms. It should be possible to work out where they came from. But how?

The answer was to come from a most unexpected source. The ancient Greeks were well acquainted with measly pork, i.e. pig meat containing little cysts.

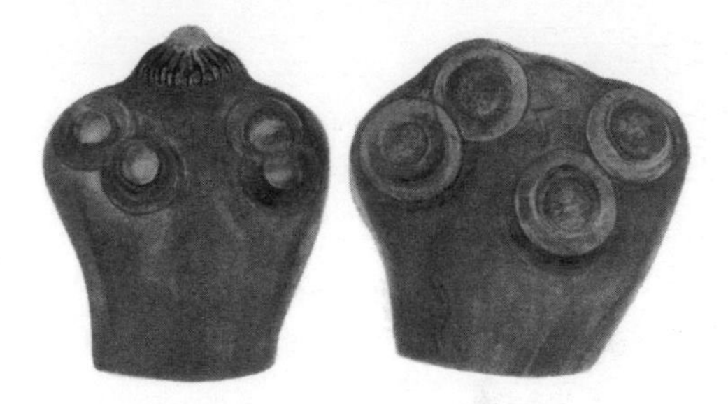

Figure 6. Head, also known as a scolex, of *T. solium* on the left and the scolex of *T. saginata* on the right. Both have four suckers but only *T. solium* has hooklets.

Aristotle (384–332 BC) compared them with hailstones and others with pearls. But they had no idea of what these lesions were. Nearly 2,000 years later, surgeons began to find similar cysts in people. Johann Udalric Rumler in 1558 in Augsburg, Germany, found some attached to the inside of the skull of a person who had suffered from epilepsy. Domenico Panaroli (1587–1657) in Rome found them in the brain of an epileptic priest in 1652 and the Englishman Thomas Wharton (1614–1673) in 1656 found large numbers in the muscles and fatty tissues of a soldier. As with large cysts (called hydatids), which had also been known for millennia, they were thought to be growths produced by collections of serum or mucus or else were enlarged and degenerate glands. No-one had any idea that these were living animals.

The first clue came with Francisco Redi (p.8). In 1684 he found some cysts in the abdominal cavity of a hare (a larger relative of a rabbit). When he examined them closely, he noticed a little nubbin inside each one like a retracted neck, and what is more, the whole cysts moved about independently as if they were animals. In the following year, Philip Hartmann (1648–1707), a doctor in what was then Königsberg, Prussia (now Kaliningrad, Russia), dissected a goat and found white, globular cysts in the abdominal cavity. When a cyst was incised and turned inside out: 'the rounded tail of a protruding intestinal worm became evident'.[132] Furthermore, when he placed a cyst in a container of warm water, it moved about:

> the vesicular body began movement in a marvellous manner; it moved with a singular form of undulation, exhibiting contraction and expansion with rising and falling of its parts.[132]

Later that year he found many cysts in the heart of a pig. These cysts, which were later given the name *Cysticercus cellulosae* (Figure 7), will turn out to be very important for our story. Again, when he opened the capsule (outer covering) of a cyst Hartmann found that 'a peculiar skin or membrane could be removed. This covered both a clear liquid and a white filament coiled like a thread—itself a small worm.'[132]

Figure 7. Drawing of multiple cysts of *Cysticercus cellulosae* in pork. The mounds within each cyst indicate the neck and head inside the cyst which become turned out (everted) in the stomach when ingested.

By 1800 when the German naturalist Johann Zeder published his textbook on helminthology, it was clear to him and many others that cysticerci were worms, albeit rather unusual worms. He recognized five different classes of worms: roundworms, flukes (a form of flatworm we will discuss in the chapter on schistosomiasis), tapeworms, cystic worms, and 'hookworms' (not the hookworms we shall discuss in the next chapter—we can forget about this class).

There was no Internet in those days and it took a while for these ideas to filter into the consciousness of most of the medical profession. Thus the proceedings of the London Medical Society in 1833 have been recorded:

> Mr Stephens entered the room in some haste to exhibit to the Society 3 or 4 hydatids [actually cysticerci] which he procured in the course of the afternoon affording the members to witness the existence of life in these 'imperfect animalcules'. The President immediately drew attention to the exhibition. 'This is very interesting' he observed 'as it settles a point greatly in dispute. Here is a head as clearly a head can be. Get a little water and see if it will revive the movements.'[12]

So it was now clear that there were such things as cystic worms. But what does all this have to do with tapeworms? There were some suggestions. Carl von Siebold (Figure 8) was sure he knew the answer. As we shall see, this professor of zoology in Erlangen, Germany, was a rather dogmatic academic. In typical fashion he thundered in 1844 that he had arrived

> at the most decided conviction that the cystic worms are strayed tapeworms which have remained undeveloped and become degenerated, and of which the body grew out in foreign soil into a vesicle [a small cyst], without developing sexual organs.[313]

Figure 8. Carl von Siebold (1804–1885).

And anyone who disagreed with him was obviously wrong.

But it was von Siebold who was wrong. The first step towards showing this was provided by a gifted Danish biologist named Johannes Steenstrup (1813–1897), who in 1842 published a book about a remarkable biological phenomenon that he had discovered. Steenstrup was not studying either tapeworms or cystic worms but he was particularly interested in coelenterates (jellyfish), tunicates (sea squirts), and trematodes (flatworms of a leafy shape called flukes) of the seashore. These words of his summarize what he found:

> An animal bears young which are, and remain dissimilar to the parent, but bring forth a new generation whose members themselves, or their descendants, return to the original form of the parent animals.[296]

In other words, there exist some animals within the animal kingdom which produce offspring that don't look anything like the parents. But when these different-looking offspring produced their own offspring, this third generation (or sometimes the fourth generation) once more look like the original first generation. This was completely unexpected and quite extraordinary. Steenstrup called it an 'alternation of generations'. You might be able to guess where we are heading.

The second requirement was to look carefully at the heads of various tapeworms and at the minute heads in cysticerci. They were clearly similar. But no-one had put it all together.

At this point Gottlieb Heinrich Friedrich[1] Küchenmeister (Figure 9) burst on the scene. Küchenmeister was able to think laterally, and like Redi, he brought *experiment* to bear to confirm or dispose of theories in helminthology. He was born at Buchheim in Germany, the son of a protestant pastor. Initially he studied theology but then switched to medicine. He began to practise in Zittau in

[1] In this book, the given or Christian name by which a person was usually called is underlined if it is not the first name.

Germany in 1846 and then in 1859 he moved to Dresden, where he died in 1890. All his experiments were done at his own behest and without the benefits and facilities of a university environment. He was a man of catholic interests, writing a textbook of parasitology, inventing surgical instruments and a novel gynaecological operation, investigating cholera and consumption, studying the behaviour of bees and the toxins of mushrooms, popularizing cremation, and he had a desk littered with theological texts.

Figure 9. Friedrich Küchenmeister (1821–1890).

Küchenmeister knew of Steenstrup's work on the alternation of generations and of the similarities between tapeworm heads and those of cysticerci. He put two and two together. Cysts were known in rabbits and mice. What eats rabbits? Foxes. What eats mice? Cats. So in 1851 Küchenmeister fed 40 *Cysticercus pisiformis* from rabbits to a fox. And guess what he subsequently recovered from the fox's intestine? Adult tapeworms (now called *Taenia pisiformis*)! Then he fed *C. fasciolaris* obtained from mice to a cat and recovered adult tapeworms (now called *T. taeniaeformis*). In the following year he fed *C. tenuicollis* of sheep and cattle to dogs and recovered adult tapeworms (now called *T. hydatigena*) then he gave *Coenurus cerebralis* (a peculiar cyst with multiple heads inside it) of sheep to dogs and recovered adult worms (now called *T. multiceps*). Küchenmeister was very thorough.

But now began the battle between academic and clinician. Von Siebold was enraged that he, an eminent university professor, had been upstaged by a mere general practitioner who had demolished his cherished theory of the dropsical degeneration of strayed tapeworms which were being turned into fluid-filled sacs. Von Siebold repeated some of Küchenmeister's experiments, confirming the results, but unaccountably held stubbornly to his own theory. He then attacked Küchenmeister personally and tried to claim credit for himself for the discovery of the phenomenon of metamorphosis (change of appearance). He went on to castigate Küchenmeister, saying that Küchenmeister had been so confused over the taxonomy (identification and classification) of adult tapeworms that if he, von Siebold, had not come to Küchenmeister's aid, then the whole theory of metamorphosis would have been thrown into such a state of confusion that it could hardly be corrected. This did not go down at all well with Küchenmeister who wrote:

> No-one has so grievously offended in the study of cestodes [i.e. tapeworms and cystic worms] as the professor of zoology.[170]

Von Siebold could hardly believe his eyes. Küchenmeister had dared to stand up to him. He riposted in print:

> Küchenmeister has been led away by his zeal, to depart from that calmness of tone which becomes scientific controversy.[314]

A clear case of the pot calling the kettle black.

But Küchenmeister was not to be distracted. In 1853 he undertook some experiments in order to complete the full circle of a life cycle. First he fed some coenuri from sheep to dogs and obtained tapeworms. Then he gave some proglottids from these tapeworms to another sheep. Sixteen days later the sheep developed 'staggers' and after 19 days he found 15 small coenuri in the brain at autopsy.

Küchenmeister had incontrovertibly shown that a number of veterinary infections underwent the cycle of egg–cyst–tapeworm–egg. But what about human tapeworms and the *Cysticercus cellulosae* that had been discovered in both humans and pigs and had similar-looking heads? In that same year, 1853, a Belgian zoologist, Pierre van Beneden (1809–1894), gave *Taenia solium* eggs derived from a human to a pig. Four and a half months later, the pig was slaughtered and he found a large number of cysticerci in its muscles. So far so good.

A number of investigators then tried to complete the life cycle by giving *C. cellulosae* from pigs to dogs: nothing happened. It was left to the redoubtable Dr Küchenmeister to try it in humans. In 1855, Küchenmeister with Dr D and Dr Z (who perhaps understandably did not wish to be named) gave cysts of *C. pisiformis* (from rabbits) and *C. tenuicollis* (from a pig) intermingled with noodles in a soup, cooled to blood temperature, to a convicted murderer without his knowledge some time before his execution. It was not for nothing that Küchenmeister's surname means 'master cook' in German! Küchenmeister was not to know that this experiment would have been a failure. Fortunately for his experiment, his wife found some *C. cellulosae* in pork obtained from a nearby restaurant three days before the execution; 61 cysticerci were given in sausages or soup over the next two days. Forty-eight hours after execution, an autopsy was performed and Küchenmeister wrote that he:

> was successful in finding a small *Taenia* which was tightly attached with its projected proboscis [i.e. snout] to a piece of duodenal mucosa…and proved clearly to be…*T. solium*.[171]

Küchenmeister attempted to head off any adverse criticism by pleading:

> that surely the harmless experiment of bladder-worm feeding be allowed to be repeated on criminals under sentence of death; so that the whole developmental cycle of *T. solium* could be observed…In the case of the subsequent pardon of the convict, the tapeworms can easily be expelled; this will calm anxious souls and serve science at the same time.[171]

There matters lay for another four years until Küchenmeister found an opportunity to repeat the experiment. On 24 November 1859, another convicted murderer was induced to swallow some cysticerci cellulosae from pigs. This was followed by some more on 16 January 1860, making a total of 40 in all. On 31 March 1860 the prisoner was decapitated and at autopsy many tapeworms up to 5 feet long were found. This time Küchenmeister did not escape censure. A commentator in the *British Medical Journal* recounted the experiment, castigated Küchenmeister, and went on to quote William Wordsworth:

> Physician, art thou, thing of eyes,
> Philosopher, a prying knave
> A man who'd peep and botanise
> Upon his mother's grave[328]

If Küchenmeister ever saw this editorial, I doubt that he would have cared. He had made his point. Understanding of the life cycle was now complete—it involved humans and pigs (Figure 10).

There were, however, several points to tidy up. It was now clear that there were two similar human tapeworms, *T. solium* and *T. saginata*. It seemed probable that an analogous process might occur with *T. saginata*. The difficulty was to know which animal was the intermediate host containing cysticerci of this worm. The clue came when physicians noticed that sickly children who had been ordered to eat raw beef in St Petersburg, Russia, not infrequently contracted infection with *T. saginata*. What is more, Jews, who were proscribed from eating pork, were infected with *T. saginata* but not *T. solium*. Therefore, in 1861, the German zoologist, Rudolf Leuckart (p.31), gave *T. saginata* proglottids to a calf. The animal died a month later and multiple cysts each containing a head with four suckers but no hooklets, just like the head of *T. saginata*, were

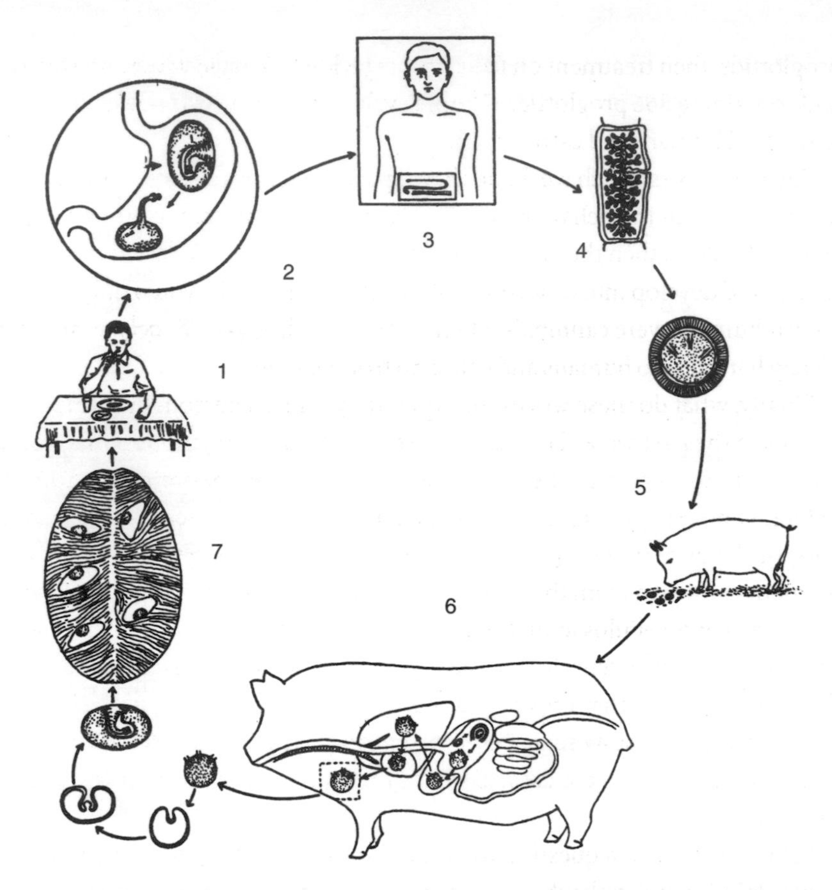

Figure 10. Life cycle of *Taenia solium*. Infection begins when a person ingests uncooked pork containing cysticerci (1). The head of the incipient tapeworm inside the cysticercus pokes outwards in the stomach (2) and this then attaches to the lining of the upper small bowel where it grows in length to several metres over the next few months by multiplying the proglottids (3). Eventually individual proglottids or a segment of them break off and are passed in the faeces (4). If they are deposited on the ground, eggs in the proglottids or free in faeces may be ingested by a pig (5). A larva hatches from each egg and they pass to the tissues of the pig (6) where they develop over a few weeks into cysticerci (7). Alternatively, if eggs are ingested by a human, e.g. in contaminated vegetables, they develop into cysticerci in the tissues just as in pigs. *T. saginata* has a similar life cycle except that cattle are the intermediate hosts and eggs cannot develop into cysticerci in humans. Not to scale.

found at autopsy; these were called *Cysticercus bovis*. Then in 1868–9 in some poorly controlled experiments, John Oliver (1837–1873), a British medical officer stationed in India, observed *T. saginata* infections in persons who ate measly beef. More definitively, the Italian veterinarian, Edoardo Perroncito (1847–1936), infected a subject with a single uncooked *C. bovis*; 54 days later he began to pass proglottids, then treatment on the 57th day yielded a *T. saginata* 4.27 metres long and containing 866 proglottids. The life cycle of *T. saginata* was now complete—it involved humans and cattle.

Further studies which we do not have space to canvass disclosed another difference between the behaviours of *T. solium* and *T. saginata*. If a human ingests eggs of *T. solium*, then they behave just as they do in pigs and larvae pass to the tissues and develop into cysticerci cellulosae; this life cycle would only be complete if humans were cannibals! On the other hand, eggs of *T. saginata* are completely harmless to humans and cause no trouble at all.

Finally, what do these worms do when they infect a person. Intestinal tapeworms in the past were claimed to cause just about any symptom imaginable but it is now clear that the main effects, if any, are abdominal discomfort (which is not surprising if you imagine a beast up to 5 metres long in your guts) or the unnerving experience of passing proglottids in the faeces or large segments of worm from the anus. Much more worrisome is human infection with cysticerci cellulosae of *T. solium*, which reach up to 1 cm in size. What symptoms, if any, occur depend partly upon how many there are and where they are located. Often they are found by chance by imaging with X-rays or computed tomography scans but most importantly they may cause epilepsy. Intestinal tapeworms are easy to treat but cysticercosis is rather resistant to therapy.

Let us finish with a question for you. In India, which tapeworm commonly afflicts Hindus and with which tapeworm are Muslims prone to be infected?

3

Hookworm anaemia

SCIENTIFIC NAMES:	*Ancylostoma duodenale, Necator americanus*
COMMON NAME:	hookworm
DISEASE NAME:	ancylostomiasis, hookworm disease
DISTRIBUTION:	widespread especially in the tropics and subtropics
TRANSMISSION:	penetration of infective larvae in water through the skin
LOCATION OF THE WORM IN A PERSON:	small intestine
CLINICAL FEATURES:	asymptomatic or symptoms of anaemia–pallor, lassitude, shortness of breath
DIAGNOSIS:	finding eggs on microscopical examination of the stool
TREATMENT:	various anthelmintics such as mebendazole and pyrantel
PREVENTION:	avoid walking on the ground bare-foot where human faeces may be deposited

When West Africans were kidnapped and transported to the Caribbean Islands and the continental Americas several centuries ago to work as slaves on sugar fields, many were to suffer a terrible affliction. The following is an extract from a book by an anonymous author of the time who appears to have been a doctor and a planter:

> When a negro is languid and listless, and so much indisposed to motion as to require to be impelled to it by threats, when he is short-breathed and unable to ascend a hill without stopping, his efforts for that purpose being accompanied with a throbbing of the temples and a violent palpitation of the heart, when he complains of giddiness, his lips being pale and his tongue white, you may know him to labour under that disorder which the French call

> mal d'estomac and the English after by the same name or 'dirt eating'.... This disorder which I observed is very common in the West Indies estates.... It disables them from effective labour for a considerable time, sometimes for years, and often terminates in dropsy (swelling of the legs with fluid).[11]

Around this time doctors were beginning to recognize these as symptoms of anaemia (low blood count). But a century or so was to pass before it became clear what caused the severe anaemia in the slaves.

Our story begins in the mortuary of the Maggiore Hospital in Milan, Italy, in 1838. Angelo Dubini (Figure 11), a young doctor, was performing a post-mortem examination. He had been born in Milan and graduated in medicine there in 1837. He was later to become head of pathology and then dermatology in the same hospital. In his retirement he wrote a popular cookbook and another work on the keeping of bees. I mention all this because he made a very important discovery during that autopsy. The patient was a peasant woman who had died from a blood clot in the lungs. When he looked in her bowels, he found 'a little worm in the small intestine, in the midst of much grey mucus...this worm impressed me as having really distinct generic characteristics'.[88] In other words, he had discovered a new worm. Perhaps because he was young, inexperienced, and unsure of himself, he did not publish news of this discovery.

Four years later in November 1842, however, he found another example of this peculiar worm in the small intestine of an old lady who suffered with dropsy. The next month he found another worm in a third patient, then on 15 December 1842 he saw a dozen worms in the small intestine of a woman who had died of a chest abscess. Whereas previously he had always found a single female worm, this time he found a mixture of male and female worms. In the course of 100 autopsies, Dubini then found these worms in 20 cadavers. Dubini described these worms as being whitish in colour, about 1 cm long and as being unisexual, i.e. either male or female. Because the worms had a mouth

Figure 11. Angelo Dubini (1813–1902).

with four hooks and the tail was slightly curved, he called them hookworms and gave them a scientific name of *Agchcylostoma duodenale*. The first (genus) name was a combination of the Greek names *ankylos* and *stoma* meaning 'hook' and 'mouth'. The second (species) name indicated their location in the duodenum (the first part of the small intestine).

Why weren't these worms discovered before? Until shortly before Dubini's time, it had not been the general practice to open the bowel at autopsy (presumably because the smell was so awful). This meant that large worms such as the giant roundworm or tapeworm could be felt through the intestinal wall but these tiny worms would be missed. When it became fashionable to open the bowels, the usual procedure was to wash out the intestines with large volumes of water, which would have carried away many worms. Finally, worms were embedded in rather opaque mucus, which would have made them very difficult to visualize (Figure 12).

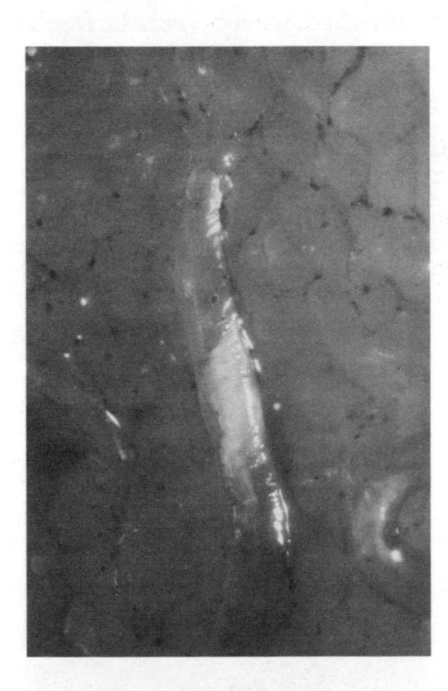

Figure 12. A hookworm with its head buried in the intestinal mucosa. The body of the worm is glistening. Parts of other worms can be seen above and below. The scattered black dots are red in real life and indicate places where hookworms have previously fed then moved on.

News of Dubini's discovery spread. In particular it reached Egypt where a condition called Egyptian chlorosis, but which was really severe anaemia, was common. There was a contingent of German doctors in Cairo, including Pruner, Bilharz (p.38), and Griesinger, all of whom verified Dubini's discovery. Twenty years later, another German doctor, Otto Wucherer (p.54), discovered the same worms in Brazil, the home of many of the slaves mentioned earlier.

Unfortunately Dubini was not a very good Greek scholar and did not get the name quite right. If there is one thing taxonomists (those who classify animals and plants) tend to get quite uptight about, it is names. Heated argument among the experts raged for 70 years until the International Commission on Zoological Nomenclature in 1915 settled the matter and re-named the worm *Ancylostoma duodenale*.

But that was not quite the end of the story. In 1868 Louis Camuset realized that the hookworm found in French Guiana in South America looked distinctly different from that seen in Europe and Africa. Unfortunately this information was buried in a thesis submitted to the University of Montpellier and no-one knew about it until the differences were rediscovered decades later. In the first few years of the twentieth century, several American doctors, thinking the hookworms they found were not quite the same as *Ancylostoma duodenale*, sent specimens to Charles Wardell Stiles (1867–1941), a parasitologist in Washington, DC. He agreed, and after some to-ing and fro-ing this worm was given a new name in 1915, *Necator americanus*, meaning the killer from America.

The next question was 'how do people become infected with these worms?' Dubini had seen eggs in the female hookworms, so perhaps it was generally assumed that a person acquired the infection by ingesting eggs. It was not until 1866 that a German medical parasitologist, Rudolf Leuckart (Figure 13), became interested in the question. Unfortunately for him, there were no human hookworms available in Germany at that time so he studied a hookworm of dogs that is now called *Uncinaria stenocephala*. He took some eggs of this hookworm and put them in damp earth. He found that the segmented embryo within the eggshell formed a larva which hatched after three or four days and then crawled around in the dirt. Over the next week, it moulted (lost its skin) twice and the interior of the larva was rearranged. Leuckart thought perhaps these worms went into some intermediate host but experiments were negative. So he introduced some of these 'third-stage' larvae in muddy water directly into the mouths of some dogs. Three weeks later, he found mature worms in their intestines. He'd done it! Or had he? Not unreasonably, Leuckart surmised that humans become infected with *A. duodenale* by drinking free-living larvae in dirty water. Leuckart was partly right in that you can get infected this way, but it turned out this was not the usual way.

Figure 13. Rudolf Leuckart (1822–1898).

In 1878, Battista Grassi (p.131) and his colleagues confirmed some of Leuckart's observations and showed that human hookworm eggs also developed into larvae, which then moulted twice in the soil. With regard to what followed, they wrote 'However, as of today, we are unable to say what happens afterwards.'[122] For some obscure reason, Grassi decided to swallow a pill containing a large number of hookworm eggs. Nothing happened. At the same time, a colleague who wished to remain anonymous, ingested a large number of newly hatched larvae (i.e. first-stage, not third-stage after two moults). Again nothing happened. Where was all this leading? Eight years later, Dr Otto Leichtenstern (1845–1900) in Cologne, Germany, undertook some experiments with volunteers (who they were we do not know). He fed them third-stage larvae of the human hookworm (A. *duodenale*) which had been prepared in soil; four to five weeks later he found hookworm eggs in the faeces! He had demonstrated the same results in humans as Leuckart had found in his animal experiments.

But how relevant was all this? Was there more to it? In the 1870s the 14-km St Gothard's railway tunnel was being built under the Swiss Alps. Large numbers of miners became heavily infected with hookworm. Why should they get infected? One might imagine that the water the miners drank was the water they took with them, but they still became infected. What else is there about mines that might be relevant? What happened when a miner felt the call of nature? He went into a quiet, dark, dank corner and deposited a pile of poo!

Figure 14. Arthur Looss (1861–1923).

It was at this point that Arthur Looss (Figure 14) in Cairo, Egypt, entered the scene and a vociferous argument ensued. Looss had been born in Chemnitz, Germany, then studied science in Leipzig, where he graduated with a PhD for his work on types of flatworms called trematodes. In 1896 he moved to Cairo as professor of biology and parasitology. He remained in Cairo, where he became a towering figure although his dogmatic manner and acrid, controversial style brought him into conflict with many famous parasitologists. At the outbreak of World War I in 1914, he was kicked out by the British and returned to Germany.

Shortly after his arrival in Cairo, however, Looss began to study hookworms. First he examined the development of eggs into larvae and wrote a paper about it in 1897, saying that very little was known about the life cycle of *A. duodenale*. Looss was savagely attacked by Leichtenstern, who said that Looss's paper was nothing more than a re-hash of the known facts combined with a curious ignorance of, if not intentional disregard for, the literature. Looss replied vigorously that for one such as him, incarcerated in Egypt and denied the scientific aids and libraries of Europe, he either worked for his own pleasure and kept his discoveries to himself, or he published them, even if they were imperfect. He did make the reasonable point:

> It is my conviction that the merits of prior authors will prevail in any scientific question, regardless of whether or not they are cited in every later paper.[193]

In any case, Looss was not at all convinced that everything had indeed been clarified about the life cycle of hookworms. In fact, he was by chance on the verge of making a momentous discovery. When doing his various experiments, Looss never had any misgivings about allowing water containing hookworm larvae (Figure 15) to settle on his hands although he made sure he kept them away from his mouth which he presumed to be the usual mode of infection. He was rather surprised in one of these experiments to find hookworm eggs in his own faeces and could not work out how they had got there. Some time later, he had an experience which gave him the clue:

> A drop of water with a high (hookworm) larva content fell on my hand one day and rolled off. I paid no attention to this moist spot which dried after a few minutes. But at the same time I felt there was an intense burning...and the spot became extremely red.[193]

What would you do in a circumstance like this? Looss put some plain water on his skin and there was no untoward reaction. So to be really sure, he put some more hookworm larvae on his skin. Exactly the same red, burning reaction recurred. What do you think had happened to the larvae? Would they still be in the drop of water? Looss scraped the moisture off his skin and had a look under the microscope and found that 'the larvae previously present in such abundance had disappeared; in their place...were found numerous empty worm skins'.[193]

What had happened? Looss came to the only logical conclusion that 'the larvae themselves could only have penetrated the skin'.[193] In fact, the next day his

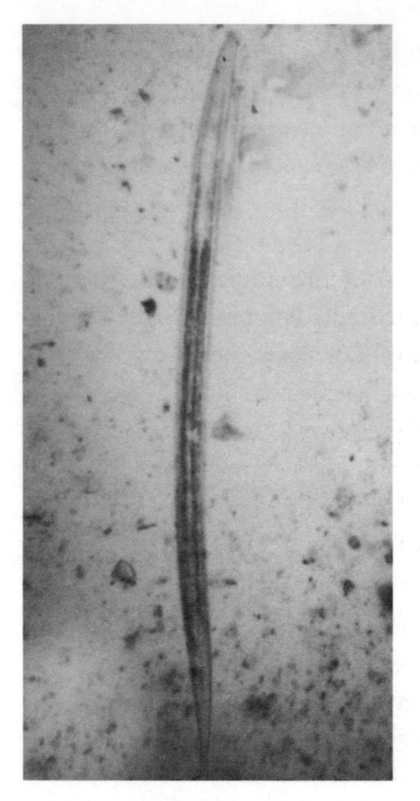

Figure 15. Third-stage infective hookworm larva.

hand became so swollen that he had to seek medical help. Two to three months later, Looss noticed a great increase in the number of hookworm eggs in his stools.

This was the first time anybody had shown that a worm was capable of penetrating the intact human skin. But he was criticized left, right, and centre so Looss decided to shut up until he could prove his point. First he put some larvae on skin which had been removed from a dead body and had been heated to 37°C, then examined it under the microscope; he found nothing. Undeterred, in about 1900, with the cooperation of Frank Madden (1873–1929), an Australian surgeon at the Kasr-el-Aini Hospital in Cairo, he put a drop of water containing many larvae on the skin of a 13-year-old boy who was going to have his leg amputated an hour later. Immediately after the amputation, he removed the infected skin and examined it microscopically. This time he had success and found that the larvae had penetrated the skin through the hair follicles, and concluded the fact 'that actual penetration of the *Ancylostoma* larva into the skin does occur can now be looked upon as unassailable'.[194]

Still there were the sceptics, including the authoritative Patrick Manson (p.58). Looss did not give up. In 1901, he persuaded a hospital attendant to volunteer for an experimental infection (there was some treatment available but it was not all that effective). First, the volunteer's stools were examined daily for six weeks and found to be free of any hookworm eggs thus proving he was naturally free of hookworm infection. Thereupon a drop of water containing hookworm larvae was placed upon his forearm and hookworm eggs began to appear in his faeces 71 days later.

Looss had to be right, and the correctness of his observations was confirmed by Charles Bentley (1873–1949), a medical officer to the Empire of India and

Ceylon Tea Company in Assam, India. Indeed, I proved it to my own personal discomfort when, in the interests of science, a colleague and I infected ourselves by each placing 1,200 larvae of *Ancylostoma ceylanicum* (a hookworm which infects dogs and sometimes humans) on our forearms and coming down with the most abominable abdominal colic within an hour of each other 29 days later.

So this raises a new question: how do larvae get from the skin to the intestines? Clearly it was going to be impossible to find out in humans so Looss decided in 1905 to infect some dogs with the dog hookworm, *Ancylostoma caninum*, and see what happened. In his first experiment, he infected some puppies percutaneously (via the skin) and when they died nine days later, he found immature hookworms in their intestines. So he repeated the experiment and killed the puppies at various times after infection. He found that some larvae entered the lymphatics (bloodless drainage tubes) in the fatty tissues under the skin which ultimately drain into the large blood vessels or else they went straight into the veins. Either way, they passed through the two chambers (atrium and ventricle) on the right side of the heart and went to the lungs. He found larvae in the capillaries (tiny blood vessels) of the lungs and in the alveoli (tiny air spaces of the lungs). The larvae had escaped from the capillaries to the alveoli then wriggled their way up the airways through the larynx to the mouth, were swallowed, and passed through the oesophagus (gullet) and stomach to the small intestine where they attached themselves to the mucosa (lining of the bowel). There they were able to develop into adult worms which could in turn produce eggs.

The same thing no doubt happened in humans and at last the life cycle was understood (Figure 16). In areas of poor sanitation, human faeces with hookworm eggs were deposited on the soil. Over a week or two, these hatched larvae which became infective. The next unsuspecting person to walk barefoot over contaminated soil or sit or lie on it became the next host of these cunning creatures.

But there was still a major question—what damage, if any, did these worms in the intestine do? In fact the beginnings of the answer to this question were given by Dubini himself when he first discovered these worms. He noticed that they sucked the intestinal lining into their mouths and this was surrounded by an area of haemorrhage. Moreover, the intestinal lining was often speckled with haemorrhages even though there were no longer any worms attached (Figure 12). It seemed that the worms caused bleeding!

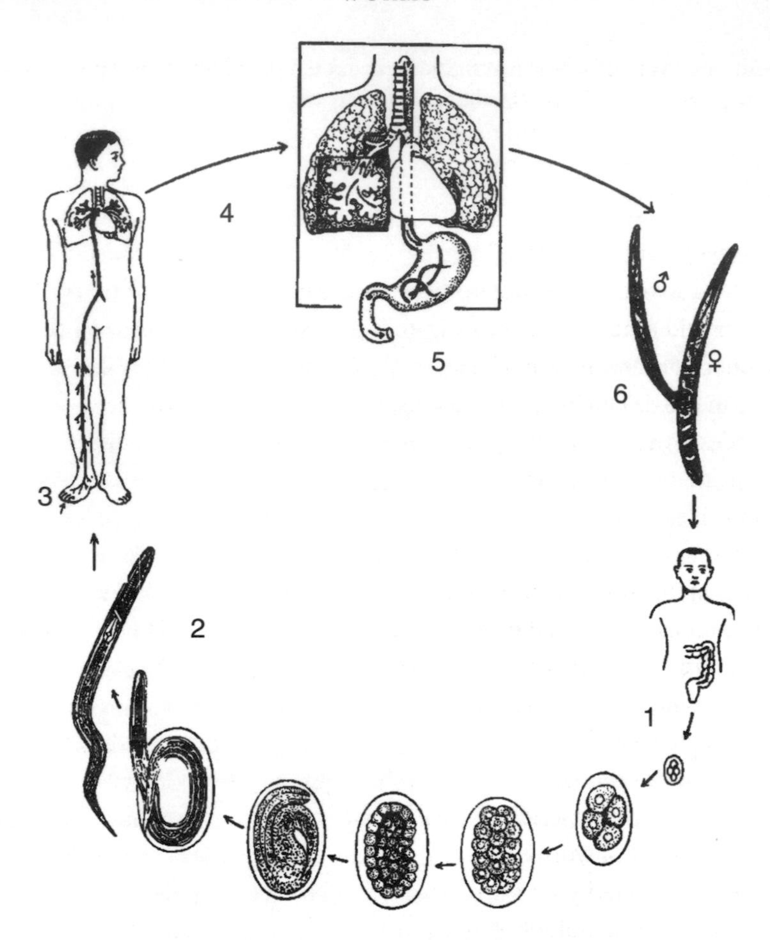

Figure 16. The life cycle of hookworm begins when eggs are passed in the stools (1) and deposited on the ground. Each egg develops into a larva which then over a week or two moults twice into an infective third-stage larva (2) which penetrates the intact skin (3). The larva passes via the lymphatics and blood vessels to the lungs (4) where it escapes from the capillaries and enters the alveoli, ascends the respiratory tree to the mouth, is swallowed, and passes via the gullet and stomach to the small bowel. Here they develop into male and female adult worms (5) and release a new series of eggs a couple of months later (6). Not to scale.

The dreadful consequences of this infection were graphically demonstrated by Wilhelm Griesinger (1817–1868) in Cairo a few years later in 1852. A 20-year-old soldier died, apparently from diarrhoea. At post-mortem examination, all his organs were extremely pale, i.e. anaemic, the condition being labelled Egyptian chlorosis, his heart was dilated, and his lungs were full of fluid. When Griesinger examined the intestines, he found that:

> the duodenum, the whole jejunum and even the upper half of the ileum [these are the first, second, and third parts of the small intestine] were completely filled with fresh, red, partly coagulated blood. Thousands of hookworms were hanging onto the mucosa of the small intestine, each with its own small petechia (area of haemorrhage) resembling the bite of a leech.[124]

Griesinger concluded, quite rightly, that Egyptian chlorosis was a severe anaemia caused by a loss of blood due to hookworm infection. Ten years or so later, Wucherer and a number of investigators in South America showed the same thing. But not everyone was a believer. When these findings were discussed at a meeting in Rio de Janeiro in 1868, a great controversy arose which was to persist for decades. There were two great questions: (1) did hookworm infection cause disease at all? and (2) how many worms were necessary to cause disease?

Loud voices were raised on both sides of these questions. We don't have time to follow the permutations but can fast-forward to the conclusions. First, it was realized that the more worms you had the more blood was sucked and the more had to be replaced. It was the ability to replace this lost blood that then became the important determinant. What were needed especially were iron and protein to make new red blood cells. Unfortunately, most areas of the world that were heavily afflicted with hookworms were just the areas in which malnutrition was rife, thus compounding the effect of the worms. If your food supply was bad, then less worms were needed to cause anaemia. And one other thing became clear; the number of worms you had was entirely dependent upon the number of larvae that entered your skin because unlike protozoa, bacteria, or viruses, most worms cannot multiply in the human body. The seriousness of hookworm infection eventually became apparent to all and this led the Rockefeller Foundation of the United States to spend an enormous amount of money in the 1920s to control, if not attempt to eradicate, global hookworm infection.

Hookworms are still with us. One final note: we expect all things American to be bigger and better. Not so with hookworms. In 1957, Marcel Roche (1920–2003) and his colleagues showed that every adult A. *duodenale* sucked 0.2 ml of blood every day whereas N. *americanus* consumed a mere 0.03 ml per day!

4

Schistosomiasis (sometimes called Bilharziasis)

SCIENTIFIC NAMES:	*Schistosoma haematobium, S. mansoni,* and *S. japonicum*
COMMON NAME:	schistosome
DISEASE NAME:	schistosomiasis formerly often called bilharziasis
DISTRIBUTION:	*S. haematobium*—Africa; *S. mansoni*—Africa and South America; *S. japonicum*—Eastern Asia
TRANSMISSION:	penetration of skin by infective larvae in water
CLINICAL FEATURES:	a. *S. haematobium* – blood in urine, b. *S. mansoni* and *S. japonicum*—enlarged liver and spleen; fluid in the abdominal cavity; bleeding from enlarged veins in the oesophagus
DIAGNOSIS:	finding eggs on microscopical examination of the urine (*S. haematobium*) or stool (*S. mansoni* and *S. japonicum*)
TREATMENT:	praziquantel
PREVENTION:	avoid walking or swimming in water in infected areas

Before he went to Cairo, Egypt, in 1850, the young German doctor Theodor Bilharz (Figure 17) sought some advice from his former teacher and mentor, Carl von Siebold (p.21), as to what branch of natural science he should particularly direct his attention. Von Siebold recommended that he concentrate on human helminths as it seemed likely that the 'strange country' would provide a fruitful field. Bilharz followed this advice and was soon in luck. In early 1851, while carrying out an autopsy on a young man he made an astounding discovery. Within the abdomen, he found a worm in a blood vessel, a location

Figure 17. Theodor Bilharz (1825–1862).

which had never before been observed in a human. With some excitement he wrote to von Siebold:

> Soon after my attention had been directed to the liver and its associated structures I found in the blood of the portal vein [the vessel that carries blood from the intestines to the liver] a number of long white helminths... A look into the microscope revealed a splendid [worm] with a flat body and a spiral tail at least ten times as long as the body.[45]

Bilharz had realized that this new worm, which was just over 1 cm long and 1 mm in diameter, was not a roundworm like *Ascaris* or hookworm or a tapeworm like *Taenia* but a flatworm belonging to the third group of worms called flukes or trematodes.

All the flukes that had been discovered in humans and animals until that time were hermaphrodites, i.e. they contained both male and female sexual organs. When Bilharz looked really carefully, he found that there were also female worms about 2 cm long which were wrapped by the shorter male worms that he had first seen (Figure 18). In a second letter to von Siebold, he wrote:

> It did not develop into a wonderful old wives' tale but into something more wonderful, a trematode with divided sex... When I searched the intestinal veins more carefully..., I soon found specimens which harboured a grey thread in the groove of their tail. You can imagine my surprise when I saw a trematode protruding from the frontal opening of the groove and moving back and forth; it was similar in shape as the first only much finer and more delicate.... It was completely enclosed in the groove-shaped half canal of the

> male posterior, similar to a sword in a scabbard. The female was easily pulled out of the male's groove and was recognised most clearly by its internal structure.[45]

Bilharz called this worm *Distomum haematobium*—*Distomum* because he believed it belonged to the previously described genus of that name, and *haematobium* from the Greek words *haema* and *bios* meaning 'blood' and 'life', respectively. In 1857 Thomas Spencer Cobbold (1828–1886), an English doctor, found a similar worm in a sooty monkey in the London zoo and thought these worms were sufficiently distinct to be placed in a new genus which in 1859 he named *Bilharzia* in honour of Bilharz. Unfortunately David Friedrich Weinland (1829–1915), a German naturalist, had in the previous year decided to place these worms in a new genus named *Schistosoma* derived from the Greek words *schistos* and *soma* for 'split' and 'body', respectively. Argument was to go on for years but eventually the International Rules of Zoological Nomenclature prevailed and the first valid name was accepted. So today we have *Schistosoma haematobium*.

But of course finding worms in the blood vessels within the abdominal cavity was one thing. The question soon arose: 'How did they get there?' Bilharz was to find the first clues. First, he saw eggs in the uterus of the female worms and noted that each had a terminal spine, i.e. a spike at its tip (Figure 19). This might not seem a big deal but make a note of it because it would turn out to be very important. This immediately led to a new question: 'What happened to these eggs and where did they go?'

Bilharz provided the answer. In 1852 he found these same eggs in the lining of the urinary bladder and saw that each one had a larva within it which was

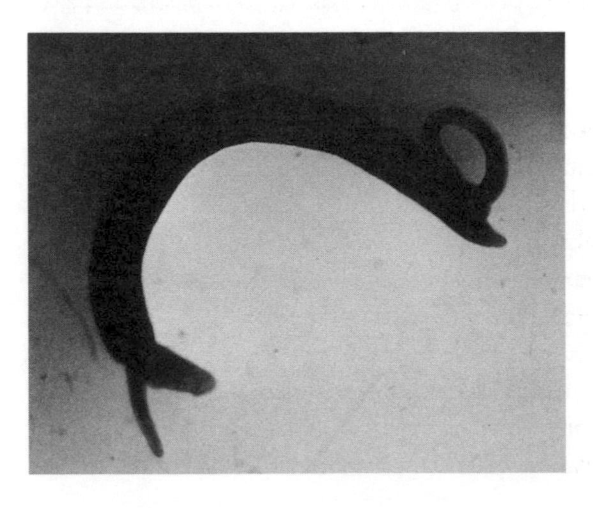

Figure 18. Adult *S. mansoni* worms. The shorter, fatter male worm is wrapped around the longer, thinner female worm.

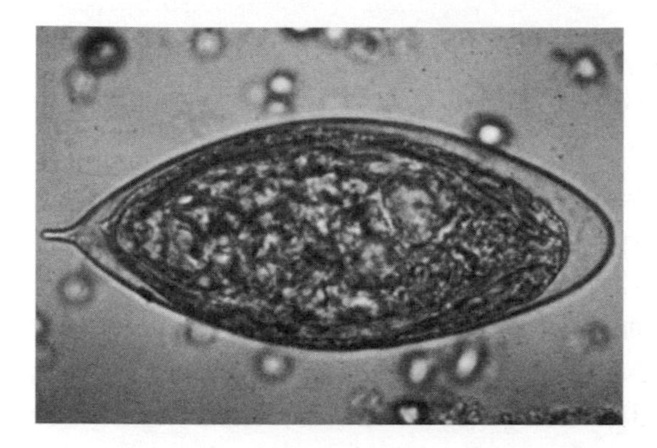

Figure 19. *S. haematobium* egg. Note the terminal spine at 9 o'clock and the larva can be seen within the eggshell.

moving about. What is more, he found these eggs in the urine of a young man who was complaining of blood in his urine. So this much was now clear: worms in the blood vessels around the bladder produced eggs which penetrated through the bladder wall into the bladder cavity and were excreted in the urine.

What then? Bilharz watched the eggs under a microscope and saw that each eggshell broke and a larva was hatched which then swam around in water with a lively rotating motion. These larvae have a technical name which we need to learn to distinguish them from a later form of larva; they are called miracidia (singular: miracidium). Everyone thought it would be easy to find out how these miracidia infected other people. How wrong they were! In the last chapter (p.36) we met Wilhelm Griesinger, who did much important work on hookworm infection in Cairo. In 1854 he wrote:

> Had I remained longer in Egypt, I would have set myself two large practical tasks. First, I would wish to discover the ways in which these entozoa [i.e. miracidia] penetrate the body. This is relatively easy considering the simple food of the people.[124]

Little did Griesinger realize that more than 60 years of confusion, concoction, and controversy would have to pass before light dawned on this problem. He pointed a finger of suspicion at three main culprits: impure water from the River Nile, contaminated bread or dates, and the half-rotten fish beloved by the local inhabitants. He thought that either eggs or larvae or mature worms ought to be able to be found in one or other of these vehicles of transmission.

The first person to undertake some experiments was John Harley (1833–1921), a London physician. This infection was present not only in Egypt but also in South Africa, where it was called 'Endemic haematuria of the Cape of Good Hope', haematuria being the technical name for blood in urine and endemic meaning that it was common in the area. Harley had an idea. Perhaps miracidia-infected snails acted as an intermediate host and then in turn infected humans? This might seem a rather weird notion but it was known at the time that some distantly related fluke worms were sometimes found in snails. So in 1869, Harley asked a medical colleague working in South Africa to send him some freshwater snails from an endemic area. He examined these snails, which were of a particular type called *Unio kaffre*, but could find no trace of schistosomes. But Harley did not give up. He asked his South African colleague to send him some schistosome eggs. In 1870 in what could be called a 'direct infection experiment' he fed the eggs in food to rabbits and dogs. When he killed the animals two to six months later, he could find no trace of schistosomes. This idea had not worked either.

Meanwhile, Cobbold was also working on the problem. He used eggs obtained from the urine of some of his patients who had been infected in South Africa to attempt to infect many species of fish and snails which could perhaps carry the worms. He reported in 1872 that he had had no luck. Cobbold was a rather cocksure and dogmatic character and began to attack Harley, who it must be admitted was often rather tortuous and circumlocutory in his writings, in rather forthright but unfair terms. This did not seem to bother Harley, who had the habit of smiling, bowing, and shaking hands with every acquaintance that he met on every possible occasion.

What neither Harley nor Cobbold appreciated was that the problem was not going to be solved in England. It was going to have to be sorted out in an area where the infection was endemic. Permeating both the 'direct' and 'snail' theories was the widely held belief that water was somehow intimately concerned with transmission of infection, either by drinking or bathing in infected water. Cobbold declared that it was quite clear to him that people in Africa were infected when they drank unfiltered water. It would turn out that Cobbold was completely wrong.

On the right track was Dr Richard Rubidge (1820–1869) of Port Elizabeth in South Africa. He observed that many boys who bathed in the river near his home began to pass blood in their urine whereas those who bathed only in the sea did not develop this problem. What is more, one of the boys who passed blood in his urine had developed a very nasty skin rash after bathing in the river. It seemed likely to Rubidge in 1869 that somehow the parasite gained entrance through the skin while a person was bathing in the fresh water of a river. Not everyone agreed, however, and many still thought the infection was most probably acquired by drinking water contaminated with the worms.

Frustrated by the inability to prove the mode of infection, three separate European missions, one from Italy, another from France, and the third from Germany, were sent to North Africa in 1893 and 1894 to settle the question once and for all. The French team put eggs directly into the stomach of guinea pigs via a tube, injected eggs intravenously into rabbits, fed eggs in food to monkeys, and kept sheep with their legs in water containing miracidia for three months. All experiments ended in dismal failure as did attempts to infect at least six different genera of snails with miracidia.

The German, Dr Arthur Looss (p.32) showed that miracidia could not survive in stomach juice or dilute hydrochloric acid so reasoned that they must therefore get in directly via the skin. No doubt influenced by his amazing discovery in 1898 of hookworm infection being acquired by direct penetration of the skin, Looss remained convinced that schistosome infection must be acquired in the same way and there was no need for a snail intermediate host, and he obstinately continued to hold this view for years to come.

The third investigator was the Italian physician, Prospero Sonsino (1835–1901). He had in fact worked in Egypt between 1873 and 1884 and had tried to infect many species of snails with *Schistosoma haematobium* miracidia without success. Moreover, he had dissected umpteen specimens of many different species of snails caught from the wild but again had found no schistosomes. He was oh so close, but when he went to Tunisia in 1893 he blotted his copybook. He wrote to the British medical journal *The Lancet* in triumph saying that he had discovered that the intermediate host of this worm was a small crustacean (rather like a mini crab) and that human infection was acquired by drinking water contaminated with infected crustacea. *The Lancet* was impressed and wrote an accompanying editorial lauding Sonsino's discovery. But it was all too soon and further studies revealed that Sonsino's 'discovery' was untenable. It was completely wrong. Sonsino retracted his claims, although he did this only in the obscure *Proceedings of the Tuscan Society of Natural Science* in Pisa rather than in *The Lancet* where everyone could see it.

All this work had come to nought. And nothing else came to light. The authoritative *prima donna* Arthur Looss held pride of place in Cairo and no-one could gainsay his insistence that infection must be acquired by miracidia penetrating the skin. But then in 1914 came the outbreak of the World War I and the British unceremoniously booted the enemy alien Dr Looss out of Egypt. The field was now clear for a fresh attempt to solve the vexed question of the manner of transmission of urinary schistosomiasis.

The British army authorities were very worried about this infection. When Napoleon had invaded Egypt in 1798, large numbers of his troops were infected. The army surgeon Renoult described it this way:

> A most stubborn haematuria [blood in urine]... manifested itself among the soldiers of the French Army.... the continual and very abundant sweats diminished the quantity of urine, the latter becoming thick and bloody. Often even, the last drops are pure blood. The sickness gives sharp pains in the region of the bladder, the last contractions of the bladder are accompanied by the most lively and piercing pains.[266]

What is more, the same problem had afflicted British troops in South Africa during the Boer War of 1899–1901. And now it was happening all over again to Imperial troops in Egypt. For example, 75 troopers of the Australian Light Horse went bathing in an oasis in Egypt. Four to ten weeks later, more than half of the soldiers suffered from fevers, shivers, sweats, headaches, muscle pains, skin rashes, and emaciation lasting for several weeks. This was followed a few weeks later by pain when passing urine then blood appeared in the urine. Something had to be done!

The British War Office gave temporary commissions to a medical parasitologist, Dr Robert Leiper, and two colleagues, Drs RP Cockin and JG Thomson, and sent them to Egypt with the brief to investigate 'bilharzia disease in that country and advise us as to the preventive measures to be adopted in connection with the troops'.[126] Leiper was not exactly a novice as far as schistosome infections were concerned. He had been to Japan!

We now have to suspend our story on urinary schistosomiasis and move to the Orient and back in time. In 1888, a Japanese doctor named Tokuho Majima performed a post-mortem examination on a man who had been suffering from fluid in his abdomen (ascites) and fluid in his legs (peripheral oedema). He found that the patient's spleen (an organ in the left side of the upper abdomen, the blood from which flows into the liver) was enlarged five times but that the liver was shrunken and contracted and full of nodules. When he looked at the liver under the microscope, he found that the nodules contained eggs, which no-one had ever seen before. He opened the bile duct (bile flows out of the liver to the gall-bladder and intestine) and the portal vein hoping to find adult worms but found nothing.

Sixteen years later in 1904, Fujiro Katsurada (Figure 20), professor of medicine at Okoyama Medical College, went to Yamanishi prefecture in Japan where many patients had enlarged spleens, ascites, peripheral oedema, anaemia, and wasting.

Figure 20. Fujiro Katsurada (1867–1946).

He looked in the faeces of these patients and found eggs that were the same as those that had been found two years earlier by a colleague, Kenji Kawanishi (1868–1927). Katsurada, however, described the eggs in detail. He concluded that the most similar eggs were those of *S. haematobium* but that the Japanese eggs, although having a miracidium within them, were rounder, shorter, and had no spine but possessed a little tiny knob on one side. Katsurada was unable to perform any autopsies on patients with the symptoms and signs described above, but he was able to review three livers and one intestine obtained at post-mortem examination a few years earlier by another couple of Japanese doctors. When he looked carefully, he found eggs in a scarred liver and in a rectal tumour (the rectum is the lowermost part of the large intestine before the anus). Then Katsurada started to look at animals. He did not find anything in dogs but in a cat he found the same eggs in its stools. When he looked in the portal vein he found some adult worms and wrote that 'they were males, probably of Bilharz's *Schistosoma haematobia* (or belonging to the same genus) and had not been seen in Japan before'.[156] Katsurada was right to hedge his bets. Events were to prove that this worm did indeed belong to the genus *Schistosoma*, but it was a new species, *japonicum*. Later in that same year, other Japanese doctors found adult worms in the portal vein of humans who had died.

All of this was unknown to Dr John Catto (1878–1908), an English medical officer at the quarantine station in Singapore. In 1904 cholera broke out in a passenger ship from China. One of the passengers died and Catto performed a post-mortem examination. The liver was scarred and the large intestine was swollen and ulcerated. Parts of these organs were preserved and prepared for examination under the microscope. Amongst other things, he found some eggs of an unknown parasite. Catto went back to England to the London School of Tropical Medicine and took all this material with him. Eventually, he found some adult worms which were clearly flukes in the intestinal blood vessels. These were sent to the International Zoological Congress in Bern, Switzerland, where they were examined by the most eminent European parasitologists, all of whom agreed that these worms were of a schistosome new to science. In 1905, one of these eminent men, Raphael Blanchard, named them *Schistosoma cattoi* in honour of Catto. But it was too late; in the previous year Katsurada had already called them *S. japonicum*.

With remarkable speed and thoroughness, Japanese investigators then set out to find out how this infection was transmitted. The Japanese had both a fact and luck on their side. The fact was that *S. japonicum* infects a wide variety of animals in nature and does so commonly, whereas African schistosomiasis in nature is primarily a disease of humans. The luck was that the Japanese managed to infect snails at almost their first attempt. What was not known then, but which is well understood now, is that only a very small number of species of snails are capable of being infected with *S. haematobium* and only a small number of different species can be infected with *S. japonicum*.

It was the ability of many different animals to be infected that enabled the route of infection to be proven. In 1909, two doctors, Akira (also known as Kan) Fujinami (Figure 21), professor of pathology at the Imperial University of Tokyo, and his colleague Hachitaro Nakamura, devised a cunning experiment with cows, large, docile animals which could be made to stand in water for hours at a time. Three cows were placed in the mud of a paddy field and three were placed in a river which received large volumes of water from the paddy fields. All became infected but much larger numbers of worms were found in the cows that had been in the rice field. In contrast, another group of seven cows had their legs washed with soap and alcohol, then they were oiled and covered with waterproof, protective leg bags. Four were put in the rice fields and three were placed in the river. Six cows remained uninfected and a solitary worm pair was found in the seventh. Clearly, infection was acquired by worms penetrating the skin. But the question remained: 'What did the worms that penetrate the skin look like?'

Figure 21. Akira Fujinami (1870–1934).

This was to be answered just a few years later by Keinosuke Miyairi (1865–1946) and Masatsagu Suzuki of Kyoto University

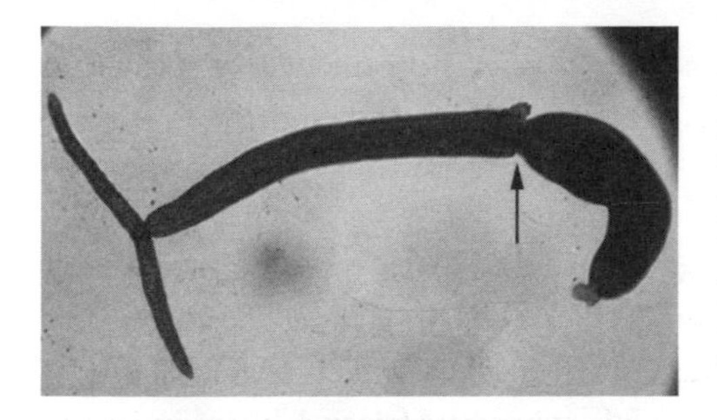

Figure 22. Cercaria stained with iodine to make it more visible. Note the forked (bifid) tail. When the larva passes through the skin, the part to the left of the arrow falls off and the part to the right known as a schistosomulum migrates through the body.

in Japan. First they found an ox with schistosomiasis and used its dung as a source of schistosome eggs. They then went to a village where human faeces left on the roadside had the same ova. Then they looked in a nearby ditch by a rice field and found some snails. They later wrote, 'as amateurs, we couldn't tell the exact name of the snail'.[219] Nevertheless they kept the snails happy by putting them in containers with water and cabbage leaves, then mixed in miracidia prepared from the ox dung. In fascination, no doubt using magnifying glasses or a special microscope, they watched what happened and recorded it thus: 'The way a fresh miracidium rushes up to its host can be compared to that of a hungry tiger coming out of his cage to hunt for something to eat.'[219]

Miyairi and Suzuki found that the miracidia penetrated into the interior of the snails, then over the next several weeks transformed themselves into a strange body called a sporocyst, which in turn developed a new generation of smaller secondary sporocysts within them, then within those, yet more larvae developed. These larvae had a distinctive shape and are called cercariae (singular: cercaria, Figure 22). These cercariae were able to escape from their snail hosts and swim around in water. Finally, when mice were placed together with infected snails, they were able to recover adult male and female worms from the mice. The main points of life cycle of *S. japonicum* were now clear (Figure 23).

At least they were to the Japanese. Miyairi and Suzuki published their findings in 1913 but only those who could read Japanese knew what they had found. A summary of their experiments was published in English in March 1914 and a German translation of their paper later that year. Thus Leiper (Figure 24) and Royal Navy surgeon Edward Atkinson (1882–1929) knew nothing of this when they left London in February 1914 with the avowed purpose of ascertaining 'the mode of spread of the trematode diseases [such as schistosomiasis] of man'.[126] They arrived in Shanghai, China, at the end of March. Imagine their chagrin, if not despair, when they eventually heard of the Japanese discoveries.

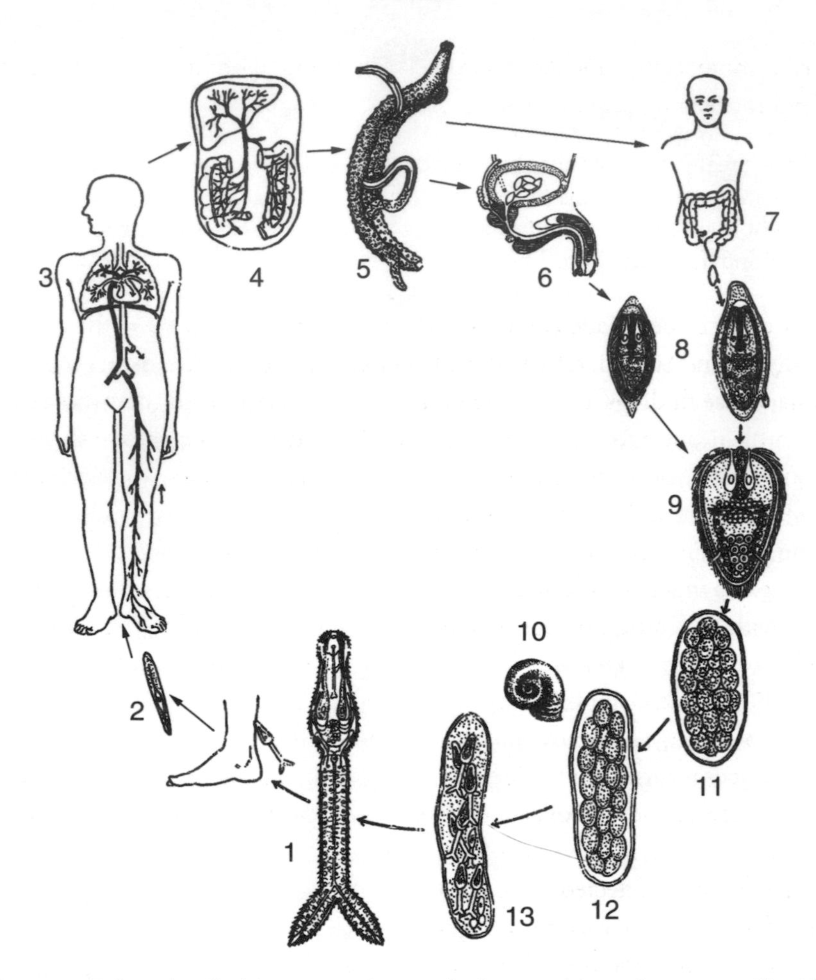

Figure 23. Life cycle of schistosomes. Start at the bottom where the cercaria (1) with a forked tail penetrates the intact skin, loses its tail, and the resultant schistosomulum (2) travels via the bloodstream through the lungs and heart (3) to the veins around the bladder (*S. haematobium*) or the veins of the intestines and liver (*S. japonicum* or *S. mansoni*) (4). Here they develop into adult worms (5) which produce eggs that enter the bladder and urine (*S. haematobium*) (6) or liver, intestines, and faeces (*S. japonicum* and *S. mansoni*) (7). When the eggs (8) are deposited in fresh water, the miracidium (9) hatches and swims around till it finds a suitable snail (10) in which it develops first into a sporocyst (11) in which develop many secondary sporocysts (12) in which develop many cercariae (13). The cercaria escapes from the snail and swims around in the water until it finds a human. Note: the relative sizes are not to scale.

Furthermore, Leiper and Atkinson had a falling out, with the latter writing to a friend concerning Leiper:

> I don't really think old chap that you can fully realise how perfectly damnable this man can be... This fellow has really been too damnable for words... I ought to give him a sound thrashing, tell him what he is for the good of his soul and then leave him.[32]

In the event Leiper headed off to Japan, collected snails, the same as those found by Miyairi and Suzuki, returned to Shanghai, then he and Atkinson confirmed the Japanese findings and infected mice with cercariae from them. Since war had broken out and Atkinson was a serving naval officer, they returned to London taking both snails and infected mice with them. The snails were identified by the British Museum as a species of a new genus of snail and given the name *Katayama nosophora* (now known as *Oncomelania nosophora*).

So it was that Leiper found himself in Egypt in February 1915. Again he was beset with difficulties. Within a month Dr Cockin fell ill and had to be invalided back to Britain, Leiper was hospitalized with scarlet fever, and the terrible casualties of the Gallipoli campaign caused Dr Thomson to join the general service of the Royal Army Medical Corps. Nevertheless, in the months until July when Leiper returned to England, they made remarkable progress. By a stroke of irony, Leiper and his colleagues set up their headquarters in the laboratories vacated by Looss in the Cairo School of Medicine. They collected almost 30 species of snails from various villages and dissected them looking for cercariae similar to those that Leiper had learnt to recognize in Japan. Three sorts of cercariae were seen and attempts were made to infect many different birds and animals. Eventually a mouse and a rat were found to have worms in the blood vessels of the liver and intestines but they were immature and it was not possible to be sure which species of worm they were. But clearly they were on the right track. They were replicating what had been found in the Orient.

Figure 24. Robert Leiper (1881–1969).

Encouraged by this success, Leiper infected mice, rats, guinea pigs, and monkeys with cercariae from a snail then called *Planorbis boissyi* (now *Biomphalaria alexandrina*) and took them back with him to England. The mice, rats, and monkeys all became infected but they were so heavily infected that they soon died. But before they died, Leiper noticed something odd. The adult worms produced eggs which instead of coming out in the urine were found in the stool. Furthermore, instead of having a spine at the end of the egg as Bilharz and others had found, these eggs all had a spine on the side (called a lateral spine, Figure 25). Such eggs had been seen before; Patrick Manson (p.58), in 1902, had described a patient of his who had become infected with schistosomiasis in the West Indies and had lateral-spined eggs in his faeces but no eggs at all in his urine.

It was beginning to look as though there may be two distinct forms of schistosomiasis hopelessly intermingled in Africa even though some such as the aforementioned Dr Looss vigorously would not countenance such a possibility, claiming that the same worm produced terminal or lateral spines depending upon where it was situated in the body. So Leiper thought he had better go back to Egypt and look into this question more closely, which he did in November 1915. To cut a long story short, Leiper showed that when he infected mice with cercariae from the snail that he called *Planorbis boissyi*, the adult worms always produced lateral-spined eggs, but when he infected them with cercariae from a different species of snails called *Bulinus*, the eggs always had a terminal spine. He then repeated these experiments in monkeys with similar findings.

It was now clear that there were two distinct forms of schistosomiasis in Egypt. The first, caused by *S. haematobium* and transmitted by the freshwater snail *Bulinus* species, affected the veins supplying the urinary system, caused bleeding into the urinary tract, and terminal-spined eggs were passed in the urine. The second was caused by a novel parasite, which Leiper named *Schistosoma mansoni* in honour of Patrick Manson and was transmitted by *Planorbis* (= *Biomphalaria*) freshwater snails. This worm lived in the veins of the intestines and the liver and its eggs damaged both these organs, with the lateral-spined eggs being excreted in the stools; patients with heavy infections might develop dysentery (diarrhoea with blood and mucus) and liver disease.

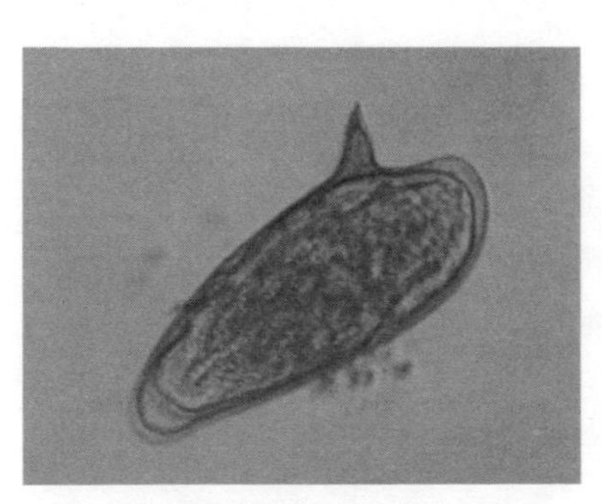

Figure 25. Egg of *S. mansoni*. Note the lateral spine.

Thus infection in a community was dependent upon several factors. First, infected people had to urinate in freshwater streams or pools (*S. haematobium*) or defecate into them (*S. mansoni*), or infected people or animals passed stools into freshwater bodies, especially paddy fields (*S. japonicum*). Secondly, snails that were susceptible to infection and would permit development through miracidia, sporocysts, and cercariae had to be present in those water bodies. This explains why, when African slaves were transported to the Caribbean and South America, only intestinal schistosomiasis mansoni could take hold; *Biomphalaria*, but not *Bulinus* snails which transmit *S. haematobium*, were present in those areas. Likewise, intestinal schistosomiasis japonica is only present in those parts of Asia where *Oncomelania* snails are present.

Only a few things remain to be said. Subsequent investigations showed that cercariae were able to penetrate intact skin, losing their spectacular bifid tail in the process. The resultant larval worm, called a schistosomulum, entered the veins and passed through the heart and lungs then arteries of the intestines or bladder, depending upon the species, to settle in the veins of those organs where it developed into an adult worm. Fertilized female worms, in turn, produced eggs which either passed via the bloodstream to the liver or entered the intestinal tract (in the case of *S. mansoni* and *S. japonicum*) or entered the bladder wall (in the case of *S. haematobium*). These adult worms can live for 5–10 years but fortunately we now have a drug called praziquantel which can kill them, although it does not repair any damage they have already done.

What is the moral of all this? Don't go wading or bathing in any water bodies where these snails and worms live, for the parasites are desperately awaiting a chance to go boring through your skin. What is good for them is not good for you.

5

Filariasis (elephantiasis)

SCIENTIFIC NAMES:	*Wuchereria bancrofti, Brugia malayi*
COMMON NAME:	filaria
DISEASE NAME:	filariasis or elephantiasis
DISTRIBUTION:	tropics and subtropics
TRANSMISSION:	bite of an infective mosquito
LOCATION OF THE WORM IN A PERSON:	lymph nodes and vessels, scrotum
CLINICAL FEATURES:	swollen, thickened legs or rarely arms (elephantiasis); fluid in the scrotal sac around the testis (hydrocoele); milky urine (chyluria)
DIAGNOSIS:	finding larvae (microfilariae) in the blood
TREATMENT:	diethylcarbamazine and ivermectin but results are unsatisfactory
PREVENTION:	avoid being bitten by mosquitoes, especially at night

Swollen legs have a number of causes, the most common being heart failure with fluid accumulating in the legs because the pumping action of the heart is impaired. If you press your finger into the foot of a patient with this problem for a few seconds then remove it, an impression of your finger is left behind; the technical term for this is 'pitting oedema'. Until recent times, with the advent of modern medical treatment, this heralded death within months or occasionally a year or two. But in tropical countries for many centuries another form of swollen legs has been recognized. People are often young when they first show signs of the illness and they live for decades. Furthermore, this fluid becomes chronically entrenched in the tissues so that if you press a leg with a finger, little impression is made; this is called 'non-pitting oedema'. Even worse, the skin becomes very thickened and sometimes warty, resembling the skin of an elephant's leg, so not surprisingly this disease came to be called elephantiasis.

This condition has been recognized for centuries. A statue of the Egyptian pharaoh Mentuhotep III (*c.* 2000 BC) shows pronounced enlargement of both his legs that was probably due to filariasis. Indeed the Roman Lucretius Carus, writing in the first century before Christ, regarded it as a characteristic disease of the Egyptians which he called 'elephas morbus' (morbus means disease and elephas refers to elephants). He thought it must be due to the climate around the Nile. Fifteen hundred years later when the Portuguese Tomes Pires went to the Malabar Coast of India (now the state of Kerala) in 1512, he wrote:

> A quarter or a fifth of the total population, including the people of the lowest castes, have very large legs, swollen to a great size, and . . . it is an ugly thing to see . . . the swelling is the same from the knees downward, and they have no pain nor do they take any notice of this infirmity.[252]

Indeed, a tradition arose that Thomas the disciple of Christ had lived and preached in this region of India and had been slain with a lance while praying in church; this in turn led to the legend that elephantiasis was the result of the curse of St Thomas.

Elephantiasis is found in both men and women. But men were sometimes afflicted with another problem, occurring either alone or in association with leg problems. The male genital organs include two oval-shaped bodies called testes (= testicles). These are each surrounded by a sac called the tunica vaginalis. This sac has two continuous layers with only a thin film of fluid separating the two. Imagine surrounding an almond with a balloon without any air in it and you have a picture of a testis surrounded by the tunica. Both sets of testes with their tunicas are surrounded by skin, fat, and muscle; this is called the scrotum. Now imagine filling the balloon around the almond with air—the whole thing swells up and the almond gets buried. In the case of the men we are talking about, however, it is not air that fills the tunica vaginalis but fluid. When this happens, the scrotum swells on one or both sides and the penis may become difficult to discern. This condition is called a hydrocoele.

Hydrocoeles may be found in men anywhere in the world, but they are much more common in areas where elephantiasis occurs and begin to afflict men at a younger age. It was especially common in West Africa and John Barbot, a French slave trader, wrote in 1732:

> Here is another unknown and foul distemper the Blacks are subject to throughout all the country about Sierra Leone . . . a wonderful swelling of, or in the Scrotum – which causes violent pains, and hinders their co-habiting with women.[37]

The relationships among elephantiasis, hydrocoeles, and inflammation of the lymph vessels (= lymphatics) were unclear and were the subject of much controversy for many years. But to this constellation of disorders, some authors gave the name 'filariasis', for reasons which will become clear shortly, a term which has stuck.

It was not until 1862 that the first glimmerings of light began to be shed. An 18-year-old man from Havana, Cuba, presented to a hospital in Paris with a tumour on the left side of his scrotum. The surgeon, Jean-Nicolas Demarquay (1814–1875), inserted a trocar (a large hollow needle) and sucked out whitish-yellow fluid similar to milk. The following year, the patient came back with a similar problem on the other side. Again 100ml of bluish-white fluid was aspirated and Demarquay demonstrated that the fluid was in the tunica vaginalis. This time, the fluid was examined by one of the house surgeons, Dr Lemoine. He found fat globules, pus cells, fibrin filaments, and many specimens of a parasite and Demarquay wrote:

> Attention was drawn above all to a little, elongated and cylindrical creature...the posterior fifth became thinner and thinner and terminated in a fine point. This worm had extremely rapid movement of coiling and uncoiling...The worm was completely transparent and did not show anything which resembled the digestive system or the genital system.[83]

The worms were about 0.3 mm long. A specimen was sent off to the parasitologist CJ Davaine (p.197) for an expert opinion but he searched in vain and could not find any worms. On the basis of drawings he was shown, he thought the worms were probably a larva (immature form) of a nematode (roundworm). Davaine did have one flash of insight, however; he thought the problem may be due to the patient having lived in Cuba.

And so matters lay without anyone paying much attention. Three years later in 1866 in Bahia, Brazil, completely unaware of the French discovery, a German doctor, Otto Wucherer (Figure 26), made similar observations except this time he found the worms in urine. He was looking for eggs of *Schistosoma haematobium*, which we discussed in the previous chapter. As it happens, that form of schistosomiasis does not occur in South America but Wucherer did not know that. One day in August he obtained a clot of milky urine from a female patient and examined it under the microscope and saw 'some threadlike worms which were very thin at one end and blunt at the other...the body was transparent and seemed to contain a granular mass'.[329] Wucherer attached no great importance to this discovery but on 9 October 1866 he found similar worms in the

Figure 26. Otto Henry Wucherer (1820–1873).

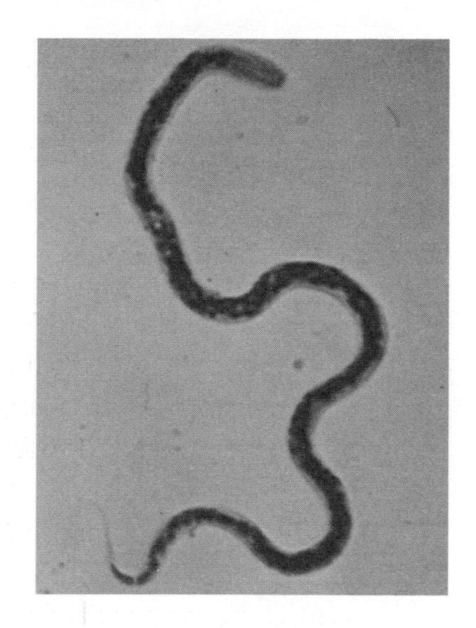

Figure 27. A microfilaria showing stained dark nuclei. Stains were not available to Lewis and he saw transparent worms.

urine of another lady, and then in the milky urine from a man. He felt he was on to something so sent specimens preserved in urine to Rudolf Leuckart (p.31) in Leipzig, Germany. Leuckart dismissed them as being of no importance whatsoever.

Unaware of all this, in 1870, Timothy Lewis (1841–1886), a British medical officer in India found the same parasite in milky urine obtained from a 25-year-old East Indian. Lewis then went on to find the same worms in another 15 or so patients. Two years later, Lewis made an even more important discovery. In July 1872 while examining the blood of a patient with diarrhoea, he was astounded to observe 'nine minute nematoid [round] worms in a state of great activity in a single slide'[186] (Figure 27). He showed them to a colleague, David Douglas Cunningham (p.150), who concurred that they were the same worms that had previously been found in milky urine. Lewis went back to check the patient but was greatly disappointed to find the patient had discharged himself one hour previously and he could not be found. Lewis heard of another woman who had milky urine, so decided to have a look at her blood. Luck was on his side this time, for he went to see her in the evening (you will see the significance of this shortly) and he found worms in her blood. Lewis kept an eye on this patient for the next two months and noticed that although they were always present in similar numbers in the urine, the numbers in the blood diminished markedly; as we shall see, this was probably because he took the samples during the day. Shortly afterwards he encountered a third patient and was amazed to find:

> that no matter what portion of his body the circulation is tapped with the point of a needle, numerous active, well-developed Haematozoa [blood-dwelling worms] are invariably obtained.[186]

He went on to calculate that one of his patients was host to some 140,000 of these parasites. He wrote up his findings and went to the Government Printing Works to examine the proofs of his report and was astounded to find that the typesetter was none other than the very first patient in whom he had found worms in his urine; he examined the man's blood and again found the same parasites. Incidentally, Lewis was later to pay for his medical curiosity, for he died in England at the early age of 44 from septicaemia following an accidental wound he received while performing a post-mortem examination.

News of Lewis's discoveries eventually filtered through to England and were reported in *The Lancet* in August 1872.[186] One of his specimens had been examined by George Busk (1807–1886), a renowned English helminthologist, who determined that they were worms which belonged to the genus *Filaria*, the name being derived from the modern Latin 'filarium', meaning 'ball of thread'. Busk named them (against all the rules which require that every organism be given only a genus name and one species name) *Filaria sanguinis hominis*, meaning filaria in human blood.

Since all of this was reported in *The Lancet*, which had an international circulation, the word spread. In 1874, Sonsino in Egypt and Rowland and Bancroft independently in Australia found the larval worms in the blood of patients. Clearly these worms were important and widespread in the tropics. But they were just immature larval worms. Where did they come from and what happened to them?

Joseph Bancroft (Figure 28) was an English medical practitioner, born in Manchester, who had moved to Brisbane, Australia, in 1864. He sent samples of his larval specimens to Spencer Cobbold (p.40) in London, the most eminent of British helminthologists. Cobbold wrote back to Bancroft, suggesting he look for adult worms which he felt must be somewhere in the body. Bancroft took up Cobbold's suggestion and on 21 December 1876 found an adult worm. Subsequently he found four more then he wrote to Cobbold in April 1877:

> I have laboured very hard to find the parental form of the parasite and am glad to tell you that I have now obtained five specimens of the worm. The worm is about the thickness of a human hair, and is from 3–4 inches long. By two loops from the centre of its body it emits the filariae [i.e. the larvae] in immense numbers. My first specimen I got on December 21st 1876 in a lymphatic abscess of the arm. Four others I obtained alive from a hydrocoele.[35]

A photograph of a worm similar to those seen by Bancroft is shown in Figure 29.

Cobbold was so impressed by Bancroft's findings that he sent Bancroft's letter together with some explanatory notes to *The Lancet*, writing 'Such Sir, is Dr Bancroft's account of his "finds", and from the brief description furnished I propose to call the adult nematode *Filaria Bancrofti*'.[35] Bancroft's discoveries were then confirmed by Lewis in India and by Antonio da Silva Araújo and by Felício dos Santos independently in Brazil later that year. In 1921, the parasitologist Léon Gaston Seurat (1872–?) examined these worms carefully and thought that they deserved to be placed in their own genus, which he named *Wuchereria* after the afore mentioned Wucherer. So, *Filaria bancrofti* became *Wuchereria bancrofti*.

So far so good. It was now clear that these larval forms, now called microfilariae, were produced by adult worms which lived in the lymph vessels near the

Figure 28. Joseph Bancroft (1836–1894).

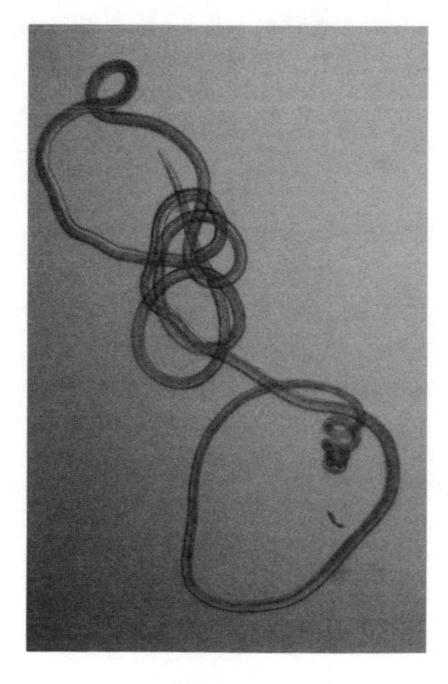

Figure 29. An adult worm. This one happens to be *Brugia malayi*.

lymph glands (= lymph nodes) or in the tunica vaginalis. But where did the adult worms come from and where did the microfilariae go?

The answer was to be provided, at least in part, by Patrick Manson (Figure 30), a Scottish doctor who went to Taiwan in 1866 and saw many patients with swollen legs and hydrocoeles. In 1875 he returned to England where he acquired a wife, a microscope, and knowledge of Lewis's discovery. The following year saw him posted to Amoy, now called Xiamen, a coastal city in south-eastern China. He soon found microfilariae in the blood of a number of his patients and deduced that their most likely means of exit was via a blood-sucking insect. He considered fleas, bedbugs, lice, mosquitoes, and sandflies. He decided to try mosquitoes first so he caught a few and on 10 August 1877 let them bite his gardener, Hin-Lo, who had many larvae in his blood (technically called a microfilaraemia). He then examined the abdominal contents of samples of the mosquitoes at daily intervals. He was later to say:

> I shall not easily forget the first mosquito I dissected. I tore off its abdomen and succeeded in expressing the blood the stomach contained. Placing this under the microscope, I was gratified to find that, far from killing the filaria, the digestive juices of the mosquito seemed to have stimulated fresh activity. And now I saw a curious thing. The little sac or bag enclosing the filaria, which hitherto had muzzled it . . . was broken through and discarded.[207]

Figure 30. Patrick Manson (1844–1922).

Manson went on to observe that over a week or so within mosquitoes, the microfilariae underwent a series of metamorphoses, i.e. increases in size, changes in shape, and development of the internal organs. Manson undertook these studies with the two mosquitoes which were common in the area—they are now called *Culex quinquefasciatus* (previously *C. fatigans*) and *Aedes aegypti*. Furthermore, he observed that only female mosquitoes ingested blood and that the replete insects flew off to nearby stagnant water and deposited eggs.

What happened next? Manson surmised that the mosquitoes died in the water and released the metamorphosed forms of the larvae (now called infective larvae) which were then ingested by humans in whom they developed into adult worms. Manson was wrong about that but he was right about microfilariae developing in mosquitoes and was the first person to indicate that an infectious agent could be transmitted by mosquitoes.

Manson did not know much about mosquitoes—their distribution or their life span—nor was he aware that female mosquitoes could bite more than once. But in this he was not alone. He wrote to the British Museum in London seeking further information about mosquitoes but the relevant authority replied that regretfully no such work existed and forwarded him a treatise on cockroaches in the hope that this would do instead! Manson sent news of his discovery to Cobbold, just as Lewis had done. Cobbold passed the information on to *The Lancet* and reported Manson's findings in more detail to the Linnean Society of London. This finally stimulated the *British Medical Journal* to publish an editorial entitled 'Is the mosquito the intermediate host of the Filaria sanguinis hominis?'[15]

Manson had squashed the whole of the posterior portion of the mosquito and had assumed that the various events all took place within the gut. Timothy Lewis in 1878 set about trying to repeat Manson's observations but did so somewhat more carefully. He removed the entire alimentary tract and found that after the third day the larvae perforated the gut lining and migrated to the thorax of the insect. So in 1883 and 1884 Manson dissected over 1,000 mosquitoes fed on the hapless Hin-Lo and showed that the most advanced forms were found one week after the blood meal. They had undergone at least two moults and were 1.5 cm long.

Meanwhile, a remarkable discovery had been made. During 1876 and 1877, Manson trained two Chinese assistants to make blood examinations. One worked during the day and the other laboured at night. Manson was struck by the fact that the night worker found more parasites than did the daytime worker, but at first the penny did not drop. In 1879 he gave directions for a particular patient's blood to be examined daily. He then realized that more microfilariae were being found on busy days when the examination was delayed until the evening. He decided to systematically examine the patient every few hours. With great excitement he wrote to Cobbold:

> The young escape into the circulation at regular intervals of 24 hours, the discharge commencing soon after sunset and continuing till near midnight, from which time till the following noon their numbers gradually decrease. By 2 or 4 o'clock till 6 they are nearly completely absent…It is marvellous how nature has adapted the habits of the filariae to those of the mosquito.

The embryos are in the blood just at the time the mosquito selects for feeding.[206]

Cobbold recounted Manson's findings, which were termed 'nocturnal periodicity', to the Quekket Microscopical Club in London in 1880. They were greeted with astonishment by some and downright disbelief by others, with one wag enquiring 'whether the filariae carried watches'.[126] Because of this phenomenon, nearly 100 years later, I had to acquire a curfew pass to allow me to go out and about at midnight when I was researching filariasis in the Philippines during the military dictatorship of Ferdinand Marcos. The scoffers were silenced later that year when Stephen Mackenzie (1844–1909) in London found a patient who had acquired filarial infection in India. He persuaded the young man to sleep by day and stay up at night and found that after a week the periodicity was reversed!

What is more, this patient died two years later in 1882 and Mackenzie was able to do an autopsy. He found that the thoracic duct (the tube into which all lymph vessels feed before it in turn runs into the bloodstream) was obliterated and embedded in inflammatory tissue and that the lymphatics below the obstruction were dilated enormously, thus confirming that the habitat of mature filariae in humans is the lymphatic system and suggesting that chyluria (milky urine) is due to rupture of distended lymphatics into the urinary tract.

And so matters rested for many years. In 1899, a man who was known to have a microfilaraemia with nocturnal periodicity committed suicide with prussic acid and died almost instantly at 8.30 in the morning. Manson concluded from a post-mortem examination that during the day, most of the microfilariae become lodged in the blood vessels of the lungs.

In that same year of 1899, Joseph Bancroft's son, Thomas (1860–1933), also a medical practitioner in Queensland, discovered that mosquitoes could be bred and kept alive in confinement for up to two months when fed upon ripe bananas. He fed them on a 15-year-old girl with microfilariae in her blood. In contrast to Manson, who had found worms in various stages of development within the one mosquito (because they had been fed on different days), Bancroft found that all of his larvae were at the same stage of development (because the mosquitoes were all fed once on the same day) and took 16 days to complete their development. In June of that year, he wrote: 'It has occurred to me that young filariae may gain entrance to the human host whilst mosquitoes bearing them are in the act of biting.'[36] In pursuance of this idea, Bancroft then sent some mosquitoes that had ingested microfilariae to Manson, who in turn passed them on to George Low (1872–1952), who was working under his direction at the London School of Tropical Medicine.

Low fixed them in celloidin (a form of cellulose), then cut very small slices called histological sections and examined them under the microscope. Low found that the larvae passed from the abdomen to the thorax and then pushed past the salivary glands into the proboscis (the biting part of a mosquito). With masterly understatement, Low wrote:

> It is difficult to avoid the deduction that the parasites so situated are there normally, awaiting an opportunity to enter the human tissues when the mosquito next feeds on man.[196]

Low was the technician who completed Thomas Bancroft's experiment.

Thomas Lane Bancroft then turned to an animal model to prove the life cycle. He used the dog heartworm, *Dirofilaria immitis,* rather similar to the human parasite. In 1901 he clearly showed that infective larvae escaped from the tip of the proboscis. Then in December 1902 he exposed a three-week-old uninfected pup to 183 filariated mosquitoes. Nine months later, microfilariae appeared in the blood. The dog was then killed and 16 male and 16 female adult *D. immitis* were recovered from the pulmonary artery and the heart. Undoubtedly a similar life cycle occurred in human filariasis (Figure 31).

Just a few of things need to be added. In 1927, A Lichtenstein and SL Brug discovered that the microfilariae in the Dutch East Indies (now Indonesia) sometimes looked different and would not develop in the mosquito *C. fatigans.* Eventually their adult worms were found and shown to differ from *W. bancrofti* and were named *Brugia malayi* in 1958. It was by now realized that not all mosquitoes are able to transmit filarial infections and that filariasis only occurs in regions where such mosquitoes (known as vectors) are present. It was also observed that while filariasis is usually a chronic illness due to obstruction of the lymph vessels over many years, sometimes an acute inflammatory reaction may occur around an infective larva when it moults into a mature worm or when an adult worm dies and disintegrates. Unfortunately there is still no good treatment for this illness. But there is some good news for the transient visitor to an endemic area. These worms can't multiply within you. Only one adult worm can develop from each infective larva. You need large numbers of adult worms to develop disease and for this to happen, you must be attacked by thousands upon thousands of night-biting mosquitoes carrying the parasite. Make sure you take insect repellent and a bed net with you!

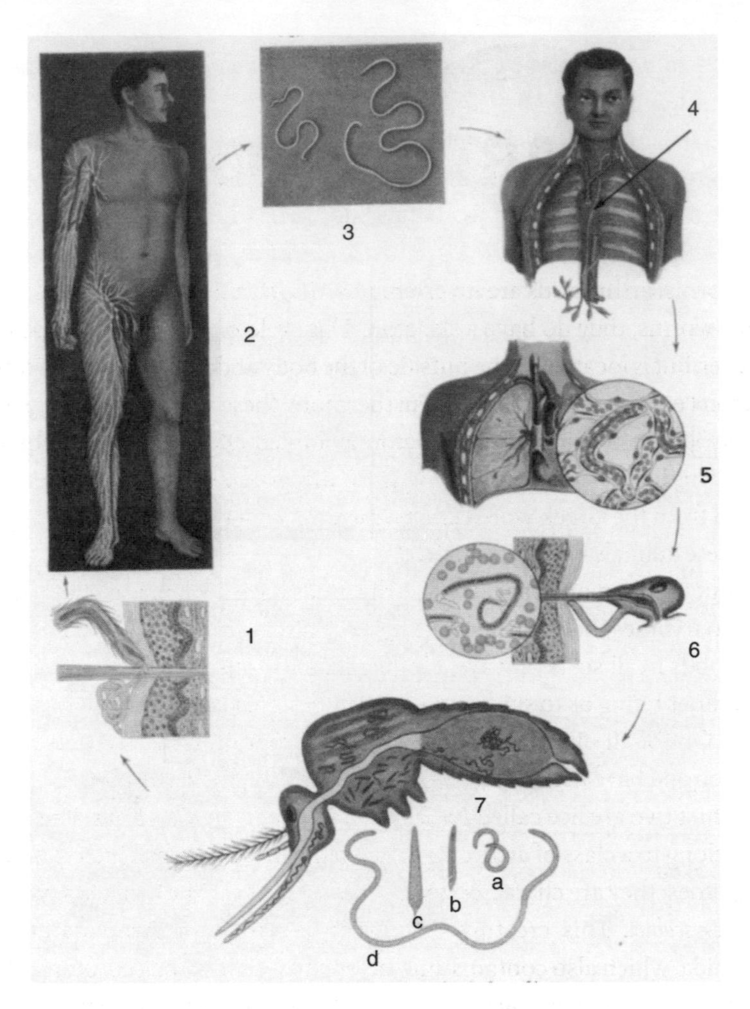

Figure 31. Life cycle of *Wuchereria bancrofti* and *Brugia malayi*. Start at (1) when a mosquito bites. One or more infective larvae escape from its proboscis then enter the hole made by the mosquito. They pass via the lymphatics (tiny tubes which contain fluid but no blood) to the region of the draining lymph nodes, e.g. in the groin or in the tunica vaginalis (2). Here they moult and mature into adult worms (3). Male worms fertilize female worms which release first-stage larvae (microfilariae) which pass via the thoracic duct (arrow) and enter the subclavian vein (4). Microfilariae enter the peripheral bloodstream at night but retire to the veins of the lungs during the day (5). When a mosquito bites and sucks blood containing microfilariae (6), the worms enter its stomach and pass over a couple of weeks via the thorax to the proboscis of the mosquito, moulting twice in the process to become infective larvae (7).

PART III

ARTHROPODS

Like worms, arthropods are invertebrates, that is, animals without backbones. Unlike worms, they do have a skeleton. This skeleton is not internal as ours is, however, but is located on the outside of the body and is therefore called an exoskeleton ('exo' meaning outside). Furthermore, these animals have appendages such as legs in which the exoskeleton is jointed allowing them to bend and move. This has given rise to the name for this group of animals. 'Arthropod' is derived from the Greek words *arthron* meaning 'joint' and *pous* indicating 'foot', thus these animals have 'jointed feet'.

About 1 million or so species of arthropods have been named but there may be up to 10 times that number waiting to be discovered. We can be bothered by arthropods in all sorts of ways, ranging from a housefly annoying us through mosquitoes biting us to spiders such as Sydney's funnel-web spiders rarely killing us. Out of all of this vast number of arthropods, however, only three are common and have had a long-standing relationship with the human race.

The first two are lice called *Pediculus humanus* and *Phthirus pubis*. These organisms belong to a class of arthropods called the Insecta, which includes flies and mosquitoes; they are characterized by having six legs. The other is a mite called *Sarcoptes scabiei*. This creature belongs to a class of arthropods called the Arachnida, which also contains animals such as spiders and scorpions; they are distinguished by having eight legs. Since these animals are parasites which live in or on the skin, they are often referred to as 'ectoparasites', meaning parasites living on the outside of the body, in contrast to 'endoparasites' such as worms and protozoa, which live inside the body. Finally, although all the other organisms in this book are said to cause infections, by a peculiarity of convention, arthropods are said to infest. This word is derived from the Latin 'infestare' meaning to make hostile, unsafe, disturbed, or troublesome. When you read about lice and mites you might agree!

6

Lice (pediculosis)

SCIENTIFIC NAMES: *Pediculus humanus capitis* and *P. h. humanus* = *P. h. corporis; Phthirus pubis*
COMMON NAMES: headlouse, body louse, pubic louse, or crab
DISEASE NAMES: pediculosis or louse infestation
DISTRIBUTION: worldwide
TRANSMISSION: contact with infected persons
LOCATION OF PARASITES: skin with eggs (nits) on hairs or clothing fibres
CLINICAL FEATURES: parasites on the scalp or body, itch, secondary infection
DIAGNOSIS: observation of lice on the skin or nits in hairs or fibres
TREATMENT: various insecticides, e.g. benzyl benzoate, permethrin, malathion
PREVENTION: avoid contact with infested people and their combs, hats, etc.

'You louse!' is a colloquial term of scorn and contempt, 'I'm feeling lousy' means you do not feel at all well, while when a person 'louses' something up they have messed it up indeed. Perversely, in American slang 'lousy with money' indicates that someone has plenty of cash. All of these expressions are derived from man's long association with lice (the plural of louse), a state of affairs which has generally, but not invariably, been viewed as being most undesirable.

Our caveman ancestors were quite familiar with these creatures. They are big enough to be seen with the naked eye, being about 3 mm in length. They are greyish or dirty brown in colour and move about on their six legs. They may excite an unpleasant itch because they are equipped with mouthparts that allow them to pierce the skin and suck blood. They only feed on humans and spend their lives living on our blood. We now recognize two species. One inhabits the head or body and is called *Pediculus humanus* (Figure 32), while the other, which

Figure 32. The body louse, *Pediculus humanus corporis.*

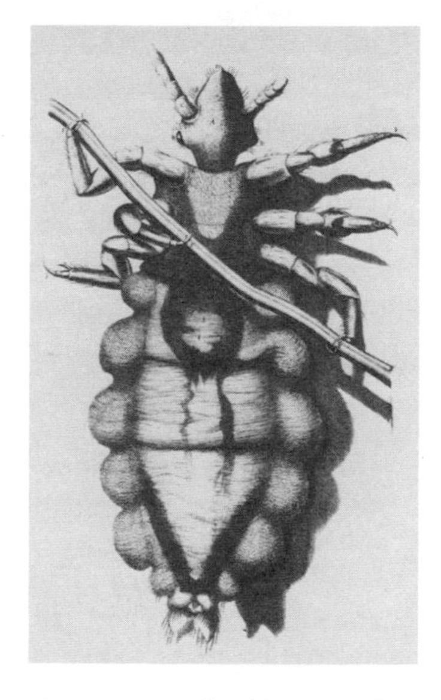

Figure 33. Hooke's drawing of a louse clutching a hair.

infests the pubic hair, is known as *Phthirus pubis.* But we are getting ahead of ourselves.

Lice have been a human annoyance throughout all human prehistory and recorded history. The Bible describes how Aaron (*c.* fifteenth century BC?) 'stretched out his hand with his rod and smote the dust of the earth and it became lice in men and beast' (Exodus 8:17, King James Version). However, this rendering may be a reflection of the preoccupation of the early seventeenth-century Englishmen who made this translation with lice since modern versions use the more innocuous and vague term 'gnat'. No-one really knows. The literature of the ancient Greeks and Romans frequently refer to lice. Most people considered them noxious vermin and often made strenuous efforts to get rid of them. Herodotus, the Greek historian (*c.* 484–*c.* 425 BC) recounts how Egyptian priests shaved their bodies completely, including their heads, in order to keep the lice away, particularly during religious ceremonies. In the 1536 edition of the late Johann Prüss of Strasbourg's book *Hortus sanitatis* (The garden of health), lice were regarded as worms with legs and it was said that 'they are called pediculi from the number of their feet'[258] (Latin: 'pedum' = foot).

Louse infestation was indeed a major problem in the Middle Ages. Thus, Ann, Countess of Dorset in England, wrote in her diary in 1603 that after a visit to see the king, they 'were all lousy by sitting in Sir Thomas Erskine's chambers'.[73] Likewise, Samuel Pepys of London in 1662 complained with irritation that he had had to go 'to my barber's, to have my Periwigg [i.e. wig] he had lately made me cleansed of nits which vexed me cruelly'.[247] Robert Hooke (1635–1703), better known to physics students for Hooke's law (which states that the extension of a spring is in direct proportion with the load added to it), published in England in 1665 a book called *Micrographia* detailing his studies with the microscope. He provided an excellent illustration of a louse clutching a hair (Figure 33) and wrote:

> This is a creature so officious, that 'twill be known to everyone at one time or another, so busie and so impudent, that it will be intruding itself into every one's company, and so proud and aspiring withal that it fears not to trample on the best, and affects nothing so much as a Crown [i.e. loves the head]; feeds and lives very high and that makes it so saucy as to pull any one by the ears that comes into its way and will never be quiet till it has drawn blood.[140]

On the other hand, not everyone viewed lice this way. The Spanish conquistador Hernán Cortes (1485–1547) saw that the Aztecs in Mexico would carry bags of lice to Montezuma, their king, to pay their respects when they could not afford gold. Centuries earlier, the Roman emperor Julian the apostate (331–363) referred with some satisfaction to his own unkempt appearance, with his shaggy beard being thick with lice that scampered about in it as though it were a thicket of wild beasts.

This excessively ascetic approach was enthusiastically adopted by some early Christian hermits. Cleanliness was viewed with abhorrence while lice were called 'pearls of poverty' and a mark of saintliness. Thomas à Becket, archbishop of Canterbury, was murdered in 1170. When his many layers of garments were removed to prepare him for burial, the undertakers found:

> finally next to the body, a tight fitting suit of coarse hair cloth, covered on the outside with linen, the first of its kind seen in England. The innumerable vermin [i.e. lice] which had infested the dead prelate were stimulated to such activity by the cold that his hair cloth garment... boiled over with them like water simmering in a cauldron and that the onlookers burst into alternate fits of weeping and laughter between the sorrow of having lost such a head and the joy of finding such a saint.[201]

This attitude did not last. The eighteenth-century evangelist John Wesley uttered those famous words 'cleanliness is, indeed, next to godliness', a phrase which resonated with the English public and was to be the watchword thereafter.

But where did these creatures come from? The Greek naturalist Aristotle (384–322 BC) considered the question and concluded that they were generated spontaneously in the flesh of animals and people, a view which was to persist for two more millennia or so. Thus Prüss, referred to above, wrote: 'this pest is, without doubt, generated from the flesh of man himself, but invisibly'.[258]

But by the seventeenth century, views began to change. As remarked above, Samuel Pepys complained about the nits in his wig. Perhaps this was because he and others recognized an association between nits and lice although the mechanism would not have been clear. Or they may simply have thought that they were excrements from lice. Robert Hooke, on the other hand, was concerned with the external appearances of lice as seen under the microscope but makes no mention of nits.

Around the same time, the Dutchman Jan Swammerdam (1637–1680) used the microscope to dissect and describe the internal anatomy of lice. He was born in Amsterdam, the son of an apothecary. He graduated in medicine but never practised it or earned a living from it, much to the annoyance of his father. Although dogged by ill-health and intermittent religious fervour, he was obsessed by insects and is one of the founders of modern entomology. Most of his writings were held in private libraries and not published until 1738, many years after his death. By an extraordinary mischance, all of the 40 lice he dissected were female (he did not know this) and Swammerdam erroneously wondered whether they were perhaps hermaphroditic, i.e. had both male and female sex organs, although he found no evidence of this. In one book printed in 1682, two years after his death, Swammerdam illustrates the nit before and after hatching, with its cap opened, and a larva. He clearly realized that lice produce eggs (nits) which hatch larvae. Perhaps he observed it happen.

His Dutch contemporary, the famous microscopist Antony van Leeuwenhoek (p.97) was also interested in observing lice. He had the problem of sustaining the creatures. He solved this difficulty by wearing a clean stocking on one leg, introducing two lice, and sealing the top of the stocking with a garter, thus allowing them to feed on himself. After 10 days he found about 90 eggs on the stocking. He had worn a black one rather than the usual white stocking in order for the eggs to be seen more easily. One louse became lost but he dissected the remaining creature and found it still full of eggs. He then wore the stocking for another 10 days and found:

> at least 25 lice of three different sizes... but I was so disgusted at the sight of so many lice that I threw the stocking containing them into the street after which

> I rubbed my leg and foot very hard in order to kill any louse that might be on it.[181]

Furthermore, Leeuwenhoek disproved Swammerdam's theory that lice were hermaphrodites by identifying male and female sex organs.

Clearly Swammerdam and Leeuwenhoek were very close to completely describing the life cycle of lice. And, of course, around the time all these things were happening, Francisco Redi (p.8) showed that maggots developed not spontaneously but from flies depositing eggs on decaying meat. Spontaneous generation was nearly off the agenda. The Italian physician Antonio Vallisnieri (1661–1730) put all this in print:

> every animal is born out of a seed (that in truth is nothing more than its egg) and I must conclude that also these lice which occur in morbid bodies within the flesh are generated thus…the Lice of the head are born from Nits and those of the shirt or clothing in the same way from their eggs.[306]

These observers realized that nits on the head were eggs cemented on to the hairs of the scalp (Figure 34), while those on the body were attached to clothing, or sometimes body hairs. They are ovoid in shape being about 0.8 × 0.3 mm in size, whitish-yellow opalescence in colour, and with an embryo inside the shell. We now know that the female louse secretes a glue when she lays an egg that makes it stick to a hair or clothing fibre. So strong is this cement that no amount of scraping will remove a nit from a hair. Our word 'nit-picking' stems from the time-consuming process of going through a person's hair attempting to remove head lice and their eggs. And like 'You louse!', 'You nit!' is also a term of almost affectionate abuse.

Leeuwenhoek was nearly right when he observed lice in three different sizes. We now know from investigators in the early twentieth century such as Gordon Nuttall (1862–1937) and Patrick Buxton (1892–1955) much more about the life history of lice. The nymph moults three times, becoming bigger after each moult but looking more or less the same, with the last moult resulting in the adult louse. Each of these larval stages is called an instar and the time from hatching to full maturation is about eight days. The reason Leeuwenhoek made an error in the number of instars is that the third instar is almost as big as the mature adult and would have been difficult for him to differentiate. Adult lice live for three to four weeks and each female louse produces about two to four eggs per day. The eggs hatch in seven to nine days and the nymphs begin feeding on blood shortly thereafter. By the time the nits are 2.5 cm (an inch) or more from the scalp, the lice have probably hatched or else are dead but the shell remains attached until the hair is cut off or falls out.

What were these lice to be called? In 1758, Carolus Linnaeus (p.7) introduced his binomial system of nomenclature and, following the usage of Prüss and others, named the human louse *Pediculus humanus* in 1758. He later realized there might be two sorts, one living on the scalp and the other on the body. In a revision in 1767 he called them *P. humanus* numbers 1 and 2—'habitat in capite' (living on the head) and 'et vestimentis hominis' (and human clothing). In the same year, Carl de Geer (1720–1778), another Swede, thinking they were different but being unsure whether they were separate species called head lice *P. humanus capitis* ('capitis', meaning 'of the head') and the body louse *P. humanus corporis* ('corporis', indicating 'of the body'). Nineteenth-century investigators decided that head lice were a separate species from body lice and called them simply *P. capitis* and *P. humanus*, respectively.

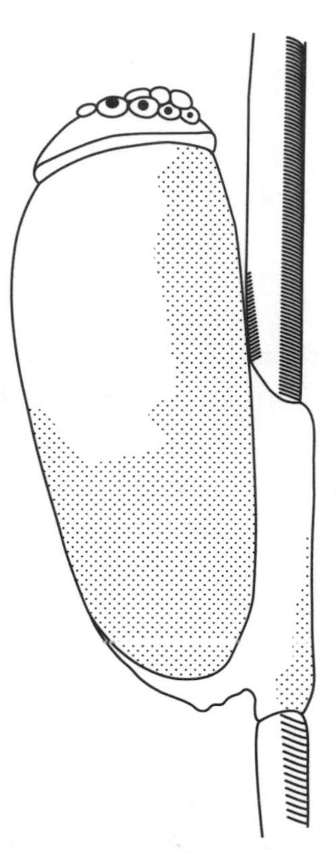

Figure 34. Nit attached to a hair.

Head lice can affect anyone. Body lice afflict those who do not change their clothes and do not wash. Remember that for many centuries, most people, especially the poor in cold and temperate climates, hardly ever changed their clothes. If they took them off at night, they put the same ones on again next day. What is more, they usually did not wash from one year to the next. Are these lice really two species? Debate has gone on ever since Linnaeus and de Geer. In the twentieth century it was finally concluded that they are probably biological races or varieties of the one species since the anatomical differences are minor and the two types when mated experimentally produce fertile progeny. Nowadays, the nomenclature of de Geer is generally used except that some call the body louse *P. humanus humanus*.

There are a couple of other things to say. In 1668, Francisco Redi described and published a figure of a distinctive louse that lives in human pubic hair. He called it 'il piattone'—the crab louse because of its crab-like appearance (Figure 35).

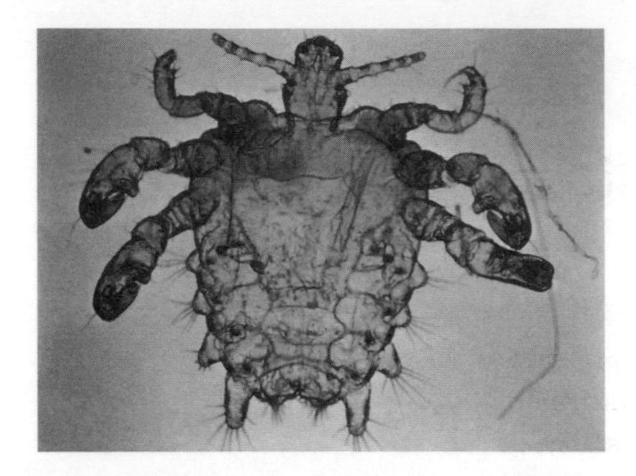

Figure 35. *Phthirus pubis*, the crab louse.

Linnaeus in 1758 named it *Pediculus pubis* but now, after many tortuous paths, it is known as *Phthirus pubis*. This name is derived from the Greek *phtheir* meaning 'louse' while *pubis* refers to the pubic area. Naturally enough, it is spread by sexual intercourse.

Head, body, and pubic lice may cause a nasty itchy rash as the body's immune system reacts to them when they feed. The holes in the skin allow bacteria which live on the skin, such as the golden staph, to enter and cause infections like boils. Finally, it was shown in the early twentieth century that lice may transmit several infections from one person to another, particularly the dreaded typhus fever that we will consider in a later chapter, as well as the bacterial infections, relapsing fever due to *Borrelia recurrentis*, and trench fever caused by *Bartonella quintana*.

7

The itch (scabies)

SCIENTIFIC NAME:	*Sarcoptes scabiei*
COMMON NAME:	itch mite
DISEASE NAME:	scabies
DISTRIBUTION:	worldwide
TRANSMISSION:	contact with infected person
LOCATION OF PARASITES:	skin
CLINICAL FEATURES:	itching, widespread rash, burrows in the skin
DIAGNOSIS:	finding mites or eggs in burrows in the skin
TREATMENT:	various insecticides, e.g. benzyl benzoate, permethrin, malathion, ivermectin
PREVENTION:	avoid direct contact with an infested person or their bedding or clothing

The dominant symptom of scabies is itch. The itch is associated with a rash which often occurs anywhere on the trunk and limbs. There are of course many causes of itch, including the lice discussed in the previous chapter, but perhaps the most important of all is infestation with a small mite. It is about 0.3 mm in length and whitish-yellow in colour, now called *Sarcoptes scabiei* (Figure 36). It is the discovery of this mite and its relationship with itch which is the subject of this chapter. This was a long involved process that took two millennia. It was made even more complicated by the fact that mites damage skin and sometimes allow bacteria such as the golden staph, *Staphylococcus aureus*, to enter and cause pustules (small lesions containing pus) or larger boils. This is known as a secondary bacterial infection or infected scabies. Such lesions are most commonly located on the back of the hands, especially around the webs of the fingers, the flexor (palmar surface) of the wrist, and the elbows. This is where the mites most commonly burrow into the skin and are easiest to find.

Itch has no doubt been a problem since long before recorded history. Some of the cases of 'leprosy' in the Old Testament of the Bible may well have really

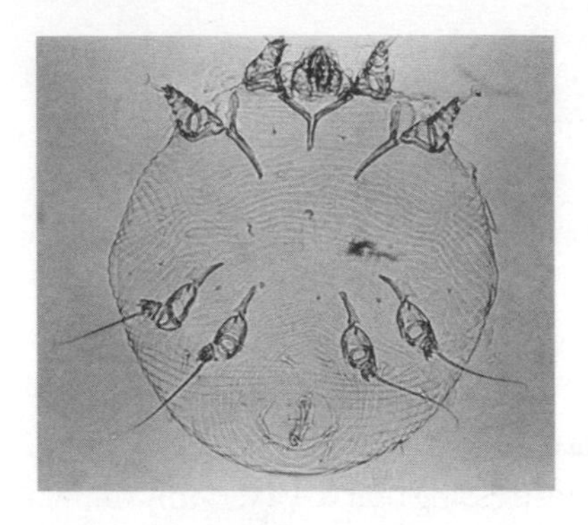

Figure 36. Female *Sarcoptes scabiei*.

been scabies. Greek physicians such as Hippocrates (p.116) and Galen (p.203) thought it was due to an imbalance in the four 'humours' (yellow bile, black bile, phlegm, and blood). Bad humours, in this case phlegm (according to Hippocrates) or black bile (in the view of Galen), were believed to escape through the skin thus causing the itch. This idea was to hold sway for well over a thousand years. Nevertheless, the Greek naturalist Aristotle (384–322 BC) may have seen the mites. He described recovering 'lice' from skin blisters but what these were precisely is unclear and he did not consider them a cause of the illness. Rather perversely, he was aware of a mite dwelling in wood which is rather similar to what we now know is the cause of scabies; he called these akari.

What happened next is quite controversial. Some hold that the discoverer of the scabies mite was Abu Hassan Ahmed ibn Mohammed al Tabari, a physician of Tabari in Persia (Iran) who lived in the tenth century AD. Although he followed the humoral doctrines of the Greek Hippocrates and Galen, he described the presence of a creature in the skin thus: 'this animalcule can be removed with the point of a needle... it moves. If crushed between the fingernails one hears it crack.'[1]

Another candidate is Saint Hildegard (1098–1179), abbess of a convent near Bingen in Germany. She wrote a book entitled *Physika* (also called in Latin *Liber Simplicis Medicinae*, i.e. *Book of Simple Medicines*) in which she described the symptoms of scabies exactly. What is more, she wrote that it was caused by what she called *suren* (medieval German for 'mite') or 'gracillimi vermiculi', which is Latin for tiny creatures that burrow into the human skin. Perhaps Hildegard did indeed find the cause of scabies. Unfortunately, people either did not know of her work or did not believe it.

Another person who met the same fate was her contemporary Avenzoar (also known as Ibn Zuhr, 1091–1161), a physician in Seville in Moorish Spain. He wrote:

> There are lice under the hand, ankle and foot like worms, and sores affecting the same areas. If the skin is removed, there appears from various parts of it, a very small animal which can hardly be seen.[33]

Some modern historians believe that although Avenzoar described what would seem to be a mite, he did not relate it to the itch.

Did any of these people discover the mite that causes scabies? We will never know for sure because we simply don't have enough documentation. Even if they were on the right track, they would have had little idea of what they were looking at because they did not have the tools to magnify any creatures that they had found. So all through the Middle Ages nearly everyone was still ascribing this illness to an imbalance in the humours of the ancient Greeks and it was so common and so persistent and so distressing that it was simply known in English as 'the itch'. Of course, the medical profession had a well-trodden path for dealing with something they did not understand. They gave it a Latin name to create an aura of mystique which physicians alone comprehended. They called it *scabies*, which was derived from the Latin word 'scabere', which simply means 'to scratch'! Even when mites were seen on people with the itch, they were considered to have been generated spontaneously as a result of corruption of the flesh and blood caused by an internal ailment of the humours.

We now need to fast-forward to the seventeenth century when Francisco Redi (p.8) was busy proving that flies resulted from eggs deposited on putrid flesh by other flies, not from spontaneous generation. In 1687 he received a letter from Giovan Cosimo Bonomo (1663–1696), a young naval physician, concerning his research into scabies in collaboration with the apothecary, Diacinto Cestoni (1637–1718), in Livorno (Leghorn), Italy. He not only describes finding mites in patients with scabies but also eggs, which flew in the face of the doctrine of spontaneous generation. Redi promptly published Bonomo's letter in a book that same year. Bonomo had seen poor women in Livorno use a needle to pull out little 'bladders of water' from the skin lesions in people with the itch. He decided to try it himself on an itchy person. He used a very fine needle to take out from a pustule (a lesion containing yellowish fluid in the skin):

> a very small white globule scarcely discernible. Observing this with a microscope, I found it to be a very minute living creature, in shape resembling a tortoise of whitish colour, a little dark upon the back with some thin and long hairs, of nimble motion, with six feet, a sharp head with two little horns at the end of the snout.[48]

Bonomo accompanied his letter with drawings of what he had seen. These are the first drawings anyone had made of these creatures and they showed a mite with six legs. We now know that the larva that hatches from an egg has six legs. After it burrows into the skin it moults and has eight legs. Bonomo repeated these observations many times in different people, of varying age and complexion as well as gender, with the same results. He was particularly interested in whether these creatures laid eggs. He was in luck and wrote:

> At last by good fortune while I was drawing the figure of one of them by a microscope, from the hinder part I saw drop a very small and scarcely visible white egg, almost transparent and oblong, like to the seed of a pineapple. I oftentimes found these eggs afterwards, from which no doubt these creatures are generated, as all others are, that is from male and female, though I have not yet been able by any difference of figure to distinguish the sex of these animals.[48]

Bonomo actually saw a mite lay an egg. He went on to say that it was now possible to give a rational explanation for the itch. It was not due to a problem of the humours or any of the other explanations that had been put forward. Scabies was simply due to:

> the continual biting of these animalcules in the skin, by means of which some portion of the serum oozing out through the small aperture in the cutis [= skin], little watery bladders are made, within which the insects continuing to gnaw, the infected are forced to scratch, and by scratching increase the mischief, and thus renew the troublesome work...From hence we come to understand how the itch proves to be a distemper [disease] so very catching, since these creatures by simple contact can easily pass from one body to another, their motion being wonderfully swift...and they multiply apace by the eggs which they lay.[48]

Furthermore, Bonomo reported that mites could stick to clothes and other objects and showed by experiment that they could live for a number of days away from a human body.

QED. Or so you might think. Unfortunately, Giovanni Maria Lancisi (1654–1720), physician to the Pope, disagreed. He recognized the presence of the parasite but believed that they were simply generated spontaneously as a result of a problem with the humours. What is more, Lancisi quoted Scripture. This was formidable opposition. Galileo had incurred the wrath of Rome by challenging accepted dogma and saying that the earth revolved around the sun instead of vice versa; rather than risk being burnt at the stake, he had recanted. Being mindful of the fate of Galileo Galilei, Bonomo was persuaded to withdraw from the debate.

There was a curious twist to this story. In 1710, Cestoni wrote a letter claiming that he had made all the discoveries and that he had written the letter to Redi under the pseudonym of Bonomo because he feared persecution. However, subsequent investigations showed that there really had been a Giovan Cosimo Bonomo, who had studied scabies in the spa of Livorno and died in 1696. Redi had also died and Cestoni probably thought he could get away with it. No-one knows why he tried.

Bonomo's letter in Redi's book was translated into English by Dr Richard Mead (1673–1754) in 1703. Mead remained convinced that mites were the cause of scabies. After a description of the symptoms and signs, he wrote in 1751:

> It may justly be called an animated disease as owing its origin to small animals. For these are certain insects, so very small as to hardly be seen without the assistance of a microscope, which deposits its eggs in the furrows of the cuticle... and the young ones, coming to full growth, into the very cutis [skin] with their sharp heads and gnaw and tear the fibres. This causes an intolerable itching.[216]

Similarly, a German doctor, Johann Ernst Wichmann (1740–1802), in 1764 published a small book entitled (in translation) *Cause of the Itch*. In this he included Bonomo's drawings as well as his own figures and wrote:

> I hope I have now thoroughly explained and proved the aetiology of scabies, or at least rendered it both plausible and logical that it is a simple skin disease caused by mites.[320]

One might well think that the cause of scabies had been established by the middle of the eighteenth century. Astonishingly, a substantial number of medical men over the next 60 years refused to accept that mites were the cause of the disease. Two examples will suffice. The Scot, William Buchan, in his *Domestic Medicine or the Family Physician* published in 1769, wrote that 'the itch... seems originally to proceed from want of cleanliness, bad air, or an unwholesome diet'. Similarly, the French doctor, Charles Lorry, wrote in 1777 in his *Tractus de Morbis Cutaneis (Book of Diseases of the Skin)*, despite being aware of Bonomo's work, that the itch was due to cold, damp air, poor and indigestible food, bad water, and filth generally. He even claimed that many fevers were cured by an attack of scabies! In true Galenic tradition, he thought the mites were secondary invaders in some way generated in the diseased tissue.

In the first part of the nineteenth century, a debate raged between believers in and sceptics of mites causing scabies. The uncertainty arose because mites could not consistently be found in the vesicles (fluid-filled lesions) in the skin. The matter was finally put to rest in 1834 by Simon-François Rénucci (1794–1884), a

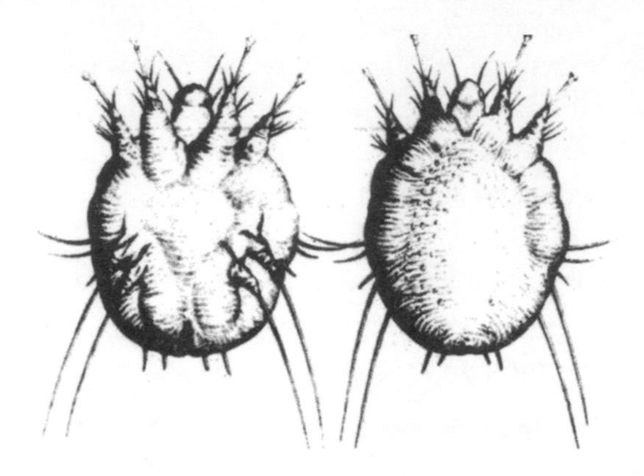

Figure 37. Rénucci's drawings of eight-legged adult mites. Front view (L) and back view (R).

Corsican medical student in Paris. He was repeatedly able to demonstrate the presence of the mites. He had learnt the trick from Corsican peasant women who used to extract mites (Figure 37) with a needle. Rénucci's success was due simply to the fact that he knew the mites were to be found in the dead ends of meandering burrows and not in the vesicles.

Encouraged by knowledge of how to reliably obtain mites, experiments on artificial infestation were begun by Rénucci and others. By 1844, the skin specialist Ferdinand Hebra (1816–1880) felt able to write that without *Sarcoptes* there can be no scabies. But it was not until 1861 when Louis Pasteur (p.182) finally dispelled the theory of spontaneous generation that everyone agreed with this statement. Lastly, during World War II (1939–45), the Briton Kenneth Mellanby and his collaborators put the matter beyond doubt when they conducted transmission experiments by bedding together infested young soldiers and conscientious objector volunteers and showed the distribution of the rash (Figure 38).

These mites were first given their present name of *Sarcoptes scabiei* in 1758 by Linnaeus. The name *Sarcoptes* was derived from a combination of the Greek words *sarx* (meaning 'flesh') and *korptein* ('to smite or cut'), while *scabiei* means 'of scabies'. In order to separate them from the morphologically indistinguishable mites of other animals, de Geer (p.69) in 1778 labelled the mite infesting humans *Sarcoptes scabiei* var *hominis*. We now know that eggs incubate in burrows and hatch in three to four days. These six-legged larvae crawl out of the burrows and migrate to the skin surface and burrow into another spot of intact skin to make short burrows called moulting pouches. Here they moult (lose their skin) to become eight-legged nymphs, which in turn moult twice more to become adults, the whole process taking about 10 days.

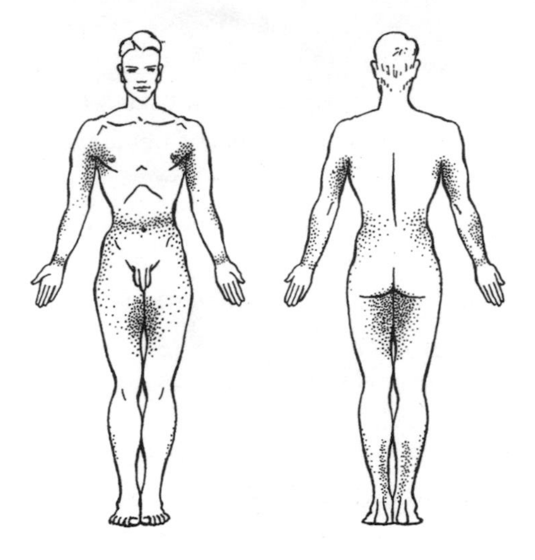

Figure 38. Mellanby's figure showing the distribution of the scabies rash. It does not correspond to the site where mite burrows are most common.

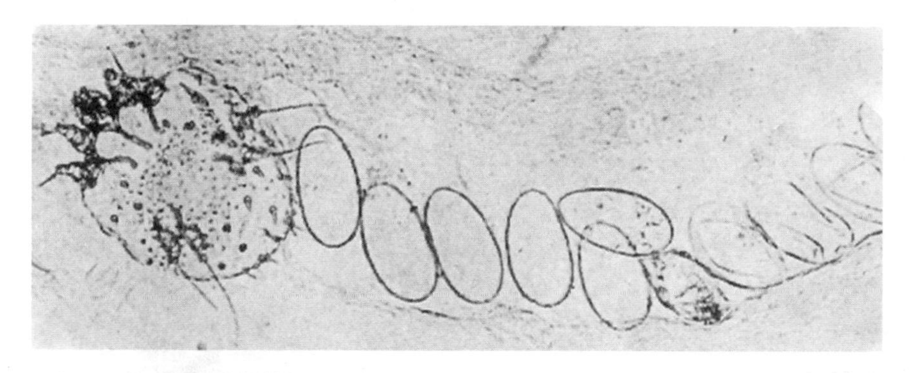

Figure 39. Wet mount of a skin scraping showing a female mite at the dead end of a burrow with a trail of eggs behind her. The oldest eggs are near the opening of the burrow.

Mating occurs when a male penetrates the moulting pouch of a female or finds a female on the skin surface. It takes place only once and the female is fertile for the rest of her life. The male soon dies but the fertilized female either continues to burrow or returns to the skin surface and crawls about at a rate of 2.5 cm/min looking for a suitable spot to burrow. She then burrows into the outer layer of the skin, biting with her jaws, digging with her legs, and using protein-digesting enzymes to dissolve the tissues of the epidermis. Within a few hours she starts to lay two or three relatively large oval-shaped eggs every day or two. She continues to lengthen her burrow at the rate of 0.5–5 mm per day, laying eggs behind her as she goes (Figure 39), for the rest of her life which lasts for one to two months.

While the mite's life cycle occurs completely on human skin, it is able to live on bedding, clothes, or other surfaces at room temperature for two to three days, yet remaining capable of infestation and burrowing. Infestations are usually asymptomatic for the first month, but then an immunological reaction to the mites, their eggs, or their faeces occurs. In reinfestation, the sensitized person develops a rapid reaction within hours with a skin eruption and intense itching (technically called pruritus). So we now recognize two sorts of lesions in scabies. The first are mite burrows with inflammation around them, especially in the finger webs, the wrists, and elbows and to a lesser degree the ankles. The second are intensely itchy, red rashes which occur elsewhere on the body (Figure 38), rarely contain a mite, result from an immunological reaction to mites elsewhere in the body, and cause much scratching.

PART IV

FUNGI

We have all seen mushrooms and toadstools and most of us recognize them as fungi, even if we cannot reliably differentiate between the two. They come out of the ground so they must be plants. Or are they? The ancient Greeks (and radio quiz shows when I was a boy) recognized three sorts of matter—animal, vegetable, and mineral, where vegetable meant plant. Linnaeus (p.7) followed this pattern in his first edition of *Systema Naturae* published in 1735, when he recognized Regnum animale, Regnum vegetabile, and Regnum lapideum, meaning the animal, plant, and mineral kingdoms. He included fungi in the Regnum vegetabile, which later became known as the Plant Kingdom. Not everybody was happy with this twofold division of all living organisms and in 1866 the German biologist Ernst Haeckel (1834–1919) created a new Kingdom, the Kingdom Protista, to house single-celled organisms such as bacteria and protozoa and the fungus-like slime moulds. Scientists argued for the next hundred years as to whether fungi should be placed in the plant or protist kingdom. It gradually became clear that there were significant differences between fungi and plants. Fungi do not have chlorophyll so they cannot photosynthesize. Instead they get their food by decomposing and degrading organic material. Unlike animals, fungi have cell walls, but unlike plants these walls are made of a material called chitin instead of cellulose. Unlike bacteria, fungi have a proper nucleus in each cell. These differences are so marked that in 1959 fungi were placed in their own kingdom, the Kingdom Fungi.

The word 'fungus' is simply the Latin word that means 'mushroom'. The plural is therefore fungi, which is a pity because no-one is quite sure how to pronounce it. I use fung-ee while others say fung-eye or funj-ee or funj-eye; so take your pick. You would think that someone who studies fungi is a 'fungologist'. Not so. The classical Greek scholars wanted a piece of the action so he or she is a 'mycologist'. This is derived from the Greek word *mykes*. You have guessed it—it means 'mushroom'.

There are thought to be about 1.5 million species of fungi and their classification is hideously complex. Fortunately for medical mycologists, only several hundred species afflict humans. What is more, they can be conveniently divided into two simple groups—moulds and yeasts. Moulds are fungi that grow in the form of microscopic multicellular filaments technically called hyphae. They sometimes grow so fast that they can soon be seen with the naked eye. An example is the green material which sometimes grows on a rotten lemon or orange; this is a mould called *Aspergillus*. Yeasts, on the other hand, are single cells. Fungal infections or diseases of humans are called either just that, fungal infections, or mycoses (singular: mycosis).

We shall consider only two types of fungal infections in humans. The first are mould infections of the skin which produce a group of clinical syndromes called tinea, but which have many different names depending upon the area of skin involved. The second is a yeast infection with fungi belonging to the genus *Candida*.

It was only in the first half of the nineteenth century that it was realized that fungi could infect humans. The scene for this was set by an extraordinary idea—that plants (as fungi were then considered) could infect animals. And this was first shown, not in humans, but in caterpillars!

Around 1805, a disease appeared in Italy in silkworms in which they became covered with a fine white powder and died. The Italians called this condition 'mal del segno' while the French (where the disease soon spread) named it 'muscardine'. The silk production industry was devastated and by 1850 most farms were abandoned. A sickly Lombard lawyer turned farmer, Agostino Bassi (Figure 40), devoted 25 years to studying this problem. He examined the white powder with a microscope and found that it was a parasitic fungus. He undertook a long series of experiments and showed that infected silkworms only became contagious to other silkworms when the fungus produced spores on the dead silkworm. A spore is a small, usually single-celled reproductive body formed asexually (that is without an egg and sperm) that is relatively resistant to drying and heat, and is capable of being dispersed then growing into a new organism. In 1834 Bassi performed a series of experiments before nine professors from the University of Pavia

Figure 40. Agostino Bassi (1773–1856).

who were convinced by him and signed a certificate to say so. He published his findings over the next two years. The professor of natural history at the University of Milan studied the fungus and named it *Botrytis* (now *Beauveria*) *bassiana* in Bassi's honour. In 1838, Victor Audouin, director of the Paris Natural History Museum, confirmed Bassi's findings. The scene was now set for finding fungal infections in humans.

8

Tinea (ringworm, etc.)

SCIENTIFIC NAMES:	*Epidermophyton, Microsporum,* and *Trichophyton* species
COMMON GENERAL NAME:	dermatophytes
DISEASE NAME:	tinea
DISTRIBUTION:	worldwide
TRANSMISSION:	contact with infected humans, animals, or contaminated environment
LOCATION IN A PERSON:	skin and hair
CLINICAL FEATURES:	scaly rash, sometimes with hair loss
DIAGNOSIS:	fungal elements seen on microscopical examination of skin scrapings then cultured on special media for specific identification
TREATMENT:	various antifungals applied directly to the skin, e.g. azoles and terbinafine, or orally, e.g. griseofulvin and azoles
PREVENTION:	avoidance of infected people or animals and wearing thongs in communal showers

In the first half of the nineteenth century, skin diseases were a bewildering array of mysterious conditions that were of unknown cause, difficult to differentiate from each other, and almost impossible to treat. Above all, a sufferer might be diagnosed with that dreadful condition 'leprosy'. We now know that they include common non-infective conditions such as eczema, dermatitis, and psoriasis as well as infections such as tinea, candidiasis, and leprosy. It was in this world that the first tentative discoveries were made that were eventually to encompass the group of conditions we now call 'tinea'. This word is the Latin word for 'moth' and was originally used to describe those infections that gave the hairs of the head a moth-eaten appearance. Another word used to describe these diseases is 'dermatophyte', being derived from a combination of the Greek words *derma* meaning 'skin' and *phyton* meaning 'plant'. A third common name is 'ringworm'. This is because the fungus on the skin of the body or the scalp tends to grow outwards more or less equally in all directions, thus producing lesions which creep in a circular or ring form like the ripple of a stone thrown into a pond.

We now know that species of three fungal genera produce this group of conditions: *Epidermophyton, Microsporum,* and *Trichophyton* species. The different species tend to affect different parts of the body although there is some overlap. Diseases of different parts of the body are given different names to reflect their location and some of them have common names as well. Some examples, with the common infecting organisms, are shown in Table 1.

Our story begins around 1840. The three key players were all German-speaking doctors who at first worked independently of each other on a particularly disfiguring form of tinea capitis then called *favus,* which is the Latin word for 'honeycomb'. This was characterized by a circular, cup-shaped, crusting followed by a scarring lesion on the scalp resembling a honeycomb.

The first observation was made by Robert Remak (Figure 41), who was born in Posen, Prussia (now Poznan, Poland). In 1836 or thereabouts when he was a medical student in Berlin, Germany, he decided to examine some material from one of these scalp lesions under the microscope. He did not think his observations

Table 1. Medical and common names of some forms of tinea showing the locations of the infections and the fungi that most commonly cause them.

Tinea barbae	Barber's itch	Beard	*Trichophyton, Microsporum*
Tinea capitis	Ringworm	Head	*Microsporum, Trichophyton*
Tinea corporis	Ringworm	Trunk, limbs	*Trichophyton, Microsporum*
Tinea cruris	Jock itch	Groin	*Epidermophyton, Trichophyton*
Tinea manuum		Hands	*Trichophyton, Epidermophyton*
Tinea pedis	Athlete's foot	Toes	*Trichophyton, Epidermophyton*
Tinea unguum	Onychomycosis	Nails	*Trichophyton, Epidermophyton*

Figure 41. Robert Remak (1815–1865).

Figure 42. Johann Schönlein (1790–1864).

were comprehensive enough to be published but he allowed his fellow student Xaver Hube to include them in his doctoral dissertation of 1837. Hube wrote:

> My friend Remak...discovered that the substance of those crusts consisted mostly of small bodies resembling in size blood corpuscles of frogs; they are smooth and rather round, are also connected to each other and have the shape of branching threads.[143]

Unfortunately, neither Remak nor Hube realized that what Remak had found was a fungus. We shall return to Remak later.

The scene then moves to Zurich, Switzerland, where Johann Lukas Schönlein (Figure 42) was professor of pathology. Schönlein had been born in Bamberg in Bavaria then graduated from the University of Würzburg. He was aware of Bassi's discovery so he obtained some infected silkworms from Milan and confirmed the findings. This stimulated him to investigate *favus* and he wrote a letter to his colleague Johannes Müller, professor of anatomy and physiology in Berlin:

> From this I was again reminded of my view as to the vegetal [plant] nature of some impetigines [skin disorders]...As I fortunately had some examples of Porrigo lupinosa [= favus] in the hospital, I examined them more closely, and the first experiments already no longer allowed any doubt as to the fungous nature of these so-called pustules. Enclosed is a microscopic picture.[284]

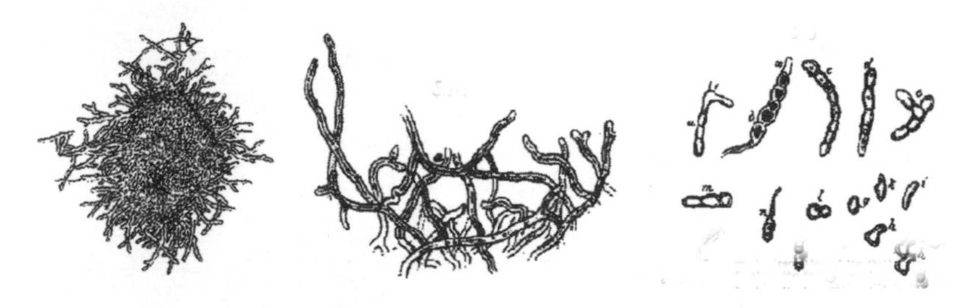

Figure 43. Drawings by Schönlein of fungal growth (left) and by Remak of fungal elements (centre and right).

Schönlein's drawing of what he had seen in 1839 is shown in Figure 43. This was the first recognition of a fungal infection of human skin.

Meanwhile Remak had not lost interest in this condition. He inoculated some crust elements from a patient into his own forearm and after a considerable delay, because of its slow growth, produced a similar lesion, thus proving the infectious nature of the crusts. He published this observation and it may have stimulated Schönlein, who had now moved to Berlin, to offer Remak a position as a private assistant in microscopy in his clinic in 1843. Working with Schönlein, Remak cultivated the agent of favus on slices of apple; it grew within six to eight days. Remak then subcultivated the fungus on more apple, described the microscopical appearances of the fungus (technically known as a thallus which looks like a tangle of filaments) on the apple (Figure 43), and related them to the appearances of the skin lesions. He published his findings in 1845 and named the fungus in honour of Schönlein as *Achorion Schönleinii*. It is now known as *Trichophyton schönleini*.

Figure 44. David Gruby (1810–1898).

Meanwhile, observations on favus were also being made in Paris, France, by David Gruby (Figure 44). Gruby had been born in the village of Kis-Kér in Hungary (now Bačko Dobro Polje, Serbia) and studied medicine in Vienna, Austria. It is well to remember that all was not plain sailing for these three investigators. Schönlein was refused citizenship of Protestant Zurich because he was a Catholic and so moved to Berlin. Remak could not be appointed to a salaried teaching position in Berlin because he was Jewish. Likewise, Gruby, who was also Jewish, was offered a position in Vienna if he would be baptised as a Christian; he refused and moved to Paris.

Shortly after his arrival as a microscopist at the Foundling Hospital (asylum for abandoned children) in 1840, Gruby began to study favus. He rapidly discovered that the crusted lesions contained a fungus and in 1841 reported to the French Academy of Sciences:

> In order to recognise the true tinea [favus]...a small amount of crust is collected and mashed in a drop of pure water. The preparation is put between two glass slides and observed at 300 fold magnification. One will see a large number of round to oblong corpuscles...One also notes small articulated filaments...Branched filaments exhibit every so often vegetable partitions forming oblong cells...The shape of these filaments does not leave any doubt about their plant nature. They belong to the group of mycodermae [skin fungi].[127]

He noted that the fungus grew between the cells of the epidermis (outer layer of the skin) and sometimes surrounded the base of the hair follicles with the result that the hairs became fragile. Gruby remarked that in view of its plant nature, this disease was likely to be contagious. This was reinforced by the observation that he could grow it on potato slices. He then tried to inoculate himself and a colleague with material from a patient's lesion. Although a little inflammation developed, he was unable to reproduce the patient's lesion. Nor could he infect plants, birds, reptiles, or mammals.

Gruby then turned his attention to other scaling skin lesions. First he looked at tinea barbae, or ringworm of the beard, also called 'mentagra' (Latin: 'menta' = 'chin'). In 1842 he found that this was associated with a fungus that lived in the hair follicles and around the roots of hairs in the skin. He did not give it a name but Charles Robin (p.92) did in 1847, when he called it *Microsporon mentagrophytes*, later to be transferred to the genus *Trichophyton*. In 1843 Gruby found a different fungus in a disease that he called porrigo decalvans, now sometimes incorrectly translated as alopecia areata, in which bald spots appear first on the scalp then may spread to other parts of the body. He found that the white dust which covered the skin was composed entirely of fungal elements. The hairs were surrounded on the outside by fungi:

> Microscopical examination of hairs...reveals a large quantity of cryptograms [fungi] which surround the hairs from all sides and which truly form a

> plant mantle which covers them upwards from the skin for a distance of 1–3 mm... These cryptograms are formed of branches, stems and spores.[128]

He noted that the fungi made the hairs extremely fragile so that they broke under the slightest pressure. The spores suggested a name to him so he called the fungus *Microsporum audouini*, with the species named after Jean-Victoire Audouin, the late director of the Natural History Museum in Paris, while the genus name reflected the small spores.

In 1844, he reported yet another fungus in a condition which he called herpes tonsurans and we now call tinea capitis. This fungus grew inside hairs. He did not give it a name but in 1859 it was described as *Trichophyton tonsurans* (after a monk's tonsure in which the scalp is shaved) by Per Henrik Malmsten (1811–1883), professor of chemistry and medicine in Stockholm. It was then that Malmsten erected the genus *Trichophyton*, the name being derived from a combination of the Greek words *thrix* (hair or thread) and *phyton* (plant).

Thus by 1844, four different fungi had been discovered. Who discovered the first one, *Trichophyton schönleini*? It all depends upon how you look at it. Remak first saw the fungal elements in favus but did not understand what he was seeing. Schönlein clearly was the first to realize it was a fungus. Then both Remak and Gruby independently described the fungus in detail, cultured it on plant material, and undertook transmission experiments to determine whether it was contagious.

Unfortunately, although Gruby was a superb mycologist, describing the first three pathogenic fungi, he was not as good a clinician and gave poor descriptions of the clinical lesions from which he had obtained them. For example, he forgot to mention that he had found *Microsporum audouini* in children. Consequently, others went looking for it in ordinary baldness in adults and found nothing. All this confused his successors and did much to discredit his work for a number of years.

So matters meandered on for the next 50 years until the appearance on the scene in Paris of Raimond Sabouraud (1864–1938), a prolific investigator of these tinea infections. Sabouraud came from a family with a long medical tradition: one of his ancestors was barber-surgeon to Louis XIII. Although he became the world's specialist on disorders of the scalp, he was bald and always wore a kind of smoking cap. Gruby's paper on *Microsporum audouini* had been forgotten by dermatologists. When Sabouraud began to study tinea in 1892, he rediscovered it. Sabouraud classified infections involving hair into two types. He called one that grew around the hair an ectothrix (outside the hair), while one that developed within hairs was labelled an endothrix (inside the hair). By 1910 when he published his book *Les Teignes (The tineas)*, he had found five different species of *Trichophyton*. It was at this time that he created the genus *Epidermophyton* meaning 'plant on the outer skin' for *E. inguinale* (later called *E. floccosum*). Thus he recognized four genera: *Achorion, Epidermophyton, Microsporum,* and *Trichophyton*. Perhaps he is best known for the material on which all medical microbiology laboratories these days grow these fungi: Sabouraud's medium.

Unfortunately, when fungi are grown in successive cultures, they tend to change their appearance. This misled many people and the 'splitters'—mycologists who create a new species for every small variation—were having their heyday, with several hundred being described. Even some of Sabouraud's species are now thought to be variants of the one species.

In 1930, Langeron and Michevich absorbed the genus *Achorion* into *Trichophyton*. Eventually in 1934, Chester Emmons published a paper in which he reduced the number of species causing tinea to only 16, all in the genera *Microsporum, Trichophyton,* and *Epidermophyton*, the last housing the species which had been described as *Trichothecium floccosum* by the German botanist Carl Otto Harz (1842–1908) in 1870. We now recognize about 30 species of these dermatophytes as affecting humans. But this will probably change with modern genetic analysis. Some species only infect humans, some infect animals as well, and some live in the soil. About two-thirds of them are found in the genus *Trichophyton*, about one-third in the genus *Microsporum*, and only one, *E. floccosum* (with which I have had the pleasure of an intimate acquaintance), in the genus *Epidermophyton*.

9

Candidiasis (thrush)

SCIENTIFIC NAMES:	*Candida albicans* and other species
DISEASE NAME:	candidiasis, thrush
DISTRIBUTION:	worldwide
TRANSMISSION:	contact, ingestion
LOCATION IN A PERSON:	mucous membranes, skin; rarely systemic
CLINICAL FEATURES:	white plaques in the mouth or vagina, vaginal discharge, skin rash
DIAGNOSIS:	fungal elements seen on microscopical examination of scrapings from the mucous membrane then cultured on special media for specific identification
TREATMENT:	various antifungals applied directly to the mucous membrane, e.g. nystatin, azoles
PREVENTION:	there is no simple preventive

A thrush is a bird but 'thrush' is also a disease. It is caused by an infection with a yeast, a unicellular fungus. It can afflict anybody but is most common in babies and women. In infants with oral thrush, there are white patches in the mouth, on the tongue, and inside the cheeks. Sometimes the organism causes a moist, red rash on the skin in the nappy area. Vaginal thrush is most common in women in the child-bearing years, especially during pregnancy. White patches appear on the lining of the vagina, a white vaginal discharge develops, and there may be red inflammation of the vulva and surrounding skin. Thrush may also be found in anyone of any age who has diabetes mellitus or some other disorder which suppresses the immune system. In such patients there is a risk of the yeast invading the tissues of the body, a potentially fatal outcome. The medical name for these conditions is candidiasis although for many years it was known as moniliasis. Indeed I well remember as a medical student in 1963 being instructed by the senior surgeon to treat a patient's moniliasis by painting his mouth with gentian violet, the favoured treatment of the day.

Some people think thrush was first described by the ancient Greek physician Hippocrates (p.116). In one of his works he talks about 'aphthae in the mouth'. They think that by this he meant 'white patches'. These days we describe an aphthous ulcer as one of those nasty, small, but very painful ulcers that may appear from time to time on the tongue or inside lining of the cheek and which last for a week or so. So what Hippocrates was referring to we will never know but it is very likely that he was familiar with thrush for it is so common. The first person to actually use the word thrush was the English Member of Parliament and naval administrator Samuel Pepys (1633–1703), who wrote in his diary of 17 June 1665 concerning a person who 'hath a fever, a thrush and a hic-cup'.[247] The origin of this word is quite obscure but some suggest that it is derived from 'torsk', an old Norwegian word for a frog and now the current Swedish word for thrush.

The cause of this disorder remained a total mystery until around 1840. The first clue, although it was not recognized as such in the usual context of oral thrush, came in 1839. Bernhard Langenbeck (Figure 45), a young university lecturer in Göttingen, Germany, performed an autopsy on a man who had died of typhoid fever. He found:

> a thick membranous coating, which covered the tonsils from a part of the pharyngeal cavity, but mainly the whole inner surface of the gullet [oesophagus] from the pharynx to the cardia [of the stomach, i.e. the whole length]...Under the microscope magnified of 200 and 400 times, the pseudomembrane did not show any of the known forms of exudates; however, it consisted of an immense number of fungi grown in confusion. The thallus of these fungi consists of tangled branched, extremely tender threads...At the outer surface of the threads, distributed over the whole length, globular or oval water bright cells of a little larger diameter than the threads and with one, rarely with two, sharply limited eccentric cell nuclei are to be seen...these cells...quite corresponded to mould spores in their form.[176]

The cause of typhoid fever was not at that time recognized. Langenbeck was aware of the muscardine disease of silkworms and canvassed the possibility that typhoid fever was a fungal disease, remarking that this would explain its known contagiousness.

All this was unknown to the three people who independently found that a fungus was the cause of oral thrush in infants. The first was Fredrik Theodor Berg (Figure 46), who made his discovery in 1841 while medical officer to the

Figure 45. Bernhard Langenbeck (1810–1887).

Figure 46. Fredrik Berg (1806–1887).

General Children's Home in Stockholm, Sweden; in 1845 he was the first person in the world to be appointed a professor of paediatrics. The second was David Gruby (p.85) at the Foundling Hospital in Paris in 1842; he would later have many famous patients including Chopin, Listz, Alexandre Dumas, and George Sand. The third was John Hughes Bennett (1812–1875) in Edinburgh, Scotland, in 1844; in the following year he would be the first person to describe leukaemia. They all described a mould-like fungus, the filaments of which spread between epithelial cells (the cells that line the tongue, palate, throat, and cheek).

Bennett thought that debilitation was an important predisposition to developing thrush. Berg, on the other hand, thought it was transmitted by unhygienic conditions and communal feeding bottles, and was therefore an unfortunate but common disease unrelated to other diseases or conditions. In an experiment that would be severely frowned upon today, Berg reported in 1846 that he had inoculated thrush membrane material into healthy babies and reproduced the disease; unfortunately, one died from candidal pneumonia!

It was then realized that thrush did not just affect the oral cavity. In 1849, JS Wilkinson in Britain described thrush in the female genital tract when he found the fungus (which he called epiphytes, i.e. plants on the outside) in the vaginal discharge of a 77-year-old woman. In 1862 L Mayer in Germany reported that he had seen six patients with vaginal thrush and was able to reproduce the

Figure 47. Charles-Pierre Robin (1821–1885).

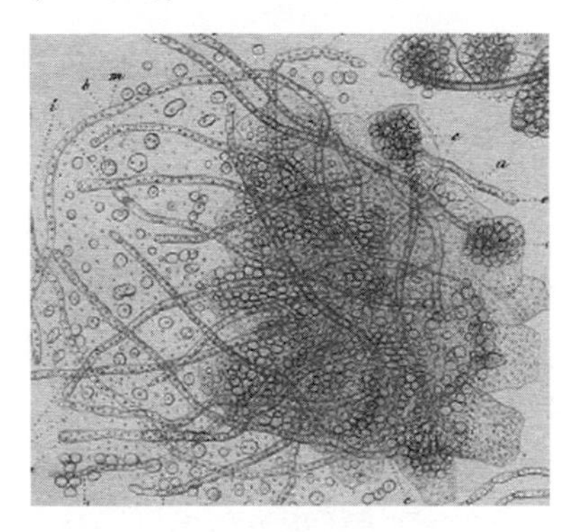

Figure 48. Robin's drawing of the fungal elements seen in oral thrush. Note the filaments (strands) and yeasts (small oval cells) on the background of the darker epithelial cells.

infection experimentally in rabbits. It is now recognized that vaginal thrush is the second most common form of this infection.

In 1853, the French doctor Charles-Pierre Robin (Figure 47) published drawings of both filaments (hyphae) and yeast cells (Figure 48). Thus it seemed that this fungus was not a simple mould but had two forms, one mould-like and the other yeast-like. The technical term now used for such organisms is that they are 'dimorphic (two forms) fungi'. Robin named what he had seen *Oidium albicans* ('oidium' reflecting the egg shape of the yeast cell). In 1864, the German physician Max Burchardt (1831–1897) working in Berlin studied epithelial (cells lining the mucous membrane) preparations of the fungus on microscope slides over two to three days. He noted the filaments were branched and were periodically blocked by internal barriers called septa. Such septate filaments are now known as hyphae. In addition, they sometimes grew on their sides small filaments which lengthened and which did not have these septa; these are called pseudo-hyphae and are a characteristic feature of *Candida* as is budding from yeast cells (Figure 49).

Five years later, in 1869 in Paris, another French doctor, Joseph Parrot (1823–1889), found that this fungus was not restricted to the mouth but could invade the linings of the stomach, intestines, larynx, trachea, and bronchi (Figure 50). He was the first person to describe what is now called invasive candidiasis.

Another important advance was made in 1877 when Paul Grawitz (1850–1932) cultivated the thrush fungus in acid conditions. He found it would then only grow as unicellular yeast cells and not as filaments, thus supporting the idea that this was a dimorphic fungus. Furthermore, Grawitz managed to experimentally infect dogs with the yeasts that he had cultivated. Nevertheless, some people believed that these different forms were different fungi, one a yeast and the other a filamentous mould. This misconception was finally put to rest by the physician Charles Audry (1865–1934), in Lyon, France, in 1887. He showed that thrush was due to a single fungus, which he called *Saccharomyces albicans*, its form varying according to the medium on which it was grown. On gelatine, it became a yeast but in broth it was sometimes filamentous.

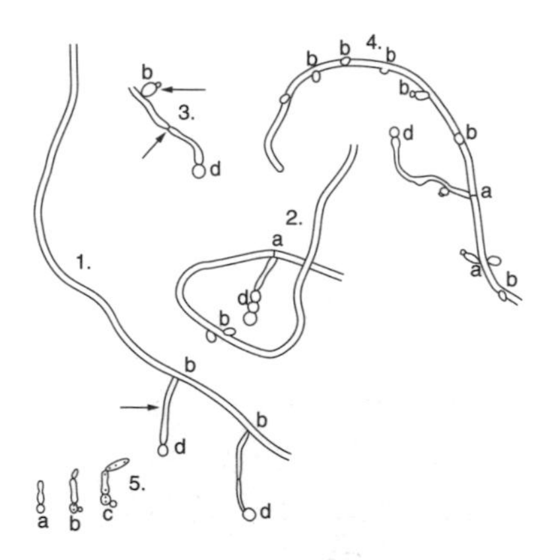

Figure 49. Burchardt's illustration. Note the budding yeast cell (upper arrow), septum (middle arrow), and pseudo-hypha (lower arrow).

Part of the confusion with the fungus that caused thrush was the multiplicity of names it was given by different investigators. In fact, 177 synonyms for *Candida albicans* have been used. The most important was *Monilia albicans* given by the German botanist Wilhelm Zopf (1846–1909) of Münster in 1890, who placed it in the genus *Monilia* erected in 1751 by Hill to describe certain fungi in rotting vegetation (they were actually *Aspergillus*). In 1923 Christine Berkhout of Delft in the Netherlands proposed a new genus, *Candida*: hence *Candida albicans*. This was finally accepted at the Eighth Botanical Congress at Paris in 1954. A number of other species of *Monilia* that infect humans had also been described over the years and these too were transferred to the genus *Candida*. The name *Candida* is derived from the Latin word 'candidus', which means 'white', while 'albicare' is the Latin verb for 'to whiten'.

Figure 50. Parrot's drawing of gastric mucosa. Note the *Candida* filaments invading the submucosa (arrow).

Most of us harbour this yeast in small numbers in our mouth or intestinal tract, as do many women in the vagina. Surveys have shown that up to 80 per cent of healthy people are infected. We now know that Bennet was right. The yeasts multiply and cause disease only when a person's defences are impaired. One such situation is when competing bacteria are suppressed by the administration of antibiotics. Another is in diseases like diabetes, AIDS, or cancer or during chemotherapy. In the more severe cases of impaired immunity, the organisms may invade the tissues, a potentially fatal disease.

PART V

PROTOZOA

Protozoa are single-celled animals. They are too small to be seen with the naked eye so their discovery had to await the invention of the microscope around the turn of the sixteenth to the seventeenth century. They generally range in size from about 1 to 50 micrometres (μm). The name is derived from the Greek words *proton* and *zoa* meaning 'first' and 'animals'. The singular is protozoon, and some taxonomists have placed them in a kingdom all of their own—the Kingdom Protozoa—a term first used by Carl von Siebold (p.21) in 1845. These cells have all the functions of an animal—they move, eat, excrete, and reproduce but are not able to photosynthesize.

There are said to be about 30,000 species of protozoa. Many are free-living in the environment, but some are parasitic and need to spend some or all of their life cycle in one or more animal hosts. Some of these cause no harm. They just simply live with us doing no damage; they are called commensals. Others not only live in us but cause disease; these are parasites. We shall review some of the more important parasitic protozoa that infect humans.

We start with the organism we now call *Giardia lamblia*, which was the first human protozoan parasite to be seen. This is acquired by ingestion as is our next subject, the amoeba *Entamoeba histolytica*, which may cause amoebic dysentery and liver abscesses. We then examine malaria parasites which were the first protozoa to be shown to be transmitted by mosquitoes. This is followed by two subspecies of trypanosomes (*Trypanosoma*), which cause sleeping sickness acquired by the bite of an infected tsetse fly. Sleeping sickness is found only in Africa whereas another trypanosomal infection, *Trypanosoma cruzi*, is transmitted by certain bugs and occurs in South America, where it is called Chagas disease after its discoverer. Finally, we examine another group of protozoal parasites called leishmaniae (*Leishmania* species), also named after their discoverer, William Leishman. They too are transmitted by insects, in this case sandflies.

10

Giardiasis

SCIENTIFIC NAME:	*Giardia lamblia* (synonyms: *G. intestinalis, G. duodenalis*
COMMON GENERAL NAME:	giardia
DISEASE NAME:	giardiasis
DISTRIBUTION:	worldwide
TRANSMISSION:	ingestion of cysts in infected food or water
LOCATION IN A PERSON:	lining of the small intestine
CLINICAL FEATURES:	diarrhoea and flatulence in acute infections
DIAGNOSIS:	parasites seen on microscopical examination of faeces
TREATMENT:	metronidazole
PREVENTION:	avoid eating uncooked vegetables or ground-fruits that may have been grown in gardens contaminated with human faeces, or drinking unboiled or untreated water

When I took a parasitology course at the School of Public Health and Tropical Medicine in Sydney in 1968, we were often asked to examine faecal specimens for parasites. These samples generally came from one of the South Pacific Islands and they almost always contained cysts of *Giardia lamblia*. The stools were solid, indicating the absence of diarrhoea, and I came to the conclusion that this parasite was almost universal and did not seem to be doing any damage, at least in chronic infections. Perhaps it was a commensal. In this idea I was not alone, as we shall see when we consider the remarks of Rendtorff later in the chapter. Several years later when I was a gastroenterology registrar at the Royal Adelaide Hospital in South Australia, I was disabused of the concept that this organism was always harmless. My consultant told me of an Australian patient he had seen who had developed severe, intractable diarrhoea after contracting acute giardiasis while visiting Fiji.

Before we go any further in our consideration of giardiasis, we need to know that this parasite exists in two forms. When it is in the lumen (inside) of the small bowel it is called a trophozoite. In this form, it has flagella, which are whip-like cords that give it the capacity to move. This is the phase that is able to divide and reproduce. As the trophozoites move down the length of the bowel, unless bowel movement is too rapid because of diarrhoea, they gradually turn into cysts. In this state, the parasite loses its flagella and develops a wall around itself, which enables it to survive more readily in the external environment when it is passed in the faeces.

Now let us go back to the beginning. This was the first protozoan parasite ever to be seen. It was found by the celebrated Dutch microscopist, Antony van Leeuwenhoek (Figure 51). He was born in Delft and took his surname, which means Lion's Corner, from the house owned by his father on a corner near the Lion's Gate (Leeuwenpoort) in that city. He began work as a shopkeeper but in 1660 received a sinecure office with the municipality as City Chamberlain, which gave him ample time to indulge in his scientific interests. He taught himself the art of lens grinding and used simple biconvex lenses to magnify the world around him. It is said that at his death he had more than 400 microscopes and magnifying glasses. He was very jealous of his instruments and never lent or sold them to anyone. He made numerous observations of scientific interest, discovering spermatozoa, red blood cells, the capillary circulation, and many minute aquatic creatures. He also discovered *Giardia*, although he did not give it any name. This discovery, as with most of his others, he announced by writing a letter in Dutch to the Royal Society in London, England, of which he became a foreign member in 1680. The Royal Society translated his letters into English or Latin and published them.

Figure 51. Antony van Leeuwenhoek (1632–1723).

In his 66th letter dated 4 November 1681, he described his discovery of the organism we now know as *Giardia lamblia*. He began by saying that he normally had one fairly thick stool each day. In the summer just finished he was attacked by frequent diarrhoea. He thought at first that this must be due to something he had eaten, but eventually he decided to examine his thin, watery stools under a microscope. Amongst the many things he saw:

> I have sometimes seen animalcules [microscopic animals] moving very prettily, some of them a little bigger, others a little smaller than a globule of blood, but all of one and the same shape. Their bodies were a little longer than broad, and their belly was flattish and furnished with several legs [i.e. flagella], with which they moved.... in such a manner that one might imagine seeing a wood-louse running up against a wall.[180]

Sometimes he saw many of these creatures, at other times only a few. When his stools returned to normal, he could no longer find them.

Unfortunately, this information disappeared into the mists of time and it was not until 1932 that an English protozoologist, Clifford Dobell (p.111), concluded that Leeuwenhoek was describing *Giardia* trophozoites. Dobell took to task with scarcely concealed insults earlier writers who had misinterpreted Leeuwenhoek's findings. Because there was a niggling doubt as to whether Leeuwenhoek could really have seen these creatures with the single-lensed microscopes at his disposal, a modern study with a microscope similar to one used by Leeuwenhoek showed that indeed he could.

Nearly 180 years were to pass before this parasite was seen again. This time it was to be the Czech physician Vilém Dušan Lambl (Figure 52), who found it in 1859 when he was working as a pathologist at the Children's Hospital in Prague. Because he found it in intestinal fluid, he named it *Cercomonas intestinalis* and wrote:

> With decided certainty I can vouch for the myriad presence of these monads in the jelly-like mucous excretion in children, a one-celled infusorium [an old-fashioned term for a microbe]... Form; similar to a tadpole. At the blunt end there is a suction bowl, in whose depth two nuclei-like formations gleam through. The tail: 0.003–0.004 mm long, 0.0008–0.0016 mm thick. The organism moves in a circular pattern like the flight of a swallow.[174]

What is more, Lambl drew what he saw (Figure 53); they are clearly *Giardia*. It is not clear though whether or not Lambl thought this parasite caused disease.

In 1888 the name for this parasite became confused and remains so to this day. The French parasitologist Raphael Blanchard (1857–1919) renamed it *Lamblia intestinalis* in honour of Lambl. Meanwhile in 1882 Johann Künstler had named a similar parasite of tadpoles *Giardia agilis*, after the French zoologist Alfred-Mathieu Giard (Figure 54). In 1902 the American parasitologist Charles Wardell

Figure 52. Vilém Lambl (1824–1895).

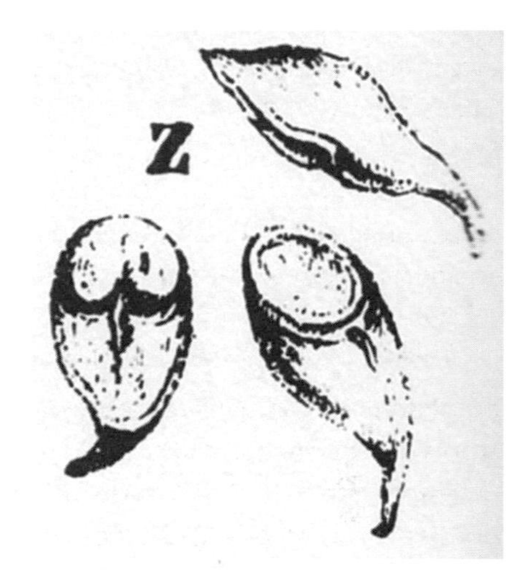

Figure 53. Lambl's drawings of trophozoites.

Stiles (p.31) thought that *Lamblia intestinalis* belonged in the genus *Giardia* (which had pre-dated Blanchard's *Lamblia*) and named it *Giardia duodenalis* because it was most common in the duodenum. Other people labelled it *G. intestinalis* and *G. enterica*, but the name that has stuck is *Giardia Lamblia* (or *lamblia*), which was proposed by Charles Kofoid (1865–1947) and Elizabeth Christiansen in 1915. Some 40 species names have been described in various animals, but cross-transmission experiments from one host to another have yielded inconsistent results. An organism is supposed to have only one name. No-one knows what the proper name is but *Giardia lamblia* is entrenched in the medical and scientific literature, is most commonly used, and has the virtue of memorializing Lambl, who first drew a picture of it and successfully brought it to medical attention. How should *Giardia* be pronounced? Some use the hard 'G', but most people, including me, say Jee-ardia, perhaps because that is how we imagine a Frenchman would say it.

Figure 54. Alfred Giard (1846–1908).

The more important question that puzzled people was whether this parasite caused significant disease or not. In 1916 Harold Fantham (1876–1937) from the Liverpool School of Tropical Medicine in England described seven soldiers who had been invalided back to Britain from either Flanders or Gallipoli with severe diarrhoea in whom *G. lamblia* was the only pathogen that could be found. He and his colleague Annie Porter then managed to infect six out of eight cats and six out of nine mice, many of which became sickly, with protozoa obtained from these soldiers. They concluded that '*Giardia (Lamblia) intestinalis* is pathogenic to man and is capable of producing diarrhoea, which may be persistent or recurrent'.[95]

Not everyone was convinced though and the next important advance occurred in 1954. In that year Robert C Rendtorff performed some experiments on prisoners in the Federal Correctional Institution in Seagoville, Texas, USA. He gave 20 prisoners between 1 and 1 million *Giardia* cysts concentrated from the stools of asymptomatic human donors in capsules. In his own words: 'The participants were prisoners who volunteered after having been fully informed of the conditions of the experiments and risks involved.'[267] Such an experiment would be severely frowned upon today because prisoners may feel they are under duress and not in a position to refuse. Rendtorff found that none of the men given only one cyst became infected but that all of those given 10 cysts or more did so after 6–15 days, with those given most cysts showing signs of infection soonest. He then gave 10 different prisoners 25 cysts in either saline or tap water but only one became infected. In the totality of his experiments, some prisoners developed looser and more frequent stools but others did not. In some subjects the parasites disappeared spontaneously but in others they persisted for months. Rendtorff did not think much of *Giardia* as a pathogen and concluded that:

> In no instance did any clinical illness or subjective complaint occur among the subjects which could be attributed with certainty or bear close relationship to *Giardia* infections.[267]

Rendtorff sought to justify his use of prisoners 24 years later at a US Environmental Protection Authority Symposium on Waterborne Transmission of Giardiasis in 1978 when he said:

> *Giardia* was thought in the 1950s to cause occasional problems of diarrhea in children but its appearance was so common and, in adults so lacking in clinical symptomatology, that most considered it a non-pathogen. As a result we felt safe in exposing prisoners to *Giardia*.[268]

So you can understand my confusion back in 1968.

But evidence for the pathogenicity of this parasite slowly accumulated. In 1955, an outbreak of water-borne giardiasis infection in Oregon, USA, infected 50,000 people. However, the sceptics were not convinced and it was not until 1969 that Dr Lyle Veazie was able to find a publisher for his account of the epidemic.

Finally Dr Theodore Elliott Nash (1943–) and his colleagues at the National Institutes of Health, Bethesda, Maryland, USA, put the matter to rest. In 1987, they reported their observations on 10 healthy volunteers who were each given 50,000 trophozoites of *G. lamblia* via a tube into the gut. The first group of five was given a preparation from one human isolate and the second group was given a preparation of a different human isolate. All of the volunteers inoculated with one of the strains were infected but not those given the other strain. Three of the five infected volunteers became ill, with two having diarrhoea and typical symptoms of giardiasis. They then gave these three people another dose of the same strain of trophozoites 12 weeks later, and at the same time administered that same strain to a third group of five volunteers who thus acted as a control group. All of the last group (who were receiving their first infections) became infected, and two developed loose stools. The rechallenged volunteers became re-infected but were asymptomatic. Thus Nash and his colleagues were able to show two things. First, there is variation in the pathogenicity for humans of different strains of *G. lamblia*. Secondly, infection produces partial immunity in that it did not protect against re-infection but it did prevent symptoms developing. This explains why chronic infections do not cause much trouble provided that the immune system is not suppressed.

So what do we know now? When cysts of *G. lamblia* in food or water are ingested, they pass through the stomach to the duodenum (the first part of the small intestine) where a trophozoite emerges from the cyst and promptly divides into two. These attach themselves to the surface of the cells lining the inside of the small bowel where they multiply by division. If there are enough parasites, they irritate and damage the bowel and impede the absorption of nutrients. This leads to abdominal discomfort, diarrhoea, and flatulence. Normally the parasites either disappear spontaneously or the immune system keeps them at such low numbers that there are no significant symptoms.

11

Amoebic dysentery and liver abscess

SCIENTIFIC NAME:	*Entamoeba histolytica*
COMMON GENERAL NAME:	amoeba
DISEASE NAME:	amoebiasis
DISTRIBUTION:	widespread in the tropics and subtropics but found elsewhere
TRANSMISSION:	ingestion of cysts in infected food or water
LOCATION IN A PERSON:	wall of the large intestine; liver; elsewhere
CLINICAL FEATURES:	diarrhoea with blood and mucus, abdominal pain
DIAGNOSIS:	parasites seen on microscopical examination of faeces
TREATMENT:	metronidazole (tissues), diloxanide furoate (bowel lumen)
PREVENTION:	avoid eating uncooked vegetables or ground-fruits that may have been grown in gardens contaminated with human faeces

Until several generations ago, the thought of contracting dysentery filled people in the first world with foreboding. It is still very common in the Third World. Dysentery is the medical term for loose, fluid stools (diarrhoea) which also contain blood and mucus. It is associated with severe abdominal pains, fever, and lassitude. Dysentery occurs in two forms. The first is acute, lasting days to weeks. The second is chronic and lasts weeks to months to years. Both can kill. We now know that the first form is usually due to infection with *Shigella*, a bacterium; we shall consider it in a later chapter. The second type is due to infection with an amoeba (plural amoebae). Most of us remember from our school days a little about an amoeba. It is a simple single-celled animal with a nucleus surrounded by cytoplasm and held in place by a cytoplasmic membrane. It eats by surrounding food particles such as bacteria with cytoplasm then ingesting them, a process known as phagocytosis. It moves by pushing projections of cytoplasm in various directions. These are known as pseudopods (or pseudopodia) meaning 'false feet'.

Figure 55. Fedor Lesh (1804–1903).

Dysentery was known to the ancients. Hippocrates (p.116) frequently alluded to it, but of course he may have been referring to either of these two forms. It was not until 1875 that the first breakthrough came in differentiating these diseases and understanding the cause of one of them. The discovery was made by Dr Fedor Aleksandrovich Lesh (also known as Friedrich Lösch, Figure 55), a Russian doctor. He graduated from the medical school in St Petersburg, where he became a lecturer at the Women's Medical College and a consultant at the Military Hospital. Although the observations were made in Russia, the case was reported in *Virchow's Archives*, a German journal, which accounts for the Russian and German versions of his name. The paper described in great detail the illness that befell a 24-year-old peasant named J Markov from Archangel in Russia who had travelled to St Petersburg in search of work. He became gravely ill with diarrhoea in 1871 but recovered spontaneously after nearly a year. In August 1873, he once more developed diarrhoea, pain on opening his bowels, abdominal pain, fever, and general weakness. His symptoms waned then waxed once more, and in November he was admitted to Lesh's hospital. At this time he had about eight foul-smelling reddish-brown evacuations containing lumps of mucus and pus each day. Lesh examined the stools (Figure 56) and found:

> large cell-like forms in great numbers, some were free, some encased in mucus. Because of their peculiar movements, they were recognised immediately as amoebae.[184]

Many treatments were tried without success. There was some improvement with quinine but the side effects of that drug, such as headache, ringing in the ears, and dizziness, became intolerable. The diarrhoea fluctuated in intensity and the numbers of amoebae in the faeces varied from day to day. By March

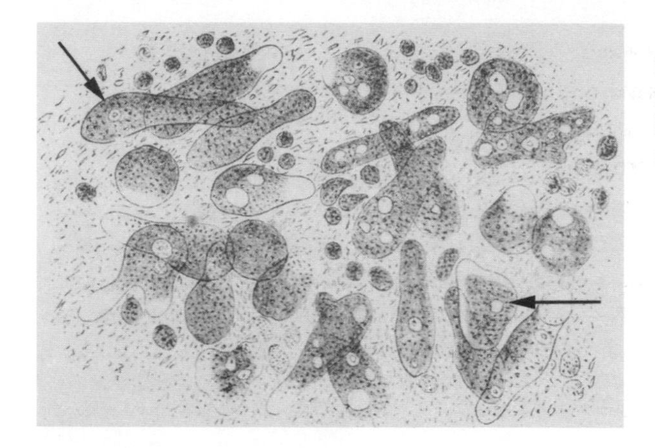

Figure 56. Lesh's illustrations of amoebae in stool—some are arrowed.

1874, the patient developed a cough, then evidence of a pleural effusion (fluid in the cavity surrounding the lungs in the chest). In April this worsened into pneumonia and the patient died from exhaustion. This illness ran its course for three years and ended in death.

Lesh then conducted an autopsy. He mentioned that the lungs had some walnut-sized cavities, the pleura (lining) of the left lower lung was coated with thick fibrinous material (from the blood), and the liver was much enlarged. He concentrated on the large intestine, which was swollen with inflammation, scarred, ulcerated, and haemorrhagic. The numerous ulcers were filled with jelly-like material. On microscopical examination of the wall of the intestine, he found infiltration with white cells and many amoebae similar to those he had seen in the stools.

Lesh then discussed at length the features of the amoebae that he had found in the patient's faeces and described their movements with finger-like extensions (pseudopods). He noted that they varied between 20 and 60 μm in size and varied in shape. He was able to see a nucleus in each parasite and granular cytoplasm which sometimes contained vacuoles (cavities filled with fluid, food, or waste) as well as fragments of ingested bacteria and red and white blood cells. He considered the parasite was similar to but different from *Amoeba princeps* (a previously described amoeba found free-living in water). He therefore named it *Amoeba coli*, after the colon (which makes up most of the large bowel) in which it was found.

He had already wondered whether the amoebae were the cause of the patient's illness, so earlier he had studied their effects on dogs. He gave these animals fresh faeces containing amoebae from the patient orally and also instilled them into the rectum daily for three days. One of the dogs became ill with dysentery and when he killed it 18 days later, Lesh found many ulcers in the rectum (the lower-most part of the large intestine) but not in the rest of the large bowel. These ulcers were heavily permeated with amoebae. His final conclusions about his patient Markow were circumspect:

> There is little doubt that in the described case the amoebae played an important part in producing the symptoms of the disease. At least they sustained the inflammation and prevented the healing of the ulcers present in the large intestine.[184]

Curiously, Lesh was unwilling to ascribe the whole illness to amoebae, writing in his final sentence:

> From this one has to assume that Markow became ill from dysentery first and that the amoebae reached the intestine later, increased there, and then sustained the inflammation.[184]

Nevertheless, Lesh had set the ball rolling in the right direction although there were to be over 100 years of confusion. The reasons for this will become clear as we read on. By publishing in Germany, Lesh had ensured widespread publication of his findings. One person who was encouraged to look further was Dr Stephanos Kartulis (1852–1920), a Greek physician practising in Alexandria, Egypt. In 1886 he published an article, again in *Virchow's Archives*, claiming to have found amoebae in 150 cases of dysentery. He too tried to transmit the infection to experimental animals but failed.

Despite Kartulis's observations, the doubt initially expressed by Lesh over the causal role of amoebae in producing dysentery persisted for many years. For example, Robert Koch (p.211) found amoebae in histological sections (thin slices of tissue which can be viewed under the microscope) of colonic ulcers while he was studying cholera in 1887 but attached no importance to this observation.

However, in that same year, Jaroslava Hlava (1855–1924), who was later to become professor of pathological anatomy at Prague, found amoebae in the stools of some patients with dysentery. He then inoculated these stools into the rectums of six cats and succeeded in producing dysentery in four of them. He published his findings in *Časopis lékařův českých* (Journal of the Czech Physicians) and this had an extraordinary consequence. He began his paper with 'O úplavici', which means 'On dysentery'. For years, those who could not read Czech mis-cited this paper as being written by one O. Uplavici!

At this point we need to step sideways and go back to 1825, to India where the British medical officer James Annesley (1780–1847) first hinted at an association between dysentery and liver abscess, writing that 'hepatic disease seems to be induced by the disorder of the bowels, more particularly when this disorder is of a subacute or chronic kind'.[6] A reason for such a connection between dysentery and liver disease was found by the earlier-mentioned Dr Kartulis. In 1887 he reported that he had made a detailed study of 20 patients with 'tropical liver abscess' and found amoebae in the livers of all of them. Three years later, Dr William Osler (1849–1919), a Canadian physician working in Baltimore in the United States, reported the sad case of a young physician who had contracted dysentery in Panama and later developed a liver abscess which led to his death. Osler found amoebae in the contents of the liver abscess as well as in the faeces of his patient but concluded 'It is impossible to speak as yet with any certainty as to the relation of these organisms to the disease.'[237] Why the uncertainty? In some patients amoebae seemed to be causing disease but other people harboured amoebae which were apparently completely innocuous.

Osler's report stimulated two of his colleagues in Baltimore, William Councilman (1854–1933) and Henri Lafleur (1865–1939), to study 14 cases of amoebic dysentery in their hospital, including Osler's case. They reported their findings in 1891. Seven of these patients developed liver (hepatic) abscesses, four of whom had lung abscesses as well, two developed peritonitis following intestinal perforation, and two suffered severe intestinal bleeding. Post-mortem examinations were performed on a number of the patients. The intestines were grossly thickened and had nodular projections on the mucous membrane with surface ulceration covered with gelatinous material containing amoebae. Amoebae invaded the submucosa (the layer under the mucous membrane), which became necrotic (dead), then the overlying surface membrane broke down producing larger ulcers. It seemed likely that amoebae travelled to the liver via the lymph or blood vessels (Figure 57). The liver abscesses were mostly in the right lobe of the liver and were sometimes multiple. The liver cells broke down, producing an opaque greyish fluid tinged brown or red with blood. Amoebae were not found in this material but were plentiful in the liver adjacent to the abscesses. Pulmonary abscesses were a consequence of amoebae invading through the diaphragm to the adjacent lung. Amoebae were found in and around the abscesses and in the sputum. Clearly amoebae could cause dreadful disease involving not only the large intestine but also the liver and lungs, but Councilman and Lafleur noted that the severity of hepatic damage did not necessarily relate to the severity of the intestinal disease.

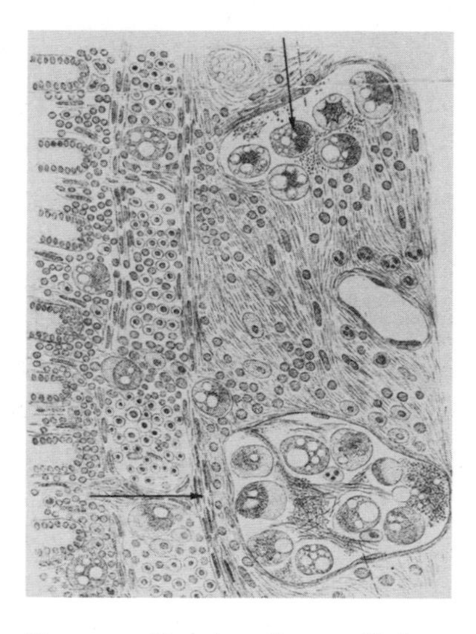

Figure 57. Drawing of a magnified view of a section of the large bowel wall. The lumen is on the left. The horizontal arrow points to the thin layer of smooth muscle that separates the mucosa (left) from the submucosa (right). The vertical arrow points to some amoebae in and entering into a blood vessel.

The next major advance was made by Heinrich Irenaeus Quincke (1842–1922), professor of medicine at Kiel in Germany, together with his colleague Ernst Roos (1866–?). In 1893, they reported two patients. The first was a 39-year-old man who had suffered for years from various nervous disorders blamed on an intensive preoccupation with music. In March 1890, together with a number of other people, he developed diarrhoea then dysentery in Palermo, Italy. This condition continued unabated for three and a half years, at which time Quincke and Roos published their observations. In his stools, they found numerous amoebae, averaging 20–25 µm in diameter, which clearly were the same as the *Amoeba coli* described by Lesh. But they made a very important discovery. Rarely, but in large numbers on one occasion after the patient had been treated with calomel, they found what they realized were cysts of the amoeba. They were smaller, being only 10–12 µm in diameter, shiny, and transparent with a distinct double contour, and the nucleus could not be clearly recognized. Furthermore, these cysts retained their form for at least three weeks.

Quincke and Roos then examined the infectivity of trophozoites (the motile forms of the amoeba) versus the cysts for cats. Faeces containing trophozoites were introduced into the rectums of eight cats via a catheter; six died within two to three weeks of amoebic dysentery. Two cats were then given the same material orally. Although they died on the fourth and tenth days, they did not contract amoebic dysentery. On the other hand, two of four cats given stools containing cysts orally developed amoebic dysentery. They wrote: 'apparently, the encysted forms are resistant to environmental and gastric conditions; therefore the transmission to man may be mediated by them'.[261]

But Quincke and Roos were not yet done. They had a second patient, a 44-year-old lady who had never left Germany and who had lived near Kiel for the past six years. She had suffered from diarrhoea for three years when she came under their care. When they examined her stools, they found amoebae. But there was a difference. These amoebae were bigger than in the first patient, being 25–30 μm in diameter. The cytoplasm was more coarsely grained and they never contained red blood cells in contrast to Lesh's amoeba. Sometimes they noticed cysts with a tough covering which were also bigger than in their first patient, averaging 16–17 μm in diameter. Furthermore, when they tried to infect cats with these parasites, there was only temporary diarrhoea and amoebae were never found.

Clearly there were two types of amoeba—*Amoeba coli* of Lesh and another one, which they called *Amoeba coli mitis* ('mitis' being Latin for 'mild'). Although their error is understandable, they would be shown to be wrong when they also wrote that the second one 'causes enteritis [infection of the bowel] in man and is not pathogenic for the cat'.[261] But Quincke and Roos were still not done. They looked in the stools of 24 people of differing ages and both sexes, some healthy and some with other medical problems. In nine of these subjects, they found amoebae identical to those seen in their second patient. They concluded that there was a third amoeba which was harmless and often found in healthy people; for it they proposed the name *Amoeba intestini vulgaris*.

I hope you are not too confused. If you are, don't worry—everyone else was at that time. From this distance of over a century later, we can say that they were in fact seeing two forms of amoebae. The second patient had another cause for her diarrhoea and had an incidental infection with Quincke and Roos's *Amoeba intestini vulgaris*, which is the same as their *Amoeba coli mitis* and which we now call *Entamoeba coli*.

In 1895, two Italian investigators, Oddo Casagrandi (1872–1943) and Pietro Barbagallo, reported the results of their biological and clinical studies of amoebae and renamed Lesh's *Amoeba coli* as *Entamoeba coli* (as we shall see, this is not the same as the *E. coli* mentioned above). This was the first time the genus name *Entamoeba* was used. Like Quincke and Roos, they also recognized two forms of amoeba, one harmless and the other pathogenic for humans.

In 1903, another German, Fritz Schaudinn (p.280), a zoologist at the Imperial Health Bureau in Berlin, entered the fray. Schaudinn believed that the erection of a new genus, *Entamoeba*, was entirely justified but then made an unfortunate error. For reasons that are hard to fathom, he wrote:

> I cannot decide whether [Lesh] had the harmless or the pathogenic kind before him. But for zoological nomenclature, the first definitive characterisation is relevant.[282]

With superb unjustified Teutonic arrogance, he went on to write:

> I designate the harmless one *Entamoeba coli* Lösch emend. [as amended by] Schaudinn . . . The other kind . . . I shall call . . . *Entamoeba histolytica* on account of its tissue-destroying ability.[282]

The species name 'histolytica' was rather a good choice because it was derived from the Greek words *histos* and *lysis* meaning 'tissue' and 'loosening' or 'destruction'. But otherwise it was all back to front. Lesh's parasite was clearly pathogenic and what we now call *Entamoeba coli* is not pathogenic. The pathogen should have been called *E. coli* and the non-pathogen given some other innocuous name.

Perhaps Schaudinn reaped his own reward. In an attempt to determine whether the parasites were a primary pathogen or of only secondary importance, he decided to infect himself. Twice he infected his large intestine with the larger parasite obtained from healthy individuals in Germany. He did not become ill but harboured amoebae which he described in detail. Most importantly, he found that there were eight nuclei in each cyst (Figure 58). He did not realize it but in fact he had infected himself with the harmless parasite now known as *Entamoeba coli*. Another German named Huber had demonstrated in the same year that there are only four nuclei in *E. histolytica*. Schaudinn did not believe this and mistakenly called this parasite *Entamoeba tetragena*, thereby causing enormous confusion for 10 years. Three years later in 1906 at the age of 34, Schaudinn tragically died. Somewhere, somehow, some say voluntarily, he appears to have become infected with *E. histolytica* and died from one of its complications, which the *British Medical Journal* carefully described as an obscure pelvic suppuration.

The term amoebiasis was introduced by William E Musgrave and Moses T Clegg in 1906 to cover any amoebic infection, whether causing disease or not. Diagnostic certainty in differentiating *E. histolytica* from *E. coli* was greatly enhanced when stains were used to demonstrate internal structures. The motile (trophozoite) stage differed in subtle ways between the two parasites. There was a major differentiator in the cystic stage, with the former having four nuclei and the latter eight. The first such stain was Lugol's iodine, a solution of elemental iodine and potassium iodide in water first made in 1829, named after the French physician Jean Guillaume Auguste Lugol (1786–1851). This took only a

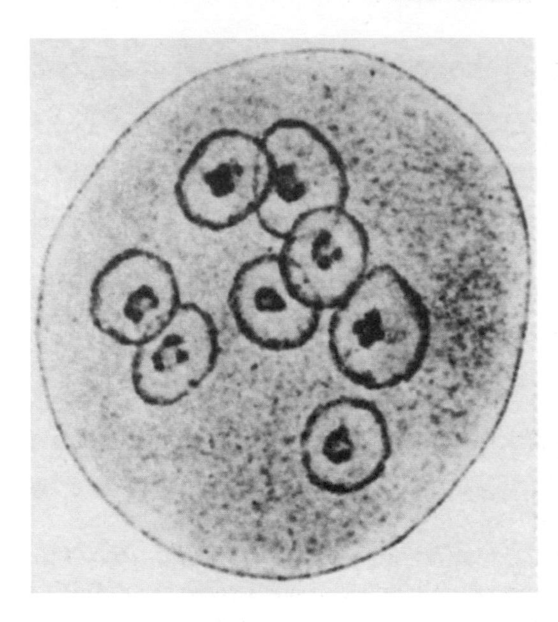

Figure 58. Drawing of a cyst of *E. coli* showing eight nuclei. In real life the nuclei are not seen in one plane and the microscope has to be focused up and down to count them.

few seconds to prepare and was easy and practical to use. Better resolution was provided by an iron–haematoxylin stain using a fixative described by Schaudinn, but this was complex to perform and took two days to complete.

The pathogenicity of *E. histolytica*, the smaller amoeba with four nuclei in its cysts, and the harmlessness of *E. coli*, the larger one with eight nuclei in its cysts, were both confirmed in a dreadful series of experiments in Manila between 1910 and 1913. Why Manila? In their war of 1898 with Spain, the United States had invaded the Philippines, expelled the Spanish, and turned the country into an American colony. Ernest Walker (1870–1952), a scientist, and Andrew Sellards (1884–1941), a medical practitioner, used the inmates of Bilibid prison for their studies.

First, they fed cysts of *E. coli* obtained from five healthy people to 20 prisoners. They did this by putting faeces containing the cysts in gelatine capsules which were then swallowed. Seventeen became infected, with amoebae appearing in the stools an average of five days later. None developed any symptoms. They wrote: 'We believe *E. coli* is non-pathogenic and consequently plays no role in the etiology of entamebic dysentery.'[315] They were right; it is harmless.

In a different experiment Sellards gave two men *E. histolytica* trophozoites and found '*E. tetragena*' cysts (with four nuclei) in the stools. He then gave one man '*E. tetragena*' cysts and recovered *E. histolytica* trophozoites from the stools. He concluded that *E. histolytica* and '*E. tetragena*' were one and the same parasite.

In a third series of complicated experiments, Sellards gave 20 prisoners *E. histolytica* from seven different sources. Fourteen of 16 men given cysts became infected, and three of four men given trophozoites became infected. Amoebae appeared in the stools after an average of six days or so. One of the uninfected men also became infected after a repeat dose of cysts. Of the 18 infected men, four developed amoebic dysentery after an incubation period averaging nine weeks. They were then treated with ipecac (which is most unreliable). *E. histolytica* therefore caused disease.

Walker then tried to justify these experiments with the following words:

> The resort to human experimentation is usually not to be recommended, but in certain infectious diseases of wide geographical distribution and prevalence...and of which an accurate knowledge of the etiology or transmission cannot be obtained, experimental infections of man have been resorted to, and the knowledge thus obtained has enabled medical science to control these diseases to a remarkable extent...Therefore, an attempt has been made to determine, once and for all the specific amoeboid organisms, if any, concerned in the production of endemic tropical dysentery by a series of carefully conducted experiments upon volunteers.[315]

These 'volunteers' were prisoners. Walker sanctimoniously added that the nature of the experiments and the possible consequences were explained to each man 'in his native dialect and each man signed an agreement'.[315] No inducements such as money or reduction in sentence were offered or given. Walker should have known better. If Dr Sellards tried this today, he would promptly be deregistered.

The next development was a technical advance. In 1925, William Boeck (1894–1972) and Jaroslav Drbohlav (1893–1946), American and Czech physicians working at Harvard University, described a medium that allowed the co-cultivation of *E. histolytica* with bacteria. This created a reliable and continuing source of parasites for further experimentation.

This was the springboard that allowed Clifford Dobell (Figure 59) to infect monkeys and describe precisely the life cycle of *E. histolytica*. Dobell worked at the Medical Research Council in England. He was a brilliant scientist and superb artist but apparently not a very nice chap. A colleague described him as being a brittle and uncompromising character, and even his wife said he was pedantic and a difficult man to know. Together with Patrick Laidlaw (1881–1940), Dobell found that he could induce encystation (turning trophozoites into cysts) by adding starch to Boeck and Drbohlav's medium.

Figure 59. Clifford Dobell (1886–1949).

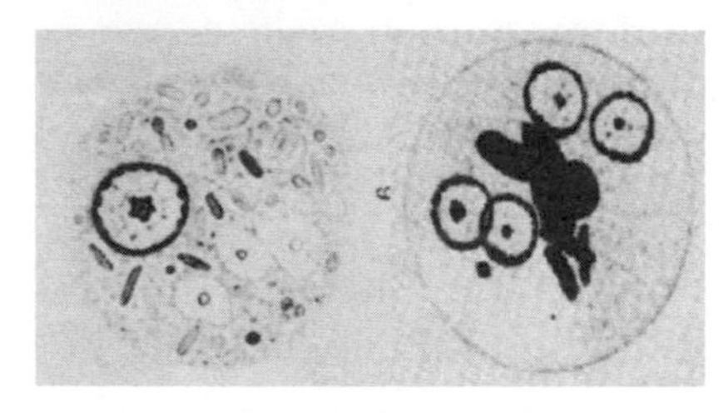

Figure 60. Dobell's illustrations of a trophozoite (left) and a cyst (right) of *E. histolytica*.

Conversely, he could induce excystation (turning cysts into trophozoites) by putting cysts in a starch-free medium. This allowed him to observe all stages in the life cycle. He concentrated on one particular strain of *E. histolytica* obtained from a macaque and which he used to infect a kitten to prove it was pure. He then serially subcultured this parasite *in vitro* (in test tubes) 220 times over 12 months; the diameter of the cysts averaged 13.5 μm. He used various stains to show the internal features of the amoebae and described them at different stages of development over the entire life cycle (Figure 60).

All was now clear. Or so he thought! The larger amoeba, *E. coli,*was not pathogenic. The smaller amoeba, *E. histolytica,* was pathogenic, and all the stages in its life cycle were known. But more and more it was found that although patients with dysentery had the typical amoebae, morphologically identical amoebae were also found in perfectly healthy individuals. What was going on? When I first became aware of all of this in the 1960s, an attempt had been made by an Englishman, Charles Wenyon, to divide *E. histolytica* into two groups—smaller ones called 'small race' or sometimes *E. hartmanni,* which was said to be non-pathogenic, and larger ones called 'large race', which was potentially pathogenic, the key word being 'potentially'. None of this was satisfactory and Ronnie Elsdon-Dew quite rightly wrote in 1968:

> No parasite has been more often wrongly labelled than has *E. histolytica* and to few organisms has such a wide gamut of clinical presentations been attributed.[92]

However, an important breakthrough which was to set the stage for the final advance had been made in 1961. Louis Stanley Diamond (1920–2009) at the

National Institutes of Health in the United States developed a medium which allowed easy cultivation of parasites in the absence of bacteria. Amoebae were now free of contamination and their biological and immunological characteristics could then be studied with confidence.

Peter G Sargeaunt (?–2008) and his colleagues John E Williams and JD Grene at the London School of Tropical Medicine and Hygiene solved the problem in 1978. They examined '*E. histolytica*' from 85 patients by using a special biochemical technique, called electrophoresis, which allowed them to distinguish the markers of three enzymes in the amoebae. Enzymes are proteins that facilitate and hasten metabolic reactions. They found they could distinguish four different enzyme patterns; they called these zymodemes. One group all came from patients with amoebic dysentery, while all of the other patterns were in parasites that had been obtained from asymptomatic people. At last a marker of pathogenicity had been discovered! After 20 years or so, this led to the availability of tests in clinical microbiology laboratories to define the significance of *E. histolytica* when it was seen on microscopical examination of faeces.

There was one final twist to the story. In 1928, Alexandre Joseph Émile Brumpt (1877–1951), a French parasitologist, had proposed that there were three morphologically similar species, two of which were identical. One was smaller, and he called this *E. hartmanni*. The other two were larger and indistinguishable except that one was pathogenic for cats and the other was not. He called the first *E. dysenteriae* (which is our present-day *E. histolytica*) and the second he named *E. dispar*. Since testing animals was impractical in a routine laboratory, no-one paid any attention to his proposal. With this in mind, however, when biochemical and immunological tests became available to distinguish the two, as Brumpt had done with his cat experiments, Diamond and CG Clark resurrected *Entamoeba dispar*, writing:

> We redescribe the invasive parasite retaining the name *Entamoeba histolytica* Schaudinn, 1903 (Emended Walker, 1911), and set it apart from the non-invasive parasite described by Brumpt, *Entamoeba dispar* Brumpt, 1925.[85]

That is now official. So we have four amoebae that commonly affect man. *E. histolytica*, the pathogen; *E. dispar*, which looks identical but is non-pathogenic; *E. hartmanni*, which looks like the first two but is small and does not cause disease; and *E. coli*, which is bigger and looks different and is also harmless. Whew!

The saga of the study of these amoebae has been peppered with powerful personalities and prima donnas who delighted in condemning the views of

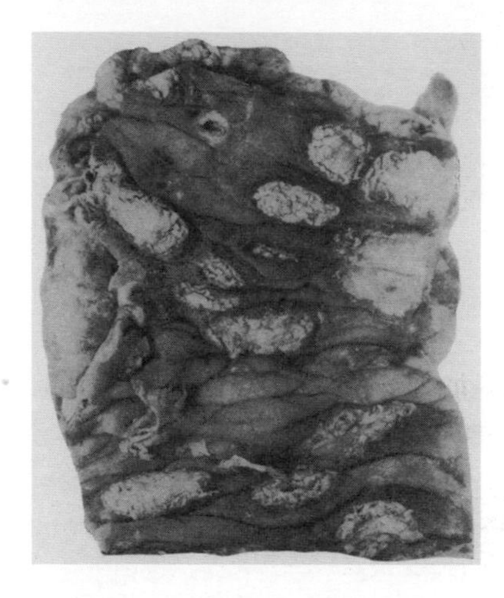

Figure 61. Section of inflamed colon showing multiple, whitish, shaggy ulcers with intact mucous membrane in between.

others and in promoting their own. So often they were wrong. We can never say 'finally' with this disease because there are still a few minor issues to resolve. However, we can understand the life cycle of *E. histolytica*. Cysts are ingested in food or water and then hatch trophozoites (motile forms) in the small intestine, which then may adhere to and invade the wall of the large intestine producing inflammation and ulcers (Figure 61). Some amoebae invade lymphatic or blood vessels and pass to other tissues, most commonly the liver, where they may produce an abscess. In chronic liver disease, trophozoites may pass to the adjacent lung and cause disease there. If the person's stools are fluid, then amoebae are excreted as trophozoites, but if they are formed, i.e. solid, then the trophozoites change into cysts which have some chance of persisting in the environment and infecting someone else. Amoebic disease is going to be around for a long time yet, especially in places where sanitation is poor.

12

Malaria

SCIENTIFIC NAMES:	*Plasmodium falciparum, P. vivax, P. malariae, P. ovale*
COMMON GENERAL NAME:	malarial parasite
DISEASE NAME:	malaria
DISTRIBUTION:	parts of the tropics and subtropics
TRANSMISSION:	bite of an infected female *Anopheles* mosquito
LOCATION IN A PERSON:	first the liver then red blood cells
CLINICAL FEATURES:	fever, shivers, abdominal pain; complications include cerebral malaria (fits and coma) and blackwater fever (kidney failure)
DIAGNOSIS:	parasites seen on microscopical examination of blood
TREATMENT:	various antimalarials—originally quinine, now drugs such as mefloquine, atovaquone-proguanil, artemether-lumefantrine
PREVENTION:	avoidance of night-biting mosquitoes with clothing and bed-nets, antimalarial drugs

In August 1966 over the space of an hour or two, I developed a fever and shivers. I did not need a blood test to know I had malaria. Six months previously, I had returned from living in a malarious area of the then Territory of Papua and New Guinea. I reached for a bottle of chloroquine that I had kept and 24 hours later I was back to normal.

Most of us have not had malaria but we have all had a fever. A fever is a raised body temperature and is often accompanied by headache, aches and pains, sweatiness, and a fast pulse. If the temperature is high enough we may become shivery, and if it is very high we may develop uncontrollable shakes called rigors. Medical science is still unsure whether fever is just the body's response to an insult of some sort or if it has a useful, protective function. As we shall see, fever occurs in malaria when red blood cells infected with malarial parasites break down.

Figure 62. Hippocrates (*c.* 460–*c.* 370 BC).

You might think that malaria is a tropical disease, and so it is. But it has not always been so. Malaria was once widespread throughout Europe and even occurred within the Arctic Circle. During the time of Hippocrates (Figure 62), malaria was endemic in Greece and he made a study of fevers. We now know of several hundred causes of fever, but malaria is unique in that it often has a distinctive, recognizable pattern. Hippocrates was the first to recognize this and he described tertian and quartan fevers. In his book on *Epidemics,* he wrote:

> In some [patients] all those described below occurred with pain. During autumn, and at the commencement of winter, there were phthisical complaints, continual fevers; and, in a few cases, ardent; some diurnal, others nocturnal, semi-tertians, true tertians, quartans, irregular fevers.[137]

A 'quartan fever' is not what you might expect, a fever that recurs every four days, but a fever that occurs on the first, fourth, seventh day, and so on, i.e. every three days (Figure 63). There is only one cause of a quartan fever and that is infection with a malarial parasite now called *Plasmodium malariae.* A 'tertian fever' is a fever that occurs on the first, third, fifth day, etc., i.e. every two days (Figure 63). This is most likely due to another malarial parasite called *P. vivax.* It can also be due to the deadly *P. falciparum,* although this parasite more commonly has a 'sub-tertian' pattern, with a cycle of somewhere between 36 and 48 hours. Incidentally, the 'quotidian fever' also described by Hippocrates is one in which fever recurs daily. There are many causes of this but in the context of malaria it can occur when a tertian fever is acquired twice, but on alternate days. Hippocrates went on to write 'the tertians were more numerous' and, quite correctly, 'the least dangerous of all, and the mildest and most protracted, is the quartan'. What is more, Hippocrates wrote in his *On Airs, Waters and Places*:

> but if there be no rivers, but the inhabitants drink the waters of fountains, and such as are stagnant and marshy, they must necessarily have prominent bellies and enlarged spleens.[137]

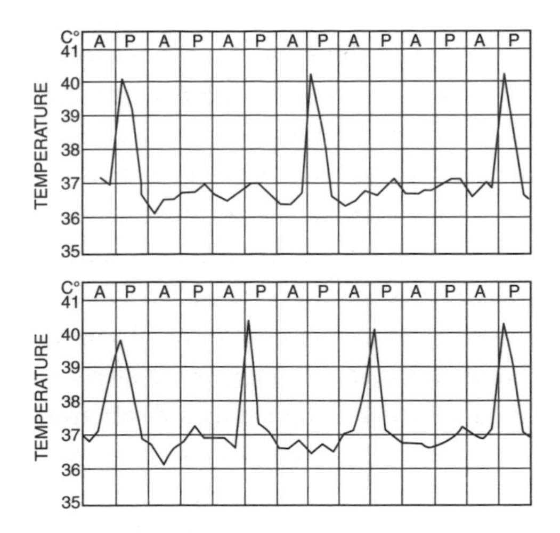

Figure 63. Quartan fever (upper) and tertian fever (lower) temperature patterns. A = a.m. P = p.m.

An enlarged spleen is a classic feature of chronic malaria, and malaria typically occurs in places where waters are stagnant and marshy. Hippocrates made an outstanding observation, although he was wrong in thinking that the affliction was acquired by drinking the water. No wonder some have called him the 'first malariologist'.

Malaria was also a problem in ancient Rome. Celsus (p.6) wrote in his *De Medicina,* with regard to quartan and tertian fevers, a description which can easily be seen to indicate a malarial paroxysm:

> Nearly always they begin with shivering, then heat breaks out, and the fever having ended there are two days free; thus on the fourth day it recurs…tertian fever…has merely this distinction, that it affords one day free and recurs on the third.[68]

A modern textbook description would be similar except that it would describe a sweating phase after the hot phase.

It is to the Italians that we owe our name for the disease. There was a widespread belief that there was a relationship between swamps and these intermittent fevers. 'Bad air' from the swamps was thought to generate vapours or miasmata. Their two words for 'bad' and 'air' were 'mala' and 'aria' and sometime in the Middle Ages the two were combined as mal'aria, which eventually became malaria. But the cause remained obscure and sometimes false trails were followed. For example, two European investigators, Edwin Klebs (p.296) and Corrado Tommasi-Crudeli (1834–1900) in the late 1870s thought that a bacterium which they isolated from swamps and called *Bacillus malariae* was the culprit. They were wrong.

Figure 64. Alphonse Laveran (1845–1922).

A big clue was the finding of a pigment. In 1716, the Italian physician Giovanni Maria Lancisi (p.74) discovered a characteristic black pigmentation of the brain and spleen in the victims of malaria. More than a hundred years later, in 1847, Heinrich Meckel von Helmsbach (1822–1856), a German pathologist, noticed some round, ovoid, or spindle-shaped protoplasmic masses containing irregular pigment granules. He was probably unsuspectingly looking at malaria parasites.

It was an interest in this pigment that led Charles Louis Alphonse Laveran (Figure 64) to make the big breakthrough. He was a French army surgeon stationed at the Military Hospital in Constantine in Algeria. Starting in 1878, he began examining post-mortem tissues from patients who had died with malaria, but had no luck for many months. He then turned his attention to looking at fresh blood taken from patients with malarial symptoms and things started to fall into place. He reported in December 1880 to the Medical Society of the Hospitals of Paris:

> On 20 October this year, while I was examining the blood of a patient suffering from malaria, I noticed, among the red corpuscles elements that seemed to me to be parasites. Since then I have examined the blood of 44 malaria patients; in 26 cases these same parasites were present... These elements were not present in the blood of patients who were not ill.[177]

One has to imagine the difficulties under which Laveran was working. He used fresh blood and would put a drop on a glass slide, cover it with a thin glass coverslip to make a thin film, then examine the slide with a not particularly powerful microscope. These parasites were colourless and transparent and were therefore difficult to see against the pale red background of a red blood cell, although sometimes associated pigment would help him find them. On the other hand, the specimens were living, so he was sometimes able to see movement. The most dramatic observation of all which confirmed beyond doubt the animal nature of these bodies was made on 6 November 1880:

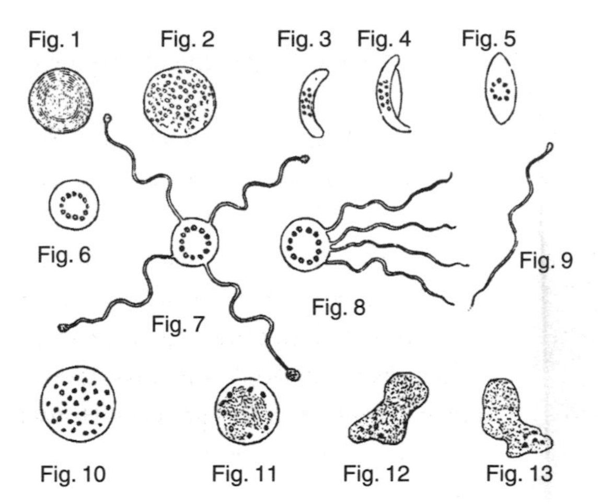

Figure 65. Laveran's drawings of the parasites he saw. See text for details.

> while examining one of the spherical pigmented elements in a preparation of fresh blood, I noticed with joy at the periphery motile elements of the animated nature of which there was no room for doubt.[177]

Actually, Laveran was incredibly lucky to see this as it does not happen in the human body. We now know that it happens when a particular sort of malarial parasite called a gametocyte enters the stomach of a mosquito. The conditions on his glass slide must have mimicked that of a mosquito's stomach.

Figure 65 shows Laveran's drawings of what he observed. He recognized various sorts of bodies. Thc first were elongated and often crescent- shaped (3–5). The second were bodies with variable shapes depending upon whether they were at rest or in motion. At rest, they were round and transparent and had pigmented granules arranged in a ring (6). In motion, they changed their shape, becoming longer and flatter then returning to spherical; in other words they behaved like an amoeba. Furthermore, this second type of body rarely had transparent filaments moving rapidly in all directions as he saw on 6 November (7–9). A third type of body was larger and had pigment granules arranged haphazardly and was sometimes distorted in shape (10–13).

Laveran named his newly found organism *Oscillaria malariae*, the genus name indicating its movements. He summed up his paper by saying that:

> Parasitic elements are found in the blood of patients who are ill with malaria. Up to now, these elements were thought incorrectly to be pigmented leucocytes. The presence of these parasites in the blood is probably the principal cause of malaria.[177]

Laveran was seeing various stages in the life cycle of a malarial parasite but he was not able to put things together in their proper sequence. However, his discovery was so important that Laveran was eventually given the Nobel Prize for it in 1907. What sort of person was he? He was certainly no money grubber. He gave the prize money to the Institut Pasteur, where he was working at the time, to build a Laboratory of Tropical Medicine. He was said to have an aloof reserve typical of his Flemish blood but he led a happy life, devoted to his family and to science.

Laveran asked his colleague Dr Eugene Richard, a military surgeon stationed 50 miles away at Philippeville, if he could replicate Laveran's findings. Richard did so, but made two important new observations. He was able to recognize an even younger form of the parasite which was just a tiny clear spot on the red cell. Secondly, he disagreed with his friend on one point; Richard believed that the parasite developed within rather than on the outside of a red cell. Time was to prove him right on both counts.

The announcements of Laveran then Richard were met with scepticism because so many similar communications had been tried and found wanting in the past. So in 1882 Laveran went to Rome to confront the sceptics, including Marchiafava, Celli, and Golgi. They were not easily persuaded, particularly Marchiafava, who had been a student of Professor Tommasi-Crudeli and had even claimed himself to have found *Bacillus malariae* in two patients with malaria. Being true scientists though, Ettore Marchiafava (Figure 66) and his colleague Angelo Celli (1857–1914) began to examine blood from their own patients (of whom there were many in Rome). Moreover, they had the advantage of better instruments called compound microscopes which gave greater magnification and clearer optics. They examined fresh blood and dried films of blood which had been stained with methylene blue (a technique introduced by Paul Erhlich in 1881) to give the parasite a blue colour which stood out against the redness of the red cell. They showed that the pigment was within the parasite and that the parasite divided by

Figure 66. Ettore Marchiafava (1847–1935).

Figure 67. Camillo Golgi (1843–1926).

fission (split in two) multiple times within red cells. Undoubtedly these pigmented bodies were alive and caused malaria. They thought this parasite was different from the one that Laveran had shown them, so they called it *Plasmodium* in 1885. This was to cause controversy and confusion for years to come because a plasmodium was and is the major component of a slime mould. But the name stuck and was eventually ratified by the International CommissiononZoologicalNomenclature. So much for Laveran's *Oscillaria*.

In 1885, Camillo Golgi (Figure 67) in Pavia in northern Italy carefully observed the cycle of development of malarial parasites in quartan fever and in 1889 reviewed his findings:

> The parasites will gradually develop within the red cells, changing from the initial unpigmented amoeboid forms to the pigmented ones which will progressively enlarge, assuming cellular substance until, having attained a certain stage of development, they undergo a series of metamorphoses [changes in shape], the end result is segmentation, occurring at the time of or just prior to the onset of fever.[117]

Golgi recognized that the parasites in different red cells underwent development at the same time and that fever erupted when the segmented red cells ruptured. What is more, a new generation of parasites thus came into being which invaded other red cells and began the cycle again. In 1886 he showed a similar sequence of events occurred in tertian fever. Furthermore, Golgi realized that the pigment was formed from the breakdown of haemoglobin in red cells.

Golgi continued to accumulate cases and in 1889 produced a definitive report describing the evolution of the parasite from a tiny body one-fifth the diameter of a red cell through amoeboid forms to segmented forms. Golgi reasoned that since there were two clinical patterns of fever, tertian and quartan, there may well be two different malarial parasites. He thought he could differentiate the amoeboid forms in these two illnesses. But it was much easier with the segmented forms. He wrote:

> The segmentation process in tertian fever is different from that of quartan fever. The differences are so clear-cut that, as a rule, the finding of certain forms may be sufficient to make a differential diagnosis between the two clinical types of intermittent fever.[117]

The key difference is that in tertian fever there are 15–20 segments per organism and it took 48 hours for them to appear, while in quartan fever there were only 6–12 segments but they took 72 hours to appear. Finally, Golgi realized that the crescents found by Laveran were found in irregular fevers and not in tertian and quartan fevers. Golgi, too was to be awarded the Nobel Prize in 1906, but for work he had done on nerve cells, which he had learnt to stain. He was married but had only one adopted daughter, which perhaps explains how he had the time to be so prolific.

It seemed there were at least two different parasites, and in 1890 Battista Grassi and Raimondo Feletti (1851–1927) named the organism causing tertian fever *Haemamoeba vivax* and gave that which caused quartan fever the name *H. malariae*. They made matters worse in 1892 by renaming them *Laverania vivax* and *L. malariae*. Not everyone agreed with this though. Many, including Laveran, maintained that there was only one parasite which had many forms.

These tertian and quartan fevers, although unpleasant, were relatively benign and did not kill people. There was, however, another form of fever which developed especially during the summer and autumn and was therefore called 'aestivo-autumnal fever'. The fever either recurred every day (a quotidian fever) or the cycle was somewhere between 36 and 48 hours (so it was called a subtertian fever, Figure 68). It was therefore considered an 'irregular fever'. But much worse, patients sometimes became extremely ill and died within several days. This type of fever was therefore also called a 'malignant fever' or a 'pernicious malaria' to reflect its nastiness.

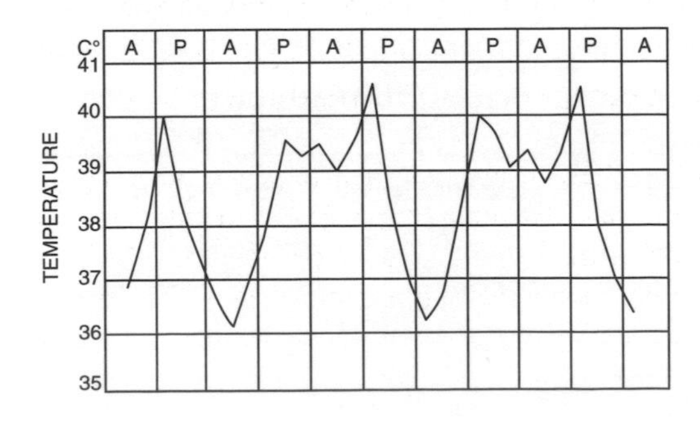

Figure 68. Irregular (sub-tertian) fever. A = a.m. P = p.m.

Such fevers were common among people living near the marshes around Rome so Marchiafava and Celli began to study them during the 1880s. One day a young girl who was severely ill was brought to them and in her blood they first recognized tiny hair-like rings, which we now know as trophozoites of a third species of malarial parasite, *P. falciparum*. In an extraordinary and potentially lethal experiment, Marchiafava and Celli then injected blood containing these parasites into another person and produced a new infection, thereby proving for the first time the transmissibility and parasitic nature of malaria. Who this recipient was, what he or she knew of the possible consequences, and what became of him or her, I am afraid I have not been able to find out. If it was Marchiafava or Celli themselves, then I suppose the experiment might be termed 'heroic'. If not, it was damnable. They also recognized that the small ring forms occurred solely in the peripheral blood but the segmented forms were found only in the blood vessels of the internal organs. This latter observation was also made by Golgi, who found segmented forms in smears made from punctures of the spleen (a rather hazardous procedure as it could be complicated by potentially fatal bleeding into the abdominal cavity).

Incidentally, around the same time in 1884, the German physician Carl Gerhardt (1833–1902), unaware of Laveran's discovery, accidentally caused a healthy person to develop clinical malaria by giving him an injection of blood from a malarial paticnt.

In 1891, Marchiafava together with his colleague Amico Bignami (1862–1929) published a 169-page book entitled (in English translation) *On Summer-Autumnal Malarial Fevers* in which they detailed all their investigations including drawings of the various phases of development of the parasite that caused this disease (Figure 69). Included among their discussion was the demonstration by Bignami and his colleague Giuseppe Bastianelli (1862–1959) that the crescent-shaped parasites of Laveran were typical of this form of fever and did not appear in the blood until the second week of the illness. They did not know what the function of these crescent forms was but they made the crucial observation:

> We consider it inexact to say that these fevers owe their origin to falciform [crescent-shaped] haematozoa [blood-dwelling animals]. It is not the crescent-shaped body, but the amoeba that has the greatest pathogenic importance—the amoeba which passes through a developmental cycle with every paroxysm of fever.[210]

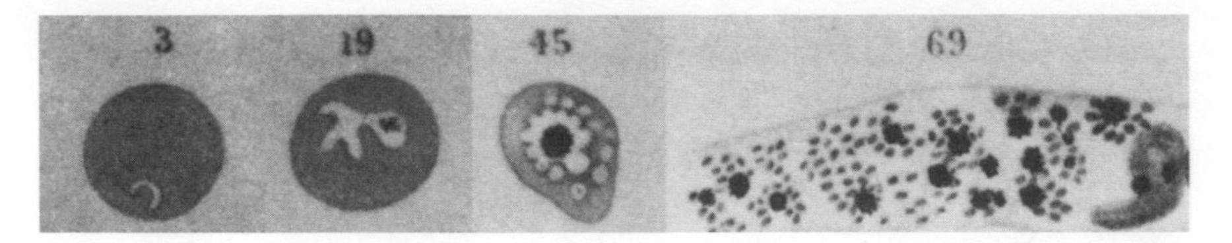

Figure 69. Marchiafava and Bignami's drawings: (3) ring trophozoite, (19) amoeboid trophozoite, (45) segmenting form, (69) segmenting forms blocking a capillary vessel in the brain.

Figure 70. Dimitri Romanowsky (1861–1921).

A few years would pass before the function of the crescent forms would become clear. What sort of a person was Marchiafava? He was a Senator of the Kingdom of Rome, a humanist, and a lover of the early Italian classics.

All these investigators had a problem—they could not see the parasites particularly well. Either they looked at transparent living parasites or they saw them stained a blue colour with methylene blue. What was needed was a way of staining them (like painting them) so that different parts of the parasites could be seen in different colours. Such a differential stain was discovered by Dimitri Leonidovich Romanowsky (Figure 70). Romanowsky graduated in medicine from the University of St Petersburg in Russia. In 1891, the very year of his graduation, he reported that malarial parasites could be dried and fixed on a glass slide then stained with a combination of the stains eosin and methylene blue. The result was that different parts of the parasite appeared in various colours. He then went on to describe in great detail the cycle of development, but first let us read how he described the colours:

> The red blood cells stain eosin-pink... The blood cells which contain the malaria parasite stain pink; however, the more pigment the parasite contains the paler is the blood cell [this makes sense because the parasite converts the pink haemoglobin of the red cell into pigment]. Therefore the red blood cell which

> contains the sporulating form [the segmenting form now known as schizonts] is hardly visible against the pale background... *The malaria parasite, no matter in which form or size it is encountered, always shows with our staining method two distinguishable parts*: one irregular form of blue colour (Prussian blue) and one unstained of constant round or oval form in the middle or on the periphery of which are seen formations of the most varied shade of carmine-violet. We have concluded that this carmine-violet formation found in every parasite is the chromatin part of the parasite nucleus... the nuclear fluid... (is) the colourless part (halo) mentioned above.[275]

Samples of illustrations of what Romanowsky saw are shown in Figure 71. Unfortunately, they do not do justice to the brilliant array of colours which a modern microscopist can see and use to differentiate stages in the life cycle and the species of parasite concerned. Romanowsky described small free forms in the blood (1) (now called merozoites) and then one entering a red cell (2). Here they evolved from a small form (4) into an amoeboid (8) then signet ring form (11) (all of these stages are now called trophozoites). Then the parasite began to divide (21) (now called a pre-schizont) then into the fully segmented form with up to 20 segments, which are in fact individual parasites (30) (the whole now called a schizont) ready to burst and release individual merozoites into the bloodstream. The word 'schizont' (generally if illogically pronounced shy-zont) is of course derived from the Greek word *schizos* meaning 'split'. In addition, Romanowsky saw other forms of the parasite which almost filled the red cell but which did not segment (34, 35). It would later be realized that these are male

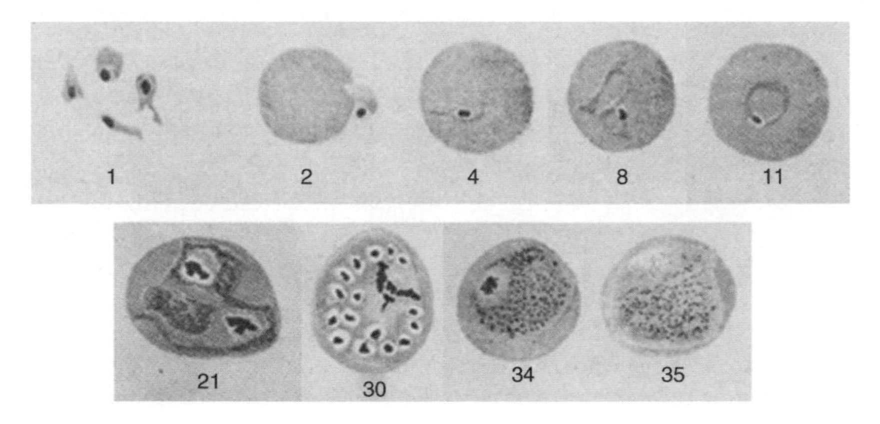

Figure 71. Romanowsky's drawings of differentially stained malarial parasites. See text for details.

and female gametocytes which are ingested by mosquitoes and continue the life cycle in those insects.

Romanowsky made an incredible contribution but his stain was a little cumbersome. Over time, the Scot William Leishman (1901), James Wright in the USA (1902), Gustav Giemsa in Germany (1904), and Jack Field (1941), an Englishman working in Malaya, each devised modifications which were easier to use and came into routine use.

In his paper, Romanowsky did not appreciate the different appearances of the parasites that caused tertian or quartan fevers or irregular fevers. This suited Laveran, who was still insisting that all these forms represented just the one species of malaria. It was becoming increasingly clear, however, that there were at least three different species. We have already seen how Grassi and Feletti ascribed tertian and quartan fever to *P. vivax* and *P. malariae*, respectively. With the use of Romanowky's stains, differences in the appearances of the two parasites were recognized and Grassi and Feletti's suggestion was justified. It was increasingly appreciated that the parasite that caused irregular fevers was rather different. It had small ring trophozoites, sometimes more than one in a red cell, segmented forms (schizonts) were not generally seen in the peripheral blood but could be found in tissues, and this form of fever was associated with the distinctive crescent-shaped organisms that Laveran had first found. Consequently, in 1897, an American, William Welch (1850–1934), proposed the name *Haematozoon falciparum* (now *Plasmodium falciparum*) for the parasite with the crescent-shaped gametocytes that caused malignant malaria. The species name was derived from the Latin 'falx' meaning 'sickle' and 'parere' indicating 'to give birth', reflecting the flagella that erupted from each crescent.

We might think we are done—there are three species of *Plasmodium* which cause malaria: not quite. In 1922 John William Watson Stephens (1865–1946), professor of tropical medicine at Liverpool in England, found a parasite in the blood of a soldier returned from East Africa which showed features similar to some half-dozen previously seen cases where difficulty arose in diagnosis as between simple tertian and quartan fevers. Since its characters seemed to be different from those of any of the three known forms of parasite, Stephens named it *Plasmodium ovale*, to reflect the oval shape that the infected red cell assumed.

To return to the 1890s, it was quite clear that malaria was caused by several species of protozoan parasites which underwent a cycle of development and multiplication inside the red cells of patients. But how did they get from one person to another? This was the next great question. Many suggestions had already been made, including miasmas and vapours from marshes and even mosquitoes had been suggested, but this was all in the realm of speculation rather than fact. You will remember that Patrick Manson (p.58) had discovered in 1877 that mosquitoes transmitted filarial worms. In 1894, five years after his return to London, he developed speculation about mosquitoes and malaria in a more concrete form and published his theory in a paper on the nature and significance of the crescentic and flagellated bodies in malarial blood. In a well-argued case, he noted that ex-flagellation (the appearance of flagella outside the parasite which had first been seen by Laveran in 1890) only occurred outside the human body and wrote that '(this) points to the conclusion that the crescentic body is intended to carry on the life of the species *outside* the human body'.[208] There had to be a way that the parasites could get outside the body, so drawing on his experience in filariasis, he suggested that mosquitoes or 'a similar suctorial insect' must be the agent. This, he said, needed to be made the subject of an experiment which could not be carried out in England so he commended it to 'the attention of medical men in India and elsewhere'.[208] His mosquito–malaria theory gained mixed reviews but obtained for him the epithet 'Mosquito Manson'.

As it turned out, this was to be of great importance for one Ronald Ross who went to see him, for Manson would guide Ross for several years in his subsequent experiments in India. Ross (Figure 72) was born in Almora, India, the son of a general in the British Indian Army. At the age of eight he was sent to England to be educated and spent much of his childhood with an aunt and uncle on the Isle of Wight. During his early years he developed lifelong interests in poetry, literature, music, and mathematics. He had no wish at all to study medicine but eventually acceded to his father's desires and joined the Indian Medical Service in 1881. In 1894 while on leave in England he met Manson, who showed him malarial parasites in the blood of a patient at Charing Cross Hospital and they discussed possible means of transmission.

Figure 72. Ronald Ross (1857–1932).

Ross returned to India in April 1895 and joined the 19th Madras Infantry Regiment in Secunderabad in south-central India. He immediately set about researching malaria but had two practical problems—a source of infected patients and a supply of mosquitoes. He found patients in the hospital but had to move quickly to get them before they were cured with quinine. Worse were the mosquitoes. Ross knew little about mosquitoes at that stage and reared any mosquito larvae that came to hand, no matter what type. It was to take him months and years to realize that not all mosquito species would bite people and not all mosquitoes would carry malaria, so at first he wasted a lot of time and effort on trying to persuade inappropriate mosquitoes to feed on patients. In May 1895, while doing one of these mosquito feeding experiments, he looked at the contents of a mosquito's stomach and saw a crescent (which he did not realize was a sexual stage of the parasite) ex-flagellate (release flagella) and round into a spherule. He was puzzled why nearly half of the crescents did not ex-flagellate and it was not until later that he realized that they were the female gametes awaiting fertilization. He did not know that he was using a mosquito that would not allow malarial parasites to complete their development, but he felt sure that Manson's hunch was correct and wrote to him to tell him so.

In September 1895, Ross was ordered to Bangalore to control an outbreak of cholera. He did manage some experiments in which water containing drowned mosquitoes that had fed on malaria patients was given to people; nothing happened. Malaria was not very common in Bangalore because it is 920 metres above sea level and the prevalent mosquitoes were what Ross called 'brown' and 'brindled', later to be shown to be species of *Culex* and *Aedes*, neither of which we now know transmit human malaria.

Ross then became aware of a third species of mosquito, which he called 'dapple-winged', but which was less common. It so happened that this was a species of *Anopheles*. On 16 August 1897, once more in Secunderabad, Ross was able to obtain some adult dapple-winged mosquitoes that emerged from pupae on the previous day. He promptly fed them on a malarious patient by the name of Husein Khan, who was paid one anna (1/16th rupee) for each engorged mosquito. Twenty-five minutes later he had earned 10 annas. Ross dissected two or three insects each day. For the first two days he saw nothing new. On the third day, one insect had 'peculiar vacuolated cells in stomach about 10 [μm] in diam'.[276] On the fourth day, 20 August, he dissected another insect and was excited to find many large single cells in the stomach wall (note it was the wall

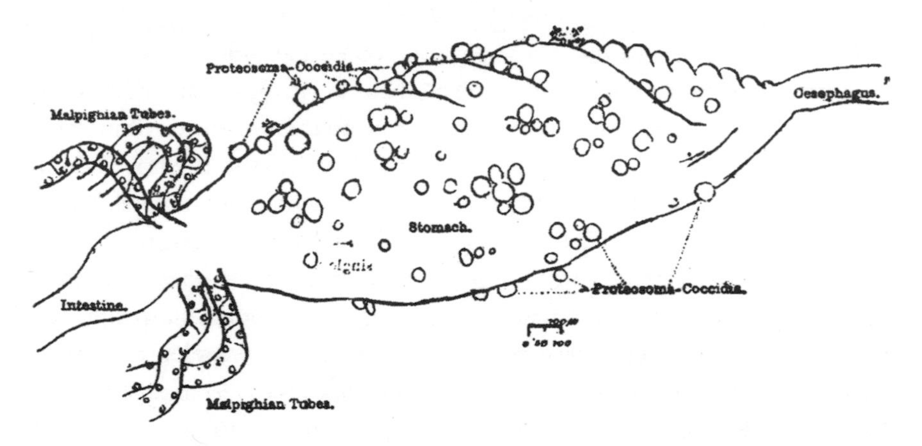

Figure 73. Ross's drawings of oocysts of bird malaria (which he called *Proteosoma coccidia*) on the stomach wall of a mosquito.

and not the lumen so they were invading the stomach wall). On the fifth day he dissected his last mosquito and found that they were even larger (see Figure 73, which shows the similar oocysts Ross later found in bird malaria). Ross was clearly on the right track but he misinterpreted his findings. He wrongly concluded that these cells were derived from the flagella. As we shall see later, they were oocysts derived from female gametes (eggs) fertilized by the flagella (sperm). But he was uplifted and, his poetic nature coming to the fore, he penned the following lines

> This day relenting God Hath placed within my hand
> A wondrous thing; and God Be praised. At his command,
> Seeking His secret deeds With tears and toiling breath,
> I find thy cunning seeds, O million-murdering Death.
> I know this little thing A myriad men will save.
> O Death, where is thy sting, Thy victory, O Grave![277]

Ross wrote to the *British Medical Journal*, which duly published his findings on 18 December 1897.[276]

Let us at this point leave Ross and move to North America. William MacCallum (1874–1944) was a Canadian medical student who was about to graduate from Johns Hopkins University in Baltimore, Maryland, USA. In 1884, Basil Danielewski in the Ukraine had discovered malaria in birds. MacCallum became interested in this subject and collected crows in Ontario, Canada, in the summer of 1897. While observing fresh blood under the microscope, he concentrated on two forms of the parasite (*Halteridium* species, now called *Haemoproteus*) which almost filled the red cell but were not segmented. One he called granular and the other he described as hyaline. First, the granular parasite escaped from the red cell and lay quietly. Then a hyaline form became agitated and burst from a red cell and threw out active flagella which swam straight to the granular parasite. The first flagellum to reach it buried into it and latecomers were repulsed. After 15 minutes or so, the fertilized parasite, now a zygote, become conical and swam away; this was to be called the travelling vermicule or ookinete, later to become an oocyst. These were what Ross had seen. MacCallum confirmed his discovery on his return to Baltimore by observing the same phenomenon on a glass slide in the blood of a woman suffering from aestivo-autumnal fever.

MacCallum had discovered sexual reproduction in which haploid organisms (with a single set of chromosomes) had been formed by meiosis. The male gametocyte (the hyaline one) released flagella which were sperm which fertilized the female gamete (the granular one) to make a fertilized oocyst. This mode of reproduction, which normally takes place in the mosquito, was in contrast to asexual reproduction in red cells where diploid organisms (with two sets of chromosomes) divide by fission (mitosis) to form segmented schizonts which burst to release diploid merozoites.

Now let us return to Ross in India. Disaster struck for Ross. In September 1897, he was ordered to remote Kherwara in north-west India where there was little malaria. He complained bitterly, and Manson moved mountains behind the scenes and Ross was transferred to Calcutta in February 1898 for six months to undertake research on malaria and kala azar. What is more, Manson drew Ross's attention to MacCallum's research. Ross was having trouble getting hold of human cases of malaria in the military compound where he worked, so he decided to see what would happen if he fed mosquitoes on birds with malaria.

Calcutta abounded with the common 'grey mosquitoes' (*Culex quinquefasciatus*) which transmitted filariasis. Ross had found it in Secunderabad and proved again in Calcutta that they would not transmit human malarial parasites. Nevertheless, he tried them with birds. He fed *C. quinquefasciatus* on larks infected with the bird malarial parasite *Proteosoma* (probably *Plasmodium relictum*) and had success. In March 1898 he found oocysts (although the term was not then current) in the mosquitoes' stomachs (Figure 73). He mistakenly called them coccidia. In June he found that these oocysts released what he called 'germinal rods' (now called sporozoites) which passed into the blood cavity of the mosquito then after a few more days to the salivary glands. This was the great clue because salivary juices would be injected whenever a mosquito fed.

Ross then allowed mosquitoes which had been infected for three to four days to feed on healthy sparrows nightly. Two weeks later the previously healthy birds were swarming with malarial parasites. He repeated his experiment with the same results and wrote to Manson on 9 July 1898, ending his letter with 'Hence I think I may now say QED and congratulate you on the mosquito theory indeed.'[277] Ross had discovered the complete life cycle of bird malaria. He had shown that human malarial parasites behaved the same way in mosquitoes in the early stages but he did not have the materials to follow it through. Without doubt, though, the same thing happened in human malaria. But Ross was not to find out. He was sent to Assam to investigate kala azar; then in February 1899 Surgeon Major Ross returned to England with his family on leave. He did not go back to India because later that year he was appointed a lecturer at the new Liverpool School of Tropical Medicine.

Meanwhile, the Italians were hot on the scent, and news of Ross's discovery gave them new impetus and direction. Giovanni Battista Grassi (Figure 74) found three species of mosquitoes that seemed to be peculiar to malarious areas, especially *Anopheles claviger*. Grassi had trained as a physician but spent most of his life working in zoology. He was widely known for his gingery character and gross short-sightedness, which perversely no doubt helped him enormously in distinguishing the 25 species or so of mosquitoes to be found in Italy. In September 1898, Grassi collected *Anopheles* mosquitoes from a highly malarious area at Maccarese about 55 km from Rome. Bignami released them in the room of Abele Sola, who was a patient in the Santo Spirito Hospital located in Rome on an *Anopheles*-free hill. Sola had volunteered. He had lived in the hospital for six years and had never had malaria. On 1 November 1898 he developed a severe attack of aestivo-autumnal fever and had several relapses. In December 1898 and January 1899, Bastianelli, Bignami, and Grassi repeated these experiments on three more volunteers with similar results.

Figure 74. Battista Grassi (1854–1922).

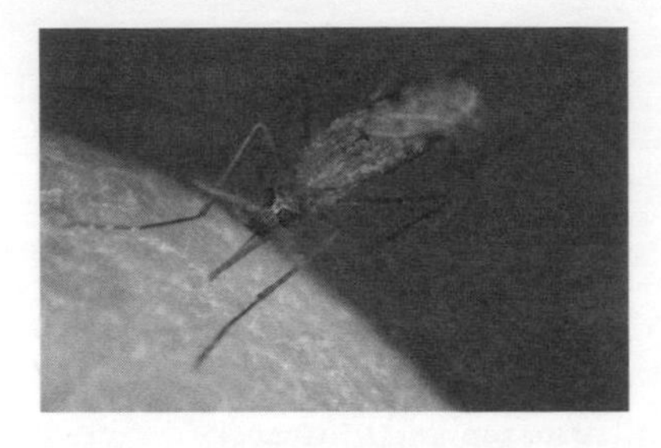

Figure 75. Biting female *Anopheles* mosquito.

Meanwhile these same three investigators were studying the malarial parasites in the mosquitoes. In November 1898 they found parasites on the gut wall of *Anopheles claviger* and in 1899 they observed the complete development of *P. falciparum* in anophelines (Figure 75). Shortly after, Bastianelli and Bignami did the same for *P. vivax* and for *P. malariae*, noted that two other species of *Anopheles* were susceptible, and confirmed that malarial parasites could not complete their cycle of development in *Aedes* and *Culex* species of mosquitoes.

There were some critics though. The volunteers, although they had no evidence of malaria, lived in a malarious area and, although it was unlikely, could have picked up the infection naturally. Manson arranged for three experimenters from the London School of Tropical Medicine, George Low (p.6), Louis Sambon (1865–1931), and the entomologist and artist Amedeo Terzi (1872–1956), to live in a wooden, mosquito-proofed hut furnished with mosquito nets prefabricated in England and set up at Fumaroli in the Roman Campagna. Because *Anopheles* only bite at night, the men went about normally by day but retired to the hut an hour before sunset until one hour after sunrise from early July to 19 October 1900. Local farmers and Red Cross workers were stricken with malaria but the three men remained well. More definitively yet, Bastianelli sent some infected *Anopheles* to Patrick Manson in London, where there was no malaria. Manson fed them on his son, Patrick Thurburn Manson, a healthy young student. Fifteen days later, he developed tertian malaria. Some mosquitoes were still living at this point, so George Warren, a laboratory assistant at the London School, thought it would be a pity to waste them and set them upon his own arm; 14 days later he too went down with tertian malaria.

There could now be no more doubt—malaria was transmitted by the night-biting *Anopheles* mosquitoes. The Italians were indeed the first to prove this in humans, but in reporting their findings, they neglected to acknowledge Ross's previous work. Ross jumped to the conclusion that the Italians were deliberately trying to steal credit that was rightfully his and wrote scathing letters to the journals, claiming that Grassi was a charlatan and a thief. Grassi wrote indignant replies, and a shameful and acrimonious feud lasted over 20 years and was continued by Ross after Grassi's death. It was Ross, not Grassi, who was awarded the Nobel Prize in 1902, was knighted in 1911, and promoted to professor of tropical medicine at Liverpool, but it never seemed enough. Ross even downplayed the role of his mentor and supporter, the man who had first given him the idea, Patrick Manson. It is difficult to decide whether Ross was just a cantankerous old man or a petulant child.

By 1900, the basic story of the life cycle of *Plasmodium* had been discovered but there were still a couple of puzzles to solve. What happened to the sporozoites (the infective forms) that were released into the human body when bitten by an infective mosquito? It took two or three weeks for people to become ill and the parasite seemed at first to disappear from view. This was clearly shown in the experiments conducted by Brigadier Neil Hamilton Fairley (1891–1966) of the Royal Australian Army Medical Corps during 1943–4 in Cairns, Queensland, during World War II. A group of Australian soldiers volunteered to face malaria-carrying mosquitoes rather than dodge Japanese bullets in the New Guinea jungle. Fairley showed that sporozoites disappeared from the blood within half an hour and that blood taken over the next week or two from these soldiers and injected into others failed to produce malaria. Once the latent period had passed, the soldiers became ill and their blood was infective if transfused into another.

Henry Edward Shortt (1887–1987) and PCC (Cyril) Garnham (1901–1994) tried to find out what happened by injecting sporozoites of *Plasmodium cynomolgi*, a malarial parasite of monkeys, into several monkeys and killed them after about a week. They meticulously searched the tissues and in 1948 eventually found segmenting parasites (schizonts) in the liver.

But what about in humans? They found a courageous volunteer, a man who had been given malaria nearly two years previously to induce a 'therapeutic fever' for the treatment of syphilis with some benefit. In March 1948 he was exposed to no less than 2,010 *Anopheles atroparvus* infected with *P. vivax* then 200 salivary glands were dissected out from the mosquitoes and injected into the patient's vein. Despite this enormous dose, he was to remain well, presumably because he had developed some immunity. A week after infection, a sample of liver was removed. Schizonts were found in the liver tissue! What was now called the 'pre-erythrocytic phase' had been discovered in humans. Sporozoites released from the mosquito passed via the bloodstream to the liver where they entered liver (hepatic) cells. Here they developed over two to three weeks into hepatic schizonts which contained thousands of parasites within each liver cell. The cells eventually burst, allowing the small parasites (merozoites) to enter the bloodstream where they penetrated red cells and started the 'erythrocytic cycle' (red cell cycle).

What about *P. falciparum*? In order to advance knowledge and benefit others, Mr CH Howard in 1949 offered himself as a volunteer to Shortt and his colleagues. He was infected with 770 mosquitoes infected with *P. falciparum*. On the sixth day, a liver sample was obtained and again hepatic schizonts were found. You will be pleased to know that Mr Howard was completely cured with the newly developed antimalarial drug, chloroquine. Finally Garnham and his colleagues, including William Cooper (1900–1964), infected Cooper with *P. ovale* sporozoites and at the beginning of 1955 found hepatic schizonts of this parasite (Figure 76). All of this explained why malaria took two to three weeks to develop into a clinical disease after biting by an infective mosquito.

But what about me who in 1966 had developed clinical malaria six months after leaving New Guinea? I had taken chloroquine as a prophylactic for one month after returning home to prevent a primary attack. What happened to me was a well-recognized phenomenon which had been described many times with *P. vivax* but not for *P. falciparum* infections. The answer was provided by Wojciech A Krotoski and his colleagues in 1982. They studied the livers of monkeys infected with *P. cynomolgi*, then those of chimpanzees infected with *P. vivax*, and used special antibodies labelled with a fluorescein dye to help locate parasites. They discovered that some sporozoites were programmed to enter liver cells but remain a single cell for many weeks after arrival (Figure 77). These

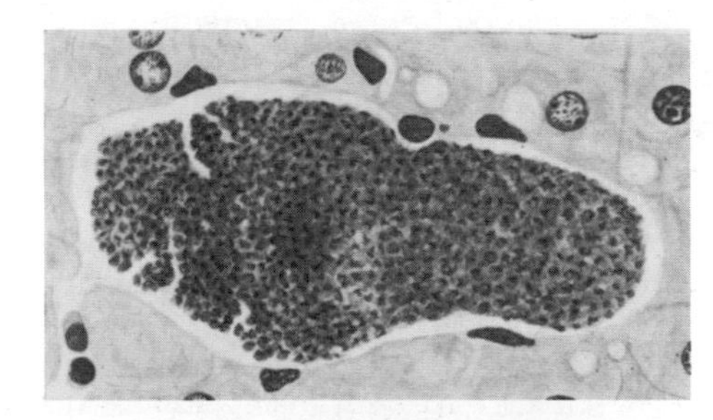

Figure 76. Hepatic schizont of *P. ovale* containing thousands of merozoites.

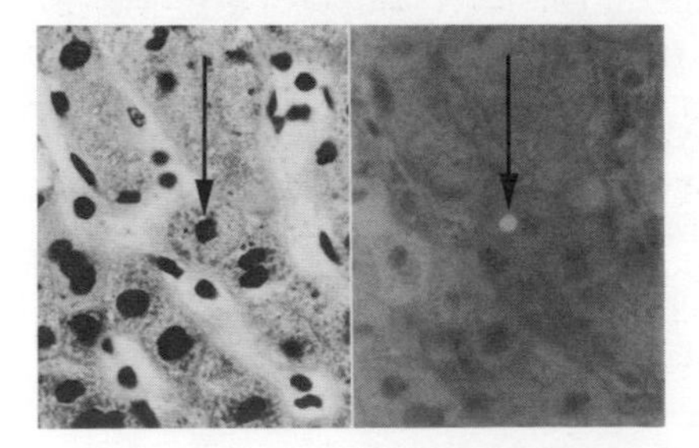

Figure 77. *P. vivax* hypnozoite in the liver of a chimpanzee. Left panel: stained with Giemsa -colophonium. Right panel: stained with a specific fluorescein-labelled antibody. You can see that you would have no idea it was there were it not for the antibody.

dormant parasites were named 'hypnozoites' because they were 'sleeping'. Eventually after a few months for unknown reasons they became active, multiplied, and seeded the bloodstream causing what is called a 'relapse' of malaria. It has been a long a tortuous process but the life cycle of malaria (Figure 78) is now finally understood (we think).

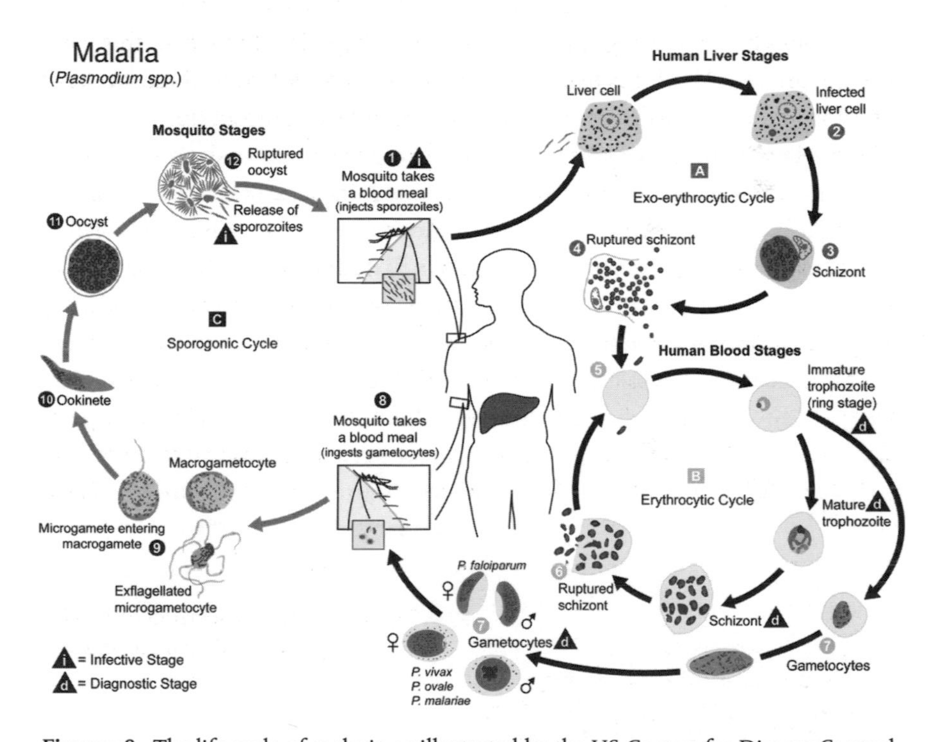

Figure 78. The life cycle of malaria as illustrated by the US Centers for Disease Control. During a blood meal, a malaria-infected female mosquito inoculates sporozoites into the human host (top middle). Sporozoites (1) infect liver cells (2) and mature into schizonts (3) which rupture and release merozoites (4). After this initial replication in the liver (A), the parasites multiply asexually in the erythrocytes (B). Merozoites infect red blood cells (5). The ring stage trophozoite matures into a schizont which ruptures releasing more merozoites (6). These stages cause the clinical disease. Some parasites differentiate into sexual stages (gametocytes) in red cells (7). Male and female gametocytes are ingested by an *Anopheles* mosquito during a blood meal (8) and undergo a sporogonic cycle (C). In the mosquito's stomach, the microgametes penetrate the macrogametes generating zygotes (9). These become motile and elongated (ookinetes) (10) which invade the midgut wall of the mosquito and develop into oocysts (11). These grow, rupture, and release sporozoites (12) which make their way to the mosquito's salivary glands. Inoculation of the sporozoites (1) into a new person perpetuates the life cycle.

13

Sleeping sickness (African trypanosomiasis)

SCIENTIFIC NAMES:	*Trypanosoma brucei gambiense* and *T.b. rhodesiense*
ORGANISMS' COMMON NAME:	trypanosome
DISEASE NAME:	African trypanosomiasis
DISEASE COMMON NAME:	sleeping sickness
DISTRIBUTION:	selected parts of Africa
TRANSMISSION:	bite of an infected tsetse fly (*Glossina* sp.)
LOCATION OF PARASITES:	blood, lymph nodes, brain
CLINICAL FEATURES:	swollen neck glands, lassitude, coma, malnutrition
DIAGNOSIS:	finding parasites in blood, lymph nodes, or cerebrospinal fluid
TREATMENT:	suramin, melarsoprol, pentamidine
PREVENTION:	avoid flies using clothing, insect repellents

The driver-guide of our safari van slapped the back of his neck and produced his kill. We were in Lake Manyara National Park in northern Tanzania. It was a tsetse fly. I had heard about these for 40 years or more and had always wanted to see one. I photographed it on my wife's hand (Figure 79). But why are we talking about tsetse flies? Let's go back to the beginning of the historical record.

The Arabian geographer Abu Abdallah Yaquf (1179–1229) visited northern Africa and found a village whose inhabitants were skin and bone and asleep. Likewise, the Sultan Mari Jata II, ruler of the empire of Mali on the upper Niger River (between modern Guinea and Mali) died in 1374 of what appears to be sleeping sickness:

> Jata had been smitten by the sleeping illness, a disease which frequently afflicts the inhabitants of that climate, especially the chieftains who are habitually affected by sleep. Those affected are virtually never awake or alert. The sickness harms the patient... until he perishes.[149]

Figure 79. Tsetse fly (*Glossina* sp.). The arrow points to the biting and feeding tube called a proboscis.

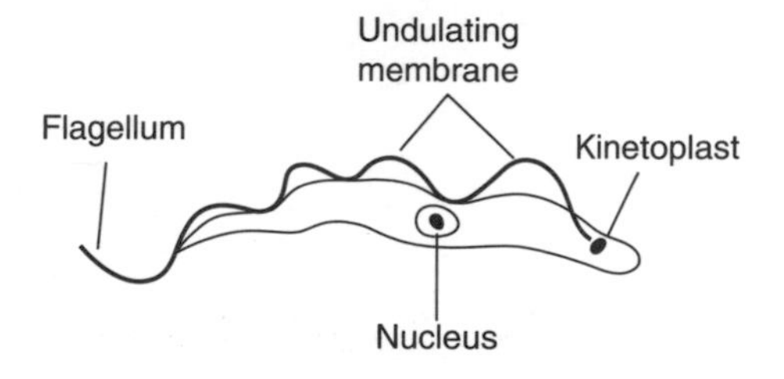

Figure 80. Drawing of the key features of a trypanosome. The flagellum starts at the kinetoplast and runs along the edge of the undulating membrane, then becomes free at the end of the body of the parasite.

The English naval surgeon, John Atkins (1685–1757), in 1742 was the first to describe the neurological features of late sleeping sickness that he saw on the Guinea Coast of West Africa:

> Their sleeps are sound and sense of feeling very little; for pulling, drubbing or whipping, will scarce stir up sense of power enough to move; and the moment you cease beating, the smart is forgot and they fall again into a state of insensibility, drivelling constantly from the mouth ... Young people are more subject to it than the old, and the judgement generally pronounced is death.[31]

Another British doctor Thomas Winterbottom (1766–1859) was for four years colonial surgeon in Sierra Leone and in 1803 gave a more detailed description:

> The Africans are very afraid of a species of lethargy, which they are much afraid of, as it proves fatal in every instance ... At the commencement of the disease the patient has a ravenous appetite ... becoming very fat ... When the disease has continued some time, the appetite declines and the patient gradually wastes away. Small glandular tumours are sometimes observed in the neck a little before the commencement of this complaint [this is Winterbottom's sign] ... Slave traders ... appear to consider these tumours as a symptom indicating a disposition to lethargy and they ... never buy such slaves ... The disposition to sleep is so strong, as scarcely to leave a sufficient respite for the taking of food ... the disease ... usually proves fatal within 3 or 4 months.[324]

Atkins, Winterbottom, and others suggested potential causes for this weird but devastating illness but no-one had any real idea. The answer was to come years later, but to understand it we have to become acquainted with some discoveries in animals.

When the nineteenth-century European explorers such as David Livingstone entered Central Africa, they noted large areas where cattle became ill and died with the result that the country was economically depressed. Indeed, in the more southern regions it was known as 'nagana', a Zulu word meaning 'depressed' or 'useless'. Understanding the cause of this was the prelude to comprehending the cause of human sleeping sickness. But even this was dependent upon previous discoveries.

In Paris, France in 1841, David Gruby (p.85) was looking at the blood of a frog when he saw a protozoon which had a peculiar corkscrew movement brought about by the contractions of an undulating membrane, which he likened to the blades of a saw (Figure 80). This motion and its location suggested a name and he called it *Trypanosoma sanguinis*, the Greek words *trypanon* and *soma* meaning 'borer' and 'body', while the Latin 'sanguinis' means 'of the blood'. His report to the French Academy of Sciences became buried in obscurity for years. In 1877, Timothy Lewis (p.55) in Bombay, India, found similar organisms in the blood of rats and, unaware of Gruby's work, illustrated them with crude woodcuts. He found the parasites in almost 30 per cent of the rats which appeared quite healthy. Later, they were to be named *Trypanosoma lewisi*. Also in India, there was a fatal, febrile (feverish) disease of horses and camels called 'surra'. In 1880, Griffith Evans (1835–1935), a Welsh veterinary surgeon in the Punjab, discovered an organism similar to that seen by Lewis in the blood of these animals. He then succeeded in transmitting it to a healthy dog and horse by injection of blood. His observation was confirmed by his colleague John Henry Steel, who thought it was a spirochaete and proposed the name *Spirochaeta evansi*. Eventually its trypanosomal nature was realized and it was renamed *Trypanosoma evansi*.

The stage was now set for David Bruce of the British Army (p.380) to solve the puzzle of nagana. In 1894, he contrived with the aid of Sir Walter Hely-Hutchinson (whom he had known in Malta), governor of the British colony of Zululand and Natal, to be sent there to investigate nagana in cattle. Together with his wife, Mary, they travelled by mule and ox-wagon for 28 days from Pietermaritzburg to Ubombo in northern Zululand (now KwaZulu-Natal in South Africa) and set up a laboratory in a primitive hut. Mary, six years his senior, was his laboratory assistant, nurse, and artist who shared all his discomforts and smoothed his path when his blustering personality made the going rough. They had no children and died within two days of each other in 1931.

Bruce was a bacteriologist but could find no evidence of bacterial infection in affected cattle. He then examined drops of fresh blood under the microscope and found a motile organism. Never having heard of trypanosomes, he initially thought it might be a microfilaria, but eventually called it 'an infusorial parasite' ('infusoria' refers to protozoa found in infusions of decaying animal or vegetable matter). He then found that if he injected blood from cattle with nagana into dogs, he produced an acute disease with abundant parasites in the blood.

'Fly disease' of domestic animals due to tsetse flies was also becoming well recognized. In the sixteenth and seventeenth centuries, the Portuguese had sent expeditions inland along the Zambesi River in search of gold. These ended in disaster because the horses succumbed to the bites of flies, then the Portuguese themselves fell prey to fever and hostile inhabitants. Livingstone even suggested a connection between nagana and the bite of tsetse flies, There were competing theories, though, including suggestions that nagana was due to animals drinking water or eating grass contaminated by large game, or else that it was a form of malaria or due to a vegetable poison.

Bruce, however, had no suspicion when he first found his 'infusorial parasites' of the connection between nagana and fly disease. It so happened that fly disease was prevalent on the lower slopes of Ubombo but not near the summit, where he resided. He then sent some dogs and two young oxen into the 'fly belt'. When they fell ill, to his surprise he found the same parasites as he had seen in the animals with nagana.

His investigations were interrupted by military duties for eight months but he then returned to Ubombo in September 1895. Bruce obtained some tsetse flies which were of the species *Glossina morsitans* from the fly belt. Flies kept in captivity failed to infect dogs. He then exposed flies to an infected dog after which he allowed them to feed on a normal dog; 10 days later he found trypanosomes in its blood. He then repeated this experiment on another dog with similar results. He concluded:

> The fly *per se* [of itself] does not give rise to any local or general disease...(but)...the fly can act readily as a carrier of the Fly Disease from affected to healthy animals.[58]

But Bruce had been lucky, although he did not know it. As we shall see later, trypanosomes undergo a cyclical development in tsetse flies over a period of several weeks, rather like malarial parasites in mosquitoes. Bruce succeeded because blood on the outside of, or still inside, the proboscis (feeding tube) of the tsetse fly that had just fed still contained living trypanosomes (they will live for up to 48 hours in this state—see later). This process is called 'mechanical transmission' because it is rather like using a syringe. He then took three healthy horses into the fly belt for several hours, preventing them from eating or drinking. Two to six weeks later, he found they were infected. Finally, he brought flies from the fly belt to Ubombo then fed them repeatedly on an uninfected horse. The horse became ill, and 22 days later he found trypanosomes in its blood. There was no doubt. It seemed that domestic animals succumbed rapidly to disease but that wildlife was unaffected. Bruce examined the blood of many species and found nothing. So, in case the numbers were too small to be seen, he injected large amounts of blood from these animals into dogs. Blood from buffalo, wildebeest, bushbuck, and hyena all proved to be infective. Here was the reservoir of infection; the animals had low levels of infection but did not become ill. The parasite was named *Trypanosoma brucei* in Bruce's honour by Henry Plimmer (1857–1918) and John Rose Bradford (1863–1935) in 1899.

All was now in readiness for the next discovery in human sleeping sickness. In May 1901, the 42-year-old English captain of the government launch on the River Gambia in West Africa was admitted with a fever (Figure 81) and a puffy face to the hospital in Bathurst. The colonial surgeon, Dr Robert Forde (1861–1948), looking for malarial parasites (for some reason in fresh blood rather than stained blood films), was confronted with the sight of many motile, worm-like bodies which he could not identify. The patient was sent home to

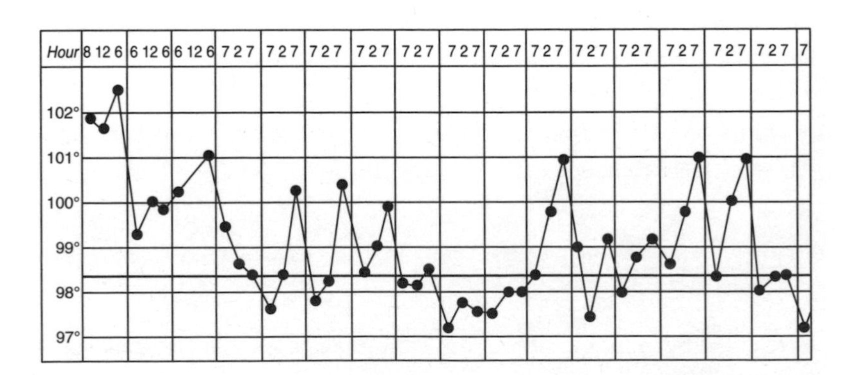

Figure 81. Chart of Dr Forde's patient with trypanosome fever. Temperatures are measured in degrees Fahrenheit. Note the irregularity.

Liverpool, England, temporarily then returned to Bathurst in December. It so happened that Dr Joseph Dutton (Figure 82) of the Liverpool School of Tropical Medicine was in Bathurst, and Dr Forde asked him to see the patient. Dutton at once recognized the parasites as trypanosomes and drew them (Figure 83). Dutton sent films to Alphonse Laveran (p.118), now very interested in trypanosomes, who pronounced them morphologically different from other known species. Dutton therefore named them *Trypanosoma gambiense* after the location in which they had been found. He then found the same parasite in the blood of an asymptomatic three-year-old African child. Sadly, Dutton died at Kasongo in the Congo at the age of 29 from African tick typhus, having just successfully transmitted that infection to monkeys. In 1902, Patrick Manson (p.58) in London saw another case in a woman from the Congo who also had a mild fever.

This disease was called 'trypanosome fever' and no-one had yet connected these parasites with sleeping sickness. In March 1903, Dr Alexander Maxwell-Adams (*c.* 1863 1924), working in the Gambia, suggested that, based upon the puffiness of the face, irritability, apathy, and huskiness of the voice, the cases with trypanosomes in the blood might just be a stage in the development of the very serious sleeping sickness. He was about to be proven right.

Figure 82. Joseph Dutton (1877–1905).

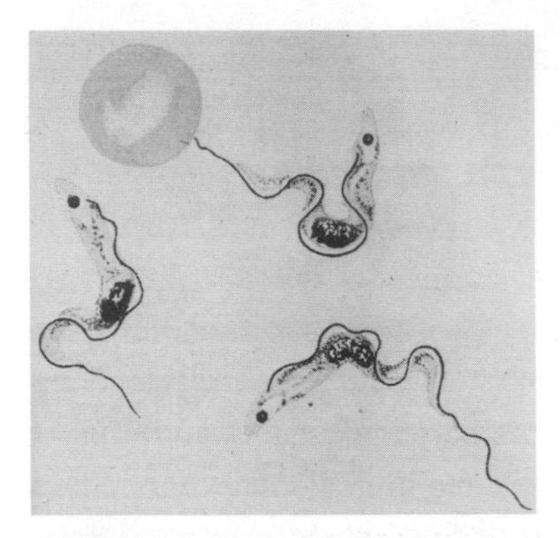

Figure 83. Dutton's drawings of trypanosomes.

Figure 84. Aldo Castellani (1874–1971).

In 1901, Albert and Jack Cook, two missionary doctors in Uganda, had reported that there was an outbreak of sleeping sickness in the Protectorate. The Royal Society in London therefore decided to sponsor a special commission to investigate the disease there. The commission consisted of three members, George Low (p.61), Cuthbert Christy (1863–1932), both graduates of Edinburgh University, Scotland, and Aldo Castellani (Figure 84) from the University of Florence, Italy. The first two had had considerable experience of tropical diseases. Christy, who was of a quarrelsome and conceited disposition, actually came to blows with Low at Mombasa because the compartment of their train was labelled 'Dr Low and party'. The commission arrived in Uganda in July 1902.

They knew nothing about sleeping sickness but had one hypothesis provided by Manson—that it was due to a filarial worm called *Dipetalonema perstans* (now known to be harmless). Christy undertook this task and travelled indefatigably from Lake Victoria all the way to Cairo. He discovered that sleeping sickness and people with enlarged lymph nodes were only found within 30 km of Lake Victoria and had no overlap whatsoever with the distribution of *D. perstans*. He died 30 years later in the Congo after being gored by a buffalo!

Castellani, meanwhile, had managed to cultivate a bacterium, the streptococcus, from blood and was inclined to think he had discovered the cause of the disease. He made the big breakthrough on 12 November 1902, when he found a few trypanosomes in some cerebrospinal fluid (fluid around the brain and spinal cord) taken from a patient with sleeping sickness. There is some doubt as to whether he immediately appreciated the significance of his find because the investigations now became extraordinary. The Royal Society sent a second commission led by David Bruce of the British Army with David Nabarro (1874–1958) as assistant which arrived in Uganda in March 1903.

There then arose the unedifying spectacle of uncooperative commissions, and jealousy between individuals was rife. Castellani refused to tell Bruce about his discovery of the trypanosome unless Bruce agreed that news of the discovery should be published under Castellani's name alone and that Nabarro should not be told about it at all. Castellani's first requirement is fair enough but his latter demand was both unethical and absurd. So Nabarro was set onto clinical matters while Castellani and Bruce worked together in the laboratory. By the time Castellani left on 6 April 1903, they had found trypanosomes in the cerebrospinal fluid of 20 out of 43 cases of sleeping sickness with none in 14 normal people. Castellani reported to the Royal Society:

> a working hypothesis on which to base further investigations is that sleeping sickness is due to the species of trypanosoma I have found in the cerebrospinal fluid of the patients in this disease and that, at least in the last stages, there is a concomitant streptococcus infection which plays a certain part in the course of the disease.[67]

Bruce honoured his agreement to keep quiet at first, but eventually could contain himself no longer, writing:

> For the sake of the future historian... At the time of the arrival of the [Army] commission, (Dr Castellani) did not consider that this trypanosome had any causal relationship to the disease, but thought it was an accidental concomitant like *Filaria perstans*... it is thought advisable to supplement his account with the above.[59]

Curiously, Nabarro, in the debate that ensued in the years to come, gave rather more credit to Castellani than did Bruce. As a later commentator put it:

> This episode is... a commentary on the unwisdom of lack of communication between scientific colleagues and on the restriction imposed upon scientific progress by jealous defence of 'priority'.[198]

But the question now was 'how do people become infected?'. In view of his experience with nagana, Bruce naturally suspected tsetse flies. He then compared the geographical distribution of tsetse flies and sleeping sickness in Uganda by sending letters requesting help and giving instructions to missionaries and officials all over the colony. By August 1903 he had received 460 collections of biting flies, both tsetse and non-tsetse. The similarity in the distributions of tsetse flies and cases of sleeping sickness was immediately obvious.

Meanwhile Bruce and Nabarro had shown that tsetse flies freshly caught near Entebbe, Uganda, were capable of infecting monkeys. Bruce, Nabarro, and Edward Grieg (1874–1950) then fed the tsetse fly, *Glossina palpalis*, on patients with sleeping sickness then allowed the flies to feed on monkeys a few hours later and again the infection was transmitted. They wrote:

> it may be considered proved that *Glossina palpalis* can convey trypanosomes from Sleeping Sickness cases to healthy monkeys up to at least 48 hours after feeding.[59]

More commissioners kept coming from the Royal Army Medical Corps. Among them was Lieutenant Forbes Tulloch, who contracted the infection after cutting his hand with an infected knife in March 1906. Knowing he was doomed, Tulloch left Uganda, travelling down the Nile. He bore his illness with cheerfulness and courage, dying in England on 20 June 1906, aged 27 years.

The next significant advance in understanding the biology of trypanosomes was made by a German, Friedrich Kleine (1861–1950) in 1909. Kleine had graduated in medicine from the University of Halle and then went to German East Africa (now Tanzania) as a member of the German Sleeping Sickness Campaign. He asked the question whether sexual propagation of trypanosomes takes place in tsetse flies as happens with malarial parasites in mosquitoes. He did not believe Bruce's contention that only mechanical transmission took place. Kleine therefore raised uncontaminated flies (*Glossina palpalis*) from pupae, a very difficult job. He infected them by feeding them on two infected sheep and one infected mule for three days. Thereafter, the tsetse flies were fed on a variety of normal animals. From days 18 to 24 they were fed on a particular sheep (sheep number 30), then from days 25 to 39 they were fed on a certain cow (cow number 2). Both of these animals became infected whereas all the animals fed before day 18 remained negative. Clearly, trypanosomes had undergone some sort of development within the flies over two and a half weeks after which they became infective.

Bruce was impressed with Kleine's research and went on himself with his colleagues Herbert Bateman (1876–1961) and Frederick Mackie (1875–1944) in 1911 to describe the full developmental cycle within tsetse flies. They found that there was no sexual stage as in malaria. Rather there was asexual multiplication with migration of trypanosomes to the salivary glands of the flies, evolving their

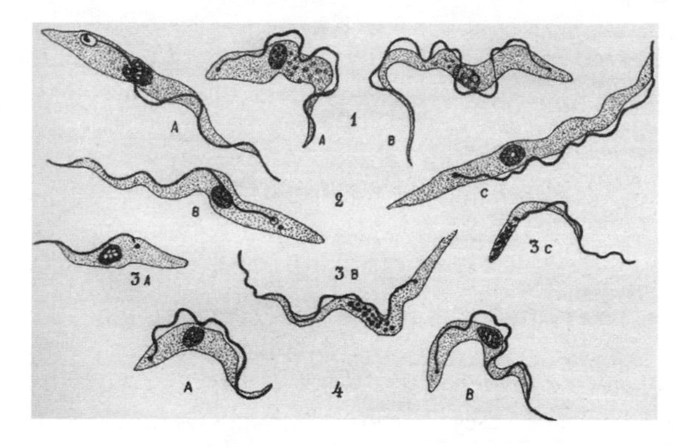

Figure 85. Laveran's reproductions in 1912 of Bruce's drawings of the varying appearances of trypanosomes: (1) in blood (the two at the top middle); (2 and 3) in the abdomen of the fly; (4) in the salivary glands of the fly (bottom two).

appearance as they went. The forms in the intestines were not infective but those in the salivary glands (now called metacyclic trypomastigotes) were infectious (Figure 85).

What actually happens in the brain of someone with sleeping sickness? The first person to examine this question was Frederick Walker Mott (1853–1923), a pathologist with the London County Asylums in England. He had studied the brain of an Englishwoman, a patient of Patrick Manson, who had died of sleeping sickness two years after returning to England. He found inflammation of the membranes surrounding the brain and in the brain itself. He did not see any parasites, but, of course, trypanosomes had not yet been related to sleeping sickness. In 1906 he was sent the brains in formalin of 24 Africans who had died in Uganda of the disease as well as those of eight monkeys experimentally infected with *T. gambiense*. Mott found the usual changes in the membranes and then described the appearances in the substance of the brain itself. There was a marked infiltration of white cells (mononuclear cells, some with an unusual appearance) around the small blood vessels and this was associated with multiplication of glial cells which are support cells in the brain. The effects of this were to destroy the neurones and hence the function of the brain. But the extraordinary thing was that trypanosomes were few and far between in the brain itself. Mott wrote:

> How the trypanosomes produce this characteristic, we may say specific, morbid change in the central nervous system we do not know. Nor has either histological examination or experiment so far solved this question.[222]

A modern writer has remarked that this is still one of the mysteries of tropical pathology. Presumably, the trypanosomes have induced an irreversible immune response in the brain and this no doubt explains why treatment with drugs that kill trypanosomes is largely useless in the late stages of sleeping sickness.

Around this time in 1910, John Stephens (p.126) in Liverpool noticed that a trypanosome in a rat which was said to be *T. gambiense* looked rather different from what he expected. Enquiry revealed that this rat had been infected with blood from an English patient in hospital who had become infected in north-east Rhodesia. He then described the appearance of this parasite and named it *T. rhodesiense*, again reflecting its area of origin. He also noticed that this parasite was more virulent (producing a quicker and more severe disease) than *T. gambiense* in rats and guinea pigs. Subsequently it was confirmed that the clinical course of sleeping sickness in humans is faster for the Rhodesian form than the Gambian form. Two years later in 1912, Allan Kinghorn (1880–1953) and Warrington Yorke (1883–1943), in experiments carried out in Northern Rhodesia (now Zambia), showed that *T. rhodesiense* obtained from humans could be transmitted to monkeys by the tsetse fly, *Glossina morsitans*. The life cycle of trypanosomes is shown in Figure 86.

Bruce meanwhile led a third commission, this time to Nyasaland (now Malawi). He compared *T. rhodesiense* with *T. brucei* obtained from the area in which he had originally found it, and was convinced they were the same species. Two Germans, Max Taute and F Huber, however, were certain they were different and in an extraordinarily dangerous experiment after World War I infected themselves and 129 local people in Tanganyika (now Tanzania) with *T. brucei* from animals. Very fortunately, no-one became infected. In more recent times, opinion has swung again and it is now considered that there are three subspecies of the one species, *T. brucei*. *T. brucei brucei* causes disease (nagana) in animals. *T. b. gambiense* primarily infects humans and is found in West and Central Africa. *T. b. rhodesiense* has both human and animal reservoirs and occurs in East and Southern Africa.

In both forms of human infection, in the early stages, the parasites are restricted to the blood and lymphatic system. Patients have fever, headache, joint pains, and itching. At this point, patients are curable. In the later, second stage, trypanosomes reach the central nervous system and progressively cause lethargy and confusion then coma. By this time it is too late and death is the inevitable outcome. Gambian trypanosomiasis is a chronic illness which lasts from

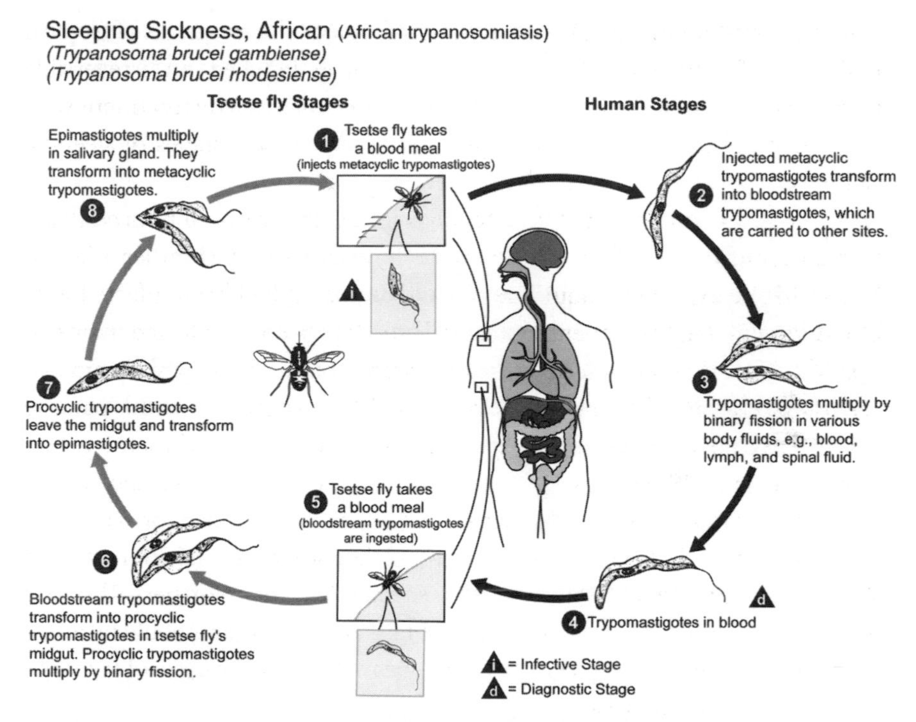

Figure 86. Life cycle of African trypanosomiasis as illustrated by the US Centers for Disease Control. During a blood meal on the mammalian host, an infected fly injects metacyclic trypomastigotes into skin tissue. The parasites enter the lymphatic system and pass into the bloodstream (1). Inside the host, they transform into bloodstream trypomastigotes (2), are carried to other sites throughout the body (lymph, spinal fluid) and multiply (3). Flies become infected with bloodstream trypomastigotes when taking a blood meal on an infected mammalian host (4, 5). In the fly's midgut, the parasites transform into procyclic trypomastigotes, multiply by binary fission (6), leave the midgut, and transform into epimastigotes (7). The epimastigotes reach the fly's salivary glands and continue multiplication by binary fission (8). The cycle in the fly takes approximately three weeks. Don't worry about these technical terms unless you are especially interested.

months to years from the bite of the tsetse fly to the death of the untreated patient. Rhodesian trypanosomiasis is a more acute disease with the untreated patient dying within a number of months. You will remember that Forbes Tulloch died within three to four months of becoming infected. He probably had the Rhodesian form. Genetic studies in the last few years have shown that Uganda (which is where the commissioners worked 100 years ago) is host to both *T. b. gambiense* in the north-east and *T. b. rhodesiense* in the central parts of the country.

So many owe so much to those who risked, and on occasion lost, their lives seeking to understand the cause of sleeping sickness. But it would be remiss of me not to mention perhaps the most shameful medical investigation in African trypanosomiasis. In 1953, the Frenchman Dr Pierre Gallais, in collaboration with seven others, reported their observations in the journal *Médecine Tropicale*. They had deliberately inoculated 12 psychiatric patients with *T. gambiense*. These patients were French, African, or Indochinese in their 20s and 30s. They were almost all schizophrenic with severe hallucinations. One was described as a delusional moron and another as mute and catatonic. Hardly material for 'informed consent', even if there were some perceived benefit. What was the justification for this apparently criminal study? Let me quote the authors' original words in French so there can be no doubt:

> Ces expériences, outre les promesses qu'elles apportent à une thérapeutique psychiatrique possible de certains états psychopathiques, ont permis des acquisitions qu'il est intéressant de confronter avec quelques données de la pathologie naturelle et expérimentale.[109]

In other words, in addition to any promise these experiments might make for a possible psychiatric treatment for certain psychopathological states, it is interesting to compare the findings with other clinical and experimental studies of the pathology of this condition. The latter seems to be all that the authors really had in mind. Did these patients live or die? The authors do not tell us. We can only assume the worst.

Fortunately, control programmes directed against tsetse flies have had considerable success. In 2009, the number of cases of human trypanosomiasis reported in sub-Saharan Africa fell to below 10,000 for the first time.

14

Cutaneous leishmaniasis (Oriental sore) and visceral leishmaniasis (kala azar)

SCIENTIFIC NAMES:	*Leishmania tropica* and *L. donovani* plus many others
DISEASE NAMES:	a. cutaneous leishmaniasis b. visceral leishmaniasis
COMMON DISEASE NAMES:	a. Oriental sore b. kala azar
DISTRIBUTION:	patchy over large areas of the world
TRANSMISSION:	bite of an infected sandfly (*Phlebotomus* sp.)
LOCATION IN A PERSON:	a. skin b. liver, spleen, bone marrow
CLINICAL FEATURES:	a. ulcerating or other lesion on the skin, b. fever, enlarged liver and spleen
DIAGNOSIS:	a. parasites seen on microscopic examination of skin biopsy, b. parasites seen on microscopic examination of liver, spleen, or bone marrow biopsy
TREATMENT:	a. difficult; often resolve spontaneously, b. amphotericin B, miltefosine, paromomycin
PREVENTION:	avoid sandfly bites, especially at night

The leishmaniases are a complex of related diseases that affect the skin, the skin plus mucous membranes, or the internal organs; they are classified as cutaneous, mucocutaneous, and visceral leishmaniasis, respectively, and there are various subtypes within each group. These infections occur in widely disparate parts of the world, with the mucocutaneous forms being endemic in Central and South America. Limitations of space have enforced a restriction to discussing only cutaneous and visceral leishmaniases of the Old World. Moreover, since the parasites causing these diseases and their modes of transmission are related, discoveries in one form influenced investigations in the other so their stories are intermingled. Why do these disorders have this unusual name? As you will see, they were named after William Leishman.

The skin disease which we now call cutaneous leishmaniasis in the Old World was recognized in the Middle East from before the time of Christ. The Arab physician Avicenna (p.7) gave a detailed description of a skin lesion called 'Balkh sore' because it was common in the once-thriving ancient city of that name in what is now northern Afghanistan. Indeed, this disease acquired a number of names depending upon where it was seen, such as Aleppo Evil (or Mal d'Aleppo), Jericho Buttons, and Delhi Boil. Today, 'Oriental sore' is used as a catch-all name for these diseases from the East (at least as seen from a European perspective).

The first description in English of this condition was given by the Scot, Alexander Russell (1715–1768), who was medical officer to the 'English factory' (i.e. trading post) in Aleppo, then part of the Ottoman Empire, now in Syria. He noticed that it was most common on the face, but could occur elsewhere, and in strangers, developed several months after their arrival. He recognized three forms of the disease:

> (The first) makes its appearance in the form of a small, red...pimple which commonly passes some weeks unregarded....afterwards it begins to encrease, and usually comes to the size of an English sixpence [*c.*20 mm]; which after some months becomes scurfy on top...forms into a thick crusty scab...till...it falls off and leaves but very little mark. The whole of its duration is seldom above eight months. (The second) begins somewhat like the former, but after a month or two becomes somewhat painful, encreases often to double the extent (of the first), discharges a good deal of ichorous matter [fluid] from under the scab and by degrees comes to have the appearance of an...ulcer with a livid circle round it...(It) is in general about a year from its first appearance before it is cured; but this is not a thing certain...After it is cicatrized, it leaves an ugly scar which remains for life and for many months has a livid colour...They seldom give much pain. (The third) seldom grows larger than about twice the size of a large pin's head and never changes its appearance, remaining...for many months without any pain, after which it throws off a few scurfy scales and disappears.[280]

Of course, neither Russell nor anyone else had any idea what the cause of these conditions was.

The first clue came in 1885 and was provided by another Scot, Surgeon Major David Douglas Cunningham (Figure 87) of the Indian Medical Service. A lesion

Figure 87. David Cunningham (1843–1914).

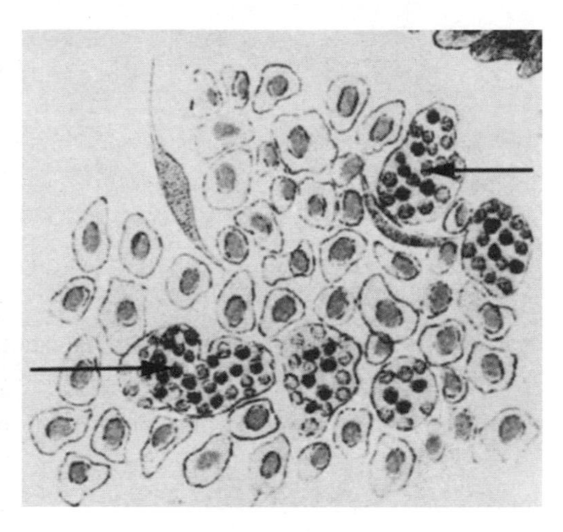

Figure 88. Cunningham's illustrations of the bodies (arrows) he saw in the Delhi boil.

was excised from a patient in Delhi and sent in alcohol to him in Calcutta. When he examined thin slices under the microscope he saw an infiltration mostly of a particular type of white cells called macrophages. However, when he examined preparations stained with gentian violet he saw something odd. After intensive washing, almost everything was transparent but:

> large numbers of peculiar violet or blue bodies appeared conspicuously among the surrounding, almost colourless masses of lymphoid elements... The average diameters of a series of measured specimens were 12.6 μ x 6.4 μ [micrometres]... Their form also varied greatly... circular... elliptical... irregularly lobate.[77]

One of his illustrations is shown in Figure 88. Cunningham did not know what these bodies were but was inclined to think that they might be Mycetozoa (slime moulds, i.e. fungus-like organisms that use spores to reproduce) and thought they probably had a role in the genesis of the lesion.

The next advance was made by Peter Borovsky (Figure 89), a surgeon in the Russian Army who was stationed in Tashkent where these skin lesions were known as Sart Sore, Sart being the name for the inhabitants of the region. In an attempt to find the cause, Borovsky and his colleagues at the end of 1895:

> began to examine the juice of papules and sores... in hanging drops [i.e. a drop was put on a microscope slide with a cover slip placed on top] and were

Figure 89. Peter Borovsky (1863–1932).

> astonished by the presence of numerous motile corpuscles, some round, some irregular while some of them possessed pseudopodia.[49]

When the fluid was dried on a slide, fixed with alcohol, and stained, they saw the same creatures:

> in most cases their dimensions are from 1½ to 2μ... These corpuscles frequently have a fine process as long as... the diameter of the corpuscle; at the end of the process there is not infrequently a small globular thickening.[49]

Moreover, if a hanging drop preparation was left for 24 hours, the round bodies disappeared and motile, fine filaments appeared.

At the end of 1897, Borovsky and his colleagues made their second major discovery. They made histological sections of tissue, stained them, and found that:

> the cells are literally packed with them [the parasites] to such an extent that the limits of the individual parasites cannot be seen, and they appear as one continuous mass in which only their nuclei, stained deeply with Loeffler's solution, are distinctly visible. This is the reason why the earlier preparations produced the impression that these accumulations consisted of cocci [bacteria]... The size of the parasites is in most cases about 1μ.[49]

So what Cunningham had measured and thought might be slime moulds were in fact white cells (now known to be macrophages) packed with these small parasites in the cytoplasm. Borovsky correctly interpreted these organisms as being protozoa, writing:

> Owing to the peripheral position of the nucleus and to the presence of processes like flagella, we are inclined to refer the parasites described to a class of protozoa.[49]

Most regrettably, Borovsky published all this wonderful new knowledge in an obscure Russian military medical journal and the world at large knew nothing about it until it was translated into English in 1938!

So it was that James Homer Wright (1870–1928) was completely unaware of this when in 1903 he examined a biopsy in Boston, Massachusetts, USA. The material had been obtained from a nine-year-old girl who had first developed the lesion in Armenia several months earlier. Wright had stained the tissue sections with his own version of Romanowsky's stain and found the bodies that Borovsky had seen. He then described the unusual structure of these bodies. They had a blue rim, a central pale area, and a larger and a smaller lilac-coloured mass which he suggested might be nuclei (it was to turn out that he was partly right—the larger one is). He thought these organisms were most likely protozoa but was uncertain as to exactly what sort of protozoa they were. He tentatively referred to them as Microsporidia (which they are not) and gave them the name *Helcosoma tropicum*. Incidentally, in 1912 Wright was charged with smuggling!

It is now 1903 so let us turn our attention to visceral leishmaniasis, for it is in that same year that a major discovery was made concerning that disease. What is visceral leishmaniasis? In retrospect, the first reasonable clinical description of this syndrome had been made by William Twining (1790–1835), a surgeon in Calcutta, India in 1827, although some of his cases may have been malaria. In 1858, the British government of Bengal in India became concerned by an epidemic of what was thought to be quinine-resistant malaria. It was decimating the population who called the disease *kala azar* meaning 'black disease'. It was characterized by a variable fever, enlarged spleen, anaemia, and, in the late stages, emaciation and darkening of the skin. Epidemics continued in Bengal on and off for the rest of the nineteenth century and there were several unenlightening investigations, including that of Ronald Ross (p.127), who thought it was a form of malaria. Between 1891 and 1901 the population was estimated to have fallen by 30 per cent in certain parts of Bengal as a result of kala azar.

The British Army had a contingent stationed at a cantonment at Dum Dum near Calcutta. It was an unhealthy place with various fevers and dysenteries and a number of soldiers were repatriated to England with obscure 'Dum Dum fever'. One of them died at the Royal Victoria Hospital at Netley in November 1900 and the post-mortem examination was performed by Major William

Figure 90. William Leishman (1865–1926).

Leishman (Figure 90), an army medical officer who had been stationed at Dum Dum 10 years earlier. The patient was grossly emaciated, the colon thickened, deeply pigmented, and with mostly scarred ulcers while the spleen was greatly enlarged and friable. The spleen, located in the upper left abdomen, is an organ of uncertain significance. One can live without it but it does play a role in resistance to infections with certain bacteria. Normally it weighs about 150 g; this soldier's was 1,100 g so it was enormous. On examination of smears made from the spleen, Leishman later wrote:

> I was struck by the curious appearance, among the spleen cells and red corpuscles, of enormous numbers of small round or oval bodies, 2 to 3μ in diameter which corresponded to nothing I had previously met with or had seen figured or described.[183]

He then stained them by Romanowsky's method and found:

> chromatin [nuclear material] in the form of definitely circular mass or ring, applied to which ... was a much smaller chromatin mass, usually in the form of a short rod set perpendicularly or at a tangent to the circumference of the larger mass. The outlines of the sphere or oval containing these masses of chromatin were only faintly visible.[183]

Leishman was completely at a loss as to the nature of these bodies. A little over two years later in 1903 a serendipitous event occurred. He happened to examine a rat which had been infected with the organism that caused nagana (*Trypanosoma brucei*) but not until 48 hours after its death. He found in its blood and tissues, organisms practically identical with those he had seen in the dead soldier. These had apparently formed from the trypanosomes which had been swarming in the rat's blood shortly before its death as a result of adverse circumstances during the ensuing 48 hours. Leishman then did a number of experiments with white rats and proved that after death of the host, the trypanosomes lose their motility and flagellate form and shrink into spherical or ovoid bodies and the chromatin assumed the appearance he had seen in the soldier's parasites. Leishman concluded that Dum Dum fever appeared to be a form of trypanosomiasis and reported these observations in the *British Medical Journal* in 1903.[183]

Figure 91. Charles Donovan (1863–1951).

Leishman's paper soon came to the attention of Surgeon Captain Charles Donovan (Figure 91) of the Indian Medical Service stationed in Madras (now Chennai). He wrote to the same journal remarking that he had also seen these parasites in the enlarged spleens of three Indian patients who had died. He had thought at first they were post-mortem degenerations of the nuclei of spleen cells but on reading Leishman's paper he realized his mistake. Thereupon on 17 June 1903 he punctured the spleen of a 12-year-old living boy to get a sample. This boy had a fever but no malarial parasites in the blood. He had identical bodies in his spleen, thus proving that they were not post-mortem artefacts. Finally, he added that there was nothing in the boy's blood that looked either like these bodies or like trypanosomes. Leishman later looked at Donovan's preparations and agreed that they were the same as what he had seen.

Ross suggested these bodies be called L–D bodies (after Leishman and Donovan). Histopathologists still use this term today. Pity about Cunningham and Borovsky! History does not always look after those who come first.

So there were now two questions. What exactly were these things? And how did they get from one person to another? As to the first question, there was considerable disagreement. Leishman thought they were trypanosomes. Donovan had sent specimens to Félix Mesnil and Alphonse Laveran (p.118) in France. The latter declared that they were piroplasms (malaria-like parasites of red blood cells, mostly seen in cattle, transmitted by ticks) and named them *Piroplasma donovani*. Ronald Ross had also seen the specimens and did not think they were either trypanosomes or piroplasms so in 1903 raised a new genus, *Leishmania*. He retained Laveran and Mesnil's species designation of *donovani*. Thus the organism was named *Leishmania donovani* in honour of both Leishman and Donovan.

Despite a new name, the precise nature of these protozoa was still unclear. In February 1904, Leishman wrote in the *British Medical Journal* that what he and Donovan had seen were similar to the organisms seen by Wright in Oriental sore but the exact relationship between these two diseases was unclear. Later in 1904, Leonard Rogers (1868–1962), professor of pathology at the Medical College Hospital in Calcutta, attempted to cultivate these L-D bodies. He took some infected blood from the spleen and put it in sodium citrate solution. When he kept the tubes at 37°C the parasites disappeared within 24 hours. He then found that if he kept the tubes at 27°C, not only did the parasites not die, but they multiplied. Next, he discovered that if he kept them at 22°C, the parasites not only survived and multiplied but turned into organisms resembling trypanosomes which would divide by undergoing longitudinal splitting. In fact, there were many intermediate forms and he concluded that L-D bodies 'must be one stage in the life-history of the organism and not degenerated forms as (Leishman) first thought them to be'.[274]

Charles Nicolle (Figure 92), director of the Pasteur Institute in Tunis, then made the next advance. He cultivated splenic blood obtained from three children in a fluid medium and then inoculated material into the peritoneal cavity of a dog twice, 79 days apart. In February 1908, 159 days after the first injection, the dog was put down and examined. L-D bodies were found in its spleen, liver, and bone marrow. Not unreasonably, Nicolle suggested that 'kala azar might be a disease of the dog transmissible to man through skin parasites (such as fleas)'.[228]

Figure 92. Charles Nicolle (1866–1936).

Later that same year of 1908, Nicolle made another important observation. A camel driver contracted Oriental sore (this time called Gafsa sore) in Algeria. Nicolle took samples of the lesions with a needle, then tried to cultivate them at room temperature. After nine days there was an abundant growth of the parasite which looked very similar to *L. donovani* cultivated under the same conditions.

Nicolle remained circumspect, remarking that 'only experimental study will permit the distinction between these two parasites,'[228] i.e. the parasite of kala azar and that of Oriental sore. The various stages that this parasite went through were beautifully illustrated in 1912 by Laveran and Mesnil (Figure 93).[178]

Most people, however, thought that the clinical presentations of Oriental sore and kala azar were so different that surely the causative parasites were different. Consequently, a multiplicity of names had been given for the parasite of Oriental sore. The one that has stuck is *Leishmania tropica* applied in 1906 to Wright's *Helcosoma tropicum* by Max Lühe (1870–1916), even though he thought it was a piroplasm. Proof that they were indeed different parasites was to come when the means of transmission was discovered.

How did these protozoa spread? As we have seen, Nicolle suggested fleas. Naturally mosquitoes were considered and others suggested bed bugs, house flies, and ticks but none withstood the glare of intensive study. Clinical as well as epidemiological investigations, i.e. the time of onset and geographical locations of both disease and potential vectors (carriers), were to provide the clue. As early as 1905, Edmond (1876–1969) and Etienne (1878–1948) Sergent, two brothers at the Institut Pasteur in Algeria, recognized a connection between the frequency of Oriental sore and the prevalence of sandflies. In 1911, Charles Wenyon (1878–1948) found parasites which could well have been *L. tropica* in the intestines of sandflies in Aleppo, Syria. In 1919, Hugh Acton (1883–1935) in Baghdad (now in Iraq) realized there was a correlation between the common locations on the body where Oriental sore was found and the sites where sandflies preferred to bite. Then in 1921 the Sergent brothers and some co-workers collected specimens of a sandfly called *Phlebotomus papatasii* from Biskra, an endemic area for Oriental sore, and sent them on the three-day journey to Algiers. Here the insects were emulsified and inoculated onto the skin of a man, even though microscopical examination

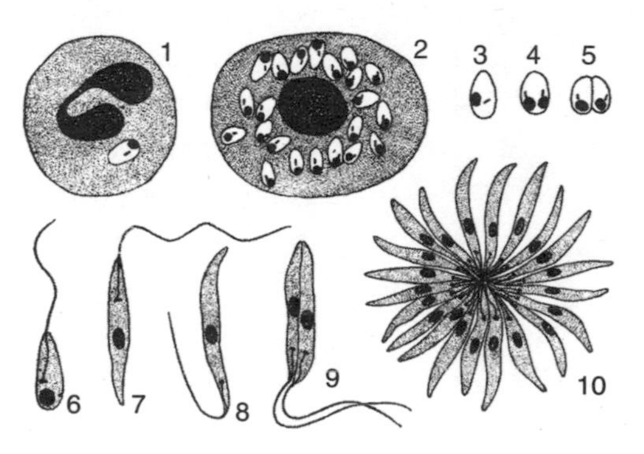

Figure 93. Laveran and Mesnil's drawings of Leishmania in culture. 1, 2: white cells containing one or more L–D bodies, 3–5: isolated parasites in the process of dividing, 6–8: flagellate forms 9: flagellate dividing, 10: rosette of flagellates.

of the emulsate failed to disclose any parasites. Over the next several months, a typical Oriental sore developed and parasites indistinguishable from *L. tropica* were seen. Then in the following year (1922), Henrique de Beaurepaire Aragão (1879–1956) in Brazil fed sandflies on sores, crushed them three days later, then inoculated a dog on the nose: a sore containing L-D bodies developed.

Grave suspicion was now turning on sandflies. What are sandflies? They are smaller than mosquitoes, being about 1.5–3 mm in size, yellowish in colour, have conspicuous black eyes, and body, wings, and the legs are hairy (Figure 94). Adult females lay eggs in batches in moist cracks and crevices in buildings. After 10 days or so, the eggs hatch larvae which feed on organic debris and after a number of moults emerge as adults in five to six weeks. The females have to feed on blood to reproduce. This accounts for its genus name—*Phlebotomus* means 'vein cutter'. A 0.3-mg fly can ingest 0.1 ml of blood but has terrible trouble flying and tends to hop about and might not be able to get out of the mosquito net it had no trouble entering.

The scene now moves to India where the School of Tropical Medicine had been established in Calcutta in 1920. Members of its staff were keen to understand kala azar, which was such a big problem in their region. Major John Sinton (1884–1956) of the Indian Medical Service (who had won the Victoria Cross for gallantry in Mesopotamia in 1916) pointed out that the distribution of kala azar in India appeared to coincide with that of the sandfly, *Phlebotomus argentipes*. Surveys of kala azar indicated that July to November was the most likely time for transmission of infection to occur. In Calcutta itself, one particular district, Ward 14, seemed to be the main focus of infection and detailed study of this ward showed that *P. argentipes* appeared here in July, with the greatest numbers in September and October, so particular attention was paid to this insect.

Figure 94. A sandfly (*Phlebotomus* sp.) feeding.

Robert Knowles (1883–1936), Leonard Napier (1888–1957), and ROA Smith fed laboratory-bred (and therefore parasite-free) *P. argentipes* on a patient with kala azar who had scanty L-D bodies in his blood. Several days later, a proportion of flies had flagellated parasites (which they called herpetomonads) in every way resembling *L. donovani* grown in a culture medium. When they fed the flies on patients with malaria and other diseases, none of the herpetomonads were found. Naturally, the authors:

> concluded that *L. donovani* passes into its flagellate form in the gut of *P. argentipes*. Any further conclusions with regard to kala azar transmission are at present carefully avoided. What is wanted is further work, which is now in progress, rather than speculation.[160]

You might think it odd that they were so cautious. As it happened they were on the right track but nearly 20 years were to pass before the question was definitively solved. The problem was that sandflies were difficult to keep in captivity and difficult to induce to feed.

Proof was first provided for *L. tropica*. Saul Adler (1895–1906), professor of parasitology at the Hebrew University of Jerusalem, managed to infect the sandfly *P. papatasii* by feeding them on a membrane over a suspension of flagellated organisms prepared in culture. He then fed them on nine people, five of whom became infected. His concluding words in 1941 were simple: '*Leishmania tropica* has been transmitted to man by the bite of *Phlebotomus papatasii*.'[2]

Experimental infection of humans with the annoying but not life-threatening *L. tropica* was one thing. What about life-threatening *L. donovani*? Would it not be better to use experimental animals? In 1932 HE Shortt (p.133), ROA Smith, CS Swaminath, and KV Krishnan reported that they had induced kala azar in a hamster on which had been fed, during a period of 17 months, 144 *P. argentipes*, the majority of which must have been infected. Had the case been sufficiently made?

Swaminath, Shortt and LAP Anderson of the Indian Kala Azar Commission evidently felt not. They too developed techniques of infecting sandflies, in this case *P. argentipes*. The 'volunteers' were healthy men of the Khasi tribe in north-eastern India. The authors tell us very little about these volunteers except that they underwent all sorts of tests to prove that they were healthy. Did they know what they were letting themselves in for? Were they later treated (and the drugs then available were both toxic and unreliable)? We are not told. Even if

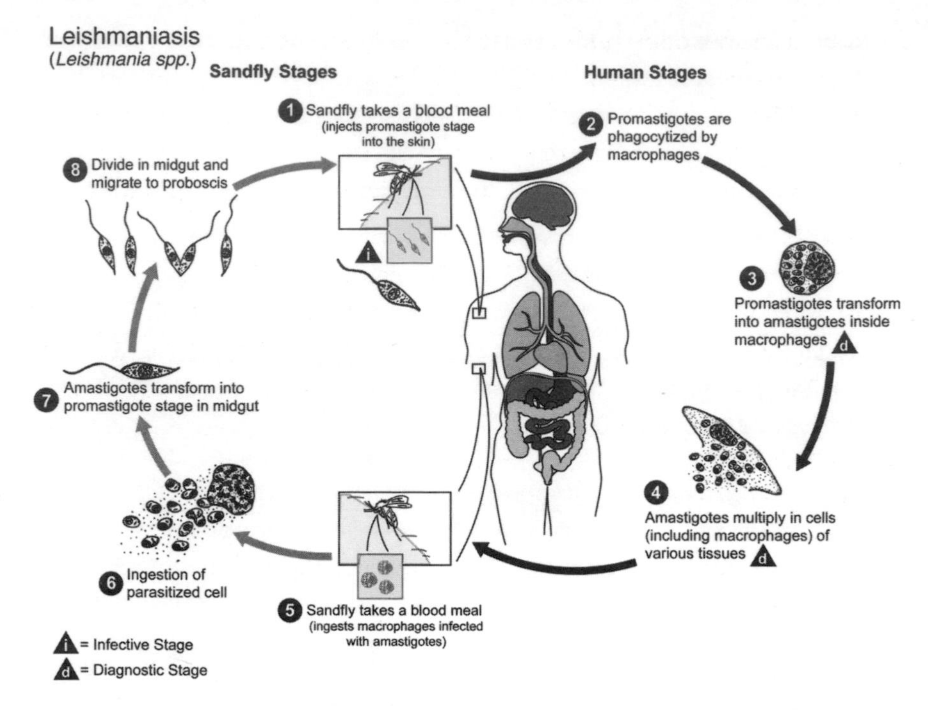

Figure 95. The life cycle of leishmaniasis as presented by the US Centers for Disease Control. Sandflies inject the infective stage (i.e. promastigotes) from their proboscis during blood meals (1) and are ingested by macrophages (white cells) (2). The parasites transform in these cells into the tissue stage of the parasite (amasitogote) (3) which multiply and infect other macrophages (4). Sandflies become infected by ingesting infected cells during blood meals (5, 6). In sandflies amastigotes transform into promastigotes in the gut (7) then migrate to the proboscis (biting tube) (8).

the volunteers had all the facts, was the study still ethical? I very much doubt that any ethics committee would pass it today. So it is with considerable discomfort we read what happened. Five men were exposed and by August 1942 all had been tested and found to be infected. The authors concluded: 'Successful transmission of Indian kala azar to man by the bites of *P. argentipes*...has been accomplished successfully in 100 per cent of cases.'[299]

In accompanying editorial comment, the editor of the *Indian Journal of Medical Research* reviewed the vicissitudes of research over the preceding 20 years and concluded thus:

> Twenty years of patient investigation have gone into forging this final link in the evidence that *P. argentipes* is the insect vector of human kala azar in India,

> and the many workers who have taken part in the investigation at different stages and have contributed to the final solution of the problem are to be congratulated on the outcome of their work.[299]

He did not say a word about volunteers or ethics. But it is true that understanding the life cycle (Figure 95) did make possible ways of controlling this infection which has killed hundreds of thousands of people. What do you think?

I mentioned at the beginning of this chapter that the leishmanial diseases cause three main clinical syndromes. Now we know that these diseases are caused by at least 13 different species of *Leishmania* which all look alike but can be separated by biochemical and immunological techniques and by DNA analysis. They are transmitted by sandflies and many have a significant animal reservoir. If you go to an endemic area you are always at risk; for example, a patient of mine contracted visceral leishmaniasis after visiting relatives in the Greek islands for several weeks.

15

Chagas disease (South American trypanosomiasis)

SCIENTIFIC NAME:	*Trypanosoma cruzi*
DISEASE NAME:	South American trypanosomiasis
COMMON DISEASE NAME:	Chagas disease
DISTRIBUTION:	patchy areas of South and Central America
TRANSMISSION:	faecal contamination by biting, infected reduviid bugs: *Triatoma, Rhodnius,* and *Panstronyglyus* species
LOCATION IN A PERSON:	many tissues, especially heart, skeletal, and smooth muscle cells
CLINICAL FEATURES:	a. acute infection—fever lasting a few weeks, swelling at the site of inoculation; rarely heart disease or convulsions, b. chronic infection—~25% of people will develop one or more of heart failure, abnormal heart rhythm, trouble swallowing, constipation
DIAGNOSIS:	positive blood tests for antibodies against *T. cruzi*; rarely parasites seen on microscopic examination of blood or tissues
TREATMENT:	difficult—benznidazole and nifurtimox
PREVENTION:	avoid the bite of reduviid bugs

Did Charles Darwin have it? Darwin suffered chronic ill-health for the rest of his life after his return as a 27-year-old from his five-year voyage around the world on the *Beagle*. Many diagnoses have been suggested, amongst them Chagas disease, which he could have acquired during his many months of travel on land in Brazil, Argentina, and Chile. We will never know whether or not he really did have South American trypanosomiasis but we do know he had an intimate acquaintance with reduviid bugs:

> We slept in the village of Luxan [Luján de Cuyo]..., in the province of Mendoza (Argentina). At night I experienced an attack of the Benchuca, a species of Reduvius, the great black bug of the Pampas. It is most disgusting to feel soft wingless insects, about an inch long, crawling over one's body. Before sucking they are quite thin, but afterwards they become round and bloated with blood, and in this state are easily crushed. One, which I caught...was very empty. When placed on a table, and though surrounded by people, if a finger was presented, the bold insect would immediately protrude its sucker, make a charge, and if allowed, draw blood. No pain was caused by the wound. It is curious to watch its body during the act of sucking, as in less than 10 minutes it changed from being as flat as a wafer to a globular form.[80]

What are reduviid bugs? They are insects which belong to the family Reduviidae, subfamily Triatominae; consequently they are also called triatomines. Colloquially, they are also called 'kissing bugs' because they tend to bite on the face, or 'assassin bugs' for reasons that will become obvious. Figure 96 shows one of these triatomines. The adult bug is about 2.5 cm long, has two pairs of long, bent legs attached to an oval-shaped abdomen, a third pair of legs that act like arms attached to the thorax, and the proboscis (biting/feeding tube), which extends from the anterior part of the head (as do two antennae). Their wings are thin and transparent and cover the top of the abdomen. Only adults have wings, which are inefficient for flying but effective for gliding and mounting their mate. They prefer nesting in roofs and ceilings and cracks in walls of houses from which they glide down upon sleeping humans to whom they are directed by radar-like heat sensors in their antennae. Triatomines glide over 100

Figure 96. Feeding *Triatoma infestans*.

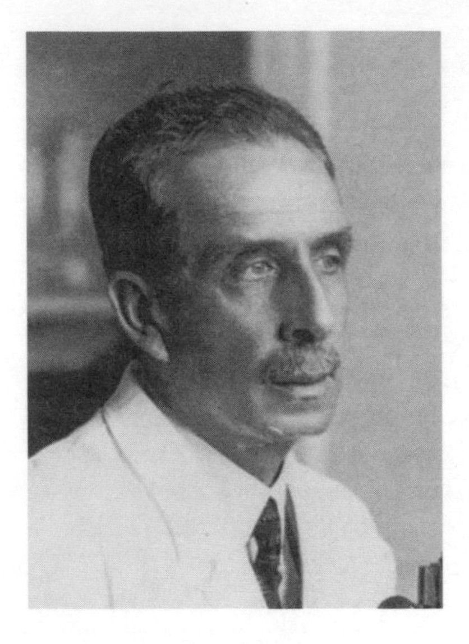

Figure 97. Carlos Chagas (1879–1934).

metres assisted by air currents and this enables them to move from house to house. The bugs pass through five wingless larval (instar) stages before they become adults and need blood at every stage of development. The bug's proboscis is coated with an anaesthetic and anticoagulant. This explains the remark of an early Spanish writer, J Gumilla, in 1702 that they suck blood with such care and sweetness that it cannot be felt. The bugs are thus able to pierce the skin and leisurely suck up to 300 mg of blood over 30 minutes. During this time, the gluttonous bug excretes faeces which are deposited at the site of the wound and the replete bug, now up to seven times its weight, waddles away with great difficulty.

You are no doubt wondering why we are talking about reduviid bugs. It is because in the case of South American trypanosomiasis, everything was done backwards. And who came up with this novel and ingenious idea? Carlos Chagas (Figure 97) did. Chagas graduated from the medical school in Rio de Janeiro, Brazil in 1902. In 1906, he joined the Instituto Manguinhos (later to be renamed Instituto Oswaldo Cruz), a medical research institute in Rio under the direction of Oswaldo Cruz, where he worked for the rest of his life.

In 1907 Chagas was sent with his travelling companion, Dr Belisário Penna (1868–1939), to the remote interior state of Minais Gerais for two years to control the malaria that was impeding the construction of an important railway. During this period he became aware that a triatomine bug, which the locals called 'barbero' (the barber), was living in the houses, particularly in huts with unplastered walls and covered with thatch roofs, and were biting people at night. Chagas was aware that mosquitoes, which are blood-sucking insects, transmitted filarial, and malarial parasites. So he wondered whether these bugs, which are also blood-sucking insects, might carry some parasite which could be passed from person to person. Chagas caught some of these bugs which he thought were probably *Conorhinus megistus* (now called *Panstrongylus megistus*). In 1909, with the ultimate in simplicity, he recorded:

> Examining the hindgut of specimens of *Conorhinus* [the bugs] collected inside human dwellings..., we discovered numerous flagellates with the morphological characteristics of crithidia.[69]

'Flagellates' of course means that the protozoa had a whip-like cord or flagellum. Crithidia are protozoa belonging to the genus of that name in the trypanosome family.

The next question was whether these parasites were pathogenic for mammals. In order to answer this, he sent bugs that had fed on humans to the Institute in Rio de Janeiro and Oswaldo Cruz (Figure 98) fed them on a monkey (*Callithrix penicillata*). This met with great success and Chagas wrote:

> After 20 or 30 days from the time of the bite, large numbers of trypanosomes were found in the peripheral blood of that monkey with morphology entirely different from that of any of the known species of the genus *Trypanosoma*.[69]

Chagas and Cruz had found a new trypanosome! They then inoculated guinea pigs, dogs, and rabbits as well as more monkeys and found that they all became infected. Monkeys and guinea pigs died a month or so after the infection but the dogs and rabbits were able to tolerate it better. In early 1909 Chagas named this new parasite *Trypanosoma cruzi*, the species name of course being given to honour Cruz.

The next obvious question was 'Does this parasite infect humans and, if so, what disease does it cause?' In order to answer this, Chagas and his colleagues visited homes in Lassance in Minais Gerais where bugs had been found and examined the inhabitants for symptoms and signs of any disease. They found that anaemia and debilitation were common, especially in children, some patients had oedema (fluid in the tissues), either localized to certain areas or generalized over the whole body, and the lymph nodes were often enlarged, especially in the neck, as were the liver and spleen. Many children were intellectually impaired and it appeared that the death rate in children was high, with most deaths being associated with convulsions.

Figure 98. Oswaldo Cruz (1872–1917).

The anaemia and oedema were suggestive of hookworm infection but examination of the stools ruled out this infection. Likewise blood examinations excluded malaria as the cause of the enlarged livers and spleens.

They took blood specimens from the inhabitants to look for parasites. Then they had some luck:

> Four days later, when we were called to treat a seriously ill child (whose blood we had examined previously), we found flagellates in the peripheral blood. In blood preparations stained with the Giemsa method, the flagellate observed was a *Trypanosoma*, with morphology identical to that of *Schizotrypanum cruzi*, whose biology we were studying in the laboratory in animals.[69]

Incidentally, this little girl whose name was Berenice survived. Fifty-two years after Chagas had seen her, she was visited by a group of doctors who confirmed that her blood tests for antibody were still positive for Chagas disease. Chagas inoculated two guinea pigs and a marmoset monkey with blood from Berenice. The guinea pigs died six days later and parasites were found in the lungs. Trypanosomes (Figure 99) appeared in the blood of the monkey after eight days. Chagas then inoculated blood into guinea pigs taken from a sickly two-year-old child who was anaemic, oedematous, debilitated, and had enlarged lymph nodes, liver, and spleen but in whom they were unable to find any parasites in the blood. One died nine days later and the same parasites were seen in the lungs. They then repeated this experiment with blood from a similar six-year-old child and this time found trypanosomes in the blood of the inoculated guinea pig 20 days later. There seemed no doubt: this parasite caused a devastating disease and there were small numbers of parasites in the blood which could be transferred to experimental animals.

The outline of the life cycle of this parasite was now clear—it passed from one human to another (or to an animal) via a reduviid (triatomine) bug. But the details needed to be worked out. Chagas observed changes in the appearance of the parasite as it passed down the length of the bug's gut and then was able to see the same sequence when parasites were cultivated on an artificial medium of blood agar. The locations of the internal organs of the parasite changed to look like crithidia and multiplied many times. After 24 hours or so, the parasites lost their

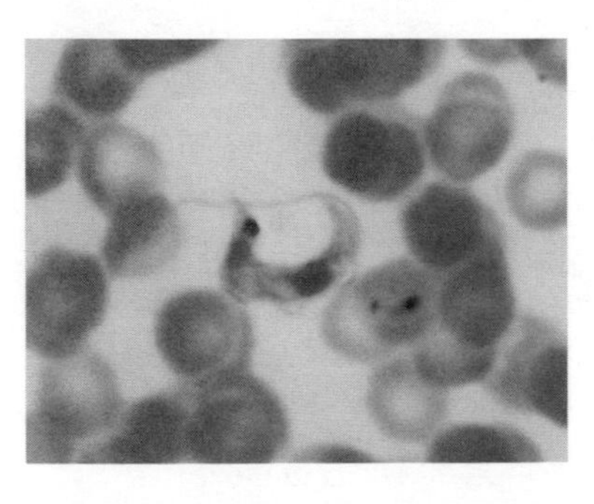

Figure 99. *T. cruzi* in blood.

undulating membrane and flagellum and became rounded in shape and looked like leishmaniae (p.149).

Chagas thought that the infective form of the parasite migrated to the proboscis of the bug and was injected at the next blood meal. He maintained this position for the rest of his life despite powerful evidence to the contrary. Émile Brumpt (p.113) reported in 1912 that the parasites 'encyst' (it is unclear what this means) in the posterior intestine of the bug and are passed as such in the excreta. Brumpt then showed that although the bite of an infected bug was not infective to mice, disease could be produced in mice by the injection of faeces containing parasites. Brumpt suggested that in humans, biting bugs deposit stool on the face which then contaminated either the conjunctivae (of the eyes) or the mucous membranes lining the mouth through which the parasite could enter. The issue was settled years later in the early 1930s when Emmanuel Dias (1908–1962) allowed 104 infected bugs to feed on 14 guinea pigs, taking precautions to ensure there was no faecal contamination; no infections resulted. On the other hand, when the mucous membranes were contaminated with faeces containing parasites, infection always developed.

The next question was 'What happens to the parasites when they get into the human body or another mammal?' Here Chagas and his colleagues made another mistake. In their studies of infected monkeys and guinea pigs at the Oswaldo Institute, they thought they saw dividing forms resembling malarial schizonts in the lungs. It was for this reason that the name of the parasite was changed late in 1909 to *Schizotrypanum cruzi*. In 1912, however, Pierre Delanoë and his wife Eugénie, students of Laveran, showed that these 'schizonts' were in fact a completely different organism which they named *Pneumocystis* (this was thought for years to be a protozoon but in the 1980s was shown to be a yeast-like fungus). Chagas paid attention to this information and changed the name of his parasite back to *Trypanosoma cruzi*.

So, if the parasite did not form schizonts in the lungs, what did it do in humans and other animals? Chagas described 27 cases of the acute form of the disease such as in the children mentioned earlier and was to perform more than 100 autopsies on patients with this infection. Not only was Chagas well aware of the acute form of the disease, but he also wondered and wrote in retrospect in 1922:

> Was it always like this or did it show well-defined chronic forms? My clinical experience and knowledge of the unusual condition of the local inhabitants led me to admit that in this trypanosomiasis, besides the acute form, other chronic ones awaited detection and description.[70]

Chagas was particularly concerned with heart disease and wrote:

> The symptoms that most deeply impressed me in the patients of that region, particularly in those that lived in houses infected by the triatomine insects, were the cardiac rhythm alterations in the form of extrasystoles [abnormal heart beats].[70]

An excellent pathological description of chronic cardiac disease was provided in 1911 by Gaspar Vianna (1885–1914) working at the Instituto Oswaldo Cruz. He found parasites inside muscle cells of the heart (Figure 100) and in skeletal muscle, and noted an inflammatory reaction around infected cells. Inside the cells, the parasites had no flagella so were called amastigotes (they look rather like leishmaniae). He wrote:

> These forms of the parasite vary numerically inside the cardiac cell and one can see them in binary fission [i.e. dividing]. Some cells have a reduced number of protozoan parasites; others have hundreds of them. In human cases, I verified the infection in large areas of the leg, and back, arm muscles and other areas . . . On reaching a certain point of saturation, when the sarcolemma (the muscle cell wall) can no longer contain the Schizotrypanosomes which have developed in its interior, it ruptures in some parts, and parasites—some still round, others already flagellated—are pushed into the interstices of the muscle fibres. There are cases in which quite peculiar degeneration of the cardiac cell is present and in which the parasites are not seen at all while the infection is positively verified in other cells of the organism. . . . Whenever a cardiac fibre ruptures, there is always a definite reaction of the connective tissue, and other cells come into this area.[309]

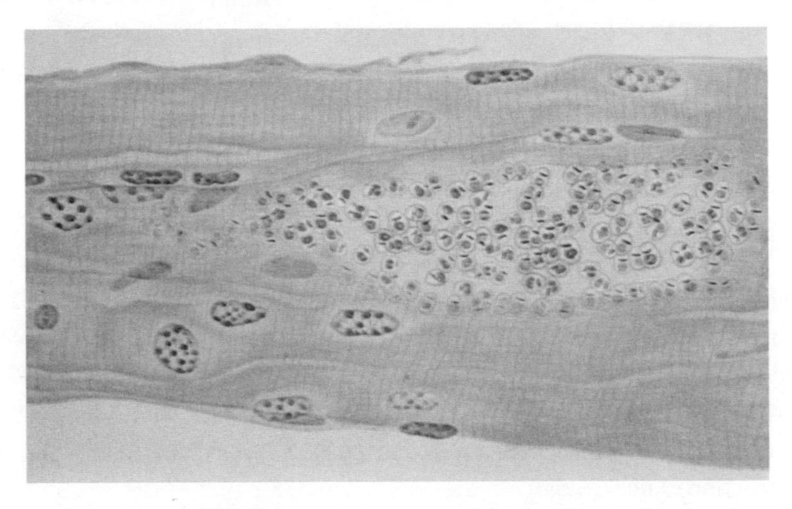

Figure 100. Drawings, probably by Vianna, of *T. cruzi* looking like leishmaniae (L–D bodies) inside heart muscle cells.

In other words, there is inflammation in heart muscle and if parasites are present in sufficient numbers, the muscle fibres die. If enough muscle fibres are affected, the heart can no longer pump blood around the body, a condition known as cardiomyopathy.

On occasion, when a cell bursts, flagellated parasites may escape into the bloodstream ready to infect another bug when it bites. The numbers of trypanosomes in the blood are usually very low and difficult to detect. It was for this reason that Brumpt in 1912 developed the technique of 'xenodiagnosis' (*xenos* is a Greek word meaning 'foreigner'). Uninfected bugs were allowed to feed on people suspected of having trypanosomiasis. Once ingested, any parasites multiplied in the gut and were easy to find in their faeces. It was now possible to describe the life cycle of this parasite (Figure 101).

But did the parasite do anything else? There were two clinical syndromes that were common in South America. One concerned difficulty with swallowing food and one practitioner wrote in 1857:

> At Limeira I became aware of the new disease...which is widely prevalent in some portions of interior Brazil.... The Brazilians call it mal de engasgo. The first indication of its existence is a difficulty in swallowing. The person thus affected appears to be in good health but in five or six years death ensues due to starvation.[13]

Another problem, called 'bicho', was chronic constipation due to gross distension of the large intestine. In 1835, José Martins da Cruz Jobim (1802–1878) in Brazil wrote that there is:

> degeneration of the layers of the stomach, duodenum and large intestine. These a few times are contracted, on occasions very dilated, above all the descendent colon, to the point of resembling a second stomach.[155]

In 1916, Chagas reported that some patients with acute Chagas disease, or while they were recovering from it, developed difficulty in swallowing. Naturally enough, he wondered whether this was the beginning of 'mal de engasgo' in which the oesophagus becomes greatly distended. There then followed decades of disagreement, with most clinicians considering that both mega-oesophagus and megacolon (bicho) were due to ill-defined non-infectious causes such as vitamin deficiency. What had been forgotten was that in 1911 Vianna had noticed that in acute Chagas disease involving the nervous system, there was often destruction of nerve cells.

Fritz Köberle (1910–1983), an Austrian-Brazilian physician revisited this issue. He showed in 1956 that the parasites destroyed the autonomic (unconscious)

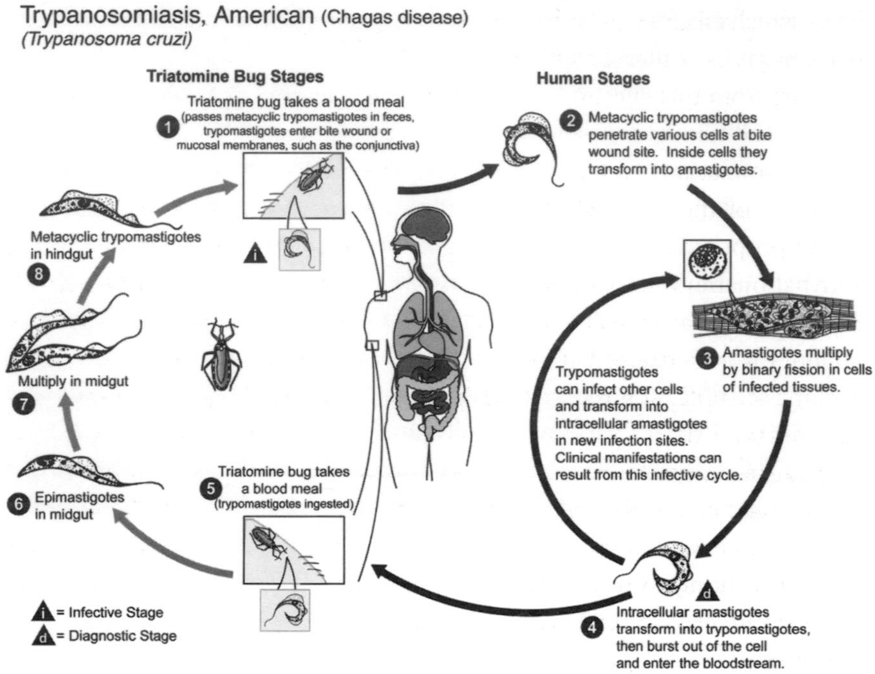

Figure 101. Life cycle of *Trypanosoma cruzi* as presented by the US Centers for Disease Control. An infected bug takes a blood meal and releases trypomastigotes in its faeces near the site of the bite wound. These enter the host through the wound or through intact mucosal membranes, such as the conjunctiva (1). Inside the host, the trypomastigotes invade cells near the site of inoculation, where they differentiate into intracellular amastigotes (2). The amastigotes multiply by binary fission (doubling of the chromosome then dividing into two) (3) and differentiate into trypomastigotes which are released into the bloodstream (4). Trypomastigotes infect cells from a variety of tissues and transform into intracellular amastigotes in new infection sites where they again divide. Clinical manifestations can result from this infective cycle. The bloodstream trypomastigotes do not replicate. The bug becomes infected by feeding on human or animal blood that contains circulating parasites (5). The ingested trypomastigotes transform into epimastigotes (Chagas's crithidia) in the vector's midgut (6) and multiply (7). In the hindgut they differentiate into infective metacyclic trypomastigotes (8) and are excreted in the bug's faeces.

nervous system that controls the muscles that cause movement in hollow tubes. Consequently, the intestinal tract was unable to contract properly and push food from top to bottom. This resulted in gross dilation of the oesophagus with patients complaining of difficulty swallowing, and similar distension of the large intestine caused constipation. Köberle also showed that apart from destruction of heart muscle, there was also disintegration of the nerves to the heart which ensure regular beating, thus explaining the abnormal heart beats that Chagas had observed 50 years before!

Apart from this question of gut disease, ten years or so after Chagas's discovery, it all seemed done and dusted. Chagas had named the disease American trypanosomiasis to distinguish it from the African form but in 1910 the President of the National Academy of Medicine in Rio de Janeiro, Miguel Couto, proposed naming the disease after its discoverer. Chagas was showered with national and international acclaim. He was twice nominated for the Nobel Prize but never awarded it (he should have been). But then the tide turned and honours were replaced with controversy and vilification. Part of this had a genuine scientific, if misplaced, basis. *T. cruzi* multiplies inside cells in the tissues and it is extremely hard to find trypanosomes in the blood so it was hard to confirm or disprove clinical diagnoses. Chagas also had to contend with jealousy and personal and political rivalries, particularly from those whom he beat when he was appointed director of the Oswaldo Cruz Institute and later director of the newly created Department of Public Health. In 1922 Chagas was viciously attacked in the Brazilian Academy of Medicine, with one opponent charging:

> You could have found some mosquitoes, you could have invented a rare and unknown disease, about which much was said, but whose victims almost no one knew, hidden in a countryside dwelling of your province, a disease that you could magnanimously distribute among your fellow countrymen, accused of being cretins.[23]

These sorts of extraordinary outbursts led to the establishment of a special commission to investigate the allegations. More than a year later, its report vindicated Chagas.

After Chagas's death in 1934, trypanosomiasis was found in Argentina. Slowly it emerged that it was present throughout Latin America and surveys using blood tests for antibodies to the parasite showed that nearly 20 million people were infected, with a huge burden of illness and premature death. Chagas was right. If anybody deserves to have a disease named after him, surely it is Carlos Chagas.

PART VI

BACTERIA

My mother is 95 years of age but she never fails to find an opportunity to tell me to 'Wash your hands—to kill the germs'. She was in her mid-twenties when penicillin arrived and, as with most members of her generation, she has a healthy respect for if not fear of germs. By 'germs', of course, she means bacteria. The word is derived from the Latin 'germen' meaning 'seed' or 'sprout' and these days tends to be applied indiscriminately by most people to both bacteria and viruses. What are bacteria? The word 'bacterium' (plural: bacteria) is derived from the Greek word *bakterion* meaning 'staff' or 'cane' and reflects the rod-like shape of many of the first bacteria that were seen through the microscope.

Bacteria are single-celled organisms. Unlike all the organisms we have considered thus far, which are eukaryotes, bacteria are prokaryotes. They have a single chromosome without a nuclear membrane around it. They do not have mitochondria (metabolic factories that supply energy) in the cytoplasm and the cytoplasmic ribosomes (that make proteins out of amino acids) are smaller in size. The chemicals making up the cytoplasmic membrane are different from those found in that of eukaryotes. Likewise, the substance that gives strength to the cell walls of bacteria is peptidoglycan whereas plant cell walls are made up of cellulose.

In 1866, the German naturalist Ernst Häckel (1834–1919) created the Kingdom Protista to house bacteria, amongst other things, in a grouping of unicellular organisms. Today, classification is in a state of flux and bacteria have been given their own Kingdom, 'Eubacteria' (although some people called it 'Prokaryota'), or have been elevated to a domain (which is bigger than a kingdom!) of their own called simply 'Bacteria'.

Medical microbiologists, however, are not concerned with such niceties. In identifying bacterial pathogens, they are more concerned with the practical features. The first is their shape. Some look round (they are actually spherical) and are called cocci (singular: coccus), from the Greek *kokkos* meaning 'seed' or 'berry', and are about 1 μm in diameter. These are then categorized as to whether they occur singly or in pairs (diplococci), chains (streptococci), or bunches (staphylococci). The other major shape of bacteria is a rod shape; these are called bacilli (singular: bacillus). This name derives from the Latin word 'baculum' meaning 'rod' or 'stick'. These bacteria are generally about 1 μm wide and 5 μm long. Others are somewhat shorter and sensibly are called 'cocco-bacilli'. Rarely, spiral bacteria may be seen.

The second most useful feature is the colour that develops after a Gram stain of the bacteria (p.192). Some are purple and are called Gram-positive while others are pink and are called Gram-negative. Rarely bacteria don't stain at all—these are usually mycobacteria, such as the organisms that cause tuberculosis and leprosy.

The third useful feature in identifying an organism is whether or not it needs or can tolerate oxygen when it grows. Those bacteria that need oxygen are called 'obligate aerobes', those that cannot grow in the presence of oxygen are called 'strict anaerobes', and those that do not care either way, somewhat confusingly, are called 'facultative anaerobes'.

When bacteria are spread out and grown on solid media such as blood agar (p.192), each individual bacterium multiplies many times to form a 'colony'. The size, shape, and appearances of these colonies, and the effects they have, if any, on the medium on which they are growing provide further clues to identification.

Bacteria have various metabolic processes such as the breakdown of different sugars or the production of certain enzymes. Sometimes one or two tests are all that is necessary to be able to identify the bacterium. On other occasions a battery of 20 different biochemical reactions is needed. The information is then fed into a computer which then spits out the identification.

Finally, different bacteria have different substances on their surface called 'antigens'. Humans and animals are able to make 'antibodies' (special proteins in the blood) that are directed specifically against these antigens. Sometimes the presence or absence of certain antigens is helpful in identifying a bacterium.

We shall begin our consideration of the discoveries of the various causes of bacterial diseases by a discussion of the 'germ theory of disease', that is, how bacteria were first recognized, how it was discovered that they cause disease, and how one could work out which bacterium causes which disease. Then follow chapters on various bacterial diseases. Clearly we cannot review them all. We shall look at those that have been most important to humanity in history and those that are the greatest problems now. They won't appear in the logical order that you might find in a textbook. Rather, we shall use a combination of textbook order with chronology relating to when the particular bacterium was discovered.

16

The germ theory of disease

The 'germ theory of disease', that is, the idea that specific bacteria could cause specific diseases burst dramatically into medical consciousness three-quarters of the way through the nineteenth century. This was not a notion that came from nowhere. Rather it was a sudden flowering resulting from the amalgam of a number of streams of thought, just as a cake rises quickly in the oven or water on the stove comes suddenly to the boil. These various factors may be summarized in this way:

1. recognition of the organisms we now call bacteria;
2. proof that certain putrefying diseases could be prevented by what we now recognize as antibacterial measures;
3. final refutation of the theory of spontaneous generation;
4. technological developments that allowed the isolation and growth in pure culture of a single bacterial type and methods to characterize those bacteria;
5. formulation of criteria, now known as 'Koch's postulates', that are necessary to prove that a specific bacterium causes a specific disease;
6. understanding of the normal bacterial flora and appreciation of the differences between extrinsic and intrinsic bacterial infections.

We shall now consider each of these in greater detail.

The first person who proposed concepts consistent with our modern understanding of infectious disease was Giralamo Fracastoro (Fracastorius) (Figure 102), a Venetian physician. In 1546 he wrote a book in Latin called (in translation) *On contagion, contagious diseases and their treatment*. Although, as noted on page 1, contagion strictly speaking refers to infection transmitted by touch, Fracastoro used it in the sense that we would now use infection. He had no idea what the nature of infectious agents was but he had great insight, writing:

> There are it seems three fundamentally different types of contagion. The first infects by direct contact only. The second does the same, but in addition leaves fomes and this contagion may spread by means of that fomes. By fomes

[we now call them fomites] I mean clothes, wooden objects, and things of that sort which though not themselves corrupted can, nevertheless preserve the original germs of contagion and infect by means of these. Thirdly, there is a kind of contagion which is transmitted not only by direct contact or by fomes as intermediary but also infects at a distance.[100]

The word which Fracastoro used to describe the infectious agent is best translated as 'germ'. He even suggested that these invisible germs generate and propagate other germs precisely like themselves. Even though he had no direct evidence to uphold such an assertion, these words were truly prophetic and apply to both bacteria, the subject of this section, and viruses, which we shall review in the next. Fracastoro was ahead of his time.

Almost certainly the first person to see bacteria was Antony van Leeuwenhoek (p.97). In 1684 he decided to have a look at what he called 'a little white matter as thick as wetted flour',[182] presumably plaque, between his teeth. He mixed some with pure rainwater which contained no 'animals' or else he mixed it with his saliva and 'then to my great surprise perceived the aforesaid matter contained very many small living Animals, which moved themselves extravagantly'.[182] Leeuwenhoek drew a diagram of the various types (Figure 103). Some (A, C, F, G) we would now recognize as rods (bacilli), while others were spherical ovoid in shape (B, E, H) and we would now view them as cocci. Only some bacteria are motile and it was those that Leeuwenhoek saw 'moved themselves extravagantly'. He calculated that there were more of these organisms in a man's teeth than there were men in a kingdom.

Figure 102. Hieronymus Fracastorius (1478–1553).

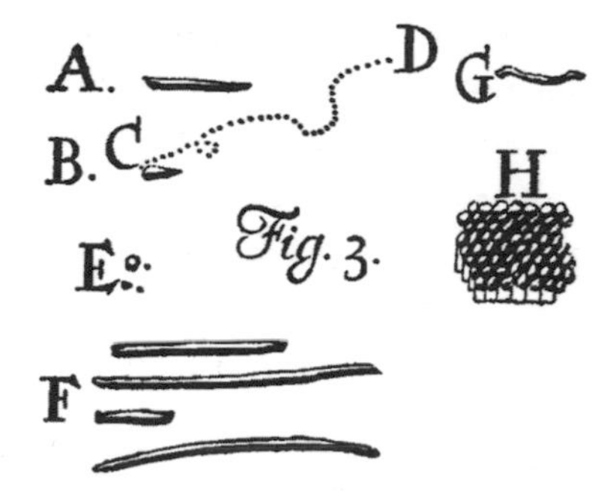

Figure 103. Leeuwenhoek's drawings of bacteria.

Naturally enough, attempts to classify these bacterial organisms began. Among the first was Otto Friderich (Friedrich) Müller (1730–1784), who divided them into two genera. The first, *Monas*, contained 10 species and were described as being inconspicuous, very simple, transparent, and dot-like organisms. The second, *Vibrio*, housed six species, which were also inconspicuous and very simple but were smooth or fine and elongated. In 1838, Christian Gottfried Ehrenberg (1795–1876) published a large book with 64 plates on organisms, which he called Infusoria. These included protozoa, rotifers, diatoms, and bacteria. Three of the 22 families, Monadina, Cryptomonadina, and Vibriona, we would now recognize as bacteria. The last family contained five genera, *Bacterium, Vibrio, Spirochaeta, Spirillum,* and *Spirodiscus,* some of which will become familiar: many of Ehrenberg's species are unrecognizable today. Soon afterwards in 1841, the French zoologist Félix Dujardin (1801–1860) published a simpler classification. All bacteria were placed in one family, Vibrioniens, and there were three genera: *Bacterium, Vibrio,* and *Spirillum.* His organisms too are often difficult to identify from the present viewpoint. In 1852, Josef Anton Maximilian Perty (1804–1884) published a new classification in which he regarded bacteria, not as animals as his predecessors had done, but as 'animal-plants'.

Concepts of the nature of bacteria moved yet further when Ferdinand Cohn (1828–1898), professor of botany at Breslau (now Wroclaw in Poland), in 1854 suggested that bacteria should be regarded as belonging to the vegetable kingdom. He was to have a major influence in the decades to come. In 1872 he asked the question as to whether or not bacteria could be classified into genera and species like animals and plants. He realized that the shape and size of bacteria were insufficient alone to classify bacteria since many similar-looking organisms may behave differently. The means at his disposal were insufficient to settle the question. He was able to divide bacteria into six genera, *Micrococcus, Bacterium, Bacillus, Vibrio, Spirillum,* and *Spirochaete* (Figure 104), but he regarded species as only provisional. Not everyone agreed with this for some thought that bacteria could change their form according to circumstances or were stages in the development of fungi. Time was to prove Cohn correct and four of his genera are still in use.

In 1874, Theodor Billroth (1829–1894), a Viennese surgeon who described the famous Billroth I and Billroth II operations on the stomach, published a book on his studies of bacteria in putrefaction and infective disorders. The main thrust of his book, that bacteria were stages in the development of an alga (a plant), was wrong but he introduced some ideas that were to have a powerful impact on medical bacteriology. He called the smallest form of bacteria 'cocci' (p.150). He then subdivided cocci into manner of arrangement such as diplococcus (pairs of cocci) and streptococcus (chains of cocci). Similarly, he divided up rod-shaped forms. Misguidedly, he thought cocci turned into rods (bacilli) and together with spores constituted the non-existent plant he called *Coccobacteria septica*, which he believed was found in putrid fluids and septic wounds. Nevertheless, some of the principles laid down by Cohn and Billroth have stood the test of time and bacteria have been classified and reclassified to the present day as new information is gathered.

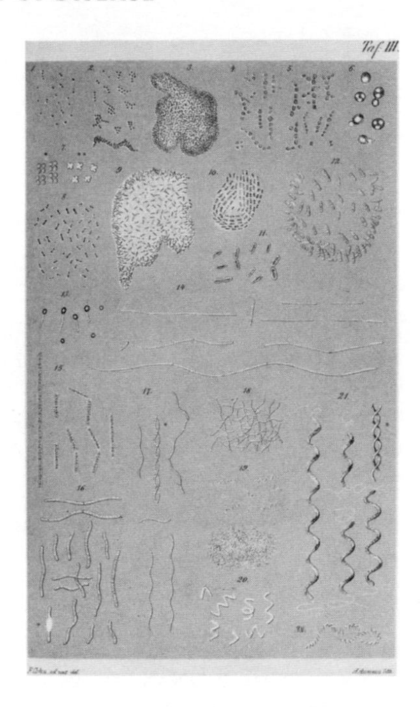

Figure 104. Cohn's drawings of bacteria.

Even though the microscope had revealed a whole new world of invisible creatures and made Fracastoro's idle speculations about invisible particles seem acceptable, there was still a fuzzy understanding of the disease process. Concepts of disease, on the one hand, and the germ or parasite that caused the disease, on the other, were often not separated. Rather, the disease itself was conceived as a parasitic organism. Jacob Henle (1808–1885), a German anatomist, in 1840 tried to clarify matters. He said that it is not the disease that reproduces itself but the cause of the disease—the 'contagium' is the disease inducer. Although he recognized that these contagious agents were so far unseen, he reckoned that they must be organic, capable of assimilating substances while parasitic in a sick body, and able to reproduce themselves. Thus far he was correct. He acknowledged that it was possible that microscopic plants and animals (infusoria) could in some instances be the culprits but he was doubtful. Furthermore, he thought it would be difficult, if not impossible, to prove that they were the cause of the disease. Fortunately, there he was wrong.

Someone who tackled the problem of infection from a more practical standpoint by preventing infectious agents from causing disease was Ignaz Semmelweis (Figure 105), even though he had no idea what the nature of those agents was. He was born in Budapest (then part of the Austro-Hungarian Empire) into a German-speaking family. He graduated from the University of Vienna in 1844 and in 1846 was appointed an assistant at the First Obstetrical

Figure 105. Ignaz Semmelweis (1818–1965).

Clinic at the Vienna General Hospital. At the time, this was the largest obstetric hospital in the world. The First Clinic was used to train male medical students, while female midwives were trained in the Second Clinic. Patients were admitted to the two clinics on alternate days. Both clinics were free but maternal and neonatal mortality from puerperal fever (childbirth fever) was appalling. Semmelweis noticed that women who gave birth in the street en route to the hospital had a much lower incidence of this fever than did those who gave birth in the hospital. Secondly, he observed that the death rate from puerperal fever in the First Clinic was twice that of the Second Clinic. He pondered various causes for this difference but could find none that was satisfying. The breakthrough came in a tragic manner. One of his mentors, Jacob Kolletschka, professor of legal medicine, had been accidentally cut with a student's scalpel while performing a post-mortem examination. Kolletschka died from an infection which spread up his arm and then into his whole body. An autopsy on Kolletschka's body revealed inflammation of the membranes within the abdomen and chest and around the brain as well as multiple abscesses. This was very similar to what Semmelweis had seen in autopsies of many women who had died from puerperal fever and in babies who had died with fever. He realized there must be a connection between cadavers and puerperal fever. As he later wrote:

> Agitated by the report of Kolletschka's death, there was forced upon my mind with irresistible clarity...the identity of this disease, of which I had seen so many hundred puerperae [women in childbirth] die...and the newborn died of the same disease as did the puerperae...The exciting cause of Professor Kolletschka's illness was known, that is to say, the wound produced by the autopsy knife was contaminated at the same time by cadaveric material...I must ask myself the question: Did the cadaveric particles make their way into the vascular systems of the individuals, whom I had seen die of an identical disease? This question I answer in the affirmative...(It) seemed to me...that decaying organic matter brought into contact with living organisms produced in them a putrefactive process.[286]

Semmelweis did not know the nature of this cadaveric material, and certainly did not equate it with bacteria, but he had made the crucial connection. It also seemed to explain why the death rate in the First Clinic was twice that in the Second Clinic because the medical students of the First Clinic attended post-mortem examinations then went to examine patients and deliver babies, whereas the student midwives of the Second Clinic did not participate in autopsies. The students neither washed their hands nor changed their clothes when they moved from the mortuary to the ward or delivery suite. Semmelweis cast around for something that might kill or neutralize this cadaveric material. He knew that chlorine had been used to fumigate the obstetric wards of the Rotunda Hospital in Dublin, Ireland. In May 1847 he introduced into the First Clinic a requirement that everyone wash their hands with liquid chlorine when returning from the mortuary. After a short while, he substituted chlorinated lime (bleaching powder, a mixture of calcium chloride hypochlorite [$CaCl(OCl)$], calcium hypochlorite [$Ca(OCl)_2$], and calcium chloride ($CaCl_2$)) for liquid chlorine because it was considerably cheaper. Semmelweis was able to embark on perhaps the first clinical trial with a control group. The mortalities in both the First and Second Clinics before and after the introduction of hand-washing are shown in Figure 106. There was a marked reduction in cases of puerperal fever in the First Clinic so that it became comparable to that in the Second Clinic.

Then he encountered another problem. A woman who probably had cancer of the cervix was nearest the door of the ward and was the first person to be examined; the students then moved up the ward in a row examining patients

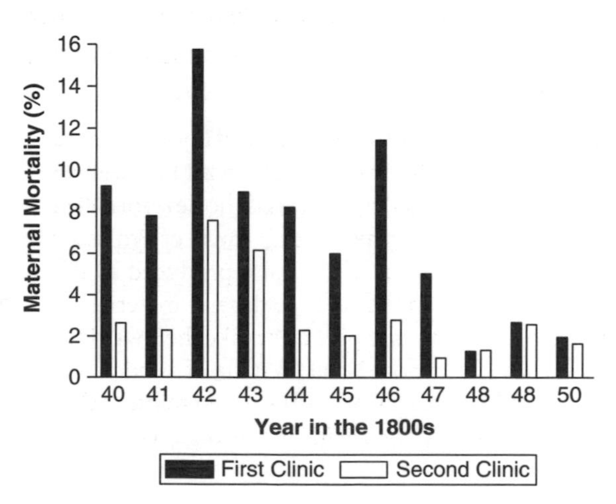

Figure 106. Graph showing the mortalities of mothers in the First and Second Clinics before and after the introduction of hand-washing with chlorine in May 1847.

one by one. Eleven of the 12 women next to the first patient died of puerperal fever. Clearly it was not just the mortuary that was a problem. Semmelweis renamed the cause of the problem as 'decomposing organic animal matter' and instituted washing of hands between examinations of every patient. Semmelweis had made a major breakthrough. Like many reformers, he encountered strong opposition. Some were offended simply by the fact that they were expected to wash their hands before examining a patient. Others had pseudo-scientific objections because Semmelweis was unable to state the precise nature of 'cadaveric material'. In 1850, in frustration, he left Vienna for home and in 1851 took the relatively insignificant, unpaid, honorary position as head of the obstetric ward of Pest's small St Rochus Hospital. Childbed fever was rampant at the clinic on his arrival but in the next five years under his supervision only 8 patients out of 933 died from the fever.

In 1857, at the age of 38, Semmelweiss married a young woman of 18. Lamenting the slow adoption of his ideas, in 1861 he published a book entitled (in translation) *The Aetiology, Concept and Prophylaxis of Childbed Fever,*[286] which outlined his methods and successes. This definitive account was still ignored so he began to write vituperative open letters such as one to Joseph Spaeth, professor of obstetrics in Vienna, concerning the 1,924 patients who had died in the Vienna hospital over the preceding 10 years in which he said that the Herr Professor had participated in the massacre and that the homicide must cease. Over the next several years, he became increasingly irrational and was eventually admitted to a mental institution in Vienna in 1865, quite possibly suffering the effects of tertiary syphilis acquired from the many patients he had examined. By an awful stroke of irony, a few weeks later, a wound on one of his fingers, apparently the result of a gynaecological operation, became gangrenous and septicaemia ensued. Semmelweis had died from the same or a similar infection to that of Kolletschka and of so many women who had died in childbirth.

While all this had been going on, the vexed question of the origin of animalcules or bacteria came to a head. A long-held view had been that, like worms, they were the product of spontaneous generation. Although Redi had shown that maggots were the result of eggs being deposited by flies rather than spontaneous generation (p.8), his experiments were largely either forgotten, disbelieved, or thought not to apply to microscopic animalcules. John Turbeville Needham (1731–1781), an English Catholic priest, undertook a number of experiments to examine this question more closely with regard to microscopic creatures. He put some mutton gravy, hot from the fire, in a vial, filling it to near the top, then closed it with a cork to prevent air which might carry insects or eggs floating in the atmosphere from entering. Then he heated the vial in hot ashes. After some time, he opened the vial and found 'my phial swarm'd with Life, and microscopical Animals of most Dimensions, from some of the largest I had ever seen, to some of the least'.[224] Needham then repeated his experiments with some dozens of different animal or vegetable infusions (liquid extracts) and always had similar results. In his report to the Royal Society in 1749, he concluded that all these micro-organisms derived from a 'vegetative force', that is, they were generated spontaneously.

In that same year, Georges Leclerc (Comte de Buffon, 1707–1788), a French naturalist, mathematician, and aristocrat, published the first of many volumes of his *Histoire Naturelle*. Buffon, who knew Needham, thought he should make a definitive statement about spontaneous generation, which he did in his second volume. Like Aristotle, he approached the problem from the point of view of logic rather than experiment and came up with this ingenious but completely wrong explanation:

> The matter of living beings preserves after death some of its vitality. The life is contained essentially in the last molecules of the body... These always active organic molecules try to move the putrefying matter; they appropriate to themselves some particles and form by this union a multitude of little organised bodies, of which some, such as earthworms, toadstools etc seem to be fairly large animals or vegetables, but others which are almost innumerable, can only be seen through a microscope. All these animals exist only by spontaneous generation, and they fill the gap which nature has made between the single living organic molecule and the animal or vegetable.[62]

In other words, when an organism is dead, it is not really dead—it preserves some 'vitality'.

Lazarro Spallanzani (1729–1799), an eminent Italian scientist, was convinced that Needham had not sealed his containers properly or heated them to a high enough temperature. Spallanzani, incidentally, is famous for discovering the means of mammalian reproduction, showing that it requires both semen and an ovum, and was the first to perform *in vitro* fertilization with frogs and artificial insemination using a dog. Spallanzani repeated Needham's experiments, put infusions of a dozen or so kinds of seeds in flagons, sealed them, then boiled them for varying lengths of time. Spallanzani noted two classes of animalcules. The first, which were larger, he termed 'ordini superiori' (these were undoubtedly protozoa), and were easily destroyed by boiling for 30 seconds. The second, which were minute, he called 'ordini ultime' (these were no doubt bacteria), and they sometimes survived boiling for 30 minutes. He concluded in 1755 that Needham's animalcules were carried into the infusions from the air or else air assisted the multiplication of those already there. This, he said, was the source of the apparent 'spontaneous generation'.

Needham disagreed with Spallanzani's conclusions, protesting that by boiling his infusions for an hour, Spallanzani had:

> treated and tortured these vegetable infusions in such a way that he has plainly not only enfeebled them and possibly entirely destroyed the vegetative force of the infusions but has also corrupted the small amount of air remaining in the phials... It is not surprising that infusions so treated should not show evidence of vitality.[225]

In other, words Spallanzani said 'You do not heat sufficiently' while Needham replied 'You heat too much and besides you are polluting the air'. Needham was sticking to his guns in favour of spontaneous generation, and it was impossible at that time to refute his argument because the composition and function of air were unknown.

The idea of spontaneous generation was still going strongly 100 years later and no-one championed it more energetically than Félix Pouchet (1800–1872,), director of the Natural History Museum in Rouen, France. In 1858 he presented a series of papers to the Academy of Sciences in Paris in which he claimed to have proved the existence of spontaneous generation by experiments using flasks, water, and hay heated to 100°C for 20 minutes. He published all his work in 1859 in a book entitled *Héterogonie*. Heterogony means spontaneous or, as it was sometimes called, equivocal generation, i.e. the birth of beings without antecedents similar to themselves. Pouchet's introduction to his book is interesting:

> When by meditation it was evident to me that spontaneous generation was one of the means employed by nature for the reproduction of living things, I applied myself to discover the methods by which this takes place.[255]

This translation perhaps does not do justice to Pouchet for he no doubt meant by 'meditation' thinking deeply about the issue, not some New Age notion of meditation according to some Eastern religion.

Nevertheless, this stirred up a hornet's nest because plenty of people did not agree with him. The French Academy of Sciences in 1859 offered a prize for 'an endeavour by means of careful experiments to throw new light on the question of spontaneous generation'. It was at this point that Louis Pasteur

(Figure 107) entered the fray. Pasteur had been born at Dôle in France, the son of a tanner and a former sergeant in Napoleon's army. After graduation, he became interested in the polarization of light and then was appointed professor of chemistry in Strasbourg following which in 1854 he became dean of science at Lille. There he developed an interest in the phenomenon of fermentation, i.e. the production of chemicals such as alcohol and lactic acid from the breakdown of sugars. He managed to separate different yeasts and showed that they produced different products. Thus 'beer yeast' generated alcohol while 'lactic yeast' gave lactic acid.

Figure 107. Louis Pasteur (1822–1895).

He then studied the fermentation of butter leading to the formation of butyric acid and discovered that this was mediated not by yeasts but by minute, motile rods about 0.002 mm in width and 0.002–0.02 mm long. Because they moved, he thought they were animals and classified them as infusoria, calling them vibrios. This contrasted with the non-motile yeasts which he thought were vegetable in nature. Furthermore, Pasteur showed that these rods could not live in the presence of air; we now classify such organisms as anaerobic bacteria. All of this led naturally to the question that he posed in his report to the Academy of Sciences in 1860:

> Among the questions raised by the researches . . . (on) . . . fermentation, none is more worthy of attention than that concerning the origin of ferments. Whence come these mysterious agents, so weak in appearance, so powerful in reality? . . . This is the problem which has led me to the study of spontaneous generation. [241]

Pasteur was thus well-placed to solve the puzzle thrown down by the Academy. In his first experiment, he drew air through cotton wool to retain dust, broke up the wad with alcohol and ether, and saw small bodies when he examined it under the microscope. He then prepared a solution of sugar, albumin (a protein), and beer yeast and placed it in a glass capsule, the neck of

Figure 108. An open swan-neck flask.

which was drawn out. Next, he boiled the solution, then allowed heated air to enter via a red-hot platinum tube after which the capsule was sealed. There was no fermentation—no spontaneous generation. He then prepared a number of flasks with an open, elongated swan's neck so that dust particles in the air could not find their way into the body of the flask but would become deposited on the damp sides of the neck (Figure 108). Fermentation fluid was then placed in the flasks. If the fluid was boiled, the solution remained clear, even though the contents of the flask were open to the atmosphere. If the fluid was not boiled, it became turbid. He then repeated his experiments with milk and wine but found that with milk, it had to be heated to 110°C under a pressure of 1.5 atmospheres to prevent turbidity, as the germs in the milk were more resistant. He then showed that if blood and urine were collected under clean conditions and stored, there was no putrefaction.

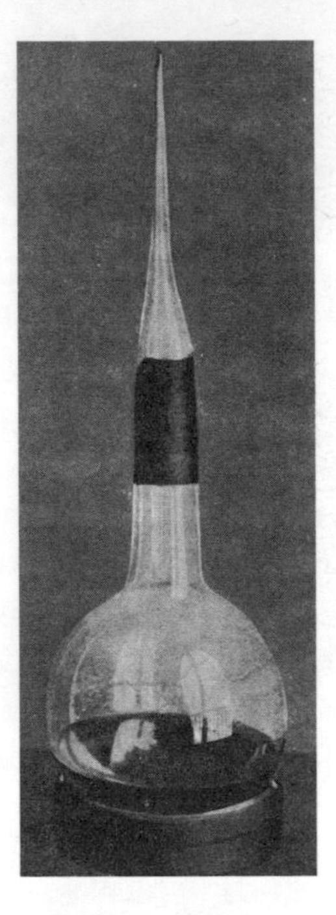

Figure 109. Flask with a vertical point.

Finally, he prepared a number of flasks with vertical points (Figure 109) with fermentation fluid, boiled them, then opened them at different locations varying in cleanliness (Table 2). They all became turbid in Paris but only one did on the Mer de Glace high up in the French Alps near Mont Blanc. In his presentation to the Academy in September 1860, Pasteur indicated:

> I have established by numerous experiments that the cause of spontaneous generation is not universally present in the atmosphere...It appears to me that the dust suspended in the air is the only origin and the first and essential condition of life in infusions.[242]

Table 2. Numbers of flasks that became turbid when they were opened in various locations

Location	Number of flasks prepared	Number turbid
Paris courtyard	11	11
Paris cellar	11	1
countryside	20	8
Mer de glace	20	1

Or as he put it in his book entitled (in translation) *On the organised bodies that live in the atmosphere. Examination of the doctrine of spontaneous generation* published in 1861:

> There exist continually in the air organised bodies which cannot be distinguished from true germs of the organisms of infusions. In the case that these bodies... are inseminated into a liquid which has been subject to boiling... the same beings appear in the liquid as those which develop when the liquid is exposed to the open air.[243]

In other words, life is in the dust of the air—there is no spontaneous generation.

Of course, Pouchet was not convinced. He prepared flasks of hay infusions which had been filtered and boiled and took them to the peaks of Aragon which, he chortled, were far higher and wilder than the ladies' favourite of Mt Blanc. This, he said, refuted Pasteur's claims because all his flasks became turbid when opened at Maladetta, 1,000 metres higher than the Mer de Glace. What Pouchet did not realize was that his hay infusions contained spore-forming organisms which were resistant to the boiling he had given them. In the event, the Academy awarded Pasteur the prize in 1862.

Still, not everyone was persuaded. Pasteur's most formidable opponent was Henry Bastian (1837–1915), professor of pathology at University College, London, who wrote a book of over 1,100 pages, *The Beginnings of Life*, in 1872 in which he championed spontaneous generation. Pasteur then had to undertake further experiments to find out where Bastian had gone wrong in his experiments. Eventually, Pasteur was able to show that all bacteria in the dry state were killed if they were heated to 180°C for 30 minutes or those in the wet state were heated to 120°C for 30 minutes.

Bastian's error was that he had not killed all the spores. Neither he nor Pouchet can hardly be blamed for this because they did not know about them. Spores were discovered by Cohn, who was intrigued by the observation that some bacteria were more resistant to boiling than others. In 1872 he began to study this phenomenon in infusions of hay. He found that a couple of days after boiling such infusions, a delicate film appeared on the surface which contained countless motile rods 0.6 μm thick and 3–7 μm long which he recognized as belonging to a single species which he called *Bacillus subtilis*. But then he noticed that the rods developed oblong, strongly refracting spores within them. These spores were viable (alive) because when they were placed in a hay infusion that had been otherwise sterilized by prolonged heating, they germinated and motile rods once more appeared. This was the explanation for all the contradictory experiments. As Cohn put it in his publication of 1876:

> The observations presented here seem to me to contain the solution to the puzzle regarding the development of organisms in boiled organic substances.[74]

In 1877, the Englishman John Tyndall (1820–1893) showed how spores could be killed. He found that while spore-forming bacteria are not killed by boiling for 10 minutes, they can be killed by boiling for just one minute then repeating the boiling twice daily for two days. What was happening in this process was that vegetative (ordinary) bacteria were killed by the one minute of heating, then over the next 12 hours, spores germinated into vegetative forms which could be killed at the next heating. And so the process was repeated until all the bacteria were destroyed.

Spontaneous generation was dead and buried for nearly everyone. Bastian did not concede defeat and went to his grave in 1915 believing in spontaneous generation. This is an excellent example of the physicist Max Planck's maxim of 1955:

> A new scientific truth does not become accepted by way of convincing and enlightening the opposition. Rather, the opposition dies out and the rising generation becomes well-acquainted with the new truth from the start.[253]

Before we leave Pasteur, you may be interested to learn a little more about him. In 1852 he was admitted to the Academy of Sciences at his third attempt by a vote of 36 to 24—in the section of mineralogy! The reason for the difficulty he had in being accepted appears to be the aggressive manner in which he dealt with those who disagreed with him. A fascinating example of such polemic is his reaction to the Franco-Prussian War of 1870–1 in which Paris fell. He returned his honorary doctorate in medicine from the University of Bonn in protest at the bombardment of the Natural History Museum writing:

> as a token of the indignation inspired in a French scientist by the barbarity and hypocrisy of one (Kaiser Wilhelm), who in order to gratify his criminal pride persists in the massacre of two great nations.[244]

Pasteur received a response:

> Sir, the undersigned, the Dean of the Faculty of Medicine in Bonn, is charged with replying to the insult which you have dared to offer to the German nation in the sacred person of its august emperor, King William of Prussia, by sending you the expression of *his greatest contempt*. Dr Maurice Naumann PS Wishing to preserve its archive from pollution, the Faculty returns herewith your libel.[223]

On the other hand, Pasteur was extremely squeamish. One of his assistants, Emile Roux, recounted how the sight of a corpse and proceedings in a mortuary revolted Pasteur and he was often obliged to leave an operating theatre overcome. Despite having a stroke in 1867, which caused a left-sided weakness, Pasteur went on to make many remarkable discoveries, some of which will be dealt with later in this book. Others include his studies on the deterioration of wine caused by bacteria (which led to the process of pasteurization), investigations of diseases of silkworms and methods to prevent them, preparation of a vaccine for swine fever, research on cholera in chickens, and production of a vaccine for anthrax. He summed up the discoveries he and others had made on bacteria in a communication to the Academy of Sciences in 1878 in a paper co-written with Jules François Joubert (1834–1910) and Charles Chamberland (1851–1908) entitled (in translation) 'Germ theory, and its application to modern medicine and surgery', remarking 'It is an alarming thought that life may be at the mercy of the multiplication of these infinitely small beings'.[245]

Figure 110. Joseph Lister (1827–1912).

Now let us return to the early 1860s, when Pasteur first prepared the way for the germ theory of disease. The next important step was taken by Joseph Lister (Figure 110). Lister was born near London. Being the son of a Quaker, he was excluded from the Universities of Oxford and Cambridge so studied medicine at the non-sectarian University College, London.

After graduation he moved to Edinburgh, where he married his boss's daughter Agnes Syme; their marriage was childless but happy. In 1860 he was appointed Regius professor of surgery at Glasgow. In 1865 he heard about Pasteur's 1863 paper *Récherche sur la putrefaction* (Researches on putrefaction) in which Pasteur showed that putrefaction was caused by living microbes. Lister therefore looked for something that would kill these organisms. He knew that the city of Carlisle used carbolic acid to get rid of the smell in the city's rubbish dump and that it also killed worms that were found nearby. He therefore decided to try carbolic acid in a case of compound fracture, that is, a break in the bone where one or more parts of the bone protrude through the skin. The standard treatment was amputation of the affected limb, and this was associated with a mortality of 50 per cent, even though anaesthesia had recently become available. On 12 August 1865, an 11-year-old boy was admitted with a compound fracture of the tibia and fibula (the large bones of the leg between the ankle and knee). Instead of performing an amputation, Lister covered the wound with lint soaked in carbolic acid, and splinted the leg (making it immobile) for several months. The boy recovered with no complications. He repeated his method in 11 patients with only one death and one amputation being required. He reported his findings in 1867:

> In conducting the treatment, the first object must be the destruction of any septic germs which may have been introduced into the wounds, either at the moment of the accident or during the time which has since elapsed.[187]

Lister then tried this treatment on abscesses that had been drained, again with excellent results. Until that time, surgeons kept a special frock coat for work in the operating theatre and in the wards. Naturally this became filthy and bloody but the filthier and bloodier it was the better, for this was an advertisement of the surgeon's busyness and therefore skill. As for washing of one's hands, that was done after the operation!—Semmelweis's practices had not yet reached Britain. Lister then went on to ensure that his surgical wards, which had been among the unhealthiest parts of the hospital, were thoroughly cleaned and kept clean, and was able to write:

> during the last 9 months, not a single case of pyaemia [blood poisoning with widespread abscesses], hospital gangrene or erysipelas [skin infection] has occurred in them.[187]

The editor of the *Lancet* wrote that if Lister's findings were confirmed elsewhere, it would be difficult to overrate his discovery. Not everyone was so supportive. James Simpson, professor of obstetrics at Edinburgh (who had introduced anaesthesia with chloroform for childbirth), writing under the pseudonym of 'Chirurgicus', needled away at Lister saying he did not think it would work, hospitals should be pulled down and rebuilt, and that in any case Semmelweis, not Lister, had thought of it first. Lister never descended to personal squabbles and wrote to the *Lancet*:

> I must forbear from any comment on his (Simpson's) allegations. In the forthcoming numbers of your journal I have arranged to publish, with your permission, a series of papers explanatory of the subject in question, and your readers will then be able to judge for themselves how far the present attack admits of justification.[188]

Lister let the facts speak for themselves.

In 1869 Simpson had a stroke and resigned at the same time as Lister was appointed to the chair of surgery at Edinburgh. Lister's first lecture was a demonstration of 'Pasteur's germ theory'. But still there was tremendous opposition, particularly in England, and even the editor of the *Lancet* had gone over to the other side. Lister realized that he had to enter the lion's den and in 1877 he was appointed professor of surgery at King's College Hospital in London. He was closely watched and after two years, John Wood, the other professor of surgery at King's, adopted Lister's methods. Lister was acclaimed at the International Congress of Medicine in Amsterdam in 1879 and London gave in. In 1897 he was created a baron, Lord Lister, and in 1903 the British Institute of Preventive Medicine was renamed the Lister Institute of Preventive Medicine. A fitting tribute.

Lister always gave credit to Pasteur for both stimulating his own work and for providing a scientific basis for understanding it. In 1874 Lister wrote to Pasteur:

> Allow me to take this opportunity for thanking you most heartily for having shown me, by your brilliant investigations, the truth of the germ theory of putrefaction, and having thus acquainted me with the one principle which can lead the antiseptic system to final success.[189]

At a meeting at the Sorbonne University, Paris in 1892, Lister climbed the stairs with arms outstretched, ready to embrace Pasteur on the occasion of the latter's 70th birthday. This is an example par excellence of how scientists ought to behave towards each other.

Crucial to understanding the roles of bacteria in various diseases were the technological abilities, first, to grow them in culture (that is allow them to multiply) and, secondly, when grown in culture, to separate one type of organism from another. Then it was necessary to find criteria for identifying different bacteria so they could be distinguished from each other.

At first, attempts were made to cultivate bacteria in natural fluids such as urine, milk, and blood, or extracts of vegetables such as hay, turnips, and carrots, but these were not very satisfactory. Artificial fluids were then developed. One of the first was 'Pasteur fluid' used by Pasteur in 1861 which contained water, candy sugar, ammonium tartrate, and yeast ash. Adolf Mayer (p.428) in 1869 used chemically pure solutions of the salts found in yeast ash and his fluid contained water, potassium phosphate, magnesium sulphate, and calcium fluid. Cohn's fluid (1872) was Mayer's fluid to which ammonium tartrate was added. A particularly useful medium was developed by Loeffler in 1881, who used meat extract, peptone, common salt, and sodium phosphate. Unfortunately, many cultures were mixed, that is, they contained two or more different types of bacteria. The holy grail was to obtain pure cultures of a single bacterial type. The first attempts were made by taking a minute quantity of a fluid culture, placing it in a new fluid medium, then repeating the process a number of times hoping contaminating organisms would be removed. This was not successful.

A better technique would be to grow them on solid media. In 1875, Joseph Schroeter (1837–1894) in Germany grew colonies of bacteria which produced pigment on the cut surface of cooked potatoes (Figure 111). He also made solid culture media from starch paste, egg albumin, bread, and meat. Unfortunately, these methods were not helpful for organisms that did not produce pigment and could not be seen easily.

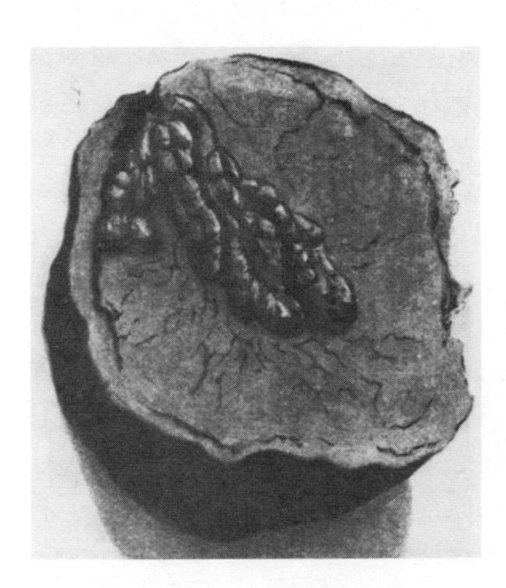

Figure 111. Pigmented bacterial colonies growing on a cut potato.

Robert Koch (p.211) had the first success in solving this problem by developing 'nutrient gelatine' in which 2.5–5 per cent gelatine was added to nutritive fluid which contained meat extract in a concentration of 1 per cent. Gelatine, which is protein made from the partial breakdown obtained from animal skins and bones, had two advantages. First, it provided solidity, as a cook knows when she is making jelly (or in America, jello). Secondly, it was transparent. This meant that if bacteria were lightly spread over nutrient media, each individual organism would multiply over one to two days to form a separate visible mound called a 'colony' which contained just the one type of organism (Figure 112). Koch placed his slabs of nutrient media on sterile glass plates which were then set under a bell jar. This technique opened the door for some of the greatest advances made in medicine, because individual organisms could be isolated, cultivated, and studied both as to their characteristics and disease-causing effects. Koch demonstrated his technique at the International Medical Congress in London in 1881 to a group that included Lister and Pasteur. Pasteur turned to Koch and said: 'C'est un grand progress, Monsieur'[55] (that is great progress, sir). Regrettably, this bonhomie between the two was not to last.

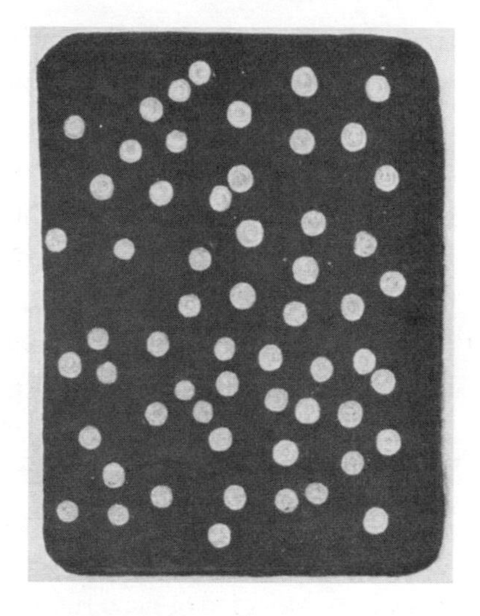

Figure 112. Bacterial colonies growing on gelatine.

One problem with gelatine is that it liquefies at body temperature and a search was made for something better. The answer was provided by Fannie Hesse, the wife of Walther Hesse (1846–1911), who had previously worked with Koch and was now studying bacteria in his home laboratory. She had obtained some agar-agar from Batavia (now Jakarta, Indonesia) from some Dutch friends, where it was used especially for making jams. It is derived from a seaweed (*Eucheuma* species), common in those waters, and had a peculiar property. It had to be heated to 90°C before it melted (and this temperature is enough to kill non-spore-forming bacteria thus making it sterile) but it does not solidify again until the temperature falls to about 40°C, when it sets into a stiff and relatively transparent solid. It was thus possible to prepare it while it was still liquid. Hesse wrote to Koch in 1881 telling him of his discovery and the latter then began to use it. One of Koch's assistants, Richard Julius Petri (1852–1921), developed rounded glass dishes with sides into which liquid agar could be poured and an overlapping lid placed on top. These 'Petri dishes' are still used today except that they are made from plastic and are thrown away after use instead of being re-used.

Over the years it was found that different bacteria had different growth requirements so different media were developed to allow them to grow. The standard culture medium in microbiology laboratories today is blood agar in which blood has been added to agar, but there are many different media including chocolate agar (in which red blood cells have been broken down) which allows certain more demanding bacteria to grow, MacConkey agar which also helps differentiate bacteria according to their ability to break down lactose, and specialized media to permit the growth of bacteria as diverse as those that cause tuberculosis, cholera, or legionnaires' disease.

When bacteria are looked at under the microscope, they are transparent and therefore difficult to see. The first person to attempt to stain bacteria to make them more visible was Hermann Hoffmann (1819–1891), a German professor of botany, who in 1869 used watery solutions of carmine (a red pigment produced by some scale insects (parasites of plants)) and fuchsin (a purplish-pink dye obtained from the flowers of *Fuchsia* plants). Koch in 1877 improved on this by first preparing thin films of bacterial solution on glass slides and then drying them. He stopped them from being washed off the slides by 'fixing them', i.e. attaching them, with alcohol. This process was improved several years later when he heated the slides as well. Koch was surprised to find that this treatment did not change the shapes or sizes of bacteria. He then tried various stains including methyl (= gentian) violet (a fuchsin) and aniline brown (isolated from coal tar). Carl Wiegert (1845–1910), an assistant of Cohn, introduced the staining of bacteria with Bismarck Brown, a diazo dye, in 1878. In 1881, Weigert's cousin, Paul Ehrlich, used methylene blue, a synthetic dye discovered in 1876 as a bacterial stain. This was then modified by making it alkaline by Loeffler in 1884.

Figure 113. Hans Christian Gram (1853–1938).

The most generally useful of all stains, the Gram stain, was developed by Hans Christian Gram (Figure 113), a Danish physician working in Berlin in 1884 in the laboratory of Karl Friedländer at the Berlin-Friedrichshain hospital. He used a stain developed by Ehrlich consisting of aniline, water, and gentian violet but then added Lugol's iodine (iodine in potassium iodide), possibly by mistake (p.256). This caused a chemical reaction to occur which turned the bacteria purple. But the big news was that when they were then treated with alcohol, the stain leached out of some types of bacteria which became transparent whereas it did not from others. A pink stain such as safranine could be added to turn the colourless bacteria pink and thus make them easily viewable. A number of modifications have been made since but this method of Gram marked a watershed. It provided a robust, easy method of dividing bacteria into two groups. Gram used it to differentiate two bacteria which were being touted as causing pneumonia—bacteria now known as *Streptococcus pneumoniae*, which stained purple (and were called Gram-positive), and *Klebsiella pneumoniae*, which stained pink (and were called Gram-negative). As the various bacteria were discovered, this method allowed Gram-positive cocci such as staphylococci and streptococci to be immediately differentiated from Gram-negative cocci such as meningococci and gonococci, all of which will be discussed in the chapters to come. We now know that these staining differences are a result of major differences in the structures of the cell walls of Gram-positive and Gram-negative bacteria.

Then, of course, new equipment was developed that played major roles in cultivating bacteria. Autoclaves were important for sterilizing equipment and preventing contamination, incubators at set temperatures (usually around 35–37°C) allowed bacteria to grow and multiply as if they were in the human body, and devices were invented for growing bacteria in the presence or absence of oxygen or with the addition of carbon dioxide.

As we shall see in the next chapter, the first disease of animals and humans that was shown to be caused by a specific organism was anthrax. Soon afterwards, Robert Koch discovered that tuberculosis was caused by an organism called *Mycobacterium tuberculosis*. Clearly, criteria were necessary to determine which specific organism caused which disease or diseases. These were first laid down by Koch in 1882 in his paper on the cause of tuberculosis and became known as Koch's postulates:

> To prove that tuberculosis is a parasitic disease, that it is caused by the invasion of bacilli and that it is conditioned primarily by the growth and multiplication of the bacilli, it was necessary to isolate the bacilli from the body; to grow them in pure culture until they were freed from any disease product of the animal organism which might adhere to them; and, by administering the isolated bacilli to animals, to reproduce the same morbid condition which, as known, is obtained by inoculation with spontaneously developed tuberculous material.[163]

Friedrich Loeffler (p.297), in a paper dated December 1883 and published in February 1884, put them this way with regard to diphtheria:

> If diphtheria is a disease caused by a micro-organism, it is essential that these three postulates be fulfilled:
>
> 1. The organism must be shown to be constantly present in characteristic form and arrangement in the diseased tissue.
> 2. The organism which, from its behaviour appears to be responsible for the disease, must be isolated and grown in pure culture.
> 3. The pure culture must be shown to induce the disease experimentally.[191]

These days, a fourth point is usually put—the same bacteria must be recovered from the experimentally infected animal (or person). Koch's name was associated with his postulates for the first time in English when the *British Medical Journal* in 1884 published an editorial on the First Cholera Conference held in Berlin:

> Of course, the whole point turns on whether Dr Koch has made out that the comma-bacillus is really the cause of the disease. In order to demonstrate that a given bacterium is the cause of a disease, it must be proved:
>
> 1. that a special bacterium with definite characteristics marking it out from other forms of bacteria, is constantly present in the parts affected
> 2. that this bacterium is present in sufficient numbers to account for the disease
> 3. that it is not similarly associated with other diseases
> 4. that this bacterium can be cultivated apart from the body, and that its introduction into lower animals is followed by the same effects as the introduction of the infective material itself.[16]

The problem with these criteria is that not all human bacterial infections, and cholera is one, can infect animals. You may prefer Loeffler's succinct criteria. They summarize the germ theory of disease in a nutshell.

At least they do for specific bacteria acquired from someone else, some animal, or from the environment which cause specific diseases in humans. These are what we might call 'extrinsic infections' because they come from outside the infected person. But there was another lesson to be learnt—there are also what we might call intrinsic infections, that is, infections of a person with his or her own bacteria. We all have bacteria on our skin and the mucous membranes of the mouth and respiratory tract and in the gut. I still remember when this concept came to me in a flash as a medical student in 1963. A fellow student explained to me that a bacterium called *Escherichia coli* was the commonest cause of urinary tract infections. 'How can that be?' I cried. 'It lives in the bowel and does not do any damage there!' (I was later to learn that that was not strictly true.) 'Ah,' replied my friend, 'these *E. coli* have got into the wrong place.' They had breached the normal defences of the body and entered the urinary tract and caused disease there. Another example is a colleague of mine who squeezed a pimple (which you should never do) and gave himself a life-threatening infection of his bloodstream with the golden staphylococcus which lives on the skin. In hospitals today, particularly following surgery when the barrier of the skin or the gut is breached, our own bacteria are the commonest cause of infections.

Now having set the scene on the germ theory of disease, we shall embark on the saga of the discoveries of the most important bacterial infections. As it happens, most of these are extrinsic infections, many of which happily have now disappeared or are rare, at least in the first world, as a result of improvements in hygiene and sanitation, and the discovery and development of immunizations and antibiotics.

17

Anthrax

SCIENTIFIC NAME:	*Bacillus anthracis*
ORGANISM'S APPEARANCE:	Gram-positive rod
DISEASE NAME:	anthrax
DISTRIBUTION:	worldwide
TRANSMISSION:	inhalation, ingestion, or contamination of cutaneous cuts and scratches with spores or bacilli
LOCATION IN A PERSON:	lungs, gastrointestinal tract, skin
CLINICAL FEATURES:	a. pneumonia, b. severe gastroenteritis, c. septicaemia, d. skin lesions
DIAGNOSIS:	recovery of the organism from sputum, blood, or tissues
TREATMENT:	penicillin or other antibiotics

When most people hear the word 'anthrax', they probably think of Saddam Hussein and weapons of mass destruction. He did not have any, but that was not enough to stop George Bush and Tony Blair going to war. No matter that the British developed anthrax as a biological weapon at the Porton Down research establishment and held it at the Scottish island of Gruinard during World War II! Or that the Americans have it at Fort Detrick, the US biodefence laboratory. Indeed, the five people who died from anthrax powder in letters posted a couple of weeks after the 9/11 Islamic terrorist attacks in New York and Washington in 2001 were victims of a home-grown terrorist. In July 2008, the FBI was about to arrest Bruce Ivins, a scientist who worked at Fort Detrick, when he committed suicide by taking an overdose of paracetamol. The reason why anthrax can be a potent biological weapon will become clear shortly.

Anthrax is primarily a disease of animals although it does infect people from time to time, particularly those in occupations that deal with sheep and cattle and their products such as hides. But none of these is the reason why we shall consider anthrax. Its importance in our narrative lies in the fact that it was the first bacterial infection in which the 'germ theory of disease' was proven—a specific disease was shown to be due to a specific bacterium, and Koch's postulates were fulfilled.

Anthrax is one of the oldest recorded diseases of grazing animals such as cattle and sheep and may be the sixth plague of the Egyptians mentioned in the Book of Exodus in the Bible. Until 100 years or so ago, thousands of animals died from it each year in Europe. The illness it causes in humans depends upon the route of infection. If the bacteria are inhaled, then severe pneumonia results. Ingestion of the bacteria leads to vomiting, abdominal pain, and bloody diarrhoea. Sometimes bacteria enter the skin through a cut or abrasion, and an abscess develops which turns into a nasty black ulcer called an eschar. Pulmonary and gastrointestinal anthrax are frequently fatal but death is less common in cutaneous anthrax.

The scientific investigation of anthrax began in France in 1850. Two Parisian doctors, Casimir-Joseph Davaine (Figure 114) and Pierre Rayer (1793–1867), examined several sheep that died in an outbreak of an illness that was called 'sang de rate' (bloody spleen). In Paris and in Chartres they inoculated blood from the spleens of dead sheep into other sheep, oxen, and horses and found that they all died after two to three days. In the blood of some of these animals Davaine and Rayer found a great number of similar-looking bacteria. Davaine wondered whether these bacteria were the consequence of putrefaction or the cause of it, but favoured the latter. Distraction by other matters put the question to the back of his mind until 1861, when he became aware of Pasteur's report of bacteria fermenting butter and he thought these bacteria looked much like those he had seen in the sheep 11 years earlier. Another two years were to pass before there was another outbreak of sang de rate that he could investigate. A farmer sent him blood from a dead sheep and in it Davaine found the same motionless bacteria he had seen in 1850. He inoculated this blood into two rabbits and a rat. The rabbits but not the rat died within several

Figure 114. Casimir Davaine (1812–1882).

days and in the rabbits' blood he again found the bacteria. The same results were found with the next 12 rabbits he injected with blood containing the bacteria. So Davaine wrote in his report to the Academy of Sciences in 1863 that the bacterium:

> is visible and palpable; an organized being, endowed with life, that develops and propagates as a living being. By its presence and rapid multiplication in the blood, it brings about, in the constitution of this liquid... those modifications from which the infected animal quickly dies.[81]

But how was it propagated? How did it get from one animal to another? More than a decade was to pass before these questions were answered. Enter Robert Koch (p.211), a medical practitioner in the small German-speaking town of Wollstein (now Wolsztyn in Poland). He began studying anthrax as a hobby in 1873, when he was aged 29. He found bacteria just as Davaine had done in the blood of sheep that died during the earlier outbreaks. In April of the following year Koch made his first major discovery when he realized that these bacteria might each form a spore within them (an endospore). A spore is a dormant, tough, and non-reproductive structure which 'hatches' under the right conditions. Bacterial spores had first been discovered by Ferdinand Cohn in 1872 (p. 176). Koch wrote:

> The bacteria swell up, become shinier, thicker and much longer. Slight bends develop. Gradually a thick felt develops. Within the long cells, cross walls appear and small transparent points develop at intervals.[161]

In December 1875, a police officer confiscated the hide of an animal that had died from anthrax and gave it to Koch, who in addition to being a general practitioner was also the public health medical officer. In the blood of this hide, Koch found large numbers of the typical bacteria. He inoculated a rabbit with this blood in the ear and the skin of its back. It died next day on Christmas Eve. He examined it at leisure on Christmas Day and found that the lymph node fluids were teeming with bacteria. He then implanted a piece of tissue containing bacteria into the eye of another rabbit. This animal died four days later and he found bacteria in the blood, spleen, lymph nodes, and the turbid fluid in the eye. This last observation gave him a brilliant idea—to use the fluid from the anterior chamber of the eye (between the cornea (the clear part) in front and the iris (which margins the pupil) and lens behind), known as aqueous humour, for the artificial culture of the bacteria causing anthrax. Thereafter Koch obtained plentiful quantities of aqueous humour from the eyes of cattle at the local abattoir.

This enabled him to study the conditions necessary for ideal growth of these bacteria. They needed temperatures of 30–35°C and they needed oxygen. Getting the right temperature might sound simple, but this was before electricity and automatically monitored electric incubators. Koch had to use a kerosene heater on which he placed dishes with moist sand on which he in turn would put his culture plates. Having achieved success with this, he was able to observe changes in the bacteria with the passage of time in this culture medium.

Koch named these bacteria *Bacillus anthracis*, the species word being a Latinised version of the Greek word *anthrakites* for 'coal' or 'carbuncle'. Whether anthracis was chosen because cutaneous anthrax has large black skin lesions or because the organism causes carbuncles (abscesses) is a moot point. Being long, fat bacilli, they are relatively easy to see. Koch noticed that when they were incubated in aqueous humour for several hours, they lengthened then split forming long chains of bacilli (Figure 115). After 24 hours, a spore developed within each bacillus near its centre giving the appearance of a chain of pearls. Then when resting spores were transferred into a fresh medium of aqueous humour, they germinated to form individual long bacilli, called vegetative forms. With some trepidation, Koch went to visit Cohn in Breslau (now Wroclaw, Poland) and demonstrated his observations to him. Cohn was enthralled and invited Koch to write his findings in a paper to be published in Cohn's journal called (in translation) 'Contributions on the

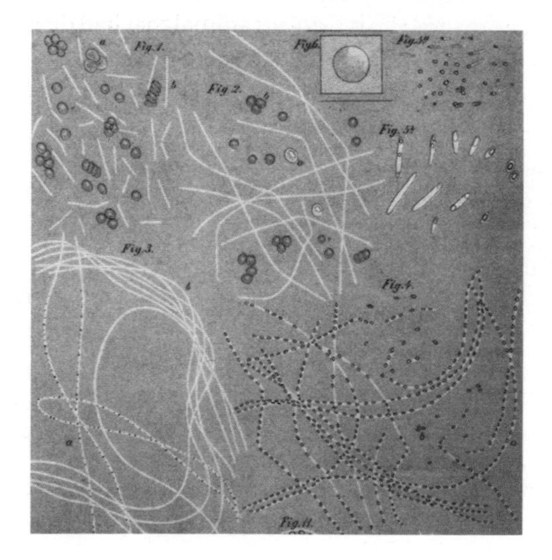

Figure 115. Koch's drawings of anthrax bacteria. At the top left (Fig. 1) are bacilli in guinea pig blood (the round-oval structures are red cells). At the top middle (Fig. 2) are chains of bacilli from a mouse spleen. At the lower left (Fig. 3) are chains of bacilli which are beginning to form spores within them after 10 hours of culture. At the bottom right (Fig. 4) are chains of spored organisms after 24 hours of culture. At the top right are individual spores (Fig. 5a) and spores turning into vegetative bacilli (Fig. 5b).

Biology of Plants'. This Koch duly did and Cohn himself wrote an introduction to Koch's paper in 1876:

> The life history of the anthrax bacillus agrees completely with that of the bacillus of hay infusions [*B. subtilis*]. Indeed, the anthrax bacillus does not have a motile stage, but otherwise the similarity with the hay bacillus is so perfect that the drawings of Koch can service without change for the clarification of my observations [on *B. subtilis*].[74]

Robert Koch had burst onto the world stage, and this was only his first mark, as we shall see in the next chapter on tuberculosis.

Unlike Pasteur and Cohn but like Davaine, Koch was not just a scientist but was also a medical practitioner. As well as the *in vitro* (test-tube) studies, he was interested to observe what happened in animals. Koch repeatedly infected mice caught in his barn, with one infected animal being used to infect the next, for a series totalling 20 mice. In every case, the spleen was engorged and full of the same bacilli but they had no spores in the living or newly dead animal.

It was the spores that made it possible to understand the epidemiology of anthrax, for many had been puzzled as to why there were occasional outbreaks. Spores were deposited on the ground and remained dormant until eaten on grass or other vegetation by herbivorous animals. In fact, it would be shown over a century later that these spores would remain viable for 100 years! Koch had proven that a specific bacterium caused a specific disease and was able to write:

> Anthrax tissues, regardless of whether they are relatively fresh, putrefying, dried or years old, can produce anthrax when these substances contain bacilli capable of developing spores of Bacillus anthracis. Thus all doubts regarding the Bacillus anthracis as the actual cause of the disease must be dispelled. Bacillus anthracis is the contagion of anthrax. The transmission of disease in fresh blood occurs rarely in nature, only in persons who come in contact with blood or tissue juices while killing, cutting and skinning animals infected with anthrax...But the great percentage of infections is produced only by the penetration of the spores of Bacillus anthracis into the animal body.[162]

In 1877, Louis Pasteur took up the investigation of anthrax. Over the next few years he confirmed Koch's observations, recommended to the French Ministry of Agriculture that animals which had died from anthrax never be buried in fields that were subsequently going to be used for grazing, and developed a vaccine by making the bacteria less virulent (dangerous). By 1882, Koch had begun to aggressively oppose Pasteur's observations, not only on anthrax but also other infections. Nevertheless, Koch refused to take part in any discussions on the matter and often had his facts wrong in the first place. Both Koch and Pasteur were strong-willed, aggressive individuals, neither spoke the other's language well, and their psyches were rooted in Franco-German antagonism. They were not above writing vitriolic comments about each other in learned journals. As the years went on, the controversy subsided although the antagonism never completely disappeared. What a contrast to the relations between Pasteur and Lister. Nevertheless, with this infection, medical and veterinary bacteriology had come of age.

18

Tuberculosis (consumption)

SCIENTIFIC NAMES:	*Mycobacterium tuberculosis* and *M. bovis*
ORGANISM'S APPEARANCE:	rod which does not stain with the Gram stain but stains pink with the Ziehl–Neelsen stain
DISEASE NAME:	tuberculosis
COMMON DISEASE NAME:	TB (previously phthisis then consumption)
DISTRIBUTION:	worldwide
TRANSMISSION:	inhalation of bacteria (usually); ingestion of contaminated milk (occasionally)
LOCATION IN A PERSON:	many tissues, most commonly the lungs
CLINICAL FEATURES:	a. pulmonary TB: cough, shortness of breath, coughing of blood, weight loss, tiredness, b. TB outside of the lungs: huge range of symptoms depending upon the organs involved
DIAGNOSIS:	isolation of *Mycobacterium* from sputum or sometimes urine, faeces, or tissues
TREATMENT:	combination of antituberculosis antibiotics, e.g. rifampicin, isoniazid, ethambutol, pyrazinamide
PREVENTION:	good ventilation and limited contact with infected people

My father shared a bedroom with his older brother. One day in about 1935, aged 20, my uncle coughed up blood. After 12 years of misery, he died of renal failure due to tuberculosis. All that my father suffered was a calcified spot on the lung seen when his chest was X-rayed. Such is the capricious nature of this awful illness.

Symptoms consistent with what we now call tuberculosis were well known to the ancient Greek physicians. The most prominent of these was marked loss of weight, so they called the illness *phthisis*, which is the Greek word for 'wasting'. Hippocrates (p.116) recognized the disease, noting that it occurred most commonly between the ages of 18 and 35 years. In his *On Epidemics* he explained that it had a gloomy prognosis:

> Early in the beginning of spring, and through the summer, and towards winter, many of those who had been long gradually declining, took to bed with symptoms of phthisis; in many cases formerly of a doubtful character the disease then became confirmed. Many, and, in fact, the most of them, died; and of those confined to bed, I do not know if a single individual survived for any considerable time...they continued throughout spitting crude matters.[137]

Of course, conditions other than tuberculosis might have produced these symptoms, but tuberculosis was undoubtedly prominent among the various causes. The Roman AC Celsus (p.6) regarded the lung form of wasting disease as being particularly dangerous and described a (useless) diagnostic test:

> The third species [i.e. form of disease], which the Greeks call phthisis, is the most dangerous by far. The malady usually arises in the head, thence it drips into the lung; there ulceration supervenes, from this a slight feverishness is produced, which even after it has become quiescent nevertheless returns; there is frequent cough, pus is expectorated, sometimes blood-stained. When the sputum is thrown upon a fire, there is a bad odour, hence those who are in doubt as to the disease employ this as a test.[68]

Galen (AD *c.* 129–*c.* 200), the most influential Graeco-Roman physician of his day and for countless succeeding generations, regarded blood in the sputum as the most important early sign and recognized fever as a poor prognostic omen. He defined phthisis as 'ulceration of the lungs, thorax or throat, accompanied by a cough, fever, and consumption of the body by pus'.[108] It was this word 'consumption' that became the lay equivalent of the medical label 'phthisis' and was commonly used until it was largely replaced by 'tuberculosis' in the nineteenth century.

The organism that causes tuberculosis can affect many different parts of the body and produce an extraordinary variety of symptoms and signs depending upon whether it is the lungs, the lymph nodes of the neck or armpit, the bowel, the brain, the kidneys, the bones and joints, or elsewhere that are involved. The most commonly involved organ is the lungs and it is disease of this organ to which Hippocrates, Celsus, and Galen were referring.

Another form of tuberculosis which excited considerable attention was the 'King's Evil'. This was because this illness was both common and easily

Figure 116. Franciscus Sylvius (1614–1672).

recognized. Lymph nodes (= glands), especially in the neck, but sometimes the armpits became swollen, red, and slightly tender. This often caused the overlying skin to break down forming an ulcer which took months or years to heal. These features led to its other name of 'scrofula', which was medieval Latin for 'swelling of the glands'. The extraordinary name of King's Evil derived from the belief in Europe from as early as the fifth century that the touch of the king could cure the malady. Philippe II Auguste of France (reigned 1180–1223) gave 1,500 touches in a single ceremony, while Edward I of England (r. 1272–1307) touched 533 sufferers in one month. Charles II of England (r. 1660–1685) touched 92,102 people during his reign. Even as late as 1886, several years after the discovery of the cause of tuberculosis, the practice was being continued by Franz Joseph I of Austria-Hungary (r. 1848–1916). Of course, it did not work and one can only consider the practice as an incredibly persistent mass delusion. Certainly it did not prevent England's Edward VI (1537–1553) and France's Charles IX (1550–1574) from being scrofulous and probably dying from tuberculosis.

It was impossible to begin to sort out the confusing array of symptoms and signs and the connections between them until patients who died from phthisis were subjected to post-mortem examinations. The person who set this ball rolling was Franz de le Boë (Latin name Franciscus Sylvius, Figure 116), a Dutch physician. He was the first to recognize what he called 'glandulous tubercles' in the lungs of patients with phthisis and noted that they increased in size with age and eventually suppurated (produced pus):

> I found more than once larger and smaller tubercles in the lungs, which on section were found to contain pus. From these tubercles I hold that not infrequently phthisis has its origin. Only the wasting caused by ulcers in the lungs is to be called phthisis [particularly when it is associated with cough, sputum and hectic fever].[301]

Here we have perhaps the first relatively modern use of the word 'tubercle', which means a small, firm, rounded swelling or nodule. It is derived from the Latin word 'tuberculum', which in turn is a diminutive (indicating small) of the

Latin word 'tuber' meaning 'hump' or 'swelling'. Sylvius said that there had to be two features present in the lungs which were affected by phthisis—tubercules or small nodules as well as ulcers. Sylvius mistook tubercles for degenerations of invisible glands which Galen had postulated were present throughout the lungs, but he did recognize similar tubercles in neck glands affected with scrofula.

Thomas Willis (1621–1675; famous for his description of the arrangement of arteries at the base of the brain still known as the Circle of Willis), an English pupil of Sylvius, showed that ulceration was not invariable. In his *Practice of Physick* published in 1684, Willis wrote:

> I have opened the dead bodies of many that have died of the disease in whom the lungs were free of any ulcers, yet they were set about with little swellings or stones or sandy matter throughout the whole... wherefore a Phthisis is better defined, That it is a withering away of the whole body arising from an ill formation of the Lungs.[321]

These little swellings, stones, or sandy matter were the tubercles of Sylvius.

Even though Willis realized that ulcers are not invariable in phthisis, ulceration, or as we would say cavitation, is a prominent feature of pulmonary (lung) tuberculosis and it is important to understand what these terms mean. The lungs are a bit like the roots of a plant. They begin with the trachea (windpipe) which breaks into two smaller branches called bronchi (singular: bronchus), which in turn break into smaller and smaller bronchi until they become the smallest tubes called bronchioles. At the end of each bronchiole are a number of air sacs called alveoli (singular: alveolus), where oxygen from the air and carbon dioxide from the body are exchanged. When Sylvius and Willis cut up lungs from patients with phthisis, they might see true ulcers in the larger airways (bronchi) in which the lining is disrupted, or they might cut across a large group of alveoli which were destroyed and appeared to be ulcerated. They did not have the microscopical techniques available to enable them to appreciate the differences. More importantly, the advent of chest X-rays would clearly show that there were tuberculous cavities in the lungs (Figure 117).

As more and more autopsies on patients with phthisis were performed, new observations were made. In 1700, Jean Jacques Manget (1652–1742), a Swiss physician, describing a case of a 17-year-old lad who had died of phthisis, first coined the term 'miliary' tuberculosis as the lesions resembled numerous millet seeds (from the Latin 'milium' meaning 'millet'). He found yellowish tubercles 1–2 mm in size throughout the lungs, liver, spleen, and mesentery (the main

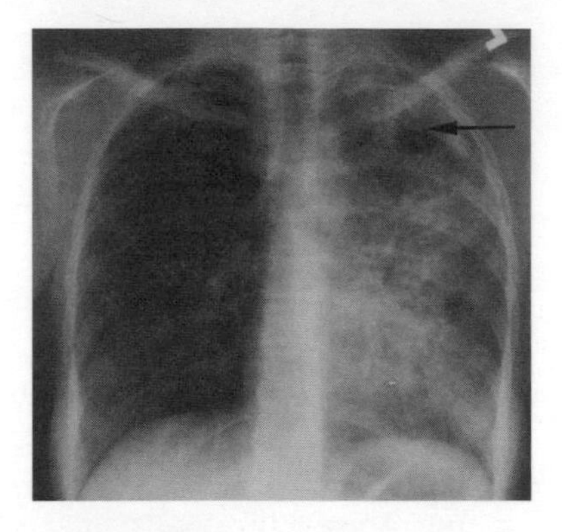

Figure 117. Chest X-ray showing infiltration with tuberculous material (the white shadows) and a cavity with air in it (arrow) formed by the breakdown of tissue in the upper lobe of the left lung.

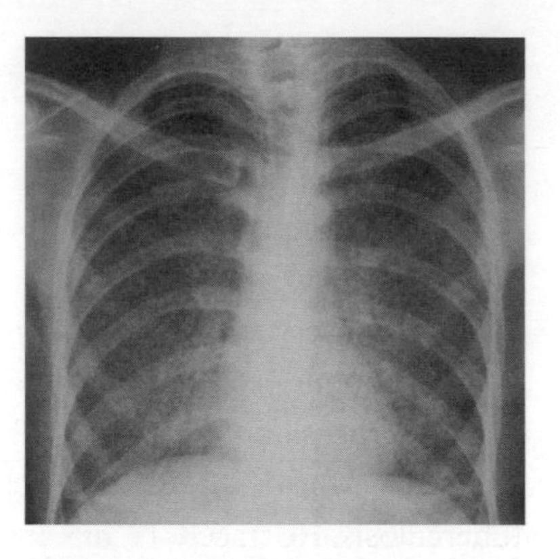

Figure 118. Chest X-ray showing tiny white opacities = tubercules scattered throughout the lungs (miliary tuberculosis).

membrane in the abdominal cavity). A chest X-ray of such a case is shown in Figure 118.

The Englishman William Stark (1741–1770) made important observations during the last five years of his short life on a mere nine patients. He noted that the appearances of the lungs in phthisis, although differing in degree in different patients, were of a similar nature, merely reflecting the speed of evolution of the pathological processes. He was the first to examine minutely the growth and development of tubercles and showed how they could lead gradually to advanced disease and death:

> In the cellular substance of the lungs are found roundish, firm bodies (named tubercles) of different sizes, from the smallest granules to about half an inch in diameter... The tubercles of a small size are always solid... of a whitish colour and... when cut through the surface appears smooth, shining and uniform... The cavities in different tubercles are of different sizes, from the smallest perceptible to... three quarters of an inch in diameter.[295]

Similar observations were made by Gaspard-Laurent Bayle (1774–1816), a French doctor who, beginning in 1802, performed 900 autopsies. He too observed millet-like tubercles and found them to be the beginnings of all the lung changes in phthisis. He said that tubercles may occur in patients prior to appearance of symptoms and correlated them with the formation of cavities. Bayle noted disseminated miliary tuberculosis and extrathoracic lesions in the intestine, lymph nodes, and larynx. Most importantly of all, he realized that the pus described by Sylvius in tubercles was not like the viscous pus seen in a boil in the skin but rather resembled cheese, and that this cheesy material, called caseation, was specific for tuberculosis in whichever organ or tissue it was found. The name caseation (adjective caseous) was derived from the Latin word 'caseus' meaning 'cheese'.

Bayle's observations had a profound influence on his friend and colleague René Theophile Hyacinthe Laënnec (Figure 119), who took the observations of Bayle on the pathology of the tubercle as the foundation for his own work. Laënnec performed more than 400 autopsies on patients who had died from tuberculosis. He traced the disease from the minute tubercle through all of its pathological aspects and, using the tubercle as a diagnostic characteristic, he recognized the disease in all the organs of the body. By finding this characteristic tubercle wherever tuberculosis existed in the body, Laënnec demonstrated that tuberculosis, whether it occurred in the lungs or outside the lungs, was really one disease. Although his conclusions were severely criticized by some, time was to prove him absolutely correct. As well as his pathological studies, Laënnec invented the stethoscope so that by listening to breath sounds with his instrument, he was able to develop a correlation between the clinical signs heard during life in the lungs with those found at post-mortem examination in death. Unfortunately, he was to fall victim to the disease he studied so assiduously.

Figure 119. René Laënnec (1781–1826). He looks gaunt perhaps presaging his death from TB.

To many people, the tubercle (Figure 120) was viewed as the fundamental pathological basis of phthisis so in 1839 Johan Schönlein (p.84), professor of medicine in Zurich, suggested that phthisis should be renamed tuberculosis. This suggestion has stood the test of time although it had to withstand the attacks of Rudolf Virchow (1821–1902), Germany's foremost pathologist, who did not believe that caseating tubercles were peculiar to tuberculosis. Virchow eventually had to concede defeat when his fellow countryman, Robert Koch, demonstrated the bacterial cause of this disease.

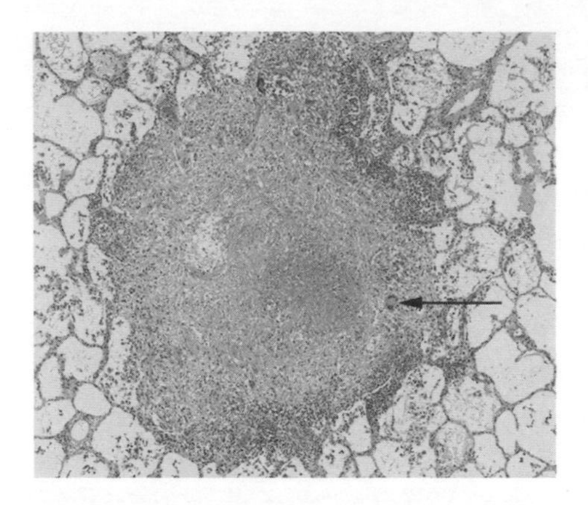

Figure 120. Histological section of a tubercle. The open chambers around it are normal alveoli (air sacs). The tubercle itself consists of a mass of inflammatory cells with dead tissue (caseation) in the centre. The arrow points to a Langhan's giant cell which is characteristic of tuberculosis.

In Laënnec's time, something approaching one in a hundred people in Europe died from tuberculosis each year. Famous people who appear to have succumbed to tuberculosis include writers the Brontë sisters (aged 29–39), DH Lawrence (44), George Orwell (46), and Robert Louis Stevenson (44); poets Elizabeth Barrett Browning (55), Robert Burns (37), John Keats (25), and Alexander Pope (56); musicians Frédéric Chopin (39) and Henry Purcell (36); and political leaders Louis XVII of France (10), Pedro I of Brazil (35), and Simon Bolivar (47).

But how was this disease acquired? Transmission by infection seemed so likely that in 1751 Ferdinand VI of Spain issued a decree requiring the reporting of all cases of pulmonary phthisis to health authorities, with rigid disinfection of houses and burning of clothes and bedding after the deaths of patients. Similar laws were soon enacted throughout southern Europe and remained in effect for most of the remainder of the eighteenth century. The public became so oppressed by the strict enforcement of such laws that a school of thought which believed in the inheritance of phthisis began to flourish and the laws were revoked. Indeed Laënnec wrote that in France at least, phthisis did not appear to be contagious.

In this, Laënnec would be proven to be comprehensively wrong. The first step was taken by examining the distribution of phthisis. Everyone knew that phthisis tended to afflict members of particular households and this supported the inheritance theory. On the other hand, differences in the distribution between different populations suggested the opposite. William Budd (p.330), a physician in Bristol, England, learnt by letter from the celebrated African explorer Dr David Livingstone that phthisis appeared to be unknown in the interior of the continent but that on the seaboard where Africans came into contact with whites, the disease became rampant among them. Africans working aboard British ships and who contracted pulmonary phthisis were often sent to Bristol for treatment where Budd encountered them. Whether or not Livingstone's assertion was actually correct, Budd believed it and came up with an alternative proposition. One day in 1856 while out walking, he pondered these things and came up with the answer. Being a cautious man, though, he waited 11 years before he wrote to *The Lancet*:

> phthisis . . . is disseminated through society by specific germs in the tuberculous matter cast off by persons already suffering from the disease.[60]

Perhaps Budd had heard about the work of Villemin (to be discussed shortly) and it was this that stimulated him to put pen to paper. Even so, the reception to Budd's suggestion was frosty.

A more important and definitive step than Budd's was made by transferring tuberculosis from one animal to another, even if the cause of the tuberculosis itself was unknown. This appears to have first been achieved in 1846 in Berlin, Germany when Philipp Klencke (1813–1881) injected tissue containing miliary tubercles into the jugular vein of a rabbit. Six weeks later extensive tuberculosis was present in the spleen and liver thus showing the disease could be transmitted experimentally. Unfortunately, no-one paid any attention to Klencke's work, even when it was confirmed in France 20 years later by Villemin.

Jean Antoine Villemin (Figure 121) had trained for three and a half years to become an army doctor. This allowed him to treat soldiers and horses but not civilian patients. Villemin noticed the similarities between tuberculosis and glanders, a disease of horses with which he was quite familiar. This illness, now known to be due to a bacterium called *Burkholderia mallei*, occurred in both an acute and a chronic form (called farcy) which affected the lungs, bones, joints, lymph nodes, and skin. Moreover, it was known that glanders could be transmitted by handling diseased horses and inoculated from one horse to another. This gave Villemin his brilliant idea: 'Could tuberculosis be transmitted experimentally?'

Major Villemin approached his colonel, who allocated him a corner of a laboratory and a few animals for experimentation. In March 1865, he took some caseating pus from the lung of a tuberculous patient who had died the day before and introduced the material through incisions in the skin of two rabbits. He understood the importance of controlling his experiments so he inoculated blister fluid from a burn in a non-tuberculous patient into another two rabbits.

Figure 121. Jean Villemin (1827–1892).

The rabbits were killed two months later. There was nothing wrong with the control rabbits but the first two rabbits had widespread tuberculosis in the lungs and lymph nodes. Thus encouraged, he repeated his experiments on a number of occasions. In addition, he undertook another series of similar experiments using tuberculous material derived from the udder of an infected cow. He found that the bovine material produced a more acute and severe disease in rabbits than did tuberculous tissue of human origin. Villemin was left in no doubt that tuberculosis was a specific infection transmitted by an agent present in tuberculous tissue.

Villemin was a meticulous and careful investigator. In December 1865 he presented his results to an audience of elderly and successful clinicians at the French Académie de Médecine. You would think that everything was cut and dried and he would be received with acclaim and approbation. The reality was tragically different. The audience listened in polite silence to the 'horse doctor' then merely offered a tepid vote of thanks for an intriguing contribution to veterinary science! Their minds were closed and Villemin's talk went completely over their heads. All they were thinking about was the next item on the agenda, which was the presentation of a 10,000-franc prize to Professor Jean Pidou, who had published a large work called *Studies on Phthisis*, which the Academy believed established beyond any reasonable doubt that phthisis was inherited. To add insult to injury, the Academy refused to publish Villemin's paper.

Undeterred, Villemin continued his researches. Over several years he accumulated evidence that tuberculosis could be transmitted to rabbits, guinea pigs, dogs, and cats. He found that caseous material, tubercles, sputum, and sometimes blood contained the virulent principle. Moreover, material taken from scrofulous lymph nodes caused tuberculosis in guinea pigs and rabbits. Thus Villemin demonstrated beyond doubt that tuberculosis was caused by a germ of uncertain nature and he proved experimentally that Laënnec was correct when he said that scrofula and phthisical lung disease were just different presentations of the one condition. In 1868, Villemin published privately his book entitled (in translation) *Studies on Tuberculosis.*[312] This earned him the mild rebuke from his commanding officer that he was a little too inclined to seek the limelight!

Villemin might have been ignored in France but not so in England. *The Times* newspaper considered his claims bizarre but nevertheless declared that they could not be wholly ignored. The British government sent Dr John Burdon-Sanderson (1826–1905), medical officer to the Privy Council, to meet Villemin. He was sufficiently impressed to repeat Villemin's experiment in collaboration with Dr John Simon (1816–1904). They too succeeded in transmitting tuberculosis to a large number of rabbits and guinea pigs by inoculating tuberculous material from a patient who had recently died. Unfortunately, for reasons that are unclear (no doubt a laboratory error of some sort), one of the control animals injected with tissue fluid from a non-tuberculous patient also became infected. This was a serious setback because Burdon-Sanderson and Simon reported back to the Privy Council that the disease could not be regarded as being necessarily dependent upon a tuberculous source. It needed the experiments of Robert Koch to undo this damage.

In the middle of 1881, Robert Koch (Figure 122) began to look for a bacterial cause for tuberculosis. The extraordinary thing was not only that he isolated the organism, but the incredible speed with which he did it. Heinrich Hermann Robert Koch had been born in Clausthal, Germany, the son of a mining official. He studied medicine at the University of Göttingen, graduating in 1866, served in the Franco-Prussian War of 1870, and then became district medical officer in Wollstein (now Wolsztyn) in Prussian Poland. Following his successful investigations of anthrax, he was appointed to the Imperial Health Office in Berlin in 1880.

Figure 122. Robert Koch (1843–1910).

Now Koch turned his attention to tuberculosis. He had access to extensive pathological material from the 'phthisis ward' of the Berlin Charité Hospital. His original material came from a 32-year-old labourer, Heinrich Günther, who died after a rapid one-month illness; at autopsy, his body was found to be riddled

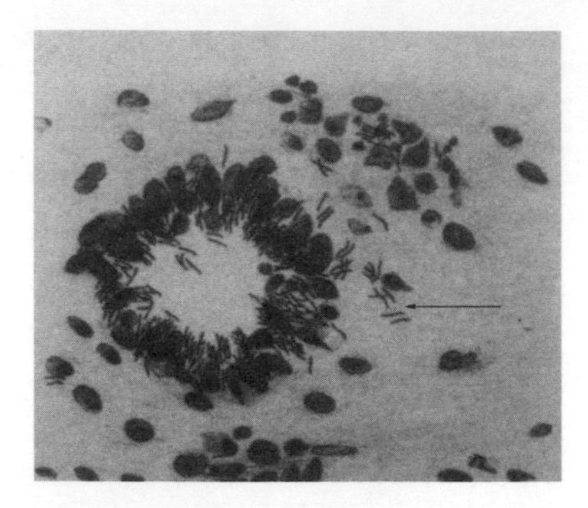

Figure 123. Koch's drawings of tubercle bacilli (arrow) stained with methylene blue. The circle of round structures is formed by the nuclei of a Langhan's giant cell.

with tubercles. First, Koch reproduced the findings of Villemin that the disease was transmissible. He inoculated numerous guinea pigs then rabbits with tuberculous material and saw the same pathological lesions develop in these experimental animals as in humans.

The next logical step was to try to stain any bacteria that were in these tissues so he spread tissue out on glass slides then stained them with methylene blue. He saw very tiny thin rods in tuberculous tissue but not in control material (Figure 123). He then stained the tissues a second time with another dye, vesuvin (Bismarck Brown), and found that the bacteria retained their blue colour and were easy to see against the brown background. What is more, other bacteria did not take up this stain, making it a good differentiator for bacteria in tuberculous material. He realized immediately that there must be something special about the properties of the wall of this particular bacterium and found that methylene blue can only penetrate it when an alkali (in the form of ammonia) is associated with the dye. Koch wrote:

> The bacteria...are very thin and are only about one-fourth to one-half as long as the diameter of a red blood cell, but can occasionally reach a length as long as the diameter of a red cell. They possess a form and size which is surprisingly like that of the leprosy bacillus.[163]

Not only did Koch find these bacteria in experimental animals but also in human tuberculous tissue. He noted that it must seem surprising that they had not been found before but explained that the bacilli were small in size, sparse in numbers, and difficult to see without special stains. He therefore concluded:

> On the basis of my extensive observations I consider it proven that in all tuberculous conditions of man and animals there exists a characteristic bacterium

> which I have designated as the tubercule bacillus which has specific properties which allow it to be distinguished from all other organisms.[163]

Koch realized that this did not prove that the bacterium was the cause of the disease. The next obvious step was to try to culture the organism in the laboratory. Koch had spent six years developing culture methods for bacteria in wound infections and anthrax, and earlier that year had perfected his plate technique and probably thought this would be a simple matter. Unfortunately, the bacterium grew very slowly, would not grow at all at room temperature, and the gelatine liquefied at body temperature. After many trials, he found that if he used serum from blood and coagulated it, then the serum provided a source of nutrients as well as a solid medium in test tubes which he kept in a slanted position. When he inoculated these slants with tuberculous material and incubated them at 37°C, he found that after two weeks, very tiny colonies appeared. At last, he was able to obtain pure cultures of the bacilli.

The next step was to infect experimental animals and see what happened. Koch was fortunate that tubercle bacilli, unlike some other bacteria, infect and produce disease in a variety of animals. He used guinea pigs, which as it happens are very susceptible. He found that when they were injected with tuberculous material, they became progressively weaker and died after four to six weeks. When their organs were examined, the recognizable features of tuberculosis were present. Koch was thus able to report:

> The results of a number of experiments with bacillus cultures inoculated into a large number of animals, and inoculated in different ways, all have led to the same results... All of these facts which are present in the tuberculous substances not only accompany the tuberculous process, but are the cause of it. In the bacillus we have, therefore, the actual tubercule virus.[163]

In the space of seven short months, Koch injected tuberculous material into guinea pigs and found the characteristic pathological changes of tuberculosis, he identified a novel bacterium with unusual staining properties in both human and animal specimens, he grew the bacilli in laboratory test tubes in pure culture, then used these organisms to once again produce tuberculosis in experimental animals. He announced his discoveries to the Berlin Physiological Society on 24 March 1882 in a lecture entitled 'On Tuberculosis'. He brought with him his supporting materials—microscope, 200 microscopic preparations, test tubes, plates, flasks, and pathological material preserved in alcohol. His colleague, Loeffler, recorded of that evening:

> Koch was by no means a dynamic lecturer who would overwhelm his audience with brilliant words. He spoke slowly and haltingly, but what he said was clear, simple, logically stated – in short, pure, unadulterated gold. With increasing excitement, the audience followed every step of his work, examined his excellent preparations, and followed the logic of his experiments. It was quite an evening![55]

Paul Ehrlich, one of the most illustrious of the attendees, was later to remark 'I hold that evening to be the most important in my life'.[55] Three weeks later, Koch's manuscript was published in the *Berliner klinische Wochenschrift*.

It was a sensation. Less than two weeks later, John Tyndall (p.186) published an English summary of Koch's work in a letter to *The Times* in London. Within a few weeks, Tyndall's letter crossed the Atlantic and was republished in the *New York Times*. Koch had predicted that it would take a year of hard work to convince the medical world of the validity of his work. He was wrong. It took only a few weeks. Pathologists who had described the various manifestations of tuberculosis in the human body were now able to find the same bacterium in all of these conditions.

The very night after Koch's lecture, Ehrlich showed that the staining properties of methylene blue were greatly improved if aniline (= phenylamine = aminobenzene, the simplest of the aromatic (smelly) amines) instead of a mineral alkali such as sodium carbonate were added. Ehrlich then found that the dye could not be washed out with acids, i.e. the dye was stuck fast to the organisms. This led to the term acid-fast bacilli, or in short-hand jargon AFB, which is still used today, to describe any organism with this property and includes not only the bacterium we now know as *Mycobacterium tuberculosis*, but also related bacilli including the bacteria that cause leprosy. Shortly thereafter, Franz Ziehl (1857–1926), a doctor in Heidelberg, Germany, found that carbolic acid worked better than did the aniline used by Ehrlich. A little later, Friedrich Neelsen (1854–1894), a medical practitioner in Rostock, Germany, introduced the use of the dye basic fuchsin as the first stain, then Ziehl's preparation of methylene blue as the counterstain. This became known as the Ziehl–Neelsen stain used in microbiology laboratories all over the world to the present day.

Naturally, Koch canvassed the means by which tuberculosis was transmitted. Correctly, he remarked:

> The great majority of all cases of tuberculosis begin in the respiratory tract, and the infectious material leaves its mark first in the lungs or in the bronchial

> lymph nodes. It is therefore, very likely that tubercle bacilli are usually inspired with the air, attached to dust particles. There can hardly be any doubt about the manner by which they get into the air, considering in what excessive numbers tubercle bacilli present in cavity contents are expectorated by consumptives and scattered everywhere.[163]

It is right and proper to add that Paul von Baumgarten (1848–1928), a physician in Königsberg, Prussia (now Kaliningrad, in the exclave of the Russian Federation), saw tubercule bacilli in unstained preparations of tuberculous tissue and actually reported his observation to the Faculty of Medicine in Königsberg on 18 March 1882, four days before Koch's lecture. But von Baumgarten did not stain the bacilli, did not grow them in pure culture, and did not reproduce disease by injecting experimental animals. The credit rightly belongs to Robert Koch.

In June 1882, Kaiser Wilhelm I appointed Koch as an Imperial Privy Councillor, increased his salary, augmented research funding, and gave him additional assistants in the Imperial Health Office. This was indeed the high point of his career and was to be reflected in him being awarded the Nobel Prize in 1905 for his discovery. In 1883, Koch found the bacterium that caused cholera, although he was never able to prove that it caused disease by using 'Koch's postulates'. In 1885 he was appointed professor of hygiene in the University of Berlin. His big let-down came in 1890, when he caused great excitement by announcing that he had discovered the cure for tuberculosis: an extract of the bacteria called 'tuberculin'. It was a total fizzer. In 1891, he became director of the new Prussian Institute for Infectious Diseases (eventually renamed the Robert Koch Institute). He retained this position until 1904, when he travelled the world studying infectious diseases in German East Africa, South Africa, India, the Dutch East Indies, and German New Guinea.

Koch was not as successful in his personal life. In 1890, at the same time as the tuberculin disaster, he began a passionate affair with a 17-year-old girl, Hedwig Freiburg, 30 years his junior. In 1893 he divorced his long-suffering wife, Emmy née Fraatz (who had agreed to marry him in 1867 provided he jettisoned his secret ambitions of exploring the South Sea Islands and winning the Iron Cross in battle), then married Hedwig. What sort of a man was Robert Koch? Naturally, like everyone else, he had his champions and his detractors. Perhaps a reasonable balance has been provided by the eminent malariologist, Cyril Garnham:

> Robert Koch had a hard life; he possessed an arrogant spirit which led to quarrels and he was intensely jealous of his contemporaries, and particularly of Pasteur. But his travels in the tropics in the last years of his life were, no doubt, both a solace and a novelty. He died, prematurely, of coronary thrombosis at the age of 67.[112]

There is no doubt that Robert Koch, whatever personal failings he may have had, was a great benefactor of mankind, for 60 years or so after his seminal discovery, the first effective antituberculous drugs began to appear and millions upon millions of lives have been saved.

There are several points left to tidy up. First, in 1883, the German botanist Friedrich Wilhelm Zopf (1846–1909) named 'Koch's bacillus' as *Bacterium tuberculosis,* the species name referring to the organism's ability to produce tubercles. In the following year Klein renamed it *Bacillus tuberculosis.* Finally in 1896, Karl Lehmann (1858–1940) and Rudolf Neumann (1868–1952) raised a new genus, *Mycobacterium,* and transferred this organism into it as *Mycobacterium tuberculosis.* The genus name is derived from the Latin prefix 'myco-', which means both 'fungus' and 'wax', and here refers to the waxy materials that compose parts of the bacterial cell wall and give its acid-fast nature.

Second was clarification of the way in which tuberculosis develops. Not only can *M. tuberculosis* affect different organs, but it tends to appear in two different age categories, the young and the elderly. When a person is first infected with tubercle bacilli, he or she either develops a progressive clinical disease, which is called primary tuberculosis, or a small tubercle develops which gradually calcifies. This is called a Ghon's focus after Anton Ghon (1866–1936), an Austrian pathologist who described the phenomenon. It was long believed that in the elderly, a new reinfection from outside the body was necessary to cause clinical progressive lesions. In the 1920s in France, Lėon Bernard (*c.* 1872–1934) realized that in the elderly, tuberculosis was mostly due to a reactivation of an old tubercle which had lain dormant for decades but was reactivated as the immune system which had kept it in check for decades gradually waned.

Third was the delineation as a separate species of *Mycobacterium bovis* derived from cattle. You will remember that Villemin infected guinea pigs and rabbits with tuberculous material derived from the udder of a cow and showed that this infection developed more quickly. In 1868 Auguste Chauveau (1827–1917) in France and Edwin Klebs (p.296) in Prague proposed that meat could transmit bovine tuberculosis. In the 1870s, the veterinarian Andreas Gerlach (1811–1878) conducted many experiments that demonstrated that milk and flesh from tuberculous cows could transmit tuberculosis to other animals to which they were fed. In 1898 Theobald Smith (1859–1934) in Boston, USA, showed that the bovine bacillus could be clearly differentiated from the human variety both in its microscopic and cultural characteristics and in the disease it produced in different animals. It was thus apparent that humans could acquire tuberculosis, particularly of the gut, by drinking cow's milk infected with this organism. In 1907, Lehmann and Neumann named it as a type of M. *tuberculosis*, *Mycobacterium tuberculosis typus bovinus*, then in 1934 David Bergey (1860–1937) and colleagues called it *Mycobacterium tuberculosis var. bovinus*. Finally, in 1970, Alfred Karlson and Erwin Lessel gave it its own specific status as *Mycobacterium bovis*.

Bovine tuberculosis was dealt with speedily by the introduction of pasteurization (gentle heating) of milk which killed the bacilli. The introduction of mass chest X-ray campaigns to detect infected people, immunization with a vaccine called BCG (an avirulent form of M. *bovis* developed by Albert Calmette (1863–1933) and Camille Guérin (1872–1961) which is controversially partially effective), and the development of effective antituberculous drugs led to the decline of tuberculosis in Western countries, at least as a public health problem, and the closure of sanatoria. With the appearance of AIDS in more recent times, which impairs or destroys the abilities of the immune system, tuberculosis on a global scale has once more reared its ugly head.

19

Leprosy (Hansen's disease)

SCIENTIFIC NAME:	*Mycobacterium leprae*
DISEASE NAME:	leprosy or Hansen's disease
DISTRIBUTION:	mostly in the tropics
TRANSMISSION:	uncertain—probably mostly by inhalation
LOCATION IN A PERSON:	skin, nerves, and internal organs
CLINICAL FEATURES:	skin lesions, nerve paralyses, anaesthesia, loss of fingers and toes, eye damage
DIAGNOSIS:	bacteria and/or characteristic lesions seen on biopsy, usually of skin lesions
TREATMENT:	rifampicin + dapsone ± clofazimine
PREVENTION:	no reliably effective preventive measures are available

I looked at the little Catholic nun with admiration. We were in Cebu in the Philippines in 1976, standing in the leprosarium's basket weaving workshop where a dozen young people, relatives of patients, were making baskets out of bamboo. She had come from Belgium 38 years previously and had devoted her life to lepers and their families. Perhaps she had been inspired to do this by reading accounts of lepers in the Bible. Immediately, though, we have a problem. Considerable uncertainty surrounds the Hebrew word transliterated as *tsaraath* and translated as 'leprosy'. The King James Version of the Bible first mentions this condition in Leviticus; chapter 13, verse 3 says:

> And the priest shall look on the plague in the skin of the flesh: and when the hair in the plague is turned white, and the plague in sight be deeper than the skin of his flesh, it is a plague of leprosy: and the priest shall look on him, and pronounce him unclean.

This clinical description is not typical of leprosy and modern translations are sometimes more circumspect, translating *tsaraath* as 'an infectious skin disease'

(New International Version (NIV)) or 'serious skin condition' (New Living Translation) although the English Standard Version retains 'leprous disease'. The truth of the matter is that no-one knows and it seems best to consider the condition as indicating one or more skin disorders. As we shall see, leprosy is indeed a skin disorder, but it is much more than that and there is no skeletal evidence of leprosy in Old Testament times in Palestine. Similar uncertainty concerns both the New Testament and the Greek translation of the Hebrew Old Testament (the Septuagint), which both use the word *lepra* for leprosy. The NIV adds a footnote to Matthew 8:3 where Jesus heals a man with leprosy: 'The Greek word was used for various diseases affecting the skin—not necessarily leprosy.'

We are in similar deep water when we consider when and where leprosy began, for different experts have different views. These days we tend to think of leprosy as a tropical disease, and that is largely true. But it was not always the case. Some think that the origins of leprosy are shrouded in the mists of time and cannot now be unravelled. Others have suggested that the earliest written records of leprosy emanate from India about 600 years BC. From there the disease spread east to China and Japan as well as westwards to Europe, perhaps being introduced by the soldiers of the Persian kings Darius and Xerxes in the fifth century BC and/or by the troops of Alexander the Great returning home from Asia 100 years later. Hippocrates (p.116) in Greece seems unaware of leprosy as we know it. Aretaeus the Cappadocian, a Greek physician of the first century AD, is said to be the first person to accurately describe the disease although he called it *elephantos*. He described nodules in the skin, the absence of sensation in hands and feet, and the loss of fingers and toes, all cardinal features of leprosy. Italy was free of leprosy until Pompey's soldiers returned from the East in 62 BC. Over the ensuing centuries, leprosy spread throughout Europe and reached its height between 1000 and 1400 AD.

In the early part of the thirteenth century, an English physician, Gilbertus Anglicae (*c.* 1180–*c.* 1250), wrote a comprehensive compendium of medicine which contained 20 pages about leprosy. This is the first reasonably accurate description of the disease in the Christian West. He defined leprosy as being a malignant (i.e. nasty, not cancerous) disease due to the dispersion of 'black bile' throughout the whole body which corrupted the complexion and the appearance of body parts, sometimes in fact causing the loss of fingers and toes or worse, hands and feet. In particular, Gilbert described a permanent loss of sensation affecting especially the fingers and toes and extending to the forearm, the arm, or the knees. The affected parts were cold and the skin was translucent with the loss of its natural folds, and looked as if it was tightly stretched or polished. He classified leprosy into four varieties depending upon the appearances of the skin: *elephantia* (swollen like an elephant's hide), *leonina* (resembling the face of a lion), *tyria* (looking like the skin of a snake), and *allopicia*, meaning loss of hair. Gilbert thought that loss of the eyebrows and eyelashes was one of the worst prognostic features and in that he was quite right. Often there was also hoarseness of the voice and an obstruction of the nostrils. Remarkably, Gilbert considered that lepers were unduly devoted to sexual pleasures. He wondered whether this may be a means of transmission of the disease but he did not rule out it being sometimes congenital, arising from conception during the menstrual period. Or, he thought, it may be the result of a corrupt diet, foul air, or of the breath from another leper.

In medieval times in Europe, lepers who roamed the streets were required to wear distinctive attire, ring bells, and, following the injunction of Leviticus 13:45, keep their distance and cry, 'Unclean! Unclean!'. The extreme disfigurement caused by this disease led to segregation of afflicted persons, who were regarded as social outcasts and made to live in special dwellings, called leprosaria, lazarettos, or lazar houses, which had but the barest of facilities. It has been estimated that at one time there were nearly 20,000 such places, many housing only a few individuals.

Whether or not it was due to such practices of isolation, leprosy declined in Europe over the last several centuries, then disappeared except in Norway, where it was common in the nineteenth century. If you have ever been to Bergen in that country at latitude 60°N, you will know it can be very cold there. Yet it was in Bergen that the cause of leprosy was eventually found. In 1840, two medical officers, Daniel Cornelius Danielssen (Figure 124) and Carl Wilhelm Boeck (1808–1875) embarked on research into leprosy. In 1847 they published a book *Om Spedalskhed* (On leprosy) in which they distinguished two forms. The first they called 'nodular' to indicate the nodular (small tumour-like) lumps of variable size and shape in the skin. The second type they labelled 'maculo-anaesthetic' to reflect loss of sensation due to disease of the nerves which became swollen and thickened and which was often associated with anaesthetic, depigmented patches (= macules) in the skin. They also recognized that some patients had mixtures of both forms of the disease or one form could transform into the other. All of this foreshadows the modern classification of leprosy, which we shall review at the end of the chapter.

Figure 124. Daniel Cornelius Danielssen (1815–1894).

But how did a person get leprosy? Clearly, throughout most of recorded history the treatment of lepers indicates that most people regarded leprosy as being contagious. But it was not as dramatically infectious as, for example, smallpox or lice infestation. No doubt with this in mind, in a series of experiments which can only be described as being either extremely brave or extremely foolhardy, Danielssen in 1844 inoculated material from a leprous nodule into himself and later in that year into two caretakers and a nurse at the hospital. Nothing happened. In 1846, Danielssen implanted a leprous nodule into his arm; this resulted in pus and an abscess forming which healed after a few weeks. Danielssen was nothing if not indefatigable. Ten years later, Danielssen, his medical assistant, several nurses, and a male helper at the Lungegarden Leprosy Hospital in Bergen were all inoculated with tissue from a leprosy nodule as well as blood or pleural fluid (fluid in the chest cavity around the lung) from a leprosy patient. In 1857, patients with favus or syphilis were infected with nodule tissue; then in 1858, yet again, Danielssen inoculated himself as well as a nurse. Despite 14 years of attempts, no-one had become infected by the experimental transfer of leprous material.

It was no wonder then that Danielssen cast around for another explanation. Both he and Boeck knew that leprosy tended to afflict members of families. This led them to the conclusion that leprosy was inherited rather than being acquired by contagion. So influential were their views that the British Secretary of State for the Colonies asked the Royal College of Physicians of London to adjudicate. The College set up a committee, none of whose members appears to have had any significant experience with leprosy, which duly reported back in 1862 that leprosy was not contagious or communicable to healthy persons by close proximity with the disease and that there was no evidence that would justify any measures for the compulsory segregation of lepers. Rather, they declared that leprosy is a constitutional disorder with a cachexia or depraved condition of the general system, absolutely in line with what Danielssen had said. If these words

Figure 125. Armauer Hansen (1841–1912).

seem waffly double-Dutch to you, then so they are. But we do need to remember that there was no proper understanding of genetic disorders in those days. In the event, the Colonial Secretary issued orders repealing any laws affecting the personal liberty of people with leprosy. Many leper hospitals were closed and there may well have been a consequent increase in the number of cases of leprosy.

This view of leprosy was about to change. In 1868 Gerhard Henrik Armauer Hansen (Figure 125) started working for Danielssen in Bergen. Hansen had been born in that town and had studied medicine at the University in Christiana (now Oslo) in Norway. In 1869, he undertook a microscopical study of lymph nodes (also known as lymph glands). These are small masses of tissue situated on the lymphatic vessels through which lymphatic fluid travels, and are responsible for removing foreign bodies from the fluid and for producing lymphocytes (white cells involved in the immune system). In specimens taken from patients with leprosy, Hansen noticed yellowish-brown, granular masses which he called 'infectious substance', not only in lymph nodes, but also in other organs. He was unable to say what these masses were composed of, but realizing the need to improve his knowledge of microscopical pathology, he went to Bonn and Vienna to further his education.

Returning to Bergen, Hansen continued his studies of leprosy, both in the laboratory and travelling around the country examining patients. He was convinced that the yellowish-brown, granular material he had seen earlier was specific for leprosy. This material was present in large amounts in skin nodules from these patients and he concentrated on this tissue. In 1873 he realized that these granules were made up of large numbers of rod-shaped bacteria (bacilli). He reported this to the Norwegian Medical Society and his communication was published in 1874. He wrote that he had seen:

> in every leprous tubercle extirpated from a living individual... small staff-like bodies, much resembling bacteria lying within the cells; not in all, but many of them.[131]

Hansen may seem a little diffident in his identification of bacteria, but he was a pioneer at the cutting edge of bacteriology. It would be another three years before Robert Koch gave the first description of the life cycle of a bacterial infection and the fulfilment of Koch's postulates.

Hansen was hampered in finding out more about these bacteria because of his inability to stain them. Furthermore, his attempts to grow the organism on artificial media came to nought and all his efforts to transfer the infection to rabbits were in vain. From the perspective of the twenty-first century, we should not be surprised at that either. We still cannot grow leprosy bacilli on artificial media. It was not until 1960 that Charles C Shepard (1914–1985) in the USA showed that leprosy bacilli would grow in the footpads of mice (but not in the rest of a mouse). Then in 1971, the extraordinary discovery was made by Waldemar Kirchheimer (1913–2001) and Eleanor Storrs in the USA that these organisms infect the American nine-banded armadillo (*Dasypus novemcinctus*). Finally, in 1985, Wayne M Meyers (1924–) and colleagues reported that an African sooty monkey (*Cercocebus atys*) imported into the USA from West Africa was infected naturally with leprosy.

But to return to Hansen, who had suffered a personal tragedy. He had married Danielssen's daughter in 1873 but she died of tuberculosis within a year. Hansen carried on his observations of the pattern of distribution of leprosy among people in Norway, as well as studies of bacteria-like organisms in leprous nodules. Hansen was convinced that leprosy was infectious. In 1879, he heard about Koch's studies with anthrax so he wrote to the latter, who suggested a new method of staining nodules. As a result, Hansen was convincingly able to show in the autumn of 1879 that this 'stuff' was masses of bacteria. Figure 126 shows two cells in a lepromatous nodule packed with bacilli. Figure 127 shows a Schwann cell filled with bacillary granules. The Schwann cell is in the myelin sheath and is pressing on the underlying axon damaging its function (a myelin sheath surrounds an axon like plastic surrounds a copper wire). But this did not prove that these organisms caused leprosy. He needed to transmit the infection. He resolved once more to try to transmit infection to humans. First, he tried to infect himself with no result. It was then that Hansen heaped coals upon his head. On 3 November 1879, he took tissue from a patient with the form of leprosy with nodules in the skin and injected it into the skin and conjunctivae (outside of the eyes) of two patients with the type of leprosy involving the nerves to see if the nodular variant of the disease would appear in them.

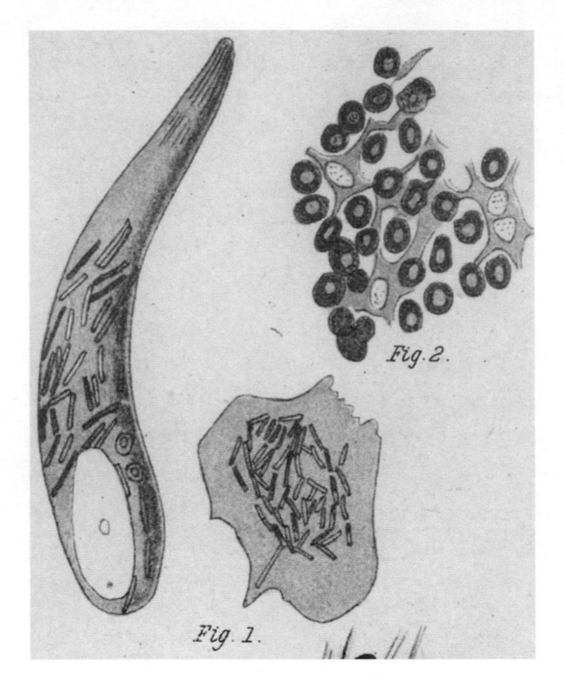

Figure 126. Hansen's drawings of two cells in a lepromatous nodule packed with bacilli inside the cytoplasm.

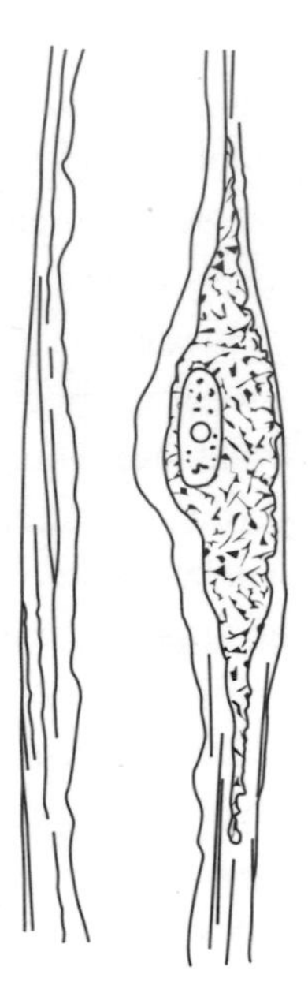

Figure 127. Hansen's drawing of a Schwann cell in a myelin sheath packed with granules; the light, oval structure is the nucleus.

Debatable though this experiment is in itself, Hansen omitted to seek permission from the patients, perhaps because he knew they would refuse. Nothing serious happened to the patients but one of them complained that he had done this without her consent, causing her much anxiety and not inconsiderable pain. This caused consternation and considerable debate over what should be done. Eventually Hansen was charged. He was found guilty and sentenced on 31 May 1880. He was deprived of his hospital post but allowed to retain his role as chief medical officer for leprosy in Norway. It is easy in retrospect to condemn Hansen. But given the state of knowledge about leprosy at the time, perhaps his sentence displays the wisdom of Solomon. He was rapped over the knuckles and his countrymen were satisfied, yet Hansen could continue his life's work, which, it has to be said, put him at considerable personal risk of contracting this dread disease.

The years 1879 and 1880 were bad years for Hansen in another sense. He was visited by Albert Neisser (p.265) from Germany, who was later to discover the cause of gonorrhoea. In a spirit of generosity, Hansen had shown Neisser all his research and given him some leprous material, which he took back to Germany. There he applied the most modern staining techniques and also identified rod-like bacteria. Unaccountably, in late 1879 Neisser published an article in which he discredited Hansen and claimed that he had discovered the microbe that caused leprosy. Neisser had stolen Hansen's discovery! Neisser probably would have received the credit had not Hansen republished his own findings in Norwegian, German, and English. In 1881, Neisser once more tried to discredit Hansen by asserting the importance of his own discoveries. Hansen had good reason to be upset as he had shown Neisser considerable hospitality and openness when Neisser had visited him, but he refrained from retaliating or even discussing the subject until he wrote his memoirs 30 years later. Extraordinarily, even at the end of his life, Neisser could not understand why his actions had been criticized and he remained wilfully obstinate and resentful that he had not received the recognition that he thought was his due.

In 1880 Hansen named the organism causing leprosy as *Bacillus leprae*. When Lehmann and Neumann (p.216) erected the new genus *Mycobacterium* in 1896, it was transferred to this genus and henceforth has been known as *Mycobacterium leprae*.

What sort of a man was Hansen? By all accounts he was well respected by his colleagues though less so by his patients whom he kept in strict isolation. He was humble and hardworking, with a lovely sense of humour, an atheist and a proponent of Charles Darwin's views. He deprecated the emancipation of women and remarried after the death of his first wife, but his new wife felt neglected by Hansen's dedication to his work and entertained the idea of divorce. He suffered from syphilis acquired during his student years and died from a heart attack at the age of 71.

So where are we today in our understanding of leprosy, and, if it is infectious, why did Danielssen and Hansen have so much trouble in transmitting the infection experimentally? In 1962, Dennis Ridley (1918–2009) and William Jopling (1911–1997) introduced a system of classification of the various sorts of leprosy which created order out of the chaos then prevalent. At its bare bones, Hansen had realized that in nodular leprosy there were large numbers of bacteria (called multibacillary leprosy), whereas there were very few bacteria to be found in maculo-anaesthetic leprosy (called paucibacillary leprosy, where 'pauci' means 'few'). Ridley and Jopling expanded on this and combined clinical features, the histological appearances of skin biopsies, the numbers of bacteria, and the immune response, to divide leprosy into five categories.

At one extreme was what they termed lepromatous leprosy. This has an insidious onset and has a progressively downhill course leading to death in untreated patients. The skin lesions are symmetrical patches of diffuse infiltration with shiny, red surfaces giving a waxy appearance and usually are not anaesthetic. Thickening is most marked on the face and the skin may become folded (known as a leonine facies), nodules appear, and the eyebrows and eyelashes may be lost. Gradual destruction of the small peripheral nerves produces a symmetrical loss of sensation beginning on the hands and feet. Loss of sensation results in repeated trauma with consequent shortening of the fingers and toes. The nose may collapse and the cornea (clear part of the eye) becomes opaque. Vast numbers of leprosy bacteria are found in the skin and other tissues and the immune system does not work properly, thus allowing the bacteria to multiply unhindered. The internal organs are affected and death results from malnutrition, kidney failure, or secondary infection with other bacteria.

At the other extreme is tuberculoid leprosy, so called because the appearances under the microscope resemble the tubercles of tuberculosis except that there is no central caseation. The skin lesions are scattered and few in number and are plaques with a raised edge and flat healing centre. There is no sensation in the skin of these lesions and they are paler and hairless. One or more large peripheral nerves are involved and these can be easily palpated through the skin. Damage to these nerves causes loss of sensation and paralysis of the muscles supplied by them. The nerve involvement is patchy and the patient may develop any of the following: a 'claw hand', inability to lift up the wrist, clawing of the feet, inability to lift up the foot, or inability to close the eyelids leading to eye damage. The loss of sensation to touch and pain may lead to ulcers on the hands and feet, infection of the underlying bones, fractures, and distortion, then loss of fingers and toes, then hands and feet. There is a marked immune response which gets rid of almost all the bacteria but at the expense of destroying nerves in the process which become infiltrated with inflammatory (white) cells. These patients may live for many years with slowly progressing disabilities.

In the middle is borderline leprosy with many skin lesions which have a great variety of size, shape, and sharpness of demarcation at the edges. There are a moderate number of bacteria in the skin lesions and there is some immune reaction. These patients have severe disease of the peripheral nerves as in tuberculoid leprosy.

Valuable though this is, it does not shed much light on the transmission of leprosy except that patients with lepromatous leprosy with vast numbers of bacilli are more likely to be a source of infection. Clearly, infection via the skin is difficult to achieve as shown by the Norwegian experiments, although cases of such a route of transmission have since been documented. Much more important is the respiratory route. Patients with lepromatous leprosy discharge abundant bacteria from their noses. It is thought that these bacteria are then inhaled by other people and lodge in the nasal passages. In most individuals nothing happens but in others bacteria enter the blood then head off to the skin and nerves where they enter particular cells in the skin called macrophages and cells around the nerves called Schwann cells. There they multiply very slowly and induce an immune response which is variable from person to person. Symptoms appear 1–20 years after infection. Danielssen was partly right. Leprosy is not inherited but susceptibility to leprosy and the form it takes probably depends upon our genes.

20

The golden staphylococcus

SCIENTIFIC NAME:	*Staphylococcus aureus*
COMMON NAME:	the golden staph
ORGANISM'S APPEARANCE:	Gram-positive cocci in clumps
DISEASE NAME:	various staphylococcal infections
DISTRIBUTION:	worldwide
TRANSMISSION:	droplet or touch
LOCATION IN A PERSON:	nostrils, skin
CLINICAL FEATURES:	boils, abscesses, wound infections, osteomyelitis, septicaemia, endocarditis, etc.
DIAGNOSIS:	recovery of the organism from pus or blood
TREATMENT:	flucloxacillin, vancomycin
PREVENTION:	not practical in daily life but in hospital, patients with MRSA (see below) are isolated

In 1965 I had to deliver 10 babies to pass my obstetric examination. But to do that, I had to be admitted to the delivery suite. And to do that, I had to demonstrate that I did not carry the golden staphylococcus, *Staphylococcus aureus*, in my nostrils. I was a carrier, so I was given antiseptic cream and re-tested two weeks later. Horror of horrors—I was still positive! What to do? An intensive course of treatment followed, including an application of cream shortly before a nasal swab was taken. Thankfully I was negative, and all went according to plan. In those days, which were shortly after *S. aureus* had gone from being usually sensitive to the wonder-drug penicillin to being usually resistant, everyone was very concerned about this organism. But it was a hopeless task and no-one worries much about ordinary *S. aureus* anymore. Instead, infection control nurses tear their hair out trying (relatively ineffectually) to control a variant of this bacterium called MRSA. But more of that anon. By the way, I still carry golden staphylococci in my nose.

Who was it that discovered the golden staphylococcus? Most textbooks will say that it was the Scottish surgeon, Alexander Ogston, while if you look at the official nomenclature it is *Staphylococcus aureus* Rosenbach 1884. But it is not as simple as all that. There was no Eureka moment such as Robert Koch had with discovering the bacterial cause of tuberculosis. Rather it was shrouded in confusion, chaos, and controversy. The golden staphylococcus is of course a coccus. We learnt on p.172 that these are spherical objects about 1 μm in diameter. We now know that there are several dozen species of cocci and we have various means, some simple, others complicated, for identifying them. But of course no-one knew that in the third quarter of the nineteenth century when pathologists started to notice them in various tissue specimens. Did they represent one species of bacterium or were they different? Did they cause disease or were they fellow travellers? Did they change their form and virulence (disease-causing capacity) according to circumstances? If they were pathogenic, did one bacterial type cause disease in one tissue type or could it cause disease in many tissues? Was a particular clinical syndrome always caused by just the one organism or could different cocci produce the same clinical effects?

Perhaps the first to see these organisms was the pathologist Friedrich Daniel von Recklinghausen (1833–1910) in Würzburg, Germany, who saw them in the kidneys of a patient who died from pyaemia in 1871. As we have seen, Ferdinand Cohn in 1872 (p.176) placed these bacteria in a genus called *Micrococcus* but that did not really help answer these questions. In 1874, Theodor Billroth (p.176) made a useful advance when he classified some cocci according to the manner in which they grouped themselves but then muddied the waters by inventing a mythical organism called *Coccobacteria septica*, which he believed was found in putrid fluids and septic wounds. Investigators such as Robert Koch tried to find the bacteria that caused wound infections in animals but the applicability of these findings to human diseases was uncertain.

Some idea of the state of confusion is conveyed in a series of articles published in 1886 entitled 'On the present state of knowledge in bacterial science in its surgical relations' by W Van Arsdale in the *Annals of Surgery*. After discussing the differing interpretations of the meanings of such words as sepsis, septicaemia, pyaemia, putridness, and metastatic abscess, he remarked:

> The most diverse theories were furthermore put forward to explain the action of the bacteria found in septic diseases. While some authors maintained that the bacteria were themselves the essential cause of all disturbance, others believed that the micro-organisms per se were harmless, but that they produced

> some alkaloid virus (ptomaine) which was the sole cause of disease, and others considered them capable of some destructive action upon the blood, thus causing disintegration of the blood-elements and occasioning disease...In fact, so great did the general confusion of opinion become, that when, even in the year 1881, E. Semmer, of Dorpat, undertook to review the state of knowledge of septic infections at that time, he could do so only by classifying the authors under different headings. He divided them all into two large groups, (1) those who experimented with putrid matter on the healthy animal body, without any regard to micro-organisms, which group comprises twenty-nine names... (2) those who paid special attention to the presence of micro-organisms. The latter group is much the larger, and includes four subdivisions...The first subdivision (a) consists of authors who attached no importance to the presence of micro-organisms (2 authors were listed)... He then (b) enumerates those who deny their influence (13 authors)... (c) The followers of theories ascribing to bacteria etiological [causative] importance in the causation of disease (14 authors)...Another (d) group consists of authors who experimented by inoculating with the blood of animals first rendered diseased. Here again we have two factions, on the one side those opposed to theories investing bacteria with etiological importance (8 authors)...On the other side those advocating these theories (18 authors).[307]

Belief, disbelief, and confusion reigned. It was in this state of ferment that we need to consider the discoveries and delineation of the golden staphylococcus as well as the streptococci that we shall consider in the next chapter.

Many people may well have seen cocci that were in fact *S. aureus* but they did not know this nor do we have any way of telling whether or not they did see them. Two people probably discovered the golden staphylococcus independently and reasonably contemporaneously. We can say this because we know that they saw the characteristic organisms in diseases that we now know are commonly caused by *S. aureus*. One person was Louis Pasteur (p.182). His laboratory was not in a hospital so he had to take whatever opportunity presented itself to him. One of his colleagues was Émile Duclaux (1840–1904), who suffered from recurrent attacks of boils. Boils are swollen, red masses in the skin and underlying tissue rather like enlarged pimples, which often break down and release pus. Duclaux, who was to become director of the Institut Pasteur after Pasteur died, himself later recorded what happened in May 1879:

> The first thing Pasteur did when I showed him one of them was prick it, or rather have it pricked, for he was not fond of operating himself, and to take therefrom a drop...in order to make a culture, in which undertaking he was successful. A second boil gave the same result.[89]

When Pasteur cultivated the pus in broth, he found the microbe presenting as a mass of rounded granules, comparable to a bunch of grapes. He injected broth containing these bacteria into animals and obtained abscesses. Shortly afterwards, Mr Lannelongue, a surgeon, asked Pasteur to examine some pus from a case of osteomyelitis (pussy inflammation of the bone). Pasteur found the same organisms and remarked 'In this case at least, osteomyelitis was a boil of the bone marrow.'[89] When he made his opinion known at the Academy of Medicine in 1880 in a paper dated 3 May, he was met with some disbelief, but he was right.

Figure 128. Alexander Ogston (1844–1929).

A person who had a lot of access to human clinical material of this sort was Alexander Ogston (Figure 128), a surgeon at the Aberdeen Royal Infirmary, Scotland. His first patient was one James Davidson, who had an abscess in his leg. When Ogston examined pus from this abscess using a microscope borrowed from a student, he found micrococci which stood out clearly when stained with an aniline-violet dye as used by Robert Koch. He was ecstatic:

> My delight may be conceived when there were revealed to me beautiful tangles, tufts and chains of round organisms in great numbers, which stood out clear and distinct among the pus cells [white blood cells] and debris.[232]

Ogston then began his research in earnest in a laboratory in his back garden in 1878; equipment for the laboratory was provided by a £50 grant from the British Medical Association (BMA). He carefully examined 88 cases of abscess which had not previously been incised and drained. Seventy of them were acute cases (i.e. of recent onset), hot and painful swellings, in various parts of the body. In all of them he found large numbers of micrococci in the pus. Fourteen cases were chronic 'cold abscesses' in which he could find no organisms; we now know that they are tuberculous in origin. In the hot cases, he calculated that there were about 2 million micrococci in every microlitre (= cubic millimetre) of pus. He noticed that in pus the cocci might be single, sometimes in pairs, at

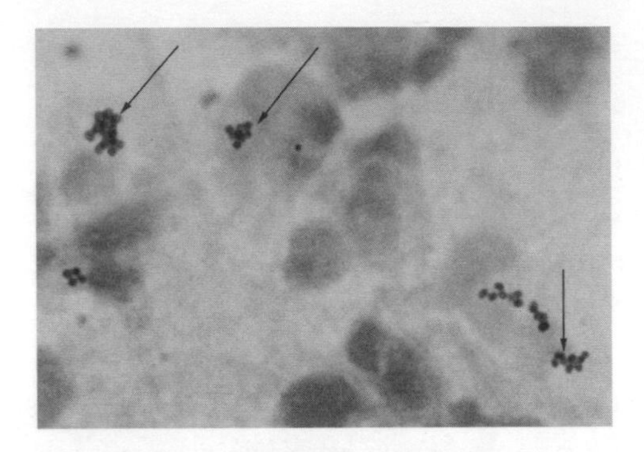

Figure 129. Clumps of staphylococci (arrows) and a chain of streptococci in pus. If you could see these in colour, they would be purple.

times in chains, and in others in clumps like a bunch of grapes or 'the roe of a fish' (Figure 129). In most cases, however, the cocci were either in clumps (31 patients) or chains (17 patients) alone. We now recognize these as being staphylococci and streptococci, respectively. When he injected pus from cold abscesses into guinea pigs or mice nothing happened, On the other hand, when he injected pus from hot abscesses an abscess always developed at the point of injection. When he examined these new abscesses, he again found the micrococci, and if he examined the blood then he often found micrococci present. If, on the other hand, he treated the pus with phenol or heated it, abscesses did not develop—the micrococci had been killed. On top of all this, Ogston was able to produce pure cultures of the organisms in hens' eggs and these always gave constant appearances on microscopy. When he turned his attention to wounds, he had similar results although sometimes there were bacilli as well as micrococci present.

But to convince the medical profession was another matter. At the Aberdeen branch of the BMA, he was received with incredulity. When he submitted a paper to the *British Medical Journal (BMJ)*, the editor, in rejecting it, asked the rhetorical question as to whether anything good can come out of Aberdeen? The bacteriological capitals of the world at the time were Paris and Berlin. So Ogston went to Berlin and presented the results of all this work in excellent German (as a student he had had a 'gap year' in Vienna and Berlin) in a paper on abscesses to the Ninth Surgical Congress in Berlin on 9 April 1880. He concluded his remarks

> On the basis of what I have said the following conclusions seem to me to be justified. (1) Micrococci are the most frequent cause of the formation of acute abscesses. (2) The occurrence of acute suppuration is everywhere very closely connected with the presence of micrococci. (3) Micrococci can produce

> blood poisoning, and (4) The individual constitution plays an important role in poisoning by micrococci and strongly influences its intensity and speed.[233]

The Germans were so impressed that Ogston was made a Fellow of the German Surgical Society. Whether or not Pasteur or Ogston first actually saw staphylococci, Ogston deserves the accolade for presenting a large series of well-documented cases.

Under some duress because the BMA had funded the research, the *BMJ* published an amplified version of Ogston's research in March 1881.[234] But a prophet has no honour in his own country. Even Ogston's hero, Joseph Lister, criticized Ogston's work in his address to the International Medical Congress in August 1881. Lister's whole antiseptic system in his mind was based on the exclusion of bacteria from wounds so as to avoid infection. With very woolly thinking, he believed that micrococci were relatively benign and that abscesses were often caused 'through the nervous system'. Reviewing Ogston's work specifically, he concluded that 'micrococci are, so to speak, a mere accident of these acute abscesses, and that their introduction depends upon the system being disordered'.[190]

Ogston had to respond, which he did in a paper entitled 'Micrococcus poisoning'. It was published in 1882 in the *Journal of Anatomy and Physiology*:[235] the *BMJ* refused to publish it. He now had 100 cases of abscesses. He dismissed Lister's nervous system hypothesis by noting that it had no evidence whatsoever to support it. He re-stated his case that septicaemia (bacteria in the blood) and pyaemia (abscesses scattered around the body) were due to a local source of infection with micrococci which then spread. Septicaemia and pyaemia were not blood diseases as such; blood was merely the vehicle by which the bacteria were spread around the body. Also in this paper he made some other important contributions. First, he showed that staphylococci and streptococci (which he called 'chain-micrococci') tended to produce different diseases. Secondly, he noted that some forms of staphylococcus produced a yellow-orange pigment in some wounds and on their bandages. Finally, it was in this paper that he gave the micrococci that occurred in grape-like bunches the name *Staphylococcus*, reserving the term streptococcus for those in chains as Billroth had done before him. This name was suggested to him by Aberdeen's Regius Professor of Greek, WD Geddes, for the Greek word *staphyle* means 'a bunch of grapes'.

Shortly thereafter, Ogston was made Regius Professor of Surgery at Aberdeen and he had to abandon research for lack of time. Lister was wrong and Ogston's research has stood the test of time. However, his name, staphylococcus, only just survived. In 1958 the International Committee of Bacteriological Nomenclature established a rule that a name is only valid if it is accompanied by a proper description. Ogston did not provide one. Fortunately, Friedrich Rosenbach, who did provide such a description, liked Ogston's name.

Anton Julius Friedrich Rosenbach (1842–1923) was a surgeon in Göttingen, Germany who was also interested in the bacteriology of wound infections. He had a great advantage over Ogston in that he was able to use Koch's new technique of culturing on solid media and obtained individual pure colonies made up of the same bacterium. From 30 abscesses, he grew five types of bacteria but by far the commonest were staphylococci. He noted two types of staphylococcal colonies distinguishable by colour. One was yellow so in 1884 he called them *Staphylococcus pyogenes aureus*, retaining Ogston's name for the genus. The middle word, 'pyogenes', is discussed on p.238. The species name reflects the Latin word *aureus* for 'golden'. The other colonies were white in colour and he named them *Staphylococcus pyogenes albus* with *albus* being the Latin word for 'white'.

Both Pasteur and Ogston had shown that staphylococci obtained from human abscesses could in turn cause abscesses when injected into animals. What would happen if staphylococci were injected into humans? Carl Garré (Figure 130), a surgeon and bacteriologist in Basel, Switzerland, determined to find out. He cultured staphylococci from both bone and skin infections. He was uncertain whether the same staphylococcal species could cause relatively harmless skin infections as well as severe osteomyelitis. To determine if the one isolate (that is a single species) could cause this spectrum of infections, he used a wire inoculated with bacteria from a patient's bone infection to scratch his own nail-bed. A very mild superficial infection developed but he was still unsure. In 1883, he scratched his forearm with a wire then smeared cultures of golden staphylococci obtained originally from a bone abscess over the

Figure 130. Carl Garré (1857–1928).

wound. He used his other arm as a control, placing sterile culture medium on a wound. By the end of the first day, the wound inoculated with staphylococci had already become red and painful. Next day he noted 'the whole thing began to be unpleasant', and he ultimately developed a carbuncle (abscess with multiple drains), swollen lymph nodes, and fever. It took him weeks to recover and he was left with numerous scars. Naturally enough, Garré concluded quite correctly that the golden staphylococcus could be responsible for both bone and skin infections. There could no longer be any doubt. *S. aureus* caused boils, wound infections, abscesses, and infections of bones.

What do we know about staphylococci today? Rosenbach's '*pyogenes*' has been dropped, being confined to *Streptococcus*, leaving the golden staphylococcus as *S. aureus*. Also, Rosenbach's name of *S. albus* has disappeared and been replaced by nearly 30 different species, *S. epidermidis* being the most common. It turned out that the colour of colonies was not always reliable in distinguishing *S. aureus* from the '*albus*' group of staphylococci. A new way of differentiating them was discovered when it was found that *S. aureus* makes an enzyme called coagulase which causes plasma (blood with the blood cells removed) to clot by turning the clotting factor, fibrinogen, into fibrin whereas all those in the '*albus*' group do not. Hence, all these latter bacteria are now called 'coagulase-negative staphylococci'. They are relatively harmless organisms and some of them, such as *S. epidermidis* and *S. capitis*, are normal inhabitants of the human skin and scalp respectively and by taking up space, protect us from worse bacteria.

The golden staphylococcus, *S. aureus*, can indeed be very pathogenic but many of us harbour it without it causing any trouble. I mentioned at the beginning that I have *S. aureus* in my nostrils. So does nearly half the population and this site is the major reservoir of this organism. From there, organisms may be deposited on the skin, especially of the hands and forearms, again without usually doing any damage. From time to time though, *S. aureus* may enter the hair follicles of the skin and cause minor infections which sometimes develop into larger boils or carbuncles. If the skin is broken by accidental injury, surgery, or tubes and catheters, then bacteria can enter the tissues and cause wound infections and abscesses. We now know that *S. aureus* can cause disease in any tissue of the body including the lungs (pneumonia), membranes around the brain (meningitis), and urinary tract. It can get into the bloodstream, sometimes for no apparent reason, and bacteria may lodge in bones causing osteomyelitis (infection of bone) or abscesses anywhere. Sometimes in staphylococcal septicaemia, bacteria lodge on the heart valves causing a dreadful disease called infective endocarditis.

At first it seemed that all these problems had been solved with the introduction of penicillin in the early 1940s. Unfortunately, important though penicillin was, and is, *S. aureus* did not abandon the field of battle. During the next 10 years, infections with this bacterium progressively became resistant to penicillin. The bacteria did this by producing an enzyme called penicillinase, a protein which has the capacity to break down the penicillin molecule (a process technically called hydrolysis). The chemists fought back by producing a variant of the penicillin molecule called methicillin which was resistant to this hydrolysis. The problem was solved—or at least for a while. The bacteria then evolved so that they changed their spots and methicillin could not attach to the bacteria and exert its effect. They were methicillin-resistant *S. aureus,* a phrase which was rapidly reduced to MRSA. Confusingly, some of these bacteria also became resistant to a variety of completely different types of antibiotics. They were multiresistant *S. aureus,* also MRSA. These organisms are no more dangerous in themselves than the original *S. aureus*—it is just that there is only a very limited array of drugs to treat them. These are some of the 'superbugs' so beloved of the media. The golden staphylococcus is going to be with us for a long time yet.

21

The pus-forming streptococcus

SCIENTIFIC NAME:	*Streptococcus pyogenes*
COMMON NAME:	group A streptococcus
ORGANISM'S APPEARANCE:	Gram-positive cocci in chains
DISEASE NAME:	various streptococcal infections, e.g. tonsillitis, cellulitis
DISTRIBUTION:	worldwide
TRANSMISSION:	droplet or touch
LOCATION IN A PERSON:	throat (pharynx)
CLINICAL FEATURES:	a. invasive infections: skin and soft tissue infections, e.g. erysipelas, cellulitis; tonsillitis; childbed fever; septicaemia, b. post-streptococcal immune diseases: rheumatic fever and acute glomerulonephritis
DIAGNOSIS:	a. recovery of the organism from pus, throat swabs, or blood, b. immunological evidence or recent streptococcal infection in the appropriate clinical context
TREATMENT:	penicillin
PREVENTION:	generally, there are no practical preventive measures

There are many species of streptococci but one, *Streptococcus pyogenes*, or as the Americans often call it, group A streptococcus, is by far and away the most important. It is the one we consider in this chapter and, as we shall see, not only does it directly cause symptoms and signs of infection, but it also produces indirect effects after the organism itself is no longer present. Before we discuss these clinical syndromes, let us dispose of the naming of these cocci.

As we have seen, Ferdinand Cohn in 1872 (p.176) placed coccal bacteria in a genus called *Micrococcus*. In 1874, Theodor Billroth (p.176) subdivided cocci according to how they were grouped and called those which formed chains streptococci. The name *Streptococcus* is derived from the Greek words *streptos* meaning 'twisted', but taken by Billroth to mean 'chain', and *kokkos* indicating 'seed' or 'berry'. In 1884, Rosenbach formally separated staphylococci from streptococci (p.234). For the chain-forming organisms, he kept Billroth's designation of *Streptococcus* for the genus name and gave the species name of 'pyogenes'; this is derived from the Greek words *puon*, meaning 'pus' (combining form 'pyo-'), and *genes*, indicating 'to form or generate'. Pus is a thick, usually creamy or yellowish fluid produced in tissue destroyed by bacteria. After Gram's stain became available in 1884 (p.192), it was found that like staphylococci, streptococci were purple in colour and are therefore called Gram-positive. But for several decades the situation became very confused and the bacterium was often referred to as *S. haemolyticus*, amongst others, until the 1930s, when the designation of *S. pyogenes* was clarified.

Childbed fever

We first encountered this organism when we read about Ignaz Semmelweis trying to prevent puerperal (childbirth) infections (p.177). He did not know it but this was the bacterium he was dealing with. The first person who probably saw it, but again he did not know what he was looking at, was Carl Mayrhofer (1837–1882) at the Vienna General Hospital in Austria. In 1865 he described various organisms that he called 'vibrions' in uterine discharges which pass through the vagina (technically known as lochia) from over 100 living and dead victims of childbed fever. The organisms differed in size and shape and in their motility, but one form was most abundant and regularly present. In view of the context in which these particular organisms were found, they may well have been *S. pyogenes* but we have no clear-cut identifying characteristics. Mayrhofer injected fluids containing these vibrions into the genital tract of newly delivered rabbits and they invariably became diseased and died; post-mortem examination revealed features similar to those seen in patients with childbed fever. Then in 1872 Heinrich Wilhelm Gottfried von Waldeyer-Hartz (1836–1921), professor of pathology in Breslau, Germany (now Wroclaw, Poland), described cocci in abscesses in the kidneys and heart and in pus in the abdominal cavity at post-mortem examination of childbed fever victims; these were almost certainly *S. pyogenes*.

But there things stood. Not much use was made of all this information until Louis Pasteur (p.182) brought the issue to the fore. On 11 March 1879, he attended a meeting of the French Academy of Medicine. A physician named Dr Hervieux gave a presentation in which he said childbed fever was not due to some hypothetical 'proto-organism' but due to 'puerperal miasma'. Pasteur immediately went to the blackboard, drew a chain of cocci, and announced that streptococci were the cause of childbed fever. When he spoke, Pasteur had made no specific study of puerperal fever, but he had seen such organisms before in 1875 when he had visited the Paris hospitals to study abscesses and had found in them multitudes of such small round organisms in pairs or in chains, like strings of beads. He was aware, too, that various German authors had described similar organisms in puerperal fever. The following day, Hervieux invited Pasteur to come to the Maternity Hospital in Paris to see a woman gravely ill with puerperal fever. From samples of blood taken from the patient's finger Pasteur was able to cultivate a micrococcus that grew in long chains. The patient died on 16 March, and Pasteur attended the autopsy. From the pus in the abdominal cavity, the blood, and from the lining of the uterus he was able to cultivate the same chain-forming micrococci. Pasteur's prediction was proven correct.

Infections of the skin and underlying tissues

So, streptococci cause puerperal fever. Do they do anything else? They can certainly cause abscesses. We have seen how Alexander Ogston in 1880 (p.231) found that about 20 per cent of abscesses were caused solely by cocci in chains. We now know that these organisms are *S. pyogenes*. An abscess is a collection of pus in any organ including the skin. As we have already noted, Rosenbach too in 1884, in his study of micrococci in abscesses, found some bacteria formed clusters while others were in chains. But *S. pyogenes* can cause other diseases of the skin. You may know that the skin has an outer layer called the epidermis, an inner layer called the dermis, and an underlying layer of subcutaneous fat. Each of these layers may become inflamed with diseases called impetigo, erysipelas, and cellulitis, respectively.

In 1881, Friedrich Fehleisen (1854–1924), a surgeon in Würzburg, Germany, began to look for a bacterial cause of erysipelas. This word means 'red skin' but the disease is characterized clinically by having a spreading but sharply demarcated border. Fehleisen first examined histological sections of skin taken either by biopsy or at autopsy. He found masses of micrococci in the lymphatic vessels (small tubes that carry fluid but not blood) in the skin. Eventually he was able to grow them on Koch's nutrient gelatine and found chain-forming cocci. Clearly they were streptococci. He then took cultures of these bacteria and inoculated them into the tips of rabbits' ears. They too developed spreading, red lesions which on histological examination were similar to the appearances seen in erysipelas in humans. So far so good. What about people? At the time, there was an extraordinary belief that erysipelas was actually a good treatment for various human tumours. This made it ethically possible for Fehleisen to see if he could transmit the infection to humans. He inoculated seven people with pure cultures and produced erysipelas in six. He had made his case! For a while Fehleisen's streptococcus was called *S. erysipelatis* but eventually it was realized that it was identical to *S. pyogenes* and the name disappeared.

Investigations continued on the other infective forms of skin inflammation. Much more common than erysipelas is cellulitis in which there is a rapidly spreading, not clearly demarcated, red swelling of the subcutaneous tissues. Impetigo of the epidermis mainly affects children and causes superficial crusting blisters on the skin. Studies over the next few decades, beginning with Max Bockhart in Germany in 1887, showed that both of these conditions could be caused by either *S. pyogenes* or *S. aureus* or both, but the relative frequencies of the two bacteria have varied from time to time and place to place.

So we now know that *S. pyogenes* can produce various diseases of the skin and soft tissues, may infect the female genital tract, and can enter the bloodstream, which may result in abscesses widely spread throughout the body. Does this bacterium cause any other types of infectious disease?

Tonsillitis and scarlet fever

If you had recurrent attacks of tonsillitis (sore throat) as a child, then it is likely that you were suffering from streptococcal sore throats. But finding the relationship between tonsillitis and streptococci was a long and tortuous process. The answer was to finally come from the delineation of scarlet fever (also called scarlatina), a particular form of streptococcal sore throat which is associated with a skin rash and redness in the throat. We now know that the form of *S. pyogenes* that causes this syndrome produces a toxin which generates the rash. But this was not known in the nineteenth century. Furthermore, with regard to the throat (pharynx), clinicians often had difficulty in distinguishing scarlet fever or what we would now call a streptococcal sore throat from diphtheria. Similarly with regard to the skin rash, they were often uncertain whether they were dealing with scarlet fever, measles, German measles, typhus, or any number of other conditions.

The first clue was found by Edward Emanuel Klein (1844–1925), born to a Jewish family in Ersek in Hungary (now Osijek, Croatia), who had moved to London in 1871. In the second half of the nineteenth century, there were a number of outbreaks of scarlet fever in Britain that seemed to be associated with cow's milk. In December 1885, an outbreak occurred which was traced to a dairy farm in Hendon in London. Klein was asked to investigate. He took samples of milk and from sores on the cows' inflamed udders (= mastitis), and from them he cultivated streptococci. Some of the scrapings from the udders were inoculated into the udders of young calves and reproduced the disease. When he performed a post-mortem examination on diseased cows or the inoculated calves, he found chains of cocci in many tissues and grew streptococci from their blood. He realized that the microscopical changes in the tissues were very similar to those he had seen in humans who had died of scarlet fever. He had fulfilled Koch's postulates. What is more, he showed that the streptococci grew rapidly in cow's milk. Klein then looked in humans with scarlet fever, both living and dead, and from their tissues he grew a streptococcus indistinguishable from that seen in the cows. He named it *Micrococcus scarlatinae*. This in turn he injected into animals and reproduced the disease.

You might say QED. The problem was that Klein's *Micrococcus scarlatinae* was indistinguishable from Rosenbach's *S. pyogenes*, which had been shown to be the cause of childbed fever, abscesses, septicaemia, and erysipelas. What is more, the economic and political implications of infected milk were profound. The result was that in Britain, Klein's work was buried. It was more convenient to say that the streptococci were fellow travellers (had just gone along for the ride) or that the animals had cowpox. So outbreaks of sore throat associated with unpasteurized milk continued over the next several decades in Britain and the United States. In 1911, a large outbreak affected 10,000 people and was associated with 200 deaths in Chicago; this was due to consumption of milk from just one dairy which had been unable to pasteurize the milk for four days.

All of these outbreaks finally led to an intensive investigation of streptococci. Around the beginning of the twentieth century it had been realized that many bacteria grew better if extra nutrients in the form of blood were incorporated into Koch's nutrient agar. Hugo Schottmüller (1867–1936) in Germany and Edward Rosenow (1875–1966) in Chicago, USA, independently found in 1903–4 that growth of colonies on blood agar proved a simple and reliable method for differentiating between *S. pyogenes* and pneumococci (p.250). *S. pyogenes* produced a large clear zone in the blood agar around each colony (Figure 131)

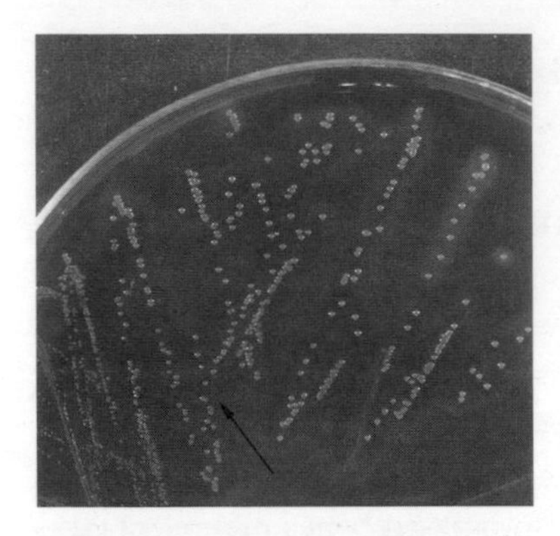

Figure 131. White colonies in streaks surrounded by beta haemolysis (arrow). If you could see this in colour, the whitish area underlying most of the arrow is red and the tip of the arrow is relatively translucent.

because the blood was completely broken down (known as haemolysis), whereas pneumococci produced a small greenish zone since the blood was only partly broken down. By the time James Howard Brown (1884–1956) and Theobald Smith (1859–1934) started their investigations of scarlet fever, it was realized that there were a number of species of streptococci. They too in 1914 divided streptococci up depending upon whether they caused a clear zone around colonies (which they called beta haemolysis), a greenish zone (alpha haemolysis), or no haemolysis at all. They realized that in outbreaks of scarlet fever, *S. pyogenes* which caused clear zones was always found in the throat of patients but that there were different strains of this bacteria in various epidemics. Things were looking suspicious but the trouble was that sometimes *S. pyogenes* could be found in the throats of healthy people.

Enter George Dick (1881–1967) and his wife Gladys (1881–1963) in Chicago, USA. They too found that β haemolytic streptococci (the ones with clear zones around colonies) were almost always present in the throats of patients with scarlet fever. They then tried to infect a variety of experimental animals with these organisms but without any success. Concluding that animals were relatively resistant to this infection, they decided to try to cause disease in human volunteers in a series of hair-raising experiments that would never pass an ethics committee today. We are not told what information was given to the volunteers about the risks they faced. In their first experiment in 1921, they studied healthy young adults who had never had scarlet fever. They collected blood from a dozen or more patients with early scarlet fever (who had negative blood tests for syphilis) and swabbed it on the tonsils of four volunteers and injected blood or serum (prepared from the blood) under the skin of another 11 volunteers. No-one got scarlet fever. They concluded that the organism was not present in the blood during the early stage of scarlet fever. They then took mucus from the throats of patients with scarlet fever and passed it through a special filter which removed any bacteria but let any viruses pass through. The filtrate was then injected into six more volunteers; again they remained well.

Scarlet fever was clearly not due to a virus. Next, they made pure cultures in the laboratory of β haemolytic streptococci from scarlet fever patients, suspended them in liquid, then painted the bacteria on the tonsils and throats of 40 volunteers. This time, seven of the volunteers developed a sore throat and a fever, but none had a rash.

They were puzzled why there was no rash. So they tried again in 1923, this time with a β haemolytic streptococcus obtained from pus from the finger of a nurse who had become infected while looking after a patient with scarlet fever and who developed scarlet fever herself. They swabbed the streptococci onto the tonsils and pharynx of five volunteers. Nothing happened in three of them but one developed a sore throat without a rash and the fifth had typical scarlet fever. This was very suggestive but could there have been an unknown virus which was travelling along with the bacteria? So the Dicks prepared a filtrate which would allow viruses to pass through but not bacteria and repeated the experiment on another five volunteers; all of these remained well. Two weeks later, four of these volunteers were swabbed again, this time with bacteria. Again, one acquired a sore throat without a rash while the fourth developed scarlet fever. The two victims were young women in their twenties; they recovered after a month or so. The Dicks were sure they had made their case, but the editor of the *Journal of the American Medical Association* clearly was not so certain because they were forced to add the following caveat to their 1923 paper in the *Journal*: 'These experiments do not justify the conclusion that all cases of scarlet fever are caused by the haemolytic streptococci described.'[86]

So the Dicks tried again. They extracted the toxin that causes the rash of scarlet fever from streptococci and showed that if it were injected into the skin of a person who had never had scarlet fever, a red rash would appear but if someone was immune by previous exposure to these streptococci, nothing happened; this became known as the 'Dick test'. They obtained two more volunteers, one of whom was shown to be immune by this test and one who was not immune, then swabbed streptococci on their throats. The non-immune subject developed scarlet fever but the immune volunteer did not. Gladys Dick felt she was able to write:

> The requirements of Koch's laws had now been fully met, and we are justified in concluding that scarlet fever is caused by the hemolytic streptococcus.[87]

It is true that the Dicks were not aware of the dangers of transmission of blood-borne viruses such as hepatitis B and C (HIV was not a problem in those days), but scarlet fever was not without its dangers and could have been fatal (p.304). Their experiments have to be seen in the context of the mores of the times. The *Journal of the American Medical Association* saw fit to publish their research thus giving it tacit approval. The Dicks were castigated, not for their methods, but because they patented a diagnostic test and vaccine that they developed!

Not everyone was convinced by their research nor did they apparently see any problems with the ethics. A number of investigators repeated their experiments without success, apart from one particularly unforgivable study. Dr Taro Toyoda and his colleagues in Japanese-occupied Manchuria infected three subjects with haemolytic streptococci recently acquired from two patients with scarlet fever. They all developed scarlet fever. However, eight subjects who had been given a strain of haemolytic streptococcus that had been repeatedly grown in the laboratory for months did not. Apparently only certain fresh strains produced scarlet fever. Who were these subjects? Children aged between three and seven years. The authors do not say so but it seems likely they were Chinese children in the occupied territory. These indefensible studies were published in the American *Journal of Infectious Diseases* in 1931. One American commentator remarked that these 'oriental investigators' had chosen their subjects 'with a disregard for human life that is impossible in the Western world'.[209] Does it not seem that this is a case of the pot calling the kettle black?

Could haemolytic streptococci from sites other than the throat cause tonsillitis? The Dicks looked into this too. In 1927 they cultured β haemolytic streptococci from three patients with erysipelas of the face. They swabbed them on the tonsils of five volunteers; three developed full-blown tonsillitis and pharyngitis (without a rash) and the other two developed a mild fever but no tonsillitis. Clearly, haemolytic streptococci are an important cause of tonsillitis.

You will notice that I have been using the term 'haemolytic streptococcus' rather than *S. pyogenes*. That is because it is the term the Dicks used. Advances needed to be made in understanding the bacteria. A major development was the discovery by Rebecca Lancefield (1895–1981) in New York in 1927 that the organism she called *S. haemolyticus* contained a particular substance known as an antigen that could be used for differentiating the bacteria. Each streptococcus had only one of these antigens, which were polysaccharides (chains of sugars). In 1933, she showed that β haemolytic streptococci belonged to one of five groups, which she named A, B, C, D, and E. Beta haemolytic streptococci recovered from humans with tonsillitis, including from patients with scarlet fever, almost always belonged to group A.

But there was still a question. Why did some group A haemolytic streptococci but not others produce scarlet fever? In 1934, Frederick Griffith (p.257) in England reported the results of an extensive study and was able to conclude:

> The haemolytic streptococci associated with scarlet fever, tonsillitis, septic conditions, etc., belong to one group or species designated *Streptococcus pyogenes*.[125]

Furthermore, he showed that there were nearly 30 different subtypes or strains of *S. pyogenes* and that in England scarlet fever was mainly due to Types 1 and 2. Strains that do not produce this erythrogenic (rash-generating) toxin are still quite capable of causing a nasty sore throat. Prophetically, Griffith wrote: 'Scarlatina is already a very mild complaint and it seems, if the present tendency is maintained, that it will disappear as a definite clinical entity.'[135] Time was to prove him right. There are now about 60 known strains of *S. pyogenes*, which vary enormously in their virulence (disease-causing capacity). These strains are determined by different antigens on their surface which are composed of proteins named M protein and T protein.

Lastly, where does *S. pyogenes* come from—where does it normally live? In fact, 10 per cent of us carry it in our throats without suffering from any ill-effects. It is only when we encounter a strain with a high disease-causing potential from someone else, that we get a sore throat.

Post-streptococcal diseases

There are two important illnesses that are the result of infection with *S. pyogenes*, even though the bacteria themselves are not present in the diseased tissues.

Rheumatic fever

When I took my examination in paediatrics (child health) in 1965, I was given an eight-year-old boy to assess and discuss. He was having his third attack of acute rheumatic fever. At that time, the hospitals were full of patients with acute and chronic rheumatic fever, the latter causing the heart to fail because of damage to heart valves. Nowadays, most medical students will finish their course without seeing a single case. Why is that? Penicillin! The administration of penicillin to children with sore throats has for all practical purposes wiped out the disease in developed countries. And unlike *Staphylococcus aureus*, not a single isolate of *S. pyogenes* has become resistant to penicillin despite 70 years of usage.

How was the association with streptococcal sore throats and rheumatic fever established? 'Rheumatic fever' is in fact rather a misnomer, for although joints are inflamed, the most damaging pathology affects the heart. 'Rheumatism' is rather obscurely derived from the Greek word *rheuma*, meaning 'flow', but has now come to signify arthritis, that is, inflammation of the joints. The French physician Guillaume de Baillou (p.292) was the first to use the term 'rheumatism' to distinguish acute (or recent onset) arthritis from gout. Thomas Sydenham (1624–1689), an English physician, noted that it chiefly affected the young and gave a graphic description of the joint changes:

> The patient is attacked by severe pains in the joints, sometimes in one and sometimes in another, sometimes in the wrist, sometimes in the shoulder, sometimes in the knee – this joint oftenest. This pain changes its place from time to time, takes the joints in turns, and affects the one that it attacks last with redness and swelling. Sometimes, during the first few days the fever and the above-named symptoms go hand in hand; the fever, however, gradually goes off whilst the pain only remains; sometimes, however, it grows worse.[300]

Sydenham also described a condition called chorea or 'St Vitus's dance':

> It first shows itself by limping or unsteadiness in one of the legs, which the patient drags. The hand cannot be steady for a moment. It passes from one position to another by a convulsive movement, however much the patient may strive to the contrary. Before he can raise a cup to his lips, he does make as many gesticulations as a mountebank; since he does not move it in a straight line, but has his hand drawn aside by the spasms, until by some good fortune he brings it at last to his mouth. He then gulps it off at once, so suddenly and so greedily as to look as if he were trying to amuse the lookers-on.[300]

Sydenham did not recognize the association of this condition with the joint disease he described but Richard Bright (see later) in 1831 realized that chorea was associated with rheumatic fever.

During the eighteenth century, inflammation of the heart muscle and lesions of the heart valves, especially the aortic valve or mitral valve which would not open or close properly and which resulted in the heart failing, were described by pathologists doing post-mortem examinations. In 1812, William Charles Wells (1757–1815) described the clinical and pathological features in a series of patients who had both acute arthritis and inflammation of the heart; he also noted that some of them had nodules under the skin called 'subcutaneous nodules'.

Involvement of the heart valves in patients with acute arthritis was dramatically described in 1836 by Jean-Baptiste Bouillaud (1796–1881), who was able to use the recently invented stethoscope:

> In auscultating the sounds of the heart in some individuals still labouring under, or convalescing from, acute articular rheumatism, I was not a little surprised to hear a strong file, saw or bellows sound such as I had often met with in chronic or organic hardness of the valves, with diminution of the orifices of the heart. Now nobody would suspect an affection of this kind amongst the majority of persons who suffered with rheumatism... Many of them were for the first time affected with articular rheumatism and had hitherto enjoyed the most perfect health.[50]

In his 1889 Harveian Lectures[71] in Britain (William Harvey described the circulation of the blood), William Butler Cheadle (1836–1910) brought together the known information and described a syndrome which could include some, or all of, inflammation of the outer layer of the heart (pericarditis) and inner valvular layer (endocarditis), acute arthritis, a skin rash (called erythema marginatum), Sydenham's chorea, subcutaneous nodules, and tonsillitis. In 1904 Ludwig Aschoff (1866–1942) described inflammation of the muscle layer of the heart (myocarditis) which, together with the valve lesions, caused the heart to stop pumping properly.

Notice the tonsillitis mentioned by Cheadle. Attention had been drawn to this in 1880 in Britain by James Kingston Fowler (1852–1934), who had himself had a severe episode of tonsillitis a month before he came down with an acute attack of rheumatic fever. Fowler described 20 patients in whom a bad sore throat preceded the appearance of acute arthritis by one to three weeks. The causation of scarlet fever, which is just one form of sore throat, by *S. pyogenes* we have just discussed. By measuring the levels of antibodies (proteins directed against *S. pyogenes*) in the blood, it was possible to show that virtually all cases of rheumatic fever had had a recent streptococcal infection. In 1944, Thomas Duckett Jones (1899–1954) in the USA put together Cheadle's signs and evidence in the blood of inflammation and increased antibodies against *S. pyogenes* to form the 'Jones Criteria', which every medical student learns, to enable a diagnosis of rheumatic fever to be made.

But how does a streptococcal sore throat cause rheumatic fever? Even today we are still not entirely sure. In 1910 an Austrian doctor, Clemens von Pirquet (1874–1929), published a paper suggesting that antibodies, as well as doing good things such as protecting from infection after immunization, sometimes also do bad things. Unfortunately he was right. It seems likely that in some genetically predisposed people, infection with certain strains of *S. pyogenes* initiates an immune response of antibodies. These attack not only the bacteria but also similar antigenic structures in the heart, joints, brain, and perhaps other tissues, a phenomenon which is known as cross-reactivity. The good news is that recurrent attacks can be prevented by taking oral penicillin over a number of years.

Post-streptococcal acute glomerulonephritis

Nephritis means inflammation of the kidneys. Glomerulonephritis means inflammation of those parts of the kidneys which contain glomeruli, the structures that filter urine from the blood. Patients with this illness have a fairly sudden onset of fever, tiredness, blood in the urine, fluid retention causing swelling of the ankles and around the eyes (oedema, previously called dropsy), and high blood pressure. Many people get better spontaneously after a few weeks.

In the eighteenth century it was recognized that in some patients fluid retention with swelling associated with the passage of scanty, dark urine, which at times was totally suppressed, developed two to three weeks after the onset of scarlet fever. The same Charles Wells we met in regard to rheumatic fever published a paper in 1812, which he had presented orally six years earlier, in which he showed that 'red matter' and 'serum of the blood' were present in the urine of patients with generalized swelling with fluid following scarlet fever.[319] He also realized that nephritis tended to afflict brothers and sisters. In 1827, Richard Bright (Figure 132) emphasized the connection between dropsy and urine that clotted (i.e. had blood in it), and this condition became known as 'Bright's disease' for well over a century. Sixty years later in 1905, Hans Reichel described the inflammatory lesions in the glomeruli when the kidneys of patients with nephritis following scarlet fever were examined under the microscope.

Figure 132. Richard Bright (1789–1859).

But what caused them? Just as the Dicks showed that scarlet fever was caused by a β haemolytic streptococcus, so various reports in the first half of the twentieth century identified cases of glomerulonephritis following streptococcal infections. These were of two types. The first was tonsillitis; John D Lyttle and his colleagues in 1938 in the USA found evidence of streptococcal infection of the throat in 109 of 116 consecutive patients with glomerulonephritis. Then Palmer H Futcher (1910–2004) and colleagues, also in America, in 1940 found that wounds of the skin infected with these bacteria also caused glomerulonephritis. But how did the streptococci cause this kidney disease? The experts are still arguing but it seems likely that in certain genetically predisposed people infected with particular strains of *S. pyogenes*, the bacteria induce the formation of antibodies. These antibodies combine with antigens that have broken off from bits of bacteria to form what are called immune complexes. These circulate in the bloodstream and are deposited in the glomeruli (which after all are a filtering mechanism) where they induce inflammation, that is a great accumulation of white cells and fluid in the tissues. Fortunately, as with rheumatic fever, penicillin has brought about a marked reduction in this disease, at least in the First World.

As we have seen, *S. pyogenes* is a remarkably dangerous bacterium which seems to have a particular propensity for humans and not animals. It often lives harmlessly in our throats, but from time to time it causes devastating infections, either with invasion of the tissues and formation of pus, or by stimulating the immune system which in turn damages our organs. It is a bacterium of which we are unlikely ever to be rid.

22

The pneumococcus and pneumonia

SCIENTIFIC NAME:	*Streptococcus pneumoniae*
COMMON NAME:	pneumococcus
ORGANISM'S APPEARANCE:	Gram-positive cocci in pairs
DISEASE NAME:	various pneumococcal infections
DISTRIBUTION:	worldwide
TRANSMISSION:	droplet (coughing and sneezing)
LOCATION IN A PERSON:	mouth and throat (nasopharynx)
CLINICAL FEATURES:	pneumococcal pneumonia, middle ear infection, sinusitis, meningitis, etc.
DIAGNOSIS:	recovery of the organism from sputum, blood, nasal secretions, ear pus, cerebrospinal fluid
TREATMENT:	penicillin, vancomycin
PREVENTION:	a vaccine is available for the elderly or those with impaired immune systems

One of my enduring memories as a teenager is watching on television one of the *Dr Kildare* movies of the 1930s in which a patient was critically ill from 'double pneumonia'. In those days before penicillin, lobar pneumonia due to a bacterium called the pneumococcus was a desperate affair. Patients either died or they made a rapid recovery when their temperature suddenly fell over 24 hours or so and they improved markedly; this was known as 'recovery by crisis' (Figure 133).

What is pneumonia? It would be better called pneumonitis—'inflammation of the lung'—but 'pneumonia' is irretrievably entrenched in both medical and lay parlance. The root word of course is the Greek *pneumon*, which means 'lung'. There are various forms of pneumonia but we are talking here about lobar pneumonia. There are three lobes (or major parts) in the right lung and two in the left

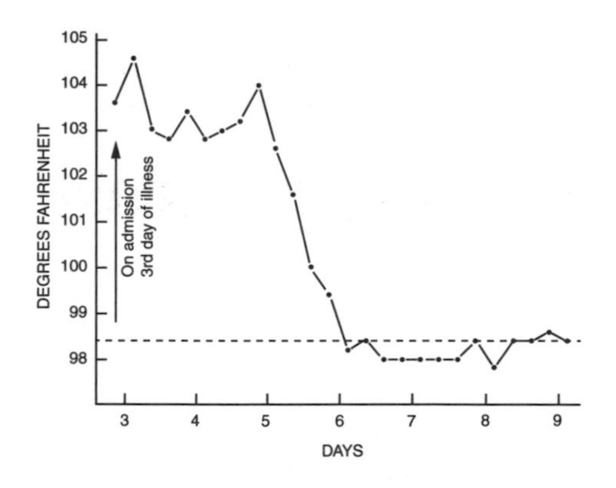

Figure 133. Temperature chart showing healing 'by crisis' from pneumococcal pneumonia in the pre-antibiotic era.

lung. In lobar pneumonia, all the tiny airspaces called alveoli in the lobe become packed with red cells, white cells, and fluid, in varying proportions depending upon the stage of the disease. The net result is that air cannot get into that lobe, which is also often inflamed on the surface (pleurisy). The patient is short of breath, has a cough, and brings up a variable amount of sputum (phlegm), has a sharp stabbing pain when coughing or inspiring deeply, fever, and prostration (being really laid low). This syndrome was well known to the ancient Greeks such as Hippocrates (p.116) and Aretaeus the Cappadocian (p.203)

A major advance in diagnostic ability was the discovery by the Austrian physician Joseph Leopold Auenbrugger (1722–1809) of the technique of percussion. You may have had your doctor percuss your chest with his or her fingers while listening to the pitch of the note produced. Auenbrugger was the son of an inn-keeper and was used to finding the level of wine in the casks; when he knocked against the part containing air he heard a higher pitch whereas it was lower when he percussed against that containing the wine. In lobar pneumonia, the affected part is filled with fluid and cells and the percussion note is lower (it is said to be 'dull') compared with the normal parts of the lung.

The next diagnostic advance was made with the invention of the stethoscope by René Laënnec (p.207). He noted that in the early stages of pneumonia when there is both fluid and air in the alveoli, one can hear the rattling of bubbles (which are called crackles or crepitations), but that when the alveoli are clogged completely, the crackles disappear:

> Such are the physical signs of pneumonia in the first degree. Of these, the most important is unquestionably the crepitous rattle... When the inflammation has reached the degree of hepatisation [so-called because lungs filled with red or white cells look solid like the liver = hepar], we no longer perceive in the infected part either the crepitous rattle or the respiratory sound... (and there

> may be) bronchophonism being nothing more than the resonance of the voice within the bronchi of the inflamed part [we now call this increased vocal resonance]...It is frequently accompanied by the sensation of blowing into the ear [we call this bronchial breathing]...When hepatisation has taken place, its resolution is invariably announced by the return of the crepitous rattle.[173]

These changes in the lung could be *seen* when X-rays were discovered in 1895 by the German physicist Wilhelm Conrad Röntgen (1845–1923), and chest X-rays were invented shortly afterwards (Figure 134).

But what causes pneumonia? There are many causes, but by far and away the most important cause of lobar pneumonia is the bacterium commonly known as the pneumococcus and officially as *Streptococcus pneumoniae*. Who discovered this organism? Who discovered what it does? It is not easy to say as a number of people had a part to play.

Edwin Klebs (p.296) thought it highly likely that lobar pneumonia was an infectious disease. In 1875 he saw in the fluid of the lungs of men dying from the disease, enormous numbers of bacteria, sometimes linked together, which he called *Monas pulmonale* and which in retrospect were almost certainly pneumococci. These organisms were not able to move about but when he tried to cultivate the causative organism by inoculating lung fluid into egg white, the bacteria became motile, and Klebs was quite confused as to exactly what he was dealing with. In 1880, Karl Eberth (p.332) examined the lung and the cerebrospinal fluid of a patient who had died from lobar pneumonia complicated by meningitis. He saw bacteria which he described as not moving and being slightly oval, almost round bodies that occurred sometimes singly but most often in pairs. These again were probably pneumococci but he did not realize that they were unique and thought they were

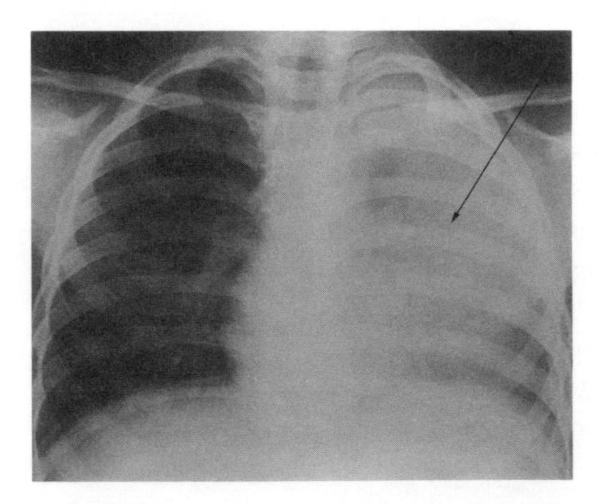

Figure 134. Pneumococcal pneumonia of the left upper lobe. It is whited out (arrow). This is called 'consolidation'.

varieties of either streptococci or staphylococci. In the same year, Dr Mátray found these cocci in the sputum (phlegm) of patients with pneumonia and also in the saliva of normal people. He called them, in German, *Pneumoniekokken*.

If either Eberth or Mátray had injected these bacteria into rabbits, they might well have been the undisputed discoverers of these organisms. This was done independently in 1880 by Louis Pasteur (p.182) and George Sternberg. Sternberg made his discovery before Pasteur but the latter reported his find first in 1881.

Pasteur in France was investigating a patient with rabies. One of the features of this disease is an inability to swallow with saliva drooling from the mouth. Naturally, Pasteur examined the boy's saliva and observed pairs of coccoid bacteria. When he injected the saliva into rabbits, they died and Pasteur recovered bacteria from the rabbit's blood with the same appearance. These were in turn injected into other rabbits to once more cause disease. As he had with staphylococci and streptococci, Pasteur avoided using proper binomial (genus + species) nomenclature and called it 'microbe septicémique du saliva' (septicaemic microbe of saliva). This organism, of course, had nothing whatever to do with rabies; Pasteur realized this even though others thought perhaps it did and mistakenly believed that Pasteur thought so as well.

George Sternberg (1838–1915), an army medical officer in the USA, also saw ovoid cocci end to end in pairs in human saliva. In 1880 he injected his own saliva into rabbits. They died and he was also able to find the same organisms in the rabbits' blood. He then found that this bacterium was present in the saliva of some people but not others. A few years later he called it *Micrococcus pasteuri*.

Although we now realize that Pasteur and Sternberg had found pneumococci, they did not call them that nor did they realize at first that they caused pneumonia. In fact, although they were obviously dangerous for rabbits, it was not clear initially that they were a problem for humans at all.

The first major study of the importance of these organisms as the cause of lobar pneumonia was published in 1882 by the German pathologist Carl Friedländer (1847–1887). At the time he was working at the Berlin-Friedrichshain Hospital and was suffering from the tuberculosis which was to take his life five years later. He reported that in the previous September he had found enormous numbers of ellipsoidal cocci, mostly occurring in pairs, in sections of lungs from eight patients with lobar pneumonia and in fluid obtained from living patients by aspirating fluid from the lungs with a needle inserted through the chest wall. In that same year, the German physician Carl Günther also found similar bacteria in specimens obtained by dangerously puncturing the lungs of living patients. Furthermore, he noticed that the pneumococci were enclosed by a capsule; this was to turn out to be so important.

In 1883, Friedländer published a more extensive study showing that he had found these encapsulated cocci in most of 50 cases of lobar pneumonia. Friedländer then set about trying to cultivate the bacteria on coagulated blood serum and upon a gelatine medium, and then to determine their ability to cause disease by injecting them into mice and rabbits. It was here that he ran into trouble. By a perverse piece of misfortune, he and a colleague, Dr Frobenius, took a specimen from a case of pneumonia in the right upper lobe of an alcoholic patient. He easily grew an organism on the gelatine medium which formed colonies looking like a 'round-headed nail'. This organism was a very short bacillus, had a striking capsule, and stained well with methylene blue. He continued to try to isolate the organism from other patients without much luck but from two, with difficulty, he grew bacteria which formed colonies on nutrient agar that looked like a flat-headed nail with a central depression, rather like a doughnut (Figure 135). He concentrated on his first isolate which was easy to grow and injected these bacteria into rabbits without ill-effect, but they killed six of nine guinea pigs and all of the mice he infected. Something was wrong here because when Steinberg injected his organism into rabbits they died. Friedländer at the time did not appreciate the significance of this point.

The year 1883 was a busy one. Drs Salvioli and Zäslein found similar bacteria in the sputum and blood of patients with pneumonia, while Strassmann found them in the sputum, and Ernst Victor von Leyden (1832–1910) in Berlin and

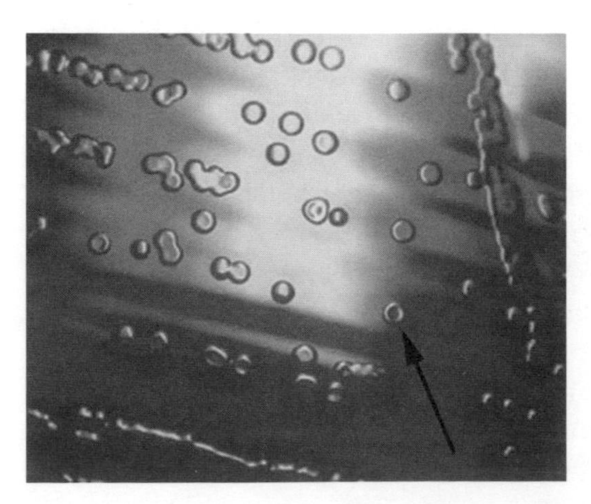

Figure 135. Colonies of pneumococci growing on chocolate (haemolysed blood) agar. Some have a central depression and look like a doughnut (arrow). Others look wet and mucoid.

Charles Talamon (1850–1929) in France both grew them from the blood of patients with pneumonia. A number of investigators were interested in studying Friedländer's organism so he distributed it to them and it became known as Friedländer's bacillus.

Friedländer's problems really began the next year in 1884 when Albert Fraenkel (1848–1916) made the first of a series of presentations on the pneumococcus. At the Congress for Internal Medicine in Berlin, he declared that he should receive some of the credit for discovering the pneumococcus. He claimed that he had begun experiments six months before Friedländer's work was made public, but because his findings were different, he delayed announcing his own results because he wished to check his findings. The cocci he isolated killed rabbits, but he too did not at first appreciate the significance of this observation. Furthermore, he reported that the appearance of colonies and the presence of a capsule were variable, and that there were other bacterial species that had capsules. Friedländer was at the meeting and in the ensuing discussion responded that there may well be two different bacteria that cause pneumonia. And so an argument was joined that was to continue for two years.

There was an easy way of resolving the issue and it was right there in Friedländer's laboratory. In 1884, Hans Christian Gram (p.192), who was working in his laboratory, had discovered the stain that bears his name. Tradition has it that Gram made his discovery by accidentally spilling Lugol's iodine on some lung sections that had been stained with methyl violet, then unsuccessfully trying to wash it off with alcohol. We would now say that these bacteria were Gram-positive but that terminology was not used at the time. Nevertheless, Gram reported his findings in 20 fatal cases of lobar pneumonia. In 19 of them he found that the bacteria stained purple (were Gram-positive) but one did not (it was Gram-negative). Gram even wrote, 'From this case stem a great part of the cultures of Dr Friedlander', i.e. it was from this last case that Friedländer distributed samples to all and sundry. And still the penny did not drop.

By 1886, Fraenkel was convinced that his coccus and those of Pasteur, Sternberg, Talamon, and Savioli were identical. After a series of careful experiments of growth in various culture media and infection of experimental animals which fulfilled all of Koch's postulates, he said that his coccus, which he called Pneumococcus, was the cause of pneumonia although it could sometimes be found in the saliva of normal people. He found that the pneumococci grew poorly on gelatine at room temperature but did well on congealed blood serum or agar at body temperature on which they developed characteristic doughnut-shaped colonies. Furthermore, he showed that if encapsulated pneumococci were cultured in broth in the laboratory, they lost both their capsules and their ability to cause disease in experimental animals. He caustically dismissed Friedländer's bacillus as an irrelevance. He was almost right.

Later in 1886, Anton Weichselbaum (p.286) described 129 fatal cases of pneumonia. He saw pneumococci in the lungs of 94 patients and grew them in 54 cases on media other than gelatine. He called these organisms *Diplococcus pneumoniae*; 'diplococcus' reflected the fact that the cocci were usually in pairs and 'pneumoniae' indicated that they were found in the lungs. But Weichselbaum showed clearly that there was a second, less common cause of pneumonia. In nine patients he saw 'Friedländer's bacillus' and grew it in six cases. He called this organism *Bacillus pneumoniae*. Weichselbaum recognized that both organisms had capsules and that *B. pneumoniae* could sometimes have variants when grown in culture which looked like cocci. Very confusing!

Then someone, and it is not clear who, realized that the answer was at hand. The two bacteria could be easily distinguished by the Gram stain—*Diplococcus pneumoniae* was Gram-positive while *Bacillus pneumoniae* was Gram-negative. In November 1886, Friedländer, in a circumlocutory fashion, acknowledged the value of the Gram stain and conceded that 'Friedländer's bacillus' was the less common cause of pneumonia but still clung to claiming credit for first seeing pneumococci on microscopy:

> From the microscopic state of affairs it is to be added that the coccus is intensely stained by the Gram procedure whereas the organism studied by me is decolorised by the Gram method.... From the foregoing investigations it emerges, therefore, that the most frequent organism in pneumonia is a capsulated coccus which was first found by me by microscopic examination...The capsulated bacterium cultivated by me (bacillus of the authors [i.e. Friedländer's bacillus]) occurs only in a minority of cases.[103]

He went on to vent his frustration with Fraenkel:

> Of the manifold personal attacks and remonstrances which Fraenkel has directed against me in different places of his work, let them cease. I do not hold them fitting.[103]

Where had Friedländer gone wrong? Almost all the bacteria he had seen microscopically were pneumococci. Unfortunately for him, *Diplococcus pneumoniae* grows poorly on nutrient agar whereas *Bacillus pneumoniae* grows much more readily. Friedländer was not to know this and he concentrated on and distributed the organism that grew more easily but would turn out to be the less important organism. We now know that they both cause pneumonia but pneumococci are much more common.

Rosenow and Schottmüller (p.241) in 1903 independently showed that pneumococci grew better on blood agar and that they could be easily distinguished from *Streptococcus pyogenes*. Pneumococci formed large greenish colonies surrounded by a narrow zone of alpha haemolysis that was not clear and had a distinct green tinge. As we have already seen in the chapter on streptococci, *S. pyogenes* formed small grey-white colonies surrounded by a large, clear zone of beta haemolysis.

The changes in names for these organisms have been very confusing. Many more have been used than those mentioned here. Weichselbaum's name of *Diplococcus pneumoniae* was officially renamed *Streptococcus pneumoniae* in 1974. Friedländer's bacillus (Weichselbaum's *Bacillus pneumoniae*) was transferred in 1887 into the newly erected genus *Klebsiella* (named in honour of the aforementioned Edwin Klebs) and has been known as *Klebsiella pneumoniae* ever since.

But much more important than wrangling over names were attempts to find out why this bacterium was able to produce such devastating disease. You will remember that the pneumococci recovered from sick patients had a large capsule. It turned out that this capsule was very important in causing disease for it made it more difficult for white cells to ingest and destroy them. On the other hand, pneumococci without capsules did not seem to cause disease and were more often recovered from healthy people. Why did some pneumococci have capsules and others not?

The British bacteriologist Frederick Griffith (1879–1941), who died in London during a German air-raid, set out to examine this issue. He found in 1928 that when living pneumococci without capsules were mixed with dead pneumococci that did have capsules and were then injected into mice, they acquired capsules. He thought there must be some 'transforming principle' which turned unencapsulated pneumococci into pneumococci with capsules but he had no idea what it might be. In 1944 Oswald Avery (1877–1955), Colin MacLeod (1909–1972), and Maclyn McCarty (1911–2005) in New York showed that this transforming principle was DNA (deoxyribose nucleic acid). This was the first report of a biological activity of a nucleic acid. Its importance was unrecognized at the time but marked the beginning of molecular genetics. Many people think they should have won the Nobel Prize.

We have learnt that pneumococci can cause deadly lobar pneumonia and can also get into the bloodstream. If that happens, they can cause abscesses anywhere, lodge on the heart valves (endocarditis—perhaps first seen by Osler in 1880), infect the membranes around the brain (meningitis—first noted by Eberth in 1880), or invade the joints (infective arthritis—first found by Weichselbaum in 1888). Indeed, in 1887 Arnold Netter showed that pneumococcal meningitis could occur whether or not there was concurrent pneumonia. Do they commonly affect other tissues?

Upper respiratory infections: sinusitis, middle ear infections, and mastoiditis

I never knew my grandfather. As a four-year-old boy on the Coolgardie goldfields in Western Australia in 1892 he developed an acute ear-ache. Someone tried to help him by putting hot olive oil in the ear canal. It didn't work. He developed chronic otitis media (middle ear infection) with a hole in the eardrum, pus pouring out of the ear canal, and progressive deafness in that ear. The infection spread to involve the mastoid bone behind the middle ear. The illness continued for his whole life. In 1939, the infection spread further from the mastoid bone to inside the skull and he died of meningitis and perhaps a brain abscess. This organism was almost certainly the pneumococcus. If he had been able to hang on for just a few more years, he may well have been cured by penicillin.

Infection of the middle ear of an adult by pneumococci was reported by Dr Emanuel Zaufel (also Emmanuel Zaufal, 1837–1910) in Prague in 1887. Middle ear infections (otitis media) occur most commonly in infants and it was eventually found that almost half of these cases are caused by pneumococci. Similarly, it was found that about a third of cases of sinusitis in children and adults are due to pneumococci. In both conditions, the bacteria generally find their way to the middle ear or sinuses from openings in the nasal passages.

Where do pneumococci normally live? A clue was given by the discoveries of Pasteur and Steinberg. We now know that 10 per cent of healthy adults and 20 per cent of healthy children carry them in the mouth and throat (nasopharynx), and that they are transmitted from one person to another mostly by coughing and sneezing. Pneumococcal infections became much less of a problem with the appearance of penicillin. Unfortunately, penicillin-resistant pneumococci appeared a couple of decades ago and these may become an increasingly important problem. There is a vaccine which has some protective value and is usually given to the elderly and others with impaired immune systems. We will always have pneumococci with us.

23

Gonorrhoea (the clap)

SCIENTIFIC NAME:	*Neisseria gonorrhoeae*
COMMON NAME:	gonococcus
ORGANISM'S APPEARANCE:	Gram-negative cocci in pairs
DISEASE NAME:	gonorrhoea
COMMON DISEASE NAME:	the clap
DISTRIBUTION:	worldwide
TRANSMISSION:	sexual intercourse
LOCATION IN A PERSON:	male—urethra; woman—cervix, urethra
CLINICAL FEATURES:	urethral discharge, vaginal discharge, pelvic inflammatory disease, conjunctivitis of the newborn, etc.
DIAGNOSIS:	recovery of the organism from urethral discharge or cervical swab or other clinical specimen
TREATMENT:	certain cephalosporin antibiotics
PREVENTION:	use of condoms and the avoidance of promiscuous sex

Urethritis means inflammation of the urethra. The urethra is the tube in the penis through which urine and semen pass. In females it is a tube, again through which urine passes, which is located in front of the vagina. The cardinal symptom of urethritis in men is an itchy, creamy discharge from the penis which stains the underpants and is often associated with painful urination. Women, too, may have urethritis. On the other hand, many have no symptoms at all or a mild vaginal discharge, at least at first, because the infection primarily involves the cervix (the lower part of the uterus). To complicate matters further, these symptoms and signs can be caused by two distinct organisms, gonococci, the subject of this chapter, and *Chlamydia*, which we will review in another.

How long has gonorrhoea afflicted mankind? Answering this question is an exercise in frustration. What is not difficult to answer is the derivation of the word 'gonorrhoea' (or as the Americans have it, gonorrhea). It comes from the Greek *gonos*, meaning 'seed', and *rhoia*, indicating 'flux' or 'flow', because it was thought to be a discharge of semen. Everything else is difficult, for what one expert says another will contradict. Leviticus 15:2–3 of the Old Testament is an example:

> Speak to the Israelites and say to them: When any man has a bodily discharge, the discharge is unclean. Whether it continues flowing from his body or is blocked, it will make him unclean.

Some think this clearly refers to a urethral discharge. Others say it could be pus discharging from any skin lesion. Some say that Hippocrates (p.116) knew of gonorrhoea. Others believe that gonorrhoea did not exist in Greece and Rome in the centuries before and around the time of Christ; this conclusion is based upon the absence of any descriptions of the complications of gonorrhoea. Whichever is true, it is clear that there were descriptions of the seeping 'excretion of semen' (or at least what was presumed to be semen) which some experts think could well indicate chlamydial urethritis. We have the same problem 1,000 years later. Some think that Persian physicians who practised during the European Dark Ages did not know of gonorrhoea but others believe that Rhazes (p.438) and Ali Abbas (AD *c.* 980) were familiar with it.

When did it come to Western Europe? Some consider that the absence of a mention of gonorrhoea in the Italian Boccaccio's bawdy collection of novels, the *Decameron*, written *c.* 1350, or in the works of the Englishman Geoffrey Chaucer (1343–1400), suggests that it was not a problem in those times. Others say that it had appeared in Western Europe by the early twelfth century. Brothels were permitted at Southwark on the south side of the River Thames in London under the jurisdiction of the Bishop of Winchester. In 1162 there was a regulation requiring 'no stew-holder [brothel owner] to keep any woman that hath the perilous infirmity of brenning [burning]'.[297]

Interpretation of the meaning of 'burning', especially in women, is particularly problematic. If it was associated with urethral discharges, that is one thing, but if it was simply the sensation of a burning pain on passing urine, that is another. It may well have represented an ordinary urinary tract infection with common bacteria such as *Escherichia coli*, a condition which affects many women, especially after sexual intercourse. On the other hand, John of Arderne (or Arden), surgeon to King Richard II of England, around 1380 described the burning of the penis as an inward heat and excoriation of the urethra which could be a tolerable description of gonorrhoea. Whatever the truth of the matter, gonorrhoea did come and the English 'burning disease' was known to the French as 'la chaude pisse' (hot piss), reflecting pain on urination.

By the late sixteenth century, this condition had acquired a new name in English—'the clap'—and was a feared consequence of promiscuous sexual intercourse. In the *Mirror for Magistrates*, a collection of English poems, the following ditty appeared in 1587:

> They give no heed before they get the clap
> And then too late they wish they had been wise[9]

Things were about to get even more complex. Syphilis (the pox) arrived on the scene and the two conditions were confused. The poet Alexander Pope penned in 1735:

> Time that at last matures a clap to pox
> Whose gentle progress makes a calf an ox[254]

So the question was: did the clap (gonorrhoea) lead to the pox (syphilis) or were they two different conditions that were apparently spread by sexual intercourse?

Enter the famous Scottish surgeon, John Hunter (Figure 136). In 1767 he performed an inoculation experiment. He collected 'venereal matter from a gonorrhoea' (in view of his later statement, probably a urethral discharge, although some have conjectured that it was purulent matter from a syphilitic sore) from a patient, dipped a lancet in it, then made two punctures in the penis of a subject. We shall discuss who this subject might have been in the next chapter. He gives no clear description of the subject developing a urethral discharge unless he is being circumlocutory with the following: 'There seemed to be a little pouting of the lips of the urethra, also a sensation in it in making water, so that a discharge was expected from it.'[146] Hunter said that discharge was

Figure 136. John Hunter (1728–1793).

'expected'. Did it actually happen? What is clear is that sores developed at the sites of inoculation of the penis after a few days, which he treated. A sore recurred on the prepuce (the part of the penis that is removed in circumcision) after four months and was associated with swelling of glands (lymph nodes) in the groin. Hunter called the ulcer of the prepuce a 'chancre', a typical lesion of syphilis. His conclusion, not unnaturally, was that 'It proves, first, that matter from a gonorrhoea will produce chancres.'[146]

In other words, he considered that gonorrhoea and syphilis were different manifestations of the same disease, thus supporting a view which had been widespread for over a hundred years. He thought a single venereal poison applied to the male urethra would provoke a discharge but if applied to the skin it would cause a syphilitic chancre. It would seem that if the subject of Hunter's attentions did indeed develop the clinical manifestations of both gonorrhoea and syphilis, then he had been given a double venereal infection. Incidentally, 'venereal' means 'pertaining to sexual intercourse' and the word is derived from Venus, the Roman goddess of love and fertility.

Not everyone considered gonorrhoea and syphilis the same. In 1793, the Scottish surgeon Benjamin Bell (1749–1806) published a two-volume work on venereal diseases.[42] In this, he marshalled his arguments, some of which are more convincing than others, that gonorrhoea and syphilis were in fact separate conditions. Among them were the following:

1. Gonorrhoea is a local disorder but syphilis is systemic disease.
2. Sometimes the two conditions can be contracted at the same time.
3. Surgeons have accidentally contracted syphilis from attending to patients with chancres or buboes but never gonorrhoea.
4. In China and Tahiti syphilis became established years before gonorrhoea.
5. Syphilis had been endemic in rural Scotland for many years but he had never seen a patient with syphilis develop gonorrhoea.
6. Medication against syphilis (mercury) is ineffective against gonorrhoea.

Bell's views gained the attention of Jean-François Hernandez (1769–1835), a French naval surgeon in Toulon who had treated Napoleon Bonaparte for injuries he received in the siege of Toulon in 1793. In 1812, Hernandez reported that he had been unable to induce syphilis by the inoculation of gonorrhoea pus into convicts under his care! Convicts apparently did not rate as human beings as far as any rights were concerned. In 1838, Philippe Ricord (p.273) clearly separated gonorrhoea from syphilis. One of the ways he did this was by observing the effects of autoinoculation. In contrast to syphilitic lesions, if gonorrhoeal material was injected into the thigh of the same patient, no local lesion developed. Perhaps this was why Ricord was convinced of the contagiousness of syphilis but not gonorrhoea. He believed that gonorrhoea was simply an inflammation of the urethra produced by a diversity of irritants, not an organic, living agent.

This misconception was rapidly disproven in the next 10 years when a number of investigators showed that if minute amounts of gonorrhoeal pus were inoculated into the urethra (as opposed to the thigh as Ricord had done), this was regularly followed by an attack of the disease. In contrast, if pus from other sources, such as an abscess, was introduced into the urethra, no discharge resulted. Naturally, people started to look for a cause and in 1836 Alfred Donné (1801–1878) found a protozoan parasite that we now call *Trichomonas vaginalis*—this can certainly cause inflammation of the vagina and a discharge but we know now that it is not the cause of gonorrhoea.

So much for men. What about women? Sometimes they developed urethritis like the men. At other times, they developed something else as well or instead. It had long been known that women sometimes suffer from a vaginal discharge, often called leucorrhoea (white flow). Some, such as Lazarus Riverius (1589–1655), described different colours and consistencies of such discharges, but no-one really had any idea what caused them or what significance they had. There are in fact five major causes of vaginal discharge. We have already encountered one: the fungus *Candida albicans*, which was delineated in the second half of the nineteenth century. The second is the *Trichomonas vaginalis* alluded to above. We are about to find the third.

In 1872 Emil Noeggerath (1827–1895), a German gynaecologist working in New York, postulated that infection could persist in men as latent gonorrhoea, i.e. the causative agent was lurking there without causing any obvious problem, even if they had apparently been cured. They were therefore capable of infecting their wives in whom it too might remain latent until it caused a problem. One such problem was inflammation throughout the woman's pelvis, a condition known as pelvic inflammatory disease, which was often associated with sterility.

There was also another problem. In fact, a big problem: 'ophthalmia neonatorum', a disease in which newborn babies developed pus in the eyes,

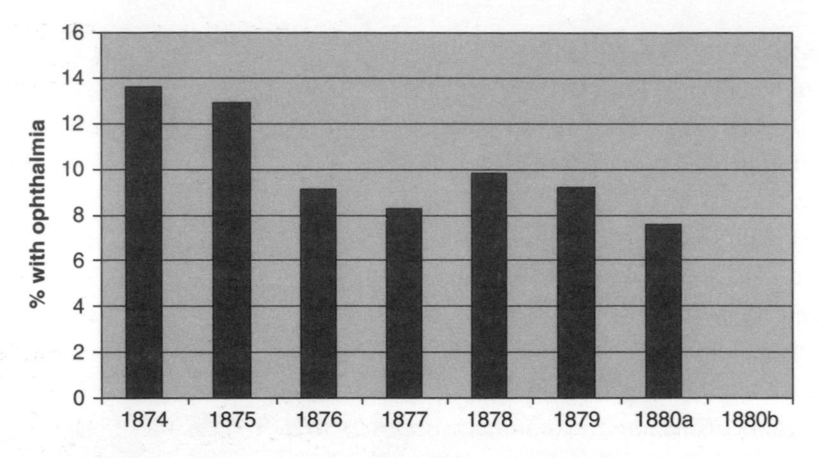

Figure 137. Graph of the percentage of babies born with ophthalmia in Credé's hospital each year. Silver nitrate drops were introduced in the second half of 1880.

often damaging the cornea, and sometimes leading to blindness. This was a major concern in maternity hospitals in Europe in the nineteenth century. Up to 35 per cent of women in maternity hospitals at the time were afflicted with leucorrhoea and many of their babies as well as those of apparently healthy women suffered from ophthalmia. Many thought it likely that the ophthalmia was acquired during passage through the birth canal. From 1874, the Berlin obstetrician Carl Credé (1819–1892) tried various ways of preventing babies acquiring the illness. He found that washing out the mother's vagina (douching) had little effect. From December 1879, vaginal douches were abandoned and a single drop of silver nitrate diluted 2 per cent in water was added to each baby's eyes. A few months later in mid-1880, this was changed to washing each baby thoroughly, wiping the eyes clean with wet cotton wool, then instilling aqueous silver nitrate into their eyes. The results were extraordinary. There were no cases in the second half of 1880 (Figure 137) and only three cases in 1,160 live births over the next three years.

The stage was now set for Albert Ludwig Sigesmund Neisser (Figure 138) to discover what was going on. We have met Neisser already when we discussed his involvement with finding the cause of leprosy. Neisser was born in Schweidnitz in Silesia (now Swidnica, Poland), the son of a Jewish physician. After graduation in 1877, he joined the Royal Dermatological University Clinic of Breslau (now Wroclaw, Poland) as an assistant. At that time in Germany, dermatology and venereology (sexually transmitted diseases) were usually combined as a single specialty. This explains Neisser's interest in leprosy,

Figure 138. Albert Neisser (1855–1916).

gonorrhoea, and syphilis. He began to look for a bacterial cause for gonorrhoea in 1879. His description of exactly who he investigated is not a model of clarity and the various numbers have been interpreted differently by various commentators. At face value, he says he found bacteria in 35 cases of urethritis, apparently male, vaginal pus from two girls exposed to a man with gonorrhoea, nine cases of purulent urethritis in women and seven children, and two adults with conjunctivitis (ophthalmia). On the other hand, he did not find any cocci in 13 cases of randomly collected vaginal secretions. Be that as it may, this is how he described the bacteria he saw, in a paper entitled (in English translation) 'A new species of coccus specific to gonorrhoea'. He wrote that the organisms were:

> seldom seen as solitary individuals; almost always they appear as micrococci packed close together so as to give the observer the impression of a single organism shaped like a figure of eight.[226]

In other words, they were a pair of cocci with the opposing sides of the bacteria often flattened like coffee beans. Neisser noted that they were often peculiarly grouped within white cells and on epithelial cells while those that were not associated with cells were usually found in groups but never in chains. However, he was unable to find them in specimens taken from patients with syphilis or balanitis (inflammation of the prepuce of the penis). All this was certainly extremely suggestive that these organisms were the cause of gonorrhoea but Neisser reserved judgement until they could be grown and transmitted experimentally.

Neisser then tried to cultivate the bacteria but had great difficulty. He reported in 1882 that they appeared to grow on peptone-gelatine medium. It is not surprising that Neisser had trouble growing these bacteria as we now recognize that the nutrients and other conditions have to be just right for them to grow (they are called 'fastidious' organisms). A number of investigators, including

L Leistikow and Friedrich Loeffler (p.194), had more success in growing the organisms on serum made from human blood, as did F Krause with serum from sheep blood as the culture medium. In 1883, the German gynaecologist, Ernst Bumm (1858–1925), grew the bacteria on a medium composed of serum obtained from the human placenta after delivery.

A number of transmission experiments of urethral pus and conjunctival pus had been undertaken during the earlier part of the nineteenth century as we have seen, but we don't know what pathogens were involved and all attempts to infect animals had been a failure. Now an avalanche of transmission experiments followed Neisser's discovery.

In 1880, Arpad Bokai (1856–1919) in Budapest confirmed Neisser's microscopic findings of micrococci in urethral pus, then instilled two drops of fluid (presumably of discharge containing these bacteria) into the urethra of three healthy volunteers, all of whom developed acute gonorrhoea. In a fourth case, discharge was placed beneath the prepuce and in two more men, secretions from a case of ophthalmia were inserted into the urethra; these three remained healthy. Bokai was uncertain what was happening and we can't be sure exactly what he was dealing with either.

In 1883, George Sternberg (p.253), after some hesitation, infected himself and two colleagues by inserting cotton that had been soaked in a 'presumed gonococcal culture' into each person's urethra. Nothing happened. Most likely, he did not have proper gonococcal cultures.

Once pure cultures had been obtained, it was possible to do transmission experiments that could be interpreted properly. This was particularly so when confusion and controversy over the nature of micrococci recovered from the urethra was at least partially resolved by use of the Gram stain. In 1886, Gabriel Roux reported that the micrococci seen in gonorrhoeal urethral discharge stained pink, i.e. were Gram-negative in contrast to staphylococci, streptococci, and pneumococci.

We have mentioned Ernst Bumm. In 1885, he reported that he had produced gonorrhoea through inoculation of pure cultures into the male urethra. This was repeated by the Viennese gynaecologist Ernst Wertheim (1865–1920) in 1891. He cultured pus from acute urethritis and from the Fallopian tubes (the passages from the ovaries to the uterus), then inoculated the cultures into the healthy urethra of five male paraplegics! All developed gonorrhoea.

In 1893, Steinschneider used Wertheim's medium to make cultures which he put in the urethra of a young medical practitioner who had apparently recovered from acute gonorrhoea three months previously and into the urethra of 'another colleague (who) offered himself most obligingly' who had no previous history of gonorrhoea. The first recipient had only minimal symptoms while the second developed typical gonorrhoea.

Even more outrageous experiments in 1893 were undertaken by Ernst Finger (1856–1939) and his colleagues in Vienna. In the first, to assess immunity to reinfection, they infected six men with a history of gonorrhoea but who had apparently recovered. Each person developed acute inflammation so the investigators concluded that previous gonococcal infection gives no immunity to reinfection. The next study seems unbelievable. In an experiment, allegedly undertaken to find out whether the presence of fever makes it less likely one can become infected, they inoculated seven men dying from a fever (one from pneumonia and seven from tuberculosis). Three of them died within 72 hours and at autopsy there was evidence of urethritis. The remaining four died at later times and no evidence of infection was found.

Could gonococci produce diseases outside of the genital tract? A Bolognese, Dr LM Petrone, in 1883 reported seeing bacteria resembling gonococci in pus taken from the knee of two young men with gonorrhoea. Were they really gonococci? Guido Bordoni-Uffreduzzi (1859–1943) in Italy in 1894 proved that such organisms could cause arthritis. He took some pus from the ankle of a young woman with gonorrhoea, cultured it then inserted material into the urethra of a healthy 23-year-old man. Two days later he developed typical gonorrhoea. And so it went on, with yet others performing transmission studies. The ethics of such experimentation seemed to bother hardly anyone. Ethically indefensible though they were, all of Koch's postulates were fulfilled and the matter was settled beyond doubt. How did these gonococci get into joints in the first place? They must have travelled there via the bloodstream. It is not surprising, therefore, that it was eventually found that these organisms could infect the heart valves (endocarditis) or cause abscesses anywhere.

What else can the organism do? Noeggerath was right about pelvic inflammatory disease. Once gonococci could be confidently diagnosed, they were found to cause about half the cases of abscesses involving the ovaries and Fallopian tubes. In men, sometimes urethral stricture (narrowing) developed as did inflammation of the testes (orchitis), epididymis, and prostate, all organs that are part of the male reproductive apparatus. Sexual behaviour has also resulted in infection of the rectum (proctitis), particularly in homosexuals, and in the pharynx (pharyngitis).

What about the name of this organism? Neisser referred to these bacteria as micrococci in his first publication in 1879 but in his second in 1882 he called it 'Gonococcus', apparently at the suggestion of his friend Paul Ehrlich, since it is the coccus that causes gonorrhoea. This name has lived on in popular parlance as 'the gonococcus' but officially it is named *Neisseria gonorrhoeae*; the genus name *Neisseria* was proposed by V Trevisan in 1885 in honour of Neisser. '*Gonorrhoeae*' is a horrible word. How do you pronounce it? Answer—gon-or-ee-eye.

The advent of penicillin ushered a dramatic change in outlook for those suffering from gonococcal infections. But it has not gone away. In the USA there were 193,000 cases in 1941 (147 cases per 100,000 people). It reached its peak in 1980 with just over a million cases (447/100,000). In 2008, there were 337,000 cases (112/100,000). But what is even more worrying is that the organism has fought back and over half the isolates are now resistant to penicillin and some are resistant to other antibiotics as well. The advice given in the 1587 edition of *Mirror for Magistrates* is still germane.

24

Syphilis (the pox)

SCIENTIFIC NAME:	*Treponema pallidum*
ORGANISM'S COMMON NAME:	spirochaete
ORGANISM'S APPEARANCE:	spiral rod which does not stain with the Gram stain but stains with silver and Giemsa stains
DISEASE NAME:	syphilis
COMMON DISEASE NAME:	the pox
DISTRIBUTION:	worldwide
TRANSMISSION:	sexual intercourse; mother to fetus
LOCATION IN A PERSON:	many tissues
CLINICAL FEATURES:	a. primary syphilis: ulcer (chancre) on the genitals and enlarged local lymph nodes, b. secondary syphilis: generalized rash and enlarged lymph nodes, fever, c. latent syphilis: apparently well, d. tertiary syphilis: tabes dorsalis, general paresis of the insane, fits, aortic aneurysm, gummas, e. congenital syphilis: bone and teeth abnormalities, deafness, eye disease, misshapen face, snuffles, enlarged liver and spleen, etc.
DIAGNOSIS:	observation of spirochaetes in primary or secondary lesions; blood tests for antibodies in latent and tertiary syphilis
TREATMENT:	penicillin
PREVENTION:	use of condoms, avoidance of promiscuous sex

In fourteen hundred and ninety two Columbus sailed the ocean blue.

On 22 February 1495, the troops of Charles XIII of France took Naples in order to enforce his tenuous claim to the Neapolitan Kingdom. The campaign had been a cakewalk and the army made merry. Shortly thereafter, a dreadful new disease broke out amongst both the invading army and its hangers-on (who included a large body of prostitutes) and the inhabitants of Naples. It did not take long for the states of the Italian peninsula to get their act together. At the Battle of Fornovo on 5 July 1495, Charles was defeated and scurried back to France taking his troops with him. His army contained mercenaries from various countries, many of whom returned to their own homes. They took this horrendous disease with them. It was so new that it did not have a name. The French called it the Neapolitan sickness and the Italians called it the French disease.

Graphic descriptions soon appeared. Alexander Benedetto (Latin name Benedictus), a Venetian doctor who served at Fornovo, wrote:

> Through sexual contact, an ailment which is new, or at least unknown to previous doctors, the French sickness has worked its way in from the West... The entire body is so repulsive to look at and the suffering is so great... that this sickness is even more horrifying than incurable leprosy or elephantiasis and can be fatal.[43]

Marcellus Cumano, another Venetian military doctor, described it in more detail:

> Several men-at-arms... had pustules on their faces and all over their bodies. They looked rather like grains of millet, and usually occurred on the outer surface of the foreskin, or on the glans [of the penis]... Sometimes the first sign would be a single pustule looking like a painless cyst, but the constant scratching provoked by the itch subsequently produced a gnawing ulceration. Some days later, the sufferers were driven to distraction by the pains they experienced in their arms, legs and feet, and by the eruption of large pustules (which) lasted... for a year or more.[76]

Jean Molinet (1435–1507), the French official historian of the House of Burgundy, described it as:

> the great pox, a violent, hideous and abominable sickness... Since no-one had heard of this awful pestilence... it was called the Neapolitan sickness, some called it the grosses pocques.[220]

This name of 'pox' or 'pockes', a Middle English word for 'pustule', indicating a small blister containing pus, was to become very popular, with 'great pox' being used to distinguish it from smallpox.

The pox did not take long to spread. France was the first to bear the brunt. By 1496 it had crossed the Rhine into Germany, reached Britain by 1497, and Russia by 1499. On 21 April 1497, the town council of Aberdeen in northern Scotland ordered that:

For the protection from the disease which had come out of France and strange parts, all light women [i.e. prostitutes] desist from their vice and sin of venery and work for their support, on pain, else, of being branded with a hot iron on their cheek and banished from the town.[260]

It did not take long for the physicians of the day to get some idea of the nature of the illness they were dealing with. Johannis (Giovanni) de Vigo (1460–1520), surgeon to Pope Julius II, published a book in 1514 in which he made many accurate observations. First, he recognized that it was acquired by sexual intercourse with an infected person, writing: 'It was and still is a contagious disease. The contagion results from sexual intercourse in which one of the parties is infected.'[311] He recognized that the first stage of the disease (now called primary syphilis) begins with ulcers on the genitals which feel hard when palpated, a characteristic called 'induration'. Such an ulcer is now called a chancre:

The first symptoms of this malady appear almost invariably upon the genital organs, that is, upon the penis or the vulva. They consist of small ulcerated pimples of a colour especially brownish and livid... These pimples are circumscribed by a ridge of callous-like hardness.[311]

Vigo also understood that there was then a second stage a few weeks later in which the lesions became widespread (now called secondary syphilis):

There then appear a series of new ulcerations on the genitalia... Then the skin becomes covered with scabby pimples or with elevated papules resembling small warts. These eruptions appear especially on the forehead, the skull, the neck, the arms, the legs, and they spread sometimes over the entire surface of the body... A month and a half, about, after the appearance of the first symptoms, the patients are afflicted with pains sufficiently severe to draw from them cries of anguish.[311]

He also knew about the even later appearance of lesions now called gummas (or gummata):

Still very much later (a year or even longer) there appear certain tumours of a scirrhus hardness [i.e. very hard] which provoke terrible suffering. The pain which they produce is characteristic in being aggravated during the night and diminished during the day... In the end, these horrible sufferings terminate most commonly in lesions of the bones... As to the secondary ulcers, they vary in form to infinity. To tell the truth, they assume all appearances, they take all kinds of masks... In fact, the multiple manifestations of this odious malady vary in an incredible manner from one subject to another.[311]

Vigo thought all these features were due to infection with, and spread of, a 'morbid virus'. Of course, he had no idea what the nature of this 'virus' was—that would take another 400 years to become apparent.

Jacques de Béthencourt, a physician in Rouen, France, took umbrage at this illness being called the 'French disease'. He hit upon a new name which he published in 1527—'morbus venereus' (= venereal disease: Latin 'morbus' = illness) meaning 'malady of Venus' since it arose from illicit love. Of course, there are many venereal diseases, so the illness we are discussing is today known by the name bestowed upon it by the Italian polymath, Hieronymus Fracastorius (p.174). In 1530, he wrote a poem, 'A Poetical History of the French Disease', in which a shepherd named Syphilis was struck with the French disease following an act which he came to regret. The poem became so famous that the word 'syphilis' soon became the near universal official term for the illness and displaced almost all other designations. The exception was 'lues venerea', derived from the Latin 'lues' meaning 'plague' which has had some currency in medical circles.

Because this disease was novel, distinct, and horrendous, the physicians of the late fifteenth and early sixteenth centuries had no difficulty in distinguishing it from urethritis, another sexually acquired illness. With the passage of time, medical men forgot this lesson and the two illnesses became merged into one. We saw in the last chapter how John Hunter in 1767 inoculated 'venereal matter from a gonorrhoea' into a subject and produced syphilis (p.261). Who was this subject? This is what Hunter wrote:

> Two punctures were made on the penis with a lancet dipped in venereal matter from a gonorrhoea; one puncture was on the glans, the other on the prepuce. This was on a Friday; on the Sunday following there was a teazing itching in those parts, which lasted till the Tuesday following. In the meantime, these parts being often examined, there seemed to be a greater redness and moisture than usual... the parts of the prepuce where the puncture had been made were redder, thickened and had formed a speck. By the Tuesday following, the speck had increased and discharged some matter.[146]

He went on to describe, on almost a daily basis, the various treatments applied to the two lesions which had become ulcerated. He added:

> Four months afterwards the chancre on the prepuce broke out again... That on the glans never did break out... While the sores remained on the prepuce and glans, a swelling took place in one of the glands of the right groin (which he called a bubo)... About two months after the last attack of the bubo... in one of the tonsils... a small ulcer was found.[146]

Speculation has been endless as to whom this subject was and we shall doubtless never know for sure. Ingenious arguments have been adduced that the subject was Hunter himself, but equally persuasive reasons have been put forward to indicate that it was not. He delayed his marriage. Some have suggested that this was to allow time for him to 'cure' his syphilis; others have maintained the delay was due to lack of money. Hunter died suddenly in 1793 during an argument at St George's Hospital over the admission of students. Some say this was a heart attack while others postulated that he had syphilitic disease of the heart and aorta. The findings at Hunter's post-mortem examination have been interpreted as refuting this idea but autopsies in those days were not altogether reliable or enlightening. But perhaps you can make your own judgement. If you were to ask me my opinion, I would say that he experimented on himself, simply on the grounds that he described many repeated examinations of the genitalia. How else would he have had such easy access?

Whoever the subject may have been, the important point was that Hunter concluded that the clap and the pox were one and the same. Such was his reputation that in many quarters this became accepted as the gospel truth. But as we have seen, not everyone agreed. Benjamin Bell (p.262) argued on epidemiological grounds (i.e. occurrence, frequency, and distribution) that the two were different, while Ricord reached the same conclusion from experimental studies.

Philippe Ricord (Figure 139) was born in Baltimore, USA, to where his parents had fled during the French Revolution but he returned to France at the age of 20 and studied medicine in Paris. In 1831, he was appointed to l'Hôpital du Midi, a hospital for venereal diseases where he was to remain for the next 30 years. The first thing he did was to reintroduce the use of the vaginal speculum (an instrument for looking inside the vagina). The then standard means of determining whether or not a prostitute was infected with syphilis was by simple inspection of the vulva and palpation of the vagina (which often infected the medical men, giving them a chancre

Figure 139. Philippe Ricord (1800–1889).

on the finger, ultimately with dreadful consequences). By being able to see, he was able to find unsuspected chancres in the vagina or on the cervix.

His second innovation was the technique of 'autoinoculation' as a diagnostic aid. Ricord believed that it was quite wrong to infect normal people with material from genital lesions but thought that it might be helpful to inoculate people with material from their own lesions. Between 1831 and 1837, he performed more than 2,500 autoinoculations in which he injected material from a genital ulcer, urethral discharge, or a lymph node which had burst, into the patient's own thigh then watched it daily for the development of lesions. He found that when material was from a chancre or from a bubo (an enlarged lymph node), then a pustule followed by an ulcer always appeared on the thigh. On the other hand, material from a urethral discharge (gonorrhoea or chlamydial urethritis), the ulcers of secondary syphilis, or certain other lesions, did not produce any ulcers on the thigh. He therefore concluded that there was just one specific cause of syphilis and that gonorrhoea, balanitis (inflammation of the prepuce), and certain warts were not due to syphilis. This re-inoculation technique, he believed, could be used to prove whether or not a venereal ulcer was syphilitic.

More importantly, Ricord made a significant advance in the classification of syphilis. He expanded de Vigo's concept of a two-stage disease into three levels. The first stage (primary syphilis) was characterized by a chancre which became prominent shortly after exposure. The next (secondary syphilis) featured generalized lesions of the skin and mucous membranes which appeared one to several months after contact. Ricord's third stage (tertiary syphilis) resulted in lesions of the bones and joints and internal organs one or more years after infection. As we shall see, more terrifying examples of tertiary syphilis would be discovered in the 1870s.

Unfortunately, Ricord made several errors. First, he thought that whether or not the inoculated ulcer was indurated (hard) depended upon individual predisposition and that both hard and soft chancres were syphilitic. In 1852, one of his former students, Edmond Bassereau, carefully studied sexual contacts and showed that there were two infections. One involved hard chancres and was associated with systemic syphilis (i.e. lesions could be found throughout the body). The other had soft chancres and was associated only with local complications; this condition was called 'chancroid' (like a chancre) and was shown in 1889 by Augusto Ducrey (1860–1940) in Naples to be due to a bacterium now called *Haemophilus ducreyi*.

Secondly, Ricord thought that secondary syphilitic ulcers were not infectious. In 1851, August Vidal de Cassis (1803–1856), Ricord's junior colleague at the hospital, not having Ricord's ethical scruples, took material from a pustule on the breast of a patient with secondary syphilis, inoculated it into a pharmacy student, and induced a typical primary chancre. Controversy continued until a Parisian committee was appointed to investigate the matter. The committee's findings were handed down in 1859. It found that the clinical, epidemiological, and experimental evidence was overwhelming that secondary syphilis was infectious. Ricord mounted the podium to acknowledge that he had made a serious error.

Then there was the question as to whether or not animals could be infected with syphilis. Ricord did not believe they could. In 1843, Joseph-Alexandre Auzias-Turenne (1812–1870) inoculated pus from chancres of humans into experimental animals including monkeys, dogs, cats, and rabbits and thought that some of these developed chancre-like lesions. His findings were greeted with derision. When Auzias-Turenne brought a monkey with facial lesions to the Society of Surgery, Ricord made the extraordinary assertion:

> One ought to have the courage of one's convictions...M Auzias should inoculate himself with pus from one of his monkeys' ulcers and wait for the appearance of symptoms.[272]

Eventually, Ritter von Welz, a young German doctor, volunteered for the crucial experiment and inoculated himself with pus from a five-day-old chancre of one of Auzias's monkeys. Five days later an ulcer appeared. Reluctantly, Ricord admitted he might have been wrong. Auzias noted that when experimental animals were inoculated repeatedly, the ulcers became progressively smaller then finally did not appear at all. He thought that this might be a way of treating intractable syphilis and called the process 'syphilization'. He was prevented from carrying out proper clinical trials on prostitutes in Paris by Ricord but the phenomenon was investigated elsewhere. Tragedy was inevitable. In 1851, a young German doctor named Lindeman repeatedly infected himself with human ulcer material and finally gave himself syphilis. Auzias treated him by syphilization, made him worse, and he died. This was the beginning of the end of syphilization as a therapeutic measure.

Another important issue was the transmission of syphilis from an affected mother to a child. This had been alluded to by various authors since the sixteenth century. Some children survived into adulthood and the person who brought it all together was the English surgeon Jonathan Hutchinson (1828–1913). At a meeting of the British Medical Association in Edinburgh in 1858 he gave a presentation entitled 'On the means of recognizing inherited syphilis in adult life'. Today, of course, we would say 'congenital', meaning present at birth, rather than 'inherited', which nowadays implies a genetic disorder. Perhaps the most famous part of his talk was his description of what are known as Hutchinson's teeth:

> The state of the teeth constitutes one of the most important signs that we possess, more especially that of the permanent teeth... The incisors... are small, rounded, peg-shaped instead of flat, presenting a broad shallow notch in their edge, at others so much worn down that this notch is not distinguishable... their colour (is) a dirty brownish hue.[147]

Meanwhile, Antoine Laurent Jessé Bayle (1799–1858), a physician working at the Royal Asylum for the Insane at Charenton in France, had made a most important observation. He was a nephew of Gaspard-Laurent Bayle (p.206) and had studied with Laënnec, both of whom had shown how important post-mortem examinations were in understanding tuberculosis. Having learnt this principle, Antoine began to do autopsies on inmates in the asylum with a disease which had been increasingly recognized since the end of the Napoleonic Wars (1815). This condition, which was known as 'general paresis of the insane' (GPI; 'paresis' means paralysis), most commonly afflicted men between the ages of 20 and 40 but also affected women. Months to years after the first appearance of symptoms, the victims became demented and died. The prevailing interpretation of the paralysis was that it was a complication of insanity.

Bayle correlated the clinical histories with post-mortem findings in these patients. He recognized three stages of the illness in life. The first was characterized by a mild paralysis, which especially affected speech, and was accompanied by a monomania in which a person was dominated by one irrational set of ideas, especially grandiose ones. This was followed by a generalized mania in which the patient became greatly excited, elated, overactive, and had extravagant delusions not based on reality, together with a worsening of the paralysis. The third and final stage was of dementia and severe paralysis. In 1822, Bayle presented the results of his first six autopsies. He had discovered a chronic inflammation of the arachnoid (arachnoiditis), the middle of the three membranes that surround the brain and spinal cord, and argued that both the mental and physical symptoms were a consequence of this chronic inflammation which affected the blood supply to the brain. He was roundly criticized and personally attacked, perhaps because he dared to suggest that a psychiatric illness had a physical basis. Undeterred, he continued doing autopsies and in 1826 published a 600-page treatise with a detailed description of 90 cases with chronic meningitis. This met with no better reception. Disgusted, Bayle gave up not only psychiatry but medicine *in toto*, and became a librarian!

Bayle had no idea that syphilis might be the cause of the disease he had been studying. In 1857, the German surgeon Johann Friedrich August Esmarch (1823–1908) and psychiatrist Peter Willers Jessen (1793–1875) described in detail three of their patients with GPI who had syphilis. They recognized the limitations of their observations but wrote:

> Three cases, naturally, could not determine the difficult question whether this is cause or complication, and it must be admitted that syphilis can also occur as a chance complication of psychic disturbance... We wished, however, only to express our belief concerning the aforementioned cases and to stimulate similar observations and also differences of opinion.[93]

Unfortunately, their hopes were not realized and their report was ignored. Indeed, many psychiatrists distinguished (quite incorrectly) between GPI and 'pseudo-general paresis of syphilitics'. At this point Alfred Fournier enters the scene. But we need to divert for a minute.

In 1840, Moritz Heinrich Romberg (1795–1873), a German physician, gave the first clinical and pathological description of a disorder he called 'tabes dorsalis' (tabes means 'wasting' while dorsalis refers to degeneration at the back of the spinal cord). This was then described in even more detail in 1858 by the French doctor Guillaume Duchenne (1806–1875), who gave it the name 'locomotor ataxia' (meaning 'difficulty in walking straight'). Duchenne noted that a few of these patients suffered from syphilis but was uncertain of the relationship.

Jean Alfred Fournier (1832–1914) had studied under Ricord and in 1875 was working at the Hôtel-Dieu de Paris. In that year he presented convincing evidence of an association between syphilis and locomotor ataxia:

> For a long time I have been struck by the fact that most patients with ataxia, or at least a great number among them, are ancient syphilitics. As this fact drew my attention, I was engaged in verifying it in numbers; and, for 30 cases of ataxia that I have observed for some years, I arrived at the result that in 24 cases syphilis was present in their past medical history. Twenty four times in 30, what a proportion![99]

Fournier's hypothesis gave rise to much controversy among neurologists (nerve specialists) but within a few years was generally accepted as others found the same relationship. Why did this association take so long to establish? The symptoms generally arose more than 10 years after the first appearance of a chancre or the rash of secondary syphilis and most patients had forgotten having had those problems. Proof of Fournier's hypothesis, of course, had to await discovery of the bacterium that causes syphilis.

Let me give you some feel for this condition. I still remember the first patient I diagnosed with tabes dorsalis in the 1960s. He was a man of about 60 who complained of progressive difficulty in walking and severe, shooting pains in his legs. When I examined him, he could not tell when his eyes were closed whether I moved his toes up or down (this is called loss of proprioception), he could not feel the vibrations of a tuning fork applied to his feet, his ability to feel cotton wool touching his feet was impaired, and he had lost the reflexes when his ankles were tapped with a hammer. When he stood up and closed his eyes, he swayed as he lost his balance (Romberg's sign). When I shone a light in his eyes, the pupils did not constrict but they did when he focused on a pencil held in front of his nose (Argyll–Robertson pupil). On top of that his memory was not too good. It was indeed a miserable condition for which there was no cure, even in the penicillin era.

Now to return to Fournier. Having made the connection between tabes and syphilis, he was convinced that the nervous system was the victim par excellence of tertiary syphilis and set out from 1879 onwards to prove that syphilis caused GPI. Over time he built up an extraordinary card system of records of 50,000 cases that were to form the basis for all his studies. He had to work prudently because he wanted to avoid confrontation with psychiatrists who might be unwilling to consider the evidence of a syphilologist. In a renowned lecture at the French Academy of Medicine in 1894, he showed that up to 90 per cent of patients with GPI had a history of syphilis

But it was not cut and dried. Many could not accept that this awful illness could be caused by an acute syphilitic condition with years of apparent good health in between and which, furthermore, did not respond to the treatments used to treat early syphilis. The question was still being bitterly debated in 1905, when the crucial discovery was made that would eventually settle the question.

But before we examine that discovery, there are a couple of other clinical syndromes we need to know about. In the late sixteenth century, the famous French surgeon Ambroise Paré (1510–1590) recognized an association between syphilis and aortic aneurysm. The aorta is the great artery by which blood leaves the heart and is distributed to the rest of the body. First it curves upwards (the ascending aorta), then becomes horizontal (the arch of the aorta) from which the arteries of the head and arms arise, then finally it turns downwards towards the pelvis and the legs. An aneurysm is a distension or ballooning outwards of the normally cylindrical aorta. If it distends too much, then it is at risk of rupturing with fatal consequences. Paré presciently wrote:

> The aneurysms that happen in the internal parts are incurable. Such as frequently happens to those who have often had the unction and sweat for the cure of the French disease . . . It (the aorta) distends to that largeness to hold a man's fist.[240]

This relationship was first documented thoroughly by Francis Welch (1839–1910), an Army surgeon at Netley in England. He performed post-mortem examinations on 34 soldiers, aged 26–42 years, who had aortic aneurysms and found that two-thirds of them occurred in the ascending aorta and the arch. He showed that the distension was caused by a loss of elasticity of the aortic wall as a result of inflammation of the small blood vessels supplying the aorta which resulted in them becoming blocked and obstructing the flow of blood (a condition called obliterative endarteritis). What is more, two-thirds of these men had syphilis. Welch thought that the case was made. He reported his findings to the Royal Medical and Chirurgical Society in 1875. None of the five eminent surgeons who discussed his presentation agreed with him. But time was to prove Welch right; aortic aneurysms are another feature of tertiary syphilis.

Welch had made an important observation in describing inflammation and obstruction of small blood vessels which led to death of the tissues they supplied (a process called infarction). Over the ensuing years, pathologists showed that this phenomenon accounted for the gummas mentioned earlier. Moreover, arteritis and infarction could affect any blood vessels anywhere. No wonder then that Hutchinson called syphilis the great 'imitator' and showed how it could mimic just about anything. He told a meeting of the British Medical Association in 1879:

> An idea has much occupied my mind during the past few years . . . It is that we ought to recognise in syphilitic forms of inflammation imitations, with certain differences, of other more common types of diseased actions which occur without any connection with specific taint.[148]

When I was a medical student, it was only 20 years or so since the introduction of penicillin and there were still plenty of cases of tertiary syphilis about. I had one very enthusiastic teacher who loved to give us lists of possible causes for every syndrome. Naturally, we were required to regurgitate these lists from time to time. We could always safely add syphilis to the list. But since this interrogation went on in front of the patient, we always had to use a pseudonym such as 'lues', 'the spirochaete' or 'WR +ve', and were gratified to see nodding approval.

Perhaps you have been a bit overwhelmed by all this information and find it a bit confusing. Don't worry. Imagine what it was like for the physicians of the nineteenth century and before. We have reached the point where it might be helpful to summarize the principal clinical syndromes of syphilis. Have a look at Table 3.

But what is it that causes syphilis? Until 1905, no-one had any idea. In February of that year, the director of the Institute of Zoology at the University of Berlin announced that an assistant in his department, Dr Siegel, had found protozoa in syphilitic chancres which he believed were the cause of syphilis. The Health Institute therefore commissioned Fritz Schaudinn and Erich Hoffmann to investigate. Schaudinn (Figure 140) was a protozoologist whose studies of amoebiasis we have encountered earlier (p.280), while Hoffman (Figure 141) was a dermatologist and syphilologist in the dermatological clinic of the University of Berlin. It is often said that Schaudinn discovered the cause of syphilis but without Hoffman there was no Schaudinn, and without Schaudinn there would be no Hoffmann. Their partnership gives a foretaste of the collaborative experiments that were to become such a feature of the rest of the twentieth century.

Hoffmann provided fluid from chancres, pustules of secondary syphilis, and from enlarged lymph nodes. It did not take Schaudinn long to realize that the protozoa Siegel had seen were saphrophytes (non-pathogens living on dead

Table 3. Summary of various stages of syphilis

- *primary syphilis*—a chancre (ulcer on the genitals), 1–4 weeks after infection, often associated with swelling of the regional lymph nodes
- *secondary syphilis*—a generalized rash on the skin and mucous membranes, generalized lymph node enlargement, and fever, 2–12 weeks after infection
- *latent period*—during which the patient appears to be normal and which can last for 5–20 years
- *tertiary syphilis*—appearing after the end of the latent period as neurosyphilis (tabes dorsalis and general paresis of the insane), aortic aneurysm, gummas, endarteritis affecting just about any organ
- *congenital syphilis*—disease of the newborn child with a huge variety of symptoms and signs

Figure 140. Fritz Schaudinn (1871–1906).

Figure 141. Erich Hoffmann (1868–1959).

tissue) and not the cause of syphilis. On 3 March, he found fine spiral organisms in an unstained smear of juice from a secondary papule. Schaudinn was not sure at this point that he had seen the cause of syphilis. But then he found them in specimens from a syphilitic lymph node which had been stained with Giemsa stain. He photographed them (Figure 142). Now he was more certain and in April 1905, Schaudinn and Hoffmann announced their findings in a joint publication. By May they had found them in primary chancres, secondary pustules, and in lymph nodes of syphilitic patients. They named this organism *Spirochaeta pallida*. *Spirochaeta* was derived from the Greek words *speira* meaning 'coil' and *chaeta* for 'hair', while *pallida* was from the Latin 'pallidus' meaning 'pale'. Later in 1905, Schaudinn renamed it *Treponema pallidum*, the genus name reflecting the Greek words *trepein* and *nema* for 'turn' and 'thread'. They described it thus:

> The length of the spirochaete varies from 4 to 10 μ [micrometres] ... the width varies from a non-measurable size to about 0.5 μ ... The number of turns of the spiral fluctuates between three and twelve ... The movements are ... of three kinds: rotation about the long axis, gliding movements forwards and backwards, and movements of flexion of the whole body.[283]

Nevertheless, Schaudinn and Hoffmann remained cautious, for as you will have appreciated, Koch's postulates had not yet been fulfilled.

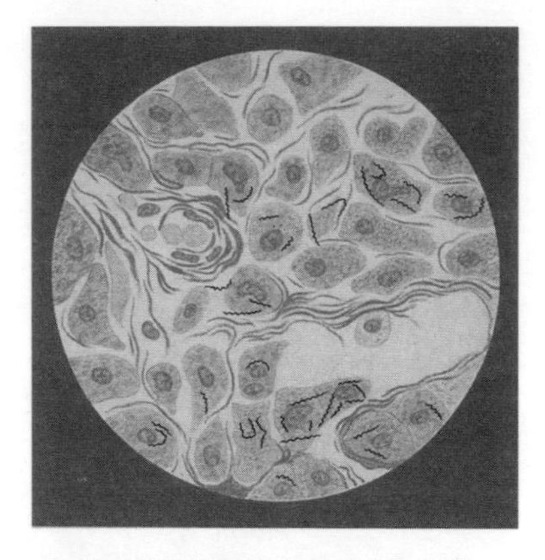

Figure 142. Schaudinn's photograph of spirochaetes stained with Giemsa stain in a syphilitic lymph node in the groin.

Hoffman therefore set about trying to prove that point. The postulates required that the organism being studied should be grown in pure culture. Unfortunately, these spirochaetes would not grow readily on artificial media so he had to proceed directly. Hoffmann rubbed blood from a young man with primary syphilis around the eyes of a rhesus monkey; a papule developed in which he found many spirochaetes. Next, he repeated the experiment with blood from another man with primary syphilis but this time infected a baboon; again he found spirochaetes. He then tried two more monkeys. They failed to become infected with blood from syphilitic patients but they did so when material from syphilitic lesions was used. The case was becoming unassailable.

What about the brain? Could spirochaetes be found in the brains of patients with GPI? Hideyo Noguchi (1876–1928) and John W Moore of the Rockefeller Institute in New York set out to find out. Noguchi, incidentally, was born in and graduated in medicine in Japan. When he was one year old, he fell into a fire and burnt off most of the fingers of his left hand. Thinking this would make clinical practice difficult, he turned to medical research for his career. This would eventually take his life, for he died in West Africa while studying yellow fever. Noguchi and Moore examined the brains of 70 patients who had died from GPI, a number of whom had a clear-cut history of syphilis. In 1913 they reported that they had found spirochaetes in the brains of 12 patients. Like Bayle many years before, they found chronic inflammation of the membranes around the brain and the underlying brain tissue was affected. It seemed that syphilis was rather like tuberculoid leprosy in which there was marked inflammation but few bacteria. There really was no longer any doubt that syphilis, including the form manifesting as GPI, was due to the bacterium *T. pallidum*.

But there were some practical problems in diagnosis. In primary and secondary syphilis, the bacteria were difficult to see and did not stain with the Gram stain (although they did with Giemsa stain). In 1906, use of a microscope modified to permit 'dark-ground illumination' (rather like the eclipse of the sun) made it easier to see the spirochaetes. And of course, how could one get a specimen from the brain or aorta in tertiary syphilis? This problem was solved, also in 1906, when August von Wasserman (1866–1925), Albert Neisser (p.224), and Carl Bruck (1877–1944) developed a blood test that showed the presence in the blood of a peculiar protein now called 'reagin' which was present in secondary and tertiary syphilis. That problem was solved. And now you understand my cryptic reference earlier to 'WR +ve', which is simply an abbreviation of 'Wasserman reaction positive'.

At the beginning of the chapter I quoted the ditty about Christopher Columbus that I am sure you learnt at school. Why? Columbus returned from his first voyage to the Caribbean in March 1493. Syphilis exploded in Europe in 1495. Naturally, this provoked suggestions that he and his crew and/or his captive Indians brought the disease back to Europe with them. Others have disputed this theory, saying that it all happened too quickly or perhaps there was evidence of syphilis in Europe before that time. The pendulum of opinion has swung back and forth over the years but now the question seems to have been settled. The evidence comes from two sources. First, detailed studies of bones (and syphilis has characteristic effects on bones) have shown that syphilis was present in the Americas including the island of Hispaniola (the present-day countries of Haiti and Dominican Republic) before Columbus's arrival there, but it was not present in Europe before that date. Secondly, DNA studies of various isolates of *Treponema* have clearly pointed to a South American origin of syphilis. There seems to be truth after all in the adage that syphilis is the Americas' revenge on Europe.

25

The meningococcus and meningitis

SCIENTIFIC NAME:	*Neisseria meningitidis*
COMMON NAME:	meningococcus
ORGANISM'S APPEARANCE:	Gram-negative coccus, often in pairs
DISEASE NAME:	meningococcal disease
DISTRIBUTION:	worldwide
TRANSMISSION:	airborne or kissing
LOCATION IN A PERSON:	pharynx (throat)
CLINICAL FEATURES:	fever, haemorrhagic rash, vomiting, stiff neck
DIAGNOSIS:	recovery of the organism from blood or cerebrospinal fluid
TREATMENT:	penicillin or cefotaxime
PREVENTION:	a partially effective vaccine is available, administration of antibiotics to those exposed to someone with meningococcal disease

The truck driver, a youngish man, started to feel a bit seedy at 10 o'clock in the morning, went to the Emergency Department of Sir Charles Gairdner Hospital in Perth, Western Australia, at 11 o'clock, was seen by a doctor at noon, and was in the Radiology Department at 1 o'clock having some X-rays taken. At 2 o'clock in the afternoon, he was dead. That is the fastest that I have heard of someone dying from overwhelming meningococcal septicaemia. He did not have time to develop meningitis. More common is meningococcal meningitis in which the course of the illness slows down to a day or so rather than a few hours. The patient is usually a child or young adult who complains of severe headache, vomits, develops a stiff neck, and a characteristic rash appears of petechial spots (which are small haemorrhages in the skin). Sometimes this progresses to block the blood supply to the fingers and toes which become gangrenous (dead). I don't know about where you live, but in Australia these infections are beloved of the media, who are forever reporting the latest case of 'meningococcal' or, worse, 'meningococcal virus'. Of course, the illness is due to a bacterium and not a virus and 'meningococcal' is an adjective which is left hanging; they should say 'meningococcal infection' or 'meningococcal disease' if not 'meningococcal meningitis'. But I suppose that is too much of a mouthful.

Meningitis means inflammation of the meninges. The meninges (Greek *meninx* = membrane) are the three layers of membranes that surround the brain and spinal cord (which together make up the central nervous system). One, the pia mater, is attached to the nerve tissue. The other two, the arachnoid on the inside and the dura mater on the outside, are attached to the bones on the inside of the skull and vertebral column. The pia mater and the arachnoid are separated by a space filled with cerebrospinal fluid; this subarachnoid space makes it easy for bacteria to traverse up and down the central nervous system in meningitis.

Whether or not the ancient Greek and later Arab physicians recognized meningitis is a matter of some debate. Pathological anatomists of the seventeenth and eighteenth centuries, such as Thomas Willis (p.205) in England and Giovanni Battista Morgagni (1682–1771) in Italy, saw pus on the surface of the brain in some patients. The condition which we would now call acute meningitis went by a number of names such as phrenitis, cephalitis, and brain fever until the French popularized the term 'meningitis' around 1800.

In the spring of 1805, a terrifying outbreak of acute meningitis struck Geneva with a total of 33 deaths during three months, mainly among the young. These outbreaks were described in separate papers by Gaspard Vieusseux (1746–1814) and Jacques-André Matthey (1778–1842) in the following year. Vieusseux described the clinical features:

> It commences suddenly with prostration of strength, often extreme, the face is distorted, the pulse feeble, small and frequent... There appears a violent pain in the head, especially near the forehead; then there comes... vomiting of greenish material, of stiffness of the spine, and in infants, convulsions. In the cases which were fatal, loss of consciousness followed. In the first case (i.e. fatal cases) the disease lasts from 12 hours to 5 days but not beyond... In most of the patients who died in 24 hours or a little after, the body is covered with purple spots... Examination of the body showed most frequently a sanguineous [bloody] engorgement of the brain.[310]

The rest of the body looked pretty normal at autopsy and both Vieusseux and Matthey, perhaps surprisingly, called it a non-contagious malignant cerebral fever (*fièvre cérébrale maligne non contagieuse*). I say surprisingly, because many of the cases occurred within the same household almost simultaneously which would surely suggest infection.

Shortly after the epidemic in Geneva, an outbreak occurred in Medfield, Massachusetts, USA, in March 1806 in which nine people died. The illness, described by Drs Lothario Danielson (1765–1841; he subsequently changed his surname to Donaldson) and Elias Mann (*c.* 1778–1822) paralleled that seen in Geneva, with patients dying within 24 hours. They performed post-mortem examinations on five of the patients. The first was on a 10-year-old boy:

> On removing the cranium and dividing the dura, there was discharged... half an ounce of a serus [*sic*] fluid. The dura and pia mater in several places adhered together, and both to the substance of the brain. The veins of the brain were uncommonly turgid... and the substance of the brain remarkably soft, offering scarcely any resistance to the finger when thrust into it.[79]

The body of his seven-year-old sister was similar:

> Between the dura mater and the pia mater was effused a fluid resembling pus, both over the cerebrum and cerebellum, the veins of the brain turgid with blood.[79]

In the remaining three cases, the only abnormality noted was that the veins of the brain were 'remarkably turgid with a very dark colored blood'.[79]

Sporadic cases and small epidemics were noted repeatedly over the next few decades in both Europe and North America but the cause remained a mystery and all attempts at treatment were futile. In 1867, the Philadelphia physician, Alfred Stillé (1836–1900), suggested calling the illness 'epidemic meningitis' or 'cerebrospinal meningitis' rather than 'spotted fever', a name which had become popular in the USA, because a skin rash was not always present. In 1884, the Baltic neurologist Woldemar (Vladimir) Kernig (1840–1917) described the sign that now bears his name and indicates meningeal irritation from infection or blood in the cerebrospinal space; the patient can't straighten the bent knee when the leg is bent at the hip because of the pain the attempt produces.

Light began to be shed on the cause of meningitis in 1880. As we have seen, Karl Eberth found bacteria, which turned out be pneumococci, in the cerebrospinal fluid of a patient who died from lobar pneumonia and meningitis. The breakthrough with regard to the epidemic disease that we have reviewed came in 1887.

In that year, the Austrian pathologist and bacteriologist Anton Weichselbaum (Figure 143) described his findings in eight patients who had died of acute cerebrospinal meningitis between 1885 and 1887. Two patients were infected with the pneumococcus. In the other six, he observed a different micrococcus in the cerebrospinal fluid. The bacteria resembled the shape of gonococci which had been described by Neisser in 1879. They occurred in pairs or fours with their adjacent surfaces flattened. The cocci stained pink with the Gram stain (they were Gram-negative) and usually lay within the cytoplasm of white cells, thus resembling gonococci, in contrast to Gram-positive pneumococci. Weichselbaum was able to grow the bacteria on agar at 37°C but not at 20°C. He then injected the cocci he grew into the chest or abdominal cavities of mice, which died within 48 hours, and the organisms were again recovered, thus fulfilling Koch's postulates.

Figure 143. Anton Weichselbaum (1845–1920).

The happy state of affairs became somewhat confused when Heinrich Jaeger of Strassburg in 1895 described finding diplococci in an epidemic of cerebrospinal meningitis which resembled those described by Weichselbaum except that Jaeger thought that after growth on agar, the bacteria became Gram-positive. It appears Jaeger's bacteriological techniques were faulty. It was eventually shown that Jaeger had contaminated his cultures with another organism, although he probably found Weichselbaum's organism in his original preparations.

Meanwhile, after an absence of a number of years, an epidemic of acute meningitis occurred in Boston between June 1896 and October 1897 in which there were 114 cases. The outbreak was described by the pathologists William Councilman (p.106), Frank Mallory (1862–1941), and James Wright (1869–1928) in 1898. Lumbar puncture was performed in 55 cases and in 38 of these Weichselbaum's diplococci were found. This technique of lumbar puncture had been introduced by the German physician Heinrich Irenaeus Quincke (p.107) in 1890. He inserted a fine needle with a cannula between the lumbar (lower back) vertebrae into the subarachnoid space to tap cerebrospinal fluid. He did this originally in infants to relieve the pressure of hydrocephalus (water on the brain) but soon realized it was an incomparable way of proving a diagnosis of meningitis and finding the infecting organism. Clearly the technique soon caught on for it was used in Boston only a few years later. Councilman and his colleagues performed post-mortem examinations on 35 cases and meningococci were found in 31 of them. There was no doubt. This organism could be a major problem.

Weichselbaum named the bacterium he saw *Diplokokkus intracellularis meningitidis* to reflect that they were cocci in pairs, were commonly found within white cells, and were seen in meningitis. Many other names were applied to this organism but it was eventually transferred to the genus *Neisseria*, which had been erected in honour of Albert Neisser (p.224), and named *Neisseria meningitidis*, retaining Weichselbaum's designation of *meningitidis* to indicate its being the cause of meningitis. Since 1890, this organism has been commonly known as the meningococcus. Everybody pronounces the 'g' in meninges and meningitis as a soft 'g', i.e. 'j'. How do you pronounce meningococcus and meningococcal? The media almost always uses the soft 'g' but since my youth I have used a hard 'g' as in 'go'. I must have been taught that way. So I looked up the *Complete Oxford English Dictionary* and was gratified to find that the official pronunciation in both British and US English is the hard 'g'. But, of course, you can say it however you want.

The question then arose as to where this organism came from. Mice could be infected relatively easily but other animals were difficult to infect experimentally and it became clear that in nature, meningococci only infect humans. A number of studies beginning in 1896 showed that about 10 per cent of the population have meningococci in their nasopharynx (behind the tonsils), usually without causing any trouble, a condition known as the carrier state. The bacteria were transmitted from one person to another through the air or by kissing. In 1934, an English microbiologist working in the USA, Geoffrey Rake (1904–1958), showed that there were three types of carriers. Some were transient carriers with infection lasting a few weeks. Others were intermittent carriers with the carrier state coming and going. Yet others carried the bacteria chronically for at least two years.

In 1909, Charles Dopter (1873–1950), who was to become Medical Inspector-General of the French Army, reported that meningococci recovered from various carriers did not all have the same antigenic structure. This means that the structures making up the surface of the bacteria had different shapes. Consequently, the different strains produced different antibodies and this partly explains why attempts to treat patients with serum therapy (the injection of serum from rabbits immunized with dead meningococci) did not work. Following intensive studies of the widespread epidemics that occurred during World War I (1914–18), these different strains were gathered into three major groups, eventually called A, B, and C. In addition, there are other groups, Y, W-135, and X, which have become problems in recent years.

Differences of opinion have been expressed by various investigators, but it seems that whether or not an epidemic occurs depends, in different circumstances, on the proportion of carriers in the population, the density of people living in confined spaces, as well as on the nature and virulence (disease-causing potential) of the bacteria that are being carried. All antigenic groups can cause epidemics although the attack rate (proportion of the population with disease) is less for group B meningococci which tends to cause sporadic (isolated) infections.

In 1933, Rake and his colleague Henry W Scherp in the USA showed that the antigens of group A meningococci were polysaccharides (many sugars) that were found in the capsule that surrounded each organism. Different polysaccharides were subsequently defined for each of the groups. Vaccines have been made for meningitis but they are only partially effective and often not administered to the population at large. Vaccines against group A meningococci are relatively effective as are the recently introduced vaccine against group C meningococci. Unfortunately, half the cases in Australia, for example, are due to group B meningococci for which no effective vaccine yet exists.

Meningococcal infection is always going to be with us. Not only is it a problem for those unfortunate enough to develop the disease, but also for the doctors who have to look after them. Antibiotics can save lives if given in time. Unfortunately, this is a horrendous illness that progresses with terrifying rapidity but which is often extremely difficult to diagnose and differentiate from more minor, non-life-threatening conditions in the early stages when it is most treatable. Community expectations are often quite unrealistic. Let me quote from the Australian Broadcasting Corporation News of 31 May 2010, with an all too often recurring theme:

> The New South Wales Coroner has found that an eight-year-old boy who died from bacterial meningitis had symptoms that were undiagnosed by a string of doctors and that hospitals could have done more to save him.[227]

All this shows is that coroners should be medical practitioners who have some understanding of what they are dealing with, not non-medical lawyers who live in an uninformed ivory tower. The operative words are 'a string of doctors' failed to make the diagnosis. Unfortunately, this is often simply not possible early in an illness which then becomes obvious but untreatable. The last thing anyone wants to do is miss this diagnosis. Being a medical practitioner working at the front line in general practice or in emergency departments can be extremely onerous and unforgiving.

26

Diphtheria

SCIENTIFIC NAME:	*Corynebacterium diphtheriae*
ORGANISM'S APPEARANCE:	Gram-positive rod; metachromatic granules when stained with methylene blue
DISEASE NAME:	diphtheria
DISTRIBUTION:	worldwide; now mostly Third World
TRANSMISSION:	aerosol droplet
LOCATION IN A PERSON:	usually throat and adjacent areas; sometimes the skin
CLINICAL FEATURES:	a. local—pseudomembrane on a severely sore throat, difficulty swallowing, foul breath, asphyxiation; skin ulcer, b. distant—paralysis, heart disease
DIAGNOSIS:	recovery of bacteria from the throat or skin lesions
TREATMENT:	penicillin, antitoxin
PREVENTION:	an effective vaccine is available

My mother had diphtheria as a small girl in 1920. Fortunately she survived, otherwise I would not be here and you would not be reading this page. Diphtheria is a distinctive clinical syndrome that seems to have come in epidemic waves, particularly in the last three centuries. Some think that Hippocrates (p.116) saw cases but the first convincing clear description is that of the Greek physician, Aretaeus the Cappadocian (AD 81–138). After considering minor ulcers in the mouth, Aretaeus described what he called Syrian or Egyptian ulcers or eschars (sloughs or scabs) of the throat:

> (Ulcers which) are broad, hollow, foul, covered with a white, livid or black concretions [inflammatory membranes] are pestilential…And if it spreads to the thorax by the windpipe, it occasions death by suffocation within the space of a day…The manner of death is most piteous; pain sharp and

> hot, ... respiration bad, for their breath smells strongly of putrefaction ... they are in such a loathsome state that they cannot endure the smell of themselves ... fever, acute thirst as if from fire and yet they do not desire to drink for fear of the pains it would occasion.[27]

Strangely, there is no record again of such awful and deadly sore throats for some 1,400 years. In the late sixteenth century in Europe there appeared epidemics of 'croup', also called 'morbus suffocans', characterized by a membrane and sore throat which was probably diphtheria. The French physician Guillaume de Baillou (Latin name Ballonius, 1538–1616) described an epidemic of this hitherto unrecognized disease in 1576. He told how people died having great difficulty in breathing after having had a growling hoarseness and swelling of the back of the mouth. A surgeon opened the body of a boy who had died from the disease and told de Baillou that he had found a tenacious and unyielding slime that stretched like a parchment across the trachea (windpipe) so that air could neither get in nor out. Such asphyxiation was the most deadly feature of the illness and in 1610, Marco Aurelio Severino (1580–1656), a surgeon in Naples, Italy, performed the operation of tracheotomy in which the windpipe was opened and a tube inserted below the obstructing membrane over the larynx, thus allowing air to pass in and out of the lungs. This was done without anaesthesia of course, but the patients were about to die and probably hardly noticed.

A community could be devastated by diphtheria. Consider the village of Hampton Falls in the British colony of New Hampshire in New England. In May 1735, a devastating 'throat distemper' came out of nowhere. The town had a population of 1,260, of whom 404 were under 10 years of age. Two hundred and ten people died from this illness that year; of these, 160 were aged less than 10. The village had lost almost half its children.

In 1818, an epidemic of diphtheria appeared in Tours, France where Pierre Fidèle Bretonneau (Figure 144) was chief physician at the hospital. The epidemic had a high mortality and over the next two years he performed post-mortem examinations on 60 people who had died. In 1821, Bretonneau presented two Memoirs to the Academy Royale de Médecine in which he detailed his findings. In the first he wrote:

> I undertake at present to prove, by the testimony of the facts, that the Scorbutic Gangrene of the Gums, Croup, and Malignant Angina are only one and the same form of phlegmasia [inflammation].[52]

Until that time, these three conditions were believed to be separate. Scorbutic gangrene of the gums afflicted 130 soldiers in Tours as well as a few other people.

The gums became swollen and the teeth loose, as in scurvy, but the disease was not arrested by the administration of sorrel juice and other anti-scorbutics, now known to contain a high concentration of vitamin C and which prevent scurvy (from which the word 'scorbutic' is derived). Furthermore, in a few patients, the disease spread backwards into the throat and looked like malignant (also called gangrenous) angina. 'Malignant' in this context means an extremely severe illness and does not imply cancer as the word usually does these days. 'Angina' means 'suffocation' or 'strangling' and is not to be confused with 'angina pectoris', which means the severe pain in the chest of heart disease. Patients with malignant angina had eschars (scabs) over the tonsils and the back of the throat. Finally, today we give the name 'croup' to an acute viral infection of the upper respiratory tract which causes a barking cough in children; this has nothing to do with Bretonneau's diphtheritic croup.

Figure 144. Pierre Bretonneau (1778–1862).

When I read Bretonneau's report nearly 200 years later, I can't say that I found his conclusions regarding scorbutic gangrene of the gums particularly convincing, but he seems to have been right about croup and malignant angina. This is how he described his autopsy findings on a seven-year-old child with what he called croup who had died:

> The walls of the pharynx [throat] were... covered with eschars; but an unexpected difference was observed in the interior of the trachea; a tube of membraniform substance [i.e. pseudomembrane] which was white, supple, elastic, consistent, adhering feebly to the mucous membrane [the normal lining of the respiratory tract]... extended from the larynx [voice-box] to the last divisions of the bronchi [the branching tubes below the trachea].[52]

Bretonneau labelled these three conditions as one—'la diphthérite', derived from the Greek word *diphthera* meaning 'skin', 'hide', or 'piece of leather', to indicate the false membrane in the throat and airways that was characteristic of the disease. This was translated into English as 'diphtheritis', which caused some dissension as words ending in 'itis' mean 'inflammation', which in this case would indicate inflammation of the skin. This was not Bretonneau's intention at all so in 1855 he changed it to 'diphthérie', which became in English 'diphtheria'. Strictly speaking, this should be pronounced diff-theria but most people find it easier to say dip-theria. Perversely, the word 'diphtheritic' rather than 'diphtheric' was retained as the adjective, as in diphtheritic sore throat. Bretonneau realized that the typical and essential lesion of diphtheria was the 'false membrane' or 'pseudomembrane' because it was not a normal anatomical structure; rather, it was composed of inflammatory material mixed with dead tissue and was stuck to the throat and respiratory tract.

Bretonneau gave further detailed reports in 1825 and 1826 in which he discussed the clinical features, diagnosis, and various therapeutic measures, including tracheotomy which he performed nearly 20 times, with about a quarter of the patients surviving. His memoirs became a landmark in the study of diphtheria. Bretonneau was something of an eccentric who did not get on particularly well with other members of his profession in Tours. They maintained that the diseases they saw were different from those he was seeing in the hospital. To prove they were wrong, with the help of his colleague Dr Velpeau, he scaled the cemetery walls at night, hastily dug up the bodies of those private patients who had died of diphtheria or typhoid fever, and made a rapid examination to prove that the diseases they had died from were the same as those he was seeing in the hospital. He worked and slept at all hours and, although a man of catholic interests, he had two particular obsessions—his garden and sore throats. If called to a patient he would tell his servant to ask if it was a case of sore throat or fever; if it was neither, he was to say the doctor was not at home. Nor were his domestic arrangements exactly normal. Bretonneau's first wife was 25 years his senior and his second wife was a girl of 18 when he married her at the age of 78!

Of course, many physicians disputed Bretonneau's conclusions and some of their objections were well founded. There could be no answer one way or another until the cause of diphtheria was found and that agent identified in individual patients. In 1855, Bretonneau returned to the fray and considered the mode of transmission of diphtheria. He was convinced that 'it is vain to deny that contagion...is the source of most epidemics',[52] but he believed that the contagion did not travel through the air. Rather, he postulated that contagion occurred when:

> the diphtheritic secretion, in the liquid or purulent state, is placed in contact with a soft or softened mucous membrane...and this application must be immediate.[52]

Investigators began to study whether diphtheria could be transferred to animals or humans and to look for contagious agents. In 1869, the German surgeon Friedrich Trendelenburg (1844–1924) reported that he had inoculated diphtheritic membrane material into 52 animals and succeeded in producing a false membrane in eight rabbits and three pigeons whereas control material caused no abnormalities. These results were confirmed in 1871 by Max Joseph Oertel (1835–1897) of Munich in a complex series of experiments in which he inoculated rabbits and pigeons either in the trachea or the muscles with diphtheritic material either directly from humans, or from animals that had been inoculated originally with human diphtheritic material. Animals died within several days whether the material was inoculated into the trachea or elsewhere, but membranes were only seen when material was implanted into the trachea. The fact that animals injected in the muscles died, but not from suffocation, is a clue to which we will return.

So much for animals. What about humans? Many doctors had contracted diphtheria after treating their patients but, on the other hand, many others had not. It was all very confusing. Some doctors made foolhardy attempts to transfer the disease to themselves. Armand Trousseau (1801–1867) in France in 1829 moistened a lancet on a false membrane then punctured his arm and his tonsils; a small blister appeared on his arm but that was all. Likewise, another French doctor and opponent of Pasteur and his theories, Michel Peter (1824–1893), was accidentally squirted in the eye by diphtheritic material while performing a tracheotomy but did not wash if off, on another occasion punctured his lip deliberately with false membrane material, and on a third occasion painted his throat with fluid from a false membrane; nothing happened. You may be wondering why? It seems most probable from this distance that they had had a prior mild attack and were immune. They were lucky.

These experiments might have been foolhardy, but hair-raising if not criminal is the word to apply to the experiments of Giuseppe Bubola in St Margherita in northern Italy. Taking his cue from the practice of vaccination for smallpox, in 1864 he inoculated diphtheritic material into the arms of 29 healthy children, half of whom belonged to families already affected by the disease, during an epidemic. Some developed a transient lesion in the skin and three children from affected families contracted diphtheria, apparently naturally and coincidentally. In the same year, another Italian, J Masotto, copied Bubola and inoculated 20 subjects, two of whom developed diphtheria.

It was all very perplexing. Animal experiments suggested that in some circumstances diphtheria could be transferred from one animal to another, but

Figure 145. Edwin Klebs (1834–1913).

the human observations were less convincing. In any case, they shed no light on the nature of the agent that caused disease. Bacteria had not yet found their place in the sun, so it is not surprising that various investigators searched particularly for fungi, as the yeast, *Candida albicans*, had been shown in the 1840s to be the cause of white plaques in the mouth called thrush.

It was at this point that Albrecht Theodor Edwin Klebs (Figure 145) first appeared on the diphtheritic scene. Klebs has been described as a restless genius with a mercurial temperament who was always dissatisfied with where he was. He had been born in Königsberg in Prussia (now Kaliningrad in the Russian exclave) and was a student at the University of Würzburg under Rudolf Virchow, Germany's pre-eminent pathologist. In 1875, Klebs was professor of pathology in Prague, when he announced he had 'discovered' the cause of diphtheria. He studied two patients who died from diphtheria. In the first, a 43-year-old woman, he examined a pseudomembrane taken from a tonsil. In the superficial layer, he saw many fungal filaments, while in the middle part of the membrane he saw masses of organisms that were micrococci. He thought that these micrococci had been produced by the superficial fungal filaments. He saw the same features in a second case, a child, and announced that a fungus was the cause of diphtheria; he named it *Microsporon diphtheriticum*. This was all rather reminiscent of Billroth's fanciful *Coccobacteria septica* (p.176). He tried to infect experimental animals but came up with uninterpretable results, presumably because his preparations were impure.

Meanwhile, his old teacher, Rudolf Virchow, had muddied the waters. Virchow showed that the pseudomembrane was cellular in nature and popularized the idea that disease was due to physical or chemical changes in the cells that make up tissues. He did not think there was a clinical entity called diphtheria. Instead, he said, any fibrous exudate in the throat was 'diphtherial', whatever its cause. Was it one disease as Bretonneau said, was it a local condition or a general state, was it due to one or more bacteria, or was their appearance in lesions completely coincidental? Such was the state of confusion that diphtheria was the main subject of the Congress for Internal Medicine held at Wiesbaden in Germany in 1883. The stormy Klebs, who had moved to Zurich in Switzerland, was the chief speaker. He confounded the audience by asserting that the diphtheria he had met in Zurich was different and was due to a bacterium. Bacilli, which were easily stained with methylene blue, could be seen in diphtheritic pseudomembrane and inside cells. Furthermore, the bacterial rods appeared to have two to four spores (Figure 146), but that was later shown to be incorrect. Klebs went on to say that he could not find these bacteria in any other tissues of patients who had died from the disease. He made no attempt to grow these bacilli in the laboratory or infect experimental animals, probably because of his inconsistent results in 1875. Nevertheless, he was convinced (again) that these bacteria were the cause of diphtheria and a diagnosis could be made by staining pseudomembranes with methylene blue. Klebs moved to the USA in 1895 and then returned to Switzerland in 1900 where he died.

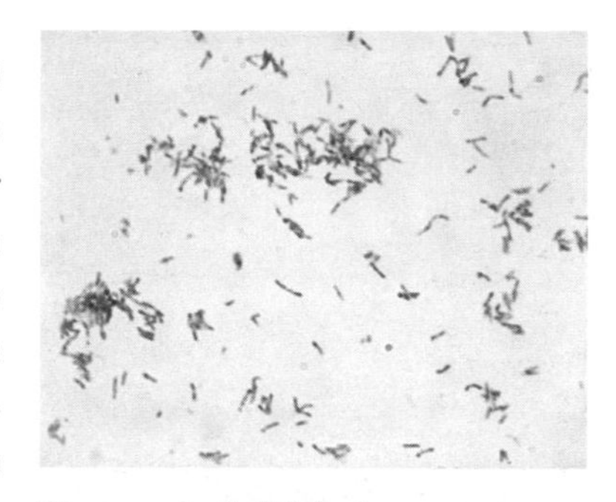

Figure 146. *C. diphtheriae* grown in culture showing granules when stained with methylene blue.

You will have appreciated that Klebs was claiming a lot because he had nowhere near fulfilled 'Koch's postulates' which were reiterated by Loeffler, who was the first to attempt to fulfil them in diphtheria. Friedrich August Johannes Loeffler (Figure 147), who had been a military surgeon, was an assistant of Robert Koch at the Imperial Health Institute in Berlin. He described his findings in a long report published in 1884.

First, Loeffler confirmed Klebs's observations and found bacilli in diphtheritic pseudomembranes that were stained with methylene blue (the formulation of which he modified and has been known ever since as Loeffler's methylene blue). He then studied the microscopical appearances in 22 cases of diphtheria and five cases of scarlet fever. He found that in the latter, the dominant microbes were streptococci. In diphtheria, although there were a few streptococci, there were large numbers of bacilli. After trial and error, he made a medium containing peptone, meat infusion, salt, grape sugar, and stiffened with blood serum, on which Klebs's bacilli would grow. After three days, he could discern two distinct types of colonies, one containing streptococci and the other made up of Klebs's

Figure 147. Friedrich Loeffler (1852–1915).

bacilli. He was then able to take individual colonies of bacilli and make pure cultures by growing them on fresh medium (a process called subculturing). He had now achieved the second of the postulates. Loeffler found that heating to 60°C killed the bacilli so what Klebs had thought were spores were not. They are known to be inclusions in the cytoplasm of uncertain function called 'metachromatic granules'.

Next, Loeffler tried to infect experimental animals with pure cultures of bacilli by injecting organisms subcutaneously. Rats and mice were resistant but guinea pigs died within several days. The lungs and kidneys were congested but bacteria could only be found at the sites of injection, which were inflamed and haemorrhagic, but not in the organs. Loeffler concluded that the animals did not die from dissemination of bacilli throughout the body but that the bacteria released a toxic poison. Similar results were obtained with various species of birds injected subcutaneously.

Loeffler then inoculated 27 rabbits. When the organisms were swabbed onto the cornea or into the trachea (17 animals), a pseudomembrane often formed. Birds, too, developed a pseudomembrane when bacteria were introduced into the throat or trachea. A pseudomembrane also formed when bacteria were introduced into the vagina of guinea pigs but not in rabbits.

Loeffler realized that it was important to determine whether or not the bacillus could be found in the mouth and throat of healthy people. Since children were the most susceptible age group, he took specimens from 20 healthy children and 10 healthy adults. In one child, he found the typical bacteria and when they were cultured and injected under the skin of two guinea pigs, they became ill and died on the third day.

Loeffler summed up the pros and cons of these bacteria being the cause of diphtheria. In favour he noted that:

> The rods were found in 13 of 27 typical cases of diphtheria with fibrinous pharyngeal exudate. The rods were present in the oldest areas of the pseudomembrane and were deeper than other organisms. Cultures of the rods were

> lethal when inoculated subcutaneously into guinea pigs and small birds. Whitish and haemorrhagic exudates developed at the site of injection... As in humans, the internal organs were free of lesions. The bacilli produced pseudomembranes on the exposed tracheae of rabbits, chickens and pigeons...[191]

Against, he said:

> The bacilli were absent in a number of typical cases of diphtheria... The pseudomembranes of rabbits and chickens were not arranged as in the pseudomembranes of man... The bacilli (only) produced disease... of animal species which were susceptible when the mucous membranes were traumatised... A typical virulent bacillus was found in the throat of a healthy child.[191]

Although Loeffler probably believed in his heart of hearts that Klebs's bacillus was the cause of diphtheria, scientific caution won the day and he concluded that 'the proof that the bacillus is the aetiological agent of diphtheria is thus not complete'.[191]

Consequently, his findings were not greeted with the enthusiasm that Koch's finding of tubercule bacilli had received. Nevertheless, over the next six years, a dozen investigators confirmed Loeffler's findings and the organism became known as the Klebs–Loeffler bacillus. When the Gram stain was discovered, these bacilli were found to stain purple (Gram-positive). It was named *Bacillus diphtheriae* by Kruse (possibly Walther Kruse, p.352) in 1886, then transferred to the genus *Corynebacterium* erected by Lehmann and Neumann (p.216) in 1896. The genus name is derived from the Greek word *koryne* meaning 'club' to indicate the club-shaped ends of the bacilli. It is pronounced rather like co-(as in cot)-rye-knee-. Loeffler, incidentally, in 1888 moved to the University of Greifswald where he remained for 25 years; at his death during World War I, he was Germany's oldest serving general.

At this point, we need to return to clinical observations in human diphtheria. It was noted that some people did not asphyxiate and appeared to recover but then became ill and sometimes died. In 1749, Jean-Baptiste Chomel (1709–1765) in Paris and Martino Ghisi (1715–1794) in Cremorna, Italy, described the occurrence of paralysis of the soft palate (the back part of the roof of the mouth) including the uvula (the bit that hangs down between the tonsils) in some cases of diphtheria; this phenomenon was observed again by Samuel Bard (1742–1821) in New York in 1771. Subsequent researchers found that paralysis supervened in about 20 per cent of cases of diphtheria. The most frequently involved area was the soft palate, but in many cases the limbs were involved and sometimes the neck and the eye muscles. In 1862, Jean-Martin Charcot (1825–1923) and Edmé Félix Alfred Vulpian (1826–1887) in France examined the affected nerves under the microscope and found that they were swollen and degenerated.

Another important organ to be affected was the heart. Symptoms of heart disease could appear early or late in the course of the disease and could cause sudden death. In 1877, J (probably Julius) Rosenbach described inflammation of the heart muscle (myocarditis). This was soon verified by others. Particular microscopical attention was paid to the conducting system of the heart (which controls the heart beat) but no consistent lesions could be found. It was not until the electrocardiograph was invented many decades later that abnormal rhythms of the heart, some potentially fatal, were found in 20 per cent or so of patients with diphtheria.

These abnormalities of the nerves and heart, and to some degree other organs, were not caused by Klebs–Loeffler bacilli being present in those tissues themselves. There seemed to be something in Loeffler's suggestion that the bacilli released a toxin which travelled through the body to cause damage. Pierre Paul Émile Roux (1853–1933) and Alexandre Yersin (p.370) at the Institut Pasteur in Paris decided to look for a toxin. First, they noted that if death did not supervene too quickly in rabbits infected with Klebs–Loeffler bacilli, they often became paralysed. In 1888, they grew the bacilli in a liquid medium and then passed the fluid through a porcelain filter that blocked bacteria. When they injected small amounts of this bacteria-free filtrate into animals, they became paralysed. When larger amounts were injected, they reproduced all the features of diphtheria with the exception of the false membrane. Their researches aroused immense interest and confirmed the role of the Klebs–Loeffler bacillus as the cause of diphtheria. This soon led to the generation of an antitoxin in immunized animals which became such an important treatment for diphtheria, especially in the pre-antibiotic days, the first child being treated in Berlin on Christmas night in 1891.

In 1931, JS Anderson and colleagues in Britain devised a new medium on which to grow *C. diphtheriae* called chocolate agar tellurite medium. When the bacteria grew on this medium they recognized three colony types. They related these types to the severity of disease in the patients from whom they were isolated. The colonies they labelled *gravis* came from the most severely ill patients, *mitis* came from the least, and *intermedius* were of intermediate severity. Subsequent investigators found that there was some relationship between colony type and virulence of diphtheria but it was by no means invariable.

In 1951, Victor Freeman (?–*c*. 2008) in the USA reported that not all *C. diphtheriae* are virulent. Only those infected with a virus that infects these bacteria called β-corynebacteriophage could produce toxin. Despite having made such an important observation, Freeman later abandoned bacteriology for psychiatry.

Not only does respiratory diphtheria persist in the Third World but so does cutaneous diphtheria acquired when organisms contaminate a cut or abrasion of the skin. So what has happened to diphtheria in the First World and the almost First World? It is a respiratory infection that only infects humans, with most cases being transmitted by aerosol droplets or by kissing. In the first half of the twentieth century, the mortality from diphtheria declined dramatically, with the fall beginning even before the introduction of preventive immunization of children and the availability of penicillin. In the second half of the twentieth century, diphtheria virtually disappeared from the First World. I have never seen a case. But the breakdown of medical services and poverty following the dissolution of the former Soviet Union saw a remarkable epidemic in the former Soviet Republics, and 3 per cent or so of healthy people carry the organism in their throats. We must always remain vigilant.

27

Whoopingcough (pertussis)

SCIENTIFIC NAME:	*Bordetella pertussis*
ORGANISM'S APPEARANCE:	short Gram-negative bacillus
DISEASE NAME:	whooping cough or pertussis
DISTRIBUTION:	worldwide
TRANSMISSION:	airborne
LOCATION IN A PERSON:	larynx (voice-box) and bronchi (lung airways)
CLINICAL FEATURES:	paroxysmal coughing, vomiting
DIAGNOSIS:	originally, recovery of the organism from sputum; in recent years, a positive chain reaction test of throat swabs
TREATMENT:	penicillin or cefotaxime
PREVENTION:	an effective vaccine is available

The *Oxford English Dictionary* defines 'whooping cough' as a 'disease chiefly affecting children, and characterized by short, violent, and convulsive coughs, followed by a long sonorous inspiration called the hoop (whoop)'. You will notice that the drawing in of a breath is called a hoop or a whoop. Indeed, 'whooping cough' and 'hooping cough' are both spellings that have been in use in England since the mid-seventeenth century and most people pronounce it as 'hooping cough'. Its derivation is from the Middle English word 'hoop' meaning a 'cry' or 'call'.

Whooping cough seems to be a relatively recent disease. Although there is some disagreement among historians, most believe that there is no mention of anything resembling this illness in the writings of the ancient Greek or Roman physicians or those of the Arabs in the centuries that followed. In his *Chronological History of France*, François Eudes de Mézeray (1610–1683) wrote that in 1414:

> A strange kind of rheum [running eyes and nose], named La Coquelúche, tormented all sorts of people during the months of February and March, and rendered their voices so hoarse that the bar, the pulpits and the colleges were mute. All the old men who were seized with it died.[217]

When this was first translated into English, 'La Coquelúche' was given as 'Hooping Cough' but this does not seem to me to fit the pattern of whooping cough which is primarily a disease of children. Rather, it suggests influenza in which laryngitis was a prominent feature.

The first clear-cut description of the disease was given by Guillaume de Baillou (p.292) in his recounting of the epidemic that hit Paris, France, in 1578. Baillou observed that the illness, which was called Quinta or Quintana, mostly afflicted young children:

> The lung is so irritated that in the effort it makes to get rid of that which affects it, it can inhale and exhale only with difficulty... The patient swells up and nearly suffocated, feels as though his breath was stopped in the middle of his throat... Some believe that this name [quinta] was made-up because of the sound of the patient in coughing. The torment of the cough is sometimes suspended for 4–5 hours after which the paroxysm returns which is often so violent that it causes blood to issue from the nose and mouth and very frequently... causes vomiting.[34]

His works were published posthumously under his Latin name of Ballonius by his nephew Jacques Thevart; this account appeared in 1640.

Thomas Sydenham (p.471) in England appears to have coined the word 'pertussis' to describe whooping cough. In his *Observationes Medicae circa Morborum Acutorum Historiam et Curationem* published in London in 1676 he wrote about 'infantum pertussis quam nostrates vocant hooping cough' which means 'the pertussis of infants which our people call hooping cough'.[300] The word 'pertussis' is derived from two Latin words: 'tussis' means 'cough'; 'per' usually means 'through', 'by', or 'from' but when used as a prefix it intensifies the following word. Thus 'pertussis' means a severe or intense cough.

Clearly there are some striking resemblances between whooping cough and diphtheria, which was discussed in the last chapter. Both afflict primarily young children. Both may occur in epidemics. Both affect the upper respiratory tract and cause marked difficulty in breathing. Both could cause death. But there were also differences. Whooping cough tended to start off rather like a cold for two weeks, during which an unremarkable cough appears (the catarrhal stage). This is followed by a two-week paroxysmal phase in which there are repetitive, short, explosive coughs (cough cough cough cough cough) ended by a huge whoop as air is breathed in. This may be followed by more coughing and vomiting. The coughing prevents blood from returning to the chest so the pressure in the blood vessels rises sometimes causing haemorrhages in the whites

Table 4. Death rates in selected countries for the five years, 1901–1905, from the four infectious diseases of the upper respiratory tract that predominantly infected children. Deaths are reported per 100,000 people per year

	Pertussis	Diphtheria	Scarlet fever	Measles
England and Wales	301	222	126	327
Germany	324	330	223	238
Italy	197	139	48	212
USA	109	296	110	90

of the eyes (subconjunctival haemorrhages), fluid accumulation around the eyes, and nosebleeds. Over another two weeks, the illness subsides (the convalescent stage). Complications include pneumonia, a permanent distortion of the airways called bronchiectasis, convulsions, and inflammation of the brain. Worse, it can be a killer (Table 4). In diphtheria, on the other hand, the illness progressed much more quickly over hours and days, the throat became sore, a pseudomembrane appeared, and there were no paroxysmal coughing spells.

Not only were the differences sufficient for clinicians to distinguish the two diseases but, once the bacterial cause of diphtheria was established in the 1880s, the similarities were enough to stimulate researchers to look for a bacterial cause for whooping cough. There was no shortage of suggestions. In fact it was utter confusion. Martha Wollstein in New York, in a paper on the bacteriology of pertussis in July 1905, began thus:

> Investigations on the etiology [cause] of pertussis by means of microscopic and cultural studies of the sputum have resulted in the finding and description of protozoa by Deichler, Kurloff, and Behla, of staphylococci by Moncorvo and Silva Aronja, of streptococci by Mircoli, of diplococci by Ritter and his pupil Buttermilch, and of bacilli. The last may be divided into three groups: First, a motile bacillus growing 'colon-like' upon all ordinary media and forming endogenous spores, described by Affanassief in 1887...Second, the bacillus found by Czeplewski and Hensel, and also by Koplik, in 1897, growing as a small, poled bacillus upon agar, Löffler's serum, gelatine and in broth...The third group comprises the influenza-like bacillus described by Spengler in 1897 and by Jochmann and Krause in 1901...They called this organism *Bacillus pertussis*, Eppendorf.[326]

And so she went on describing the findings of other authors. It was a pig's breakfast. We don't need to worry about the intricacies of all this but simply to note that she did not mention the observation made by Bordet and Gengou

Figure 148. Jules Bordet (1870–1961).

back in 1900 that would turn out to be correct! But I think she can be forgiven. David Davis in Chicago in March 1906 published a very long paper on the bacteriology of whooping cough with an extensive review of all the work that had been done until that time. He wrote that 'the etiology of whooping-cough has not as yet been determined, notwithstanding the large number of investigations upon this subject'.[82] He did not mention Bordet and Gengou either. I don't think they knew about them. I suspect Bordet and Gengou had not published their 1900 observation.

Jules Jean Baptiste Vincent Bordet (Figure 148) was born at Soignies in Belgium. He graduated in medicine from the University of Brussels in 1892 and began to work at the Institut Pasteur in Paris in 1894 where, with Eli Metchnikoff (p.335), he discovered the phagocytosis (ingestion and destruction) of bacteria by white cells, then the phenomenon of haemolysis (breaking down of red blood cells) when red blood cells were exposed to foreign serum (the fluid left after blood clots). In 1890 he moved back to Belgium to direct the Institut Antirabique et Bacteriologique du Brabant (renamed in 1903 as the Institut Pasteur) in Brussels. It was there that he worked with the Belgian bacteriologist Octave Gengou (1875–1957). In 1900 they observed a new ovoid bacillus which they had found in the sputum of a six-month-old infant with whooping cough. The bacilli which stained pink on the Gram stain (Gram-negative) were present in such abundance and purity that their association with whooping cough seemed very likely. They named it 'Microbe de la coquelúche'.

But this was nothing like proof. Bordet and Gengou had not even fulfilled the first of Koch's postulates which required that the organism must be shown to be constantly present in characteristic form and arrangement in the diseased tissue. And there was a huge problem with fulfilling the second postulate—they could not grow the organism in pure culture on an artificial medium. In fact, it was to take them six years of experimenting with different formulations of media until they came up with one on which these bacilli would grow. This was a complex witch's brew containing agar, glycerine, potato, and blood. Bordet and Gengou based their claim in 1906 that whooping cough was due to this bacillus upon two main observations. First it was present in overwhelming numbers in the early stages of the disease until the whoop appeared. Secondly, when these bacilli were mixed with serum from a patient who had recovered from whooping cough, an immunological reaction occurred in which a substance in the blood called complement was removed. This reaction did not occur with the influenza bacillus which we will mention shortly.

Unfortunately, growing the Bordet–Gengou bacillus was a difficult business, even using this new medium. In 1916, Ingeborg Chievitz and Adolph Meyer introduced the 'cough plate method' in which a child coughed directly onto a plate which was then immediately incubated (Figure 149). This yielded a larger number of positive results than did collecting sputum in a container and then spreading it onto a plate. Some children were too young to cooperate in such a manner, and it was found in them that the likelihood of growing the organism was improved if a swab was taken by passing a swab through the nostril to the

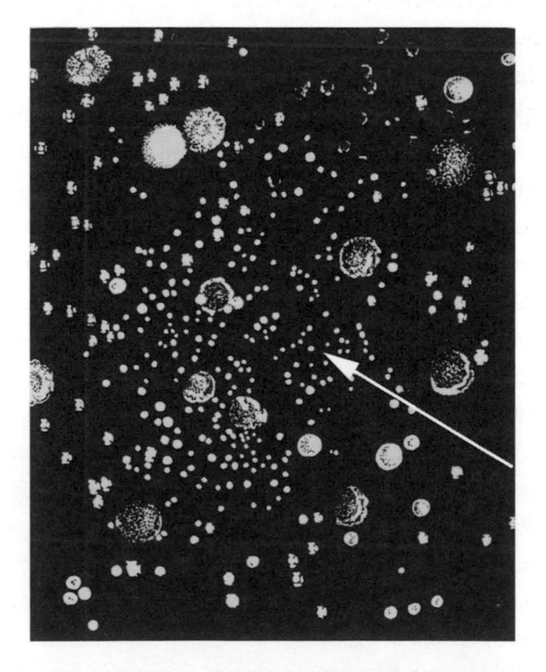

Figure 149. Cough plate of Bordet–Gengou medium. The arrow points to a clump of colonies of *B. pertussis*. The other larger colonies are mouth organisms.

back of the nasopharynx and collecting mucus. Of course, performing such a procedure on a screaming infant with whooping cough is not for the faint-hearted.

Bordet and Gengou had achieved the second of Koch's postulates. A number of investigators set about proving the third by transmitting infection. Monkeys, dogs, rabbits, guinea pigs, mice, and even chimpanzees were inoculated by swabbing the pharynx or the nasal passages. Once more, utter confusion and chaos followed with contradictory results. The problem was that although all the researchers thought they were using the Bordet–Gengou bacillus, this was not always the case. Some were using a very similar small Gram-negative bacillus, then called the influenza bacillus (now called *Haemophilus influenzae*), which was often found in large numbers in humans in upper respiratory tract infections. It would take some time before the two could be reliably differentiated. Eventually it was realized that the Bordet–Gengou bacillus grew well on Bordet–Gengou medium after three to four days, forming smooth, glistening pearl-white colonies (Figure 149). On the other hand, the influenza bacillus grew after 24 hours on a medium called chocolate agar (made by heating red blood cells to break them down before mixing them with agar) forming dull, irregular colonies. As time progressed, various immunological and biochemical tests were developed which helped separate the two types of bacteria. The other problem was that some researchers were unknowingly transferring a bacterium now called *Bordetella bronchiseptica* among animals. This organism rarely infects humans but is a natural pathogen of some animals; it causes snuffles in rabbits and distemper in dogs. On top of all this was uncertainty as to whether the Bordet–Gengou bacillus was actually pathogenic for any of these species of animals or only caused disease in humans.

The pattern became clearer from about 1930 onwards. We can be reasonably sure that researchers were really using the Bordet–Gengou bacillus, which had been renamed *Haemophilus pertussis* in 1923 by Bergey (p.217) and his colleagues. In 1929 in America, Louis Sauer (1885–1930) and Leonora Hambrecht produced an illness like whooping cough in monkeys, with paroxysmal cough appearing one to three weeks after inoculation. In 1932, Arnold Rich (1893–1968) and his colleagues in the USA generated typical whooping cough in monkeys 24–30 days after inoculation of either pure cultures of *H. pertussis* or unfiltered mucus obtained from the trachea of children with whooping cough. The pathogenicity of this bacterium was validated when they used mucus which had been passed through a filter to remove bacteria so that they could determine whether a virus caused whooping cough; all the investigators transmitted was a cold which appeared after two to three days. Others then confirmed their findings. Then in 1934 in Japanese-occupied Manchuria, Itsuyoshi Inaba and Seiichi Inamori succeeded in transmitting the disease to puppies. It was looking pretty convincing that *H. pertussis* was the cause of whooping cough. The third of Koch's postulates seemed to have been met.

But what about humans? No-one would dare to attempt to give this potentially fatal disease to humans? No-one would give it to their own children. Or would they? If you find the following hard to believe, then you can see it in print for yourself in the *Journal of Infectious Diseases* in 1933 in an article written by H MacDonald and EJ MacDonald.[204] All we are told about them in that article is that they lived in Evanston, Illinois, USA. Nor are we told precisely who the four subjects were except that they were healthy brothers and we are given the first two initials of each child. In fact, 'H' was Hugh MacDonald, a physician, and 'E' was Edith his wife. The MacDonalds indicated with regard to the subjects studied that 'unusual circumstances made it possible for them to contribute their services' but we are not told what those circumstances were. Two of them, an eight-year-old and a nine-year-old, had been immunized against whooping cough seven months earlier. A culture was made of the organism (which they called *Bacillus pertussis*) from a child with typical whooping cough, this was suspended in saline, then passed through a filter to remove bacteria but not viruses. The filtrate was instilled into the nose and throat of each child who were 'then quarantined together with a trained nurse in a rural apartment for eight weeks'.[204] Was their mother a nurse? Eighteen days into this incarceration, 140 whooping cough bacilli were instilled into the nose and throat of each child who at this point were called 'volunteers'. Can an eight-year-old child be a volunteer in these circumstances? I doubt that anyone reading this would think so. The two immunized children remained well throughout the ordeal. The other two boys, aged six and eight years, began to cough a week later. The MacDonalds reported their ordeal:

> Within a week their coughing had become increasingly severe, and at night was almost incessant... The paroxysms became more prolonged and intense with daily expectoration of mucus (by 18 days)... The temperatures ranged between 100 and 100.8°F [normal <98]; there was distention of the eyelids most of the time; the face and neck were distended after coughing; the cough plates were positive... The coughs became typically paroxysmal with vomiting of mucus and food... (32 days after inoculation) there was daily

> whooping, with vomiting of mucus and food, and headache after severe paroxysms...(39 days after inoculation) the paroxysms had become less frequent and less severe; the cough was less productive, but whooping and vomiting continued at night; the boys were more restful; their appetite improved; the temperature ranged from 98.9 to 99.8°F; the cough plates were negative.[204]

And of course, all through this, the poor children, who were indeed the MacDonalds' offspring were subjected to repeated blood tests. Their conclusion was succinct and accurate:

> A filter-passing virus plays no role in the etiology of pertussis. The disease is caused by the bacillus of Bordet and Gengou. Active immunity is conferred by the injection of B. pertussis vaccine.[204]

They might have fulfilled the last of Koch's postulates but I am very glad that they were not my parents. No editor of a medical journal would ever publish such a report these days. It is a sad commentary on the mores of those times.

Extraordinarily enough, a variation of this dreadful experiment was repeated by Dr JM Frawley in Fresno, California, in 1936. Questions had been raised, the details of which need not concern us, that there might really be a virus involved in whooping cough. The MacDonalds had used a filtrate of a bacterial culture made on artificial medium. Dr Frawley thought he would remedy this deficiency. Washings were taken directly from infected children, then passed through a filter to remove bacteria. Drops of this filtrate were then inoculated into the nasal passages every 15 minutes for eight hours into 12 healthy children. These children, who were as young as two years old, were put into strict isolation in the Fresno County Hospital. None of them contracted whooping cough although several developed colds. You might be wondering what the children and their parents thought about all this? So do I. The mind boggles. At least he did not give them whooping cough bacilli.

Let us return to a happier note. Bordet won the Nobel Prize in 1919 for his contributions to bacteriology and immunology. In a fitting memorial to him, Manuel Moreno-Lopez in 1952 erected a new genus, *Bordetella*, in his honour. Bordet's 'microbe de la coquelúche', which at that point was known as *Haemophilus pertussis*, was transferred into it as *Bordetella pertussis*. What about Octave Gengou? He seems to have largely disappeared from the pages of history, except that I can tell you he was Bordet's brother-in-law.

What has happened to whooping cough? You will remember from Table 4 that at the beginning of the twentieth century, there were about 300 deaths from whooping cough each year for every 100,000 people. In Australia in 1991, there were 1.9 cases (no deaths) per 100,000 people. This was a remarkable achievement. Partly it was due to the effectiveness of antibiotics if given early in the course of the disease, aborting attacks (and hence transmission), and when given as a preventive to children who had been exposed to whooping cough. Regrettably, antibiotic treatment is next to useless in the paroxysmal phase. Most of the credit for this dramatic reduction in incidence is probably due to immunization. Unfortunately, immunization of children against infectious diseases drew a bad press when a paper was published in 1998 in the prestigious medical journal *The Lancet* linking immunization against measles, mumps, and rubella (German measles) with autism and bowel disease. Parents became worried and vaccination rates against all childhood infections fell. In 2008 in Australia, there was a dramatic increase in whooping cough to 136 cases (again, no deaths) per 100,000 people. By September of 2009, three babies had died so far that year and by the end of the year there were 30,000 cases of whooping cough. In 2010, the editor of *The Lancet* retracted the offending paper because of irregularities in its preparation and the lead author, Dr Andrew Wakefield, was struck off the United Kingdom Medical Register for acting 'dishonestly and irresponsibly'. It was all a tragic false alarm. Immunize your children against whooping cough.

28

Cholera

SCIENTIFIC NAME:	*Vibrio cholerae*
ORGANISM'S APPEARANCE:	short, comma-shaped Gram-negative bacillus
DISEASE NAME:	cholera
DISTRIBUTION:	currently in parts of the tropics
TRANSMISSION:	usually ingestion of contaminated water
LOCATION IN A PERSON:	small intestine
CLINICAL FEATURES:	violent vomiting, profuse diarrhoea, dehydration, and shock
DIAGNOSIS:	isolation of *V. cholerae* from the stools
TREATMENT:	primary – fluid replacement (oral or intravenous) secondary – tetracycline or similar antibiotic
PREVENTION:	avoid unboiled or untreated water and uncooked foods

William Sproat, a barge-man in Sunderland, County Durham in north-eastern England, began to feel ill on Saturday 22 October 1831. He had a severe shivering fit and giddiness followed by abdominal cramps together with violent vomiting and diarrhoea. When he was seen by his surgeon, Mr Holmes, the next morning, he found him:

> evidently sinking: pulse almost imperceptible, and extremities cold, skin dry, eyes sunk, lips blue, features shrunk, he spoke in whispers, violent vomiting and purging [diarrhoea], cramps of the calves and legs, and complete prostration [loss of strength].[221]

He was seen later that afternoon by James Kell, surgeon to the 82nd Regiment then stationed in Sunderland; he had seen cholera while stationed at Mauritius, an island in the Indian Ocean. He confirmed that Sproat was suffering from Asiatic cholera. There was little his medical advisers could do and Sproat died at noon on Wednesday 26th. Within a few days, his son was dead and his granddaughter was recovering from the illness.

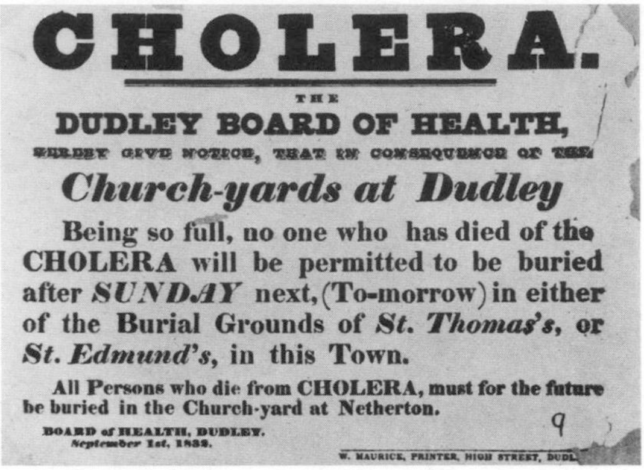

CHOLERA.

THE
DUDLEY BOARD OF HEALTH,
HEREBY GIVE NOTICE, THAT IN CONSEQUENCE OF THE
Church-yards at Dudley
Being so full, no one who has died of the CHOLERA will be permitted to be buried after *SUNDAY* next, (To-morrow) in either of the Burial Grounds of *St. Thomas's*, or *St. Edmund's*, in this Town.
All Persons who die from CHOLERA, must for the future be buried in the Church-yard at Netherton.
BOARD of HEALTH, DUDLEY.
September 1st, 1832.
W. MAURICE, PRINTER, HIGH STREET, DUDL

Figure 150. Sign erected in 1832 in the town of Dudley in the English Midlands saying that the burial grounds in the parish churches of St Thomas and St Edmund were full. My great-great-great-grandparents Grove lived through the epidemic and were married in St Thomas's five years later.

William Sproat has passed into medical history as the first recognized case of cholera acquired in the British Isles. The malevolent illness spread rapidly north and south to affect the whole island over the next 18 months. There were said to be some 22,000 deaths in England and 9,000 in Scotland, although these figures are likely to considerably underestimate the true number (Figure 150). Britain was to be hit by three more waves of cholera in 1848–9, 1853–4, and 1865–6 and it was during the last two epidemics that considerable advances were made in understanding the mode of transmission of cholera. But we are getting ahead of ourselves. Let us go back to the beginning.

As you will have seen from the description of William Sproat's symptoms and signs, the clinical features of cholera with its rapid onset of terrible vomiting, the passage of frequent, watery stools with flecks of mucus in them ('rice-water stools'), dehydration, and death within a couple of days in about half of those afflicted, are fairly distinctive. We can therefore look back at early accounts of this illness with considerable confidence. Cholera was observed by ancient Roman, Greek, Egyptian, Babylonian, Indian, and Chinese physicians. An early Chinese record (about 430 BC) reads:

> The cholera is a sudden attack of pain in the heart and abdomen, with vomiting and purging, a dread of cold and a desire of warmth. It is accompanied with pain in the head and giddiness ... When the pain is severe, the patient has spasms and when these enter the abdomen, death ensues.[138]

But the heartland of cholera was India, and it is there that the modern history of cholera begins in the early years of the nineteenth century. There were odd pockets of cholera scattered throughout the subcontinent but in 1817 it arose from its slumbers like a dragon on fire. With terrifying speed, the disease overran the whole peninsula, crossed its borders, and spread to neighbouring countries, entering China, Persia (Iran), and the Middle East then reaching the very frontiers of Europe where it halted at Astrakhan in Russia, north of the Caspian Sea. This is called the First Cholera Pandemic and lasted six years. A pandemic is an epidemic which spreads over a large area.

It was the Second Pandemic which began in 1826 and lasted 11 years that became truly global, spreading around the whole world and taking William Sproat's life in its wake. It began on the banks of the River Ganges during a religious festival, was taken home by the pilgrims, and spread along the trade routes. Thirty thousand people were said to have died in Cairo and Alexandria in a single 24 hours. The disease spread rapidly up the Danube and killed 100,000 people in Hungary between June and September 1831. Almost every major European city and port was afflicted. There were 20,000 deaths in Mauritius and 100,000 in Java (Indonesia).

The British authorities knew that it was coming and had some experience of the disease; the 'cholera morbus' had taken the lives of 3,000 of the 10,000 troops of the Marquess of Hastings (Francis Rawdon), Governor-General of British India in 1818 during the First Pandemic. The question was 'What could they do to prevent its arrival in Britain?' The government sought advice but was brought face to face with the opposing views of those who thought cholera was contagious and those who did not. In the event, the government passed an Act in 1825 which required a 14-day quarantine period for ships coming from abroad. As we have seen, ultimately this measure did not work and the government rescinded the regulations in March 1832.

Cholera was ravaging the USA as well. On 3 August 1832, the Revd Gardiner Spring, pastor of the Brick Presbyterian church in New York, preached a sermon in which he said:

> This fatal scourge is the hand of God. Already it has well nigh girt the globe, and yet the most observing among men are alike ignorant of its causes and its remedy. It is the hand of God... God will turn away his anger, when we have turned away from our wickedness. The judgement we deplore has armed its vengeance at three prominent abominations – SABBATH-BREAKING, INTEMPERANCE and DEBAUCHERY.[138]

Medicine would overtake theology by showing that cholera had an earthly (especially aquatic) basis rather than divine opprobrium. Although others had similar ideas, the man who made the definitive breakthrough was John Snow (Figure 151) in London. Snow had been born in York and had considerable

Figure 151. John Snow (1813–1858).

experience of cholera in 1832–3 while apprenticed to Mr William Hardcastle, surgeon in Newcastle-upon-Tyne. In 1849 he was practising in London, especially as an anaesthetist using the newly discovered ether and chloroform. He would later administer chloroform to Queen Victoria for the births of the last two of her nine children in 1853 and 1857, thus doing much to popularize this wonderful new medical advance. But this was by no means the end of his activities. London was once more in the throes of an outbreak of cholera as part of the Second Pandemic and Snow published in 1849 a tract entitled 'On the mode of communication of cholera'[292] which expounded his ideas on its transmission. In this paper, he argued that cholera must be a poison introduced into the alimentary tract by the mouth. But hard data to support his contentions were lacking.

This all changed in 1854 when London was facing the Third Pandemic, which had yet again spread from India in 1852. A dreadful but geographically limited outbreak of cholera occurred in the vicinity of Broad Street (now renamed Broadwick St) in the Soho district of central London. This gave Snow the opportunity to prove that water was the carrier of the cause of cholera. On 31 August 1854, an outbreak of cholera appeared which took some 500 lives in an area about 250 square yards, centred on Broad Street in London. The mortality would have been even higher if many of the population had not fled. In those days, there was not a reticulated water supply to each household in that part of London so people obtained their water from a communal well. Snow heard about the outbreak and inspected the water pump in Broad Street on 3 September and observed small, white, flocculent particles in the water over the next two days. He obtained a list of deaths reported to the General Register Office for the local subdistricts for the week ending 2 September. Eighty-three of the 89 deaths occurred on the last three days of the week. Snow found that all except 10 of the deaths took place in houses near the Broad Street pump and those who lived further away but died either used that pump because they preferred its water or their children went to a school that

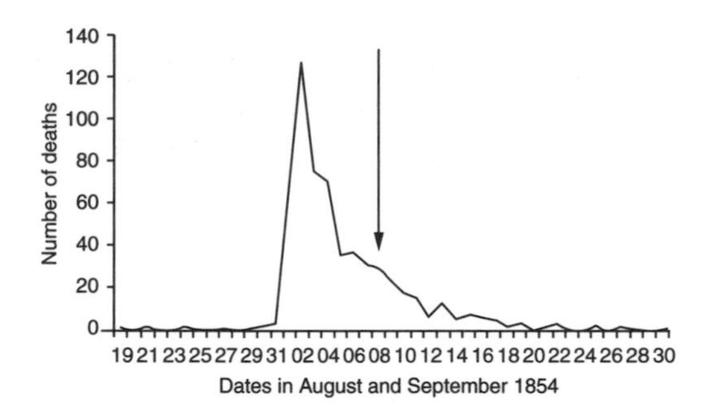

Figure 152. Graph showing the numbers of deaths in the region of the Broad Street pump. The arrow marks the date the handle was removed.

used the pump's water. There had been no similar dramatic increase in cholera deaths in other parts of London. Snow discussed his findings with the parish authorities and the pump handle was removed on 8 September making it inoperative. As you can see from Figure 152, the outbreak declined almost as suddenly as it had begun.

Snow drew a map of the area and marked all the affected houses (Figure 153). There was a workhouse in Poland Street surrounded on three sides by houses with cholera deaths but only 5 of the 535 inmates died of cholera. It turned out they had their own, separate water supply. Similarly, of the 70 men who worked in the brewery in Broad Street, none had cholera—they drank only beer! These and many other data that Snow gathered clearly incriminated the pump in Broad Street. Some patients became ill within hours of drinking the Broad Street pump water for the very first time, showing how rapidly the disease could progress. Snow could only conclude that somehow the well had become contaminated with faecal discharges from one or more cholera patients. The question then remained as to why this particular water pump was the source of cholera. Later enquiry, not by Snow, revealed that the well had probably been contaminated by an overflowing cesspool (sewage trap) which served No. 40 Broad Street, the house closest to the pump. On 28 August, a baby in that house had been attacked with cholera and the watery faeces had been deposited in the cesspool.

Snow then turned his attention to the more widespread but less sensational outbreak of cholera that occurred in London south of the River Thames between July and October 1854. This part of the city was serviced with a reticulated water supply which was provided by two companies. Some regions were supplied by the Southwark and Vauxhall Company, another by the Lambeth Company, but for some extraordinary reason, one large region was supplied by both companies (Figure 154). The pipes of both companies ran down each

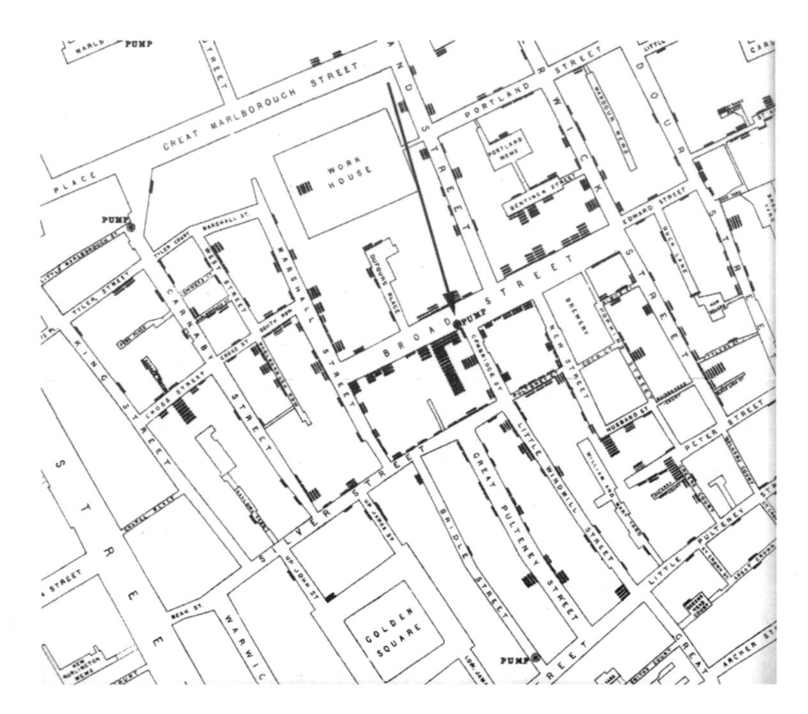

Figure 153. Snow's map of the area around Broad Street. Houses with cholera deaths are marked with black square blocks. The arrow points to the location of the pump.

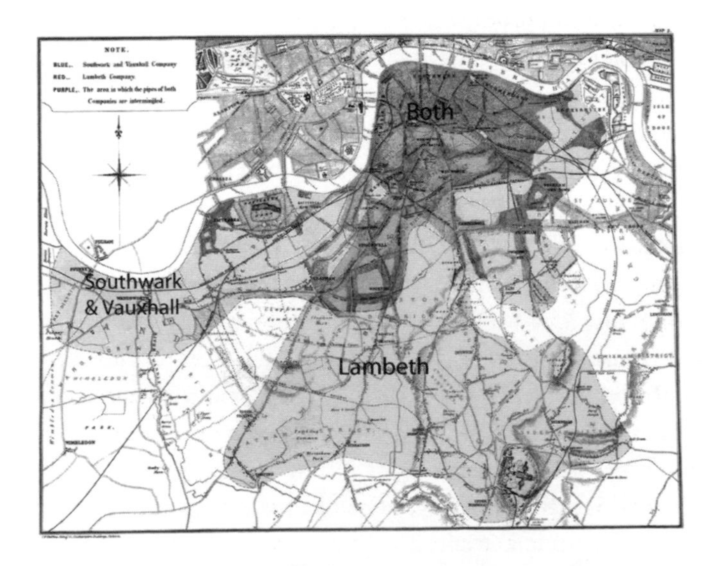

Figure 154. Map showing the area supplied by the two water companies and the area of overlap.

street and one house might be supplied by one company and its neighbours by the other. It was all totally indiscriminate and offered the opportunity for an amazing survey. Snow determined which company supplied the water for each household in which there was a cholera death. He conducted perhaps the first major epidemiological survey. The Southwark and Vauxhall Company supplied 40,000 houses and in these there were 286 cholera deaths. In contrast, the Lambeth Company supplied 26,000 houses but there were only 15 cholera deaths. The chances of contracting fatal cholera were increased 14-fold in houses supplied by the Southwark and Vauxhall Company.

Why should this be? Snow was like a bloodhound on the trail. It turned out that the Southwark and Vauxhall Company still drew its water from the River Thames at Battersea whereas the Lambeth Company had recently moved its source of supply upstream to Thames Litton where there was unquestionably less contamination with sewage. Snow published his findings in a greatly enlarged *On the mode of communication of cholera* in 1855.[292] Sadly, he died three years later at the young age of 45 from a stroke. How did one man manage to accomplish so much? He was clearly driven, but he had time—he was a bachelor. John Snow was voted in a poll of British doctors in 2003 as the greatest physician of all time.

Snow's observations that cholera was transmitted by water were conclusive but that did not mean that everyone would accept them, particularly those in authority, as the implications for public health were hugely costly. Medical opinion still had among its ranks plenty who believed that cholera was caused by some ill-defined miasma. The Fourth Pandemic that hit London was to prove Snow right. This time it fell to Dr William Farr (1807–1883), who worked in the Registrar-General's Office, to investigate. As the figures flowed in during July 1866, Farr was struck by the way in which deaths from cholera/diarrhoea were concentrated not just in East London north of the Thames, but in a particular part of it which was supplied by just one water company—the East London Water Company. This company supplied an area housing one-seventh of the population of London but was home to five-sevenths of all the deaths. The company obtained its water from two sources, but one of them, at Old Ford, had uncovered reservoirs which were home to many eels, not what one would expect to find in pure, filtered water.

Despite all this amassed evidence, there were still many who were convinced that cholera was not a water-borne disease. Indeed, at the fourth of the International Sanitary Conferences convened in 1874 to formulate uniform maritime quarantine regulations, the 21 governments present voted that the ambient air was the principal vehicle of 'the generative agent of cholera'.

Whatever the vehicle that transmitted it, what was the agent? As the bacterial cause of diseases began to be uncovered in the 1870s and 1880s, the investigative baton passed to the Europeans, especially the Germans and French. In 1883, Egypt, which was now under British control, was being swept by a severe cholera epidemic and the Egyptian government appealed for help. Following his triumphant discovery of the tubercle bacillus in 1882, Robert Koch (p.211) and his small team of the bacteriologists, George Gaffky (p.333), Bernhard Fischer, and Ludwig Treskow, a chemist, set out for Egypt to search for a bacterial cause. It was only a dozen years since the Franco-Prussian War and there was much animosity between Germany and France, so France did not want to be outdone. Pasteur was unable to lead the French mission because of ill-health so a team made up of his assistants, Emile Roux (p.300), Louis Thuillier (1856–1883), Edmond Nocard (1850–1903), and Isidore Straus (1845–1896) left for Egypt. Both teams set up shop in Alexandria in August 1883, the French arriving nine days before the Germans.

The Frenchmen dissected 24 cadavers and examined stool specimens from cholera patients. They saw so many bacterial types that they could not discern a likely pathogen. In blood smears from these patients, they found tiny bodies in blood smears which they thought might have a causal relationship. Koch was later to conclude that they had merely seen platelets, normal constituents of the blood involved in clotting. The French tried to infect guinea pigs, rabbits, mice, hens, pigeons, quails, pigs, a jay, a turkey, and a monkey but had no success. Unfortunately, Thuillier contracted cholera and died. In this tragedy, old enmities were forgotten and the Germans laid laurel wreaths upon Thuillier's coffin. On 7 October, disheartened, the three remaining members of the French mission left Egypt for home. They reported that they could not attribute a specific action to 'the microbe that we have encountered in the greatest abundance in the greatest number of cases'.[55]

Koch was likewise cautious. In his first report to the German Minister of the Interior on 17 September, he commented that faeces of patients with cholera had a multiplicity of organisms but that he had found a distinctive bacillus in the intestinal mucosa of the 10 corpses he had examined since his arrival (Figure 155) but it was too early to be sure. There were no signs of this bacterium in the blood or the organs of the body. Attempts to infect monkeys, dogs, mice, and hens with choleraic material proved fruitless. Likewise, he had been unable

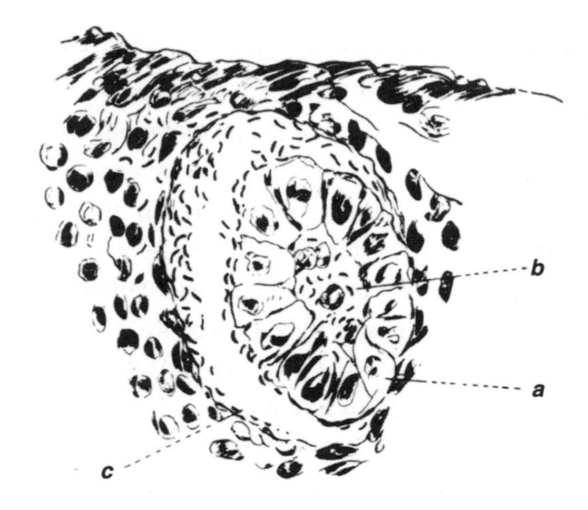

Figure 155. Koch's drawings of bacteria in the intestinal lining. His a, b, and c point to comma bacilli.

to cultivate the organism he had seen in the mucosa in pure culture. The cholera epidemic in Alexandria was now waning so Koch travelled on to Calcutta in British India where he was given access to the laboratories of the Medical College Hospital.

Here he had better luck. Within days of their arrival, Koch and his colleagues had a pure culture of the cholera organisms. The key to success was the availability of fresh material with cholera bacteria present in almost pure culture and relatively few other intestinal bacteria in 'rice-water stools'. This enabled them to separate the cholera organisms. Working all hours in December and January 1884, they examined specimens from 17 living patients and 29 corpses. The bacteria they saw were the same as they had seen in Egypt. In his despatch to the German government on 2 February Koch wrote:

> It can now be taken as conclusive that the bacillus found in the intestine of cholera patients is indeed the cholera pathogen... We have determined special properties that make it possible to definitively separate the cholera bacillus from the other bacteria.[164]

He then went on to describe those characteristics:

> The bacillus is not a straight rod, but rather is a little bent, resembling a comma. The bending can be so great that the little rods almost resemble half circles... They are very actively motile... Colonies on (nutrient gelatine) are formed which at first appear compact but gradually spread out as the gelatine is liquefied. In cultures, the colonies of the cholera bacillus can therefore be readily distinguished from other bacteria, making isolation into pure cultures easy.[164]

Within a few weeks, Koch showed that comma bacilli could be found in water tanks throughout the Ganges delta and that this was the likely source of infection. He showed that the bacillus could not be breathed in but that it could be conveyed on clothes and linen, particularly under warm, damp conditions. By the end of his investigations, he had performed 52 necropsies on those who had died from cholera, all with consistent findings.

By May, the heat was making it impossible to work with gelatine cultures as they liquefied. Besides, they were feeling homesick so the German team returned to Berlin to a hero's welcome. German science had shown its superiority over the British and French! Koch was loaded with honours including a medal, a bust of the Kaiser, 100,000 marks, an official reception at court, and a huge banquet. A major conference on cholera was held in Berlin at the end of July and even Rudolf Virchow, Koch's eminent opponent, appeared to be convinced.

Of course, you will have noticed that Koch had not fulfilled the last of his postulates that the disease should be transmitted to another animal or human from a pure culture isolated from a prior case. This meant there were plenty of sceptics. The French were rather miffed that they had not found the cause. Louis Pasteur, who had been on the receiving end of Koch's invective over other matters, was sceptical and very cross when Koch went to France when cholera broke out in Toulon in the summer of 1884 and successfully showed the comma bacilli to Roux and Straus. Pasteur wrote to Straus:

> Try to find out the fallacy in his story... He must have made some sort of great error... As much as possible, work by yourself... Keep your cadavers to yourself. The reports that tell you how great this Koch is are wrong. His knowledge on cholera is not that good. If your results agreed with his, he alone would get all the credit. Already the German newspapers are crowing.[246]

Hardly the words of an open-minded scientist seeking after the truth! One French medical journal opined: 'The great microbe hunter has followed a completely false trail. (Will he give back his decorations?).'[55]

The French might have been miffed but feelings in Britain were at first more mixed. If what Koch said was true, then that was a good thing. On the other hand, Koch's discoveries had been made in Egypt and India, territories under British suzerainty, and this reflected poorly on British scientific prowess. On 6 August 1884, a British mission consisting of Emanuel Klein (who was actually Austro-Hungarian, p.241), Heneage Gibbs, and a technician sailed for Calcutta to check Koch's findings. Unbelievably, they repudiated Koch's thesis and dismissed the role of water in the transmission of cholera. The British Secretary of

State for India appointed a committee of 13 eminent medical men to consider the Klein–Gibbs report; eight submitted memoranda endorsing their conclusions. One, John Burdon-Sanderson, stated in a public lecture that Koch's investigations had been 'an unfortunate fiasco'. The British attitude became so negative that at the Sixth International Sanitary Conference in May 1885, the British delegation successfully blocked any 'theoretical discussion of the aetiology [cause] of cholera',[55] even though Koch himself was present.

So matters dragged on until the third of Koch's postulates was fulfilled in a most unexpected manner. The main antagonist to Koch in Germany itself was Max von Pettenkofer (1818–1901), professor of hygiene in Munich. Pettenkofer had for many years been the great protagonist for the 'soil theory of cholera'. According to him, in addition to a specific microbe in the soil which he called 'x', there had to be another factor in the soil ('y') which was essential to produce the actual infectious material ('z'). He thought that the spread of cholera was dependent upon the condition of the soil and the water played no role. Von Pettenkofer was so convinced that Koch's comma bacilli were not the cause of cholera that in 1892, he swallowed 1 ml of a broth culture containing about one billion cholera bacilli. He developed a mild case of cholera (which he insisted was just diarrhoea) and comma bacilli were isolated from his stools. His assistant Rudolf Emmerich (1852–1914) was not so fortunate. A few days later he repeated the experiment, became seriously ill with cholera and almost died. Professor von Pettenkofer was lucky. He later wrote:

> Even if I had deceived myself and the experiment endangered my life, I would have looked Death quietly in the eye, for mine would have been no foolish or cowardly suicide; I would have died in the service of science.[248]

Fine words to say when you have come through relatively unscathed! He might have had a death wish—nine years later he did commit suicide.

The year 1892 was a bad one for cholera. Nearly 300,000 Russians died. In Hamburg, Germany there were 17,000 cases, with 8,500 deaths in three months. The Prussian government ordered Koch to investigate. The adjoining city of Altona, which is downstream from Hamburg and might be expected to have water contaminated by discharge of sewage from Hamburg, had only 500 cases of cholera with 328 deaths. Koch found that Hamburg used unfiltered water from the River Elbe while Altona used water, also from the Elbe, but filtered through sand-beds. Koch had repeated Snow's pioneering observations but made no mention of them. Perhaps he did not know of them. The case for the water transmission of cholera through the agency of Koch's comma bacillus was incontrovertible.

There was, however, to be a twist to the water story. In 1905, at a quarantine station at El-Tor on the Sinai Peninsula in Egypt, Felix Gotschlich (1874–1914), a German doctor, found comma bacilli in the bowels of six pilgrims returning from Mecca who died at El-Tor from other diseases, even though they had no clinical or post-mortem evidence of cholera. There was no doubt that these bacteria were cholera organisms for they reacted with antibodies that were specific for these organisms. This El-Tor biovar, for so it was called, was found to cause a relatively mild form of cholera which was rarely fatal. Furthermore, this biovar can survive in people without symptoms for much longer than can the classical comma bacilli and some people become long-term carriers (even for years). This characteristic allows carriers to infect a greater number of people. Moreover, as well as being transmitted in water, these bacteria can be acquired by consuming uncooked food fertilized with human faeces, as is not uncommon in the tropics. This organism was to be the cause of the Seventh Cholera Pandemic which began in Sulawesi, Indonesia, in 1961.

There is an extraordinary postscript to this story. While Robert Koch discovered the cholera bacilli and eventually convinced the world in the teeth of almost fanatical opposition that it really was the cause of cholera, he was not the first to find this organism. He had been anticipated by 30 years. In 1854, Filippo Pacini (Figure 156), an anatomist in Florence, Italy, published a tract entitled (in translation) 'Microscopical observations and pathological deductions on cholera'. In this paper, he described finding 'miriadi di vibrioni' (myriads of vibrios) in patients with cholera and came to the same conclusions that Koch would later come to, and for the same reasons. The word 'vibrio' is derived from the Latin word 'vibrare' meaning to vibrate, and reflects the energetic movement of these comma bacilli. It is not surprising that this work was ignored because at that time the germ theory of disease had not

Figure 156. Filippo Pacini (1812–1883).

been developed and cholera was generally thought to be due to a miasma. In five further publications over the next 26 years, Pacini developed his thesis. Unlike Koch, who misunderstood the nature of disease in cholera, Pacini correctly realized that the massive loss of water and electrolytes (salts made up of sodium, potassium, chloride, and bicarbonate) was due to a toxic action of vibrios on the intestinal lining. What is more, he recommended the intravenous injection of a salt solution of 10 g in a litre of water in serious cases. This is so close to what would later become the standard treatment. Unfortunately, few people knew of Pacini's work. Robert Koch appears to have been completely unaware of it in 1883 when he sailed for Egypt, the same year that Pacini died. Posthumous recognition eventually came. In 1965, the International Committee on Bacteriological Nomenclature adopted *Vibrio cholerae* Pacini 1854 (which means, of course, the vibrio of cholera) as the official name for the cholera bacterium.

Animals are not naturally infected with *V. cholerae*. Cholera is a human disease which travels along lines of human activity. Immunization is of little value. Personal hygiene, safe water supplies, and the proper disposal of sewage are the bulwarks on which we must rely for the control of this disease. You can never tell when it will reappear. After an absence of 50 years, cholera is spreading once more with thousands of cases in Papua New Guinea, right on Australia's doorstep.

29

Typhoid fever

SCIENTIFIC NAME:	*Salmonella enterica serovar Typhi*
COMMON NAME:	*Salmonella typhi* or typhoid bacillus
ORGANISM'S APPEARANCE:	Gram-negative rod
DISEASE NAME:	typhoid fever or enteric fever
DISTRIBUTION:	formerly everywhere, currently in parts of the tropics
TRANSMISSION:	usually ingestion of contaminated water or food
LOCATION IN A PERSON:	small intestine then spread via the bloodstream to many organs
CLINICAL FEATURES:	fever, prostration, enlarged liver and spleen, relatively slow pulse. Complications include intestinal perforation and peritonitis
DIAGNOSIS:	isolation of *S. typhi* from blood, faeces, or urine
TREATMENT:	various antibiotics but the organism is becoming more resistant and has to be tested—amoxicillin, co-trimoxazole, ciprofloxacin, ceftriaxone, chloramphenicol
PREVENTION:	a partially effective vaccine is available; avoid unboiled or untreated water or uncooked food in endemic areas

'You feel hot. Where's the thermometer?' A conversation something like this has been repeated countless times in innumerable households over the last few decades. Yet it has not always been so. Until about 130 years ago, clinicians generally had to rely on feeling the patient's skin with their own hands. This enabled them to detect marked fever with some confidence and enabled them to discern patterns of fever. Fever, by the way, simply means a body temperature increased above normal. Ancient physicians such as Hippocrates were able to discern the remarkable fevers of malaria, which recurred at regular intervals producing recognizable patterns (p.116). Primitive thermometers, called thermoscopes, were invented in the seventeenth century by Galileo Galilei (1564–1642) and Santorio Santorio (1561–1636). These instruments were susceptible to air pressure, so in the middle of the seventeenth century a sealed thermometer was made for the Grand Duke of Tuscany.

The problem was that there was no agreed standard scale. The German, Daniel Gabriel Fahrenheit (1686–1736), who for the most part lived in England and Holland, invented an alcohol thermometer in 1709 and a mercury thermometer in 1714. He used the latter to develop a scale in 1724 called the Fahrenheit scale, which became popular throughout the English-speaking world. He used three fixed points. The lowest, which he called 0 degrees, was the temperature at which either a mixture of water and ammonium chloride or salty water froze. The second was the temperature at which rain water freezes (32 degrees). The third was the reading when the thermometer was placed under the armpit (96 degrees). Others later showed that water boiled at 212 degrees. A competing scale was proposed by the Swedish astronomer Anders Celsius (1701–1744) in 1742. Rather extraordinarily, he gave zero as the boiling point of water and 100 degrees as the melting point of ice. A year later, the Frenchman Jean Pierre Cristin (1683–1755) inverted Celsius's scale to produce the Centigrade scale with a freezing point of 0° and a boiling point of 100°. By international agreement in 1948, Cristin's adapted scale became known as the Celsius Scale and is used in those countries which have adopted the metric system.

Why am I talking about all this? Fever is a hallmark of so many infections, including the one that we are discussing in this chapter, typhoid fever. The development of these instruments allowed them to be applied to clinical medicine. The person who popularized the measurement of temperatures in patients and demonstrated the value of temperature charts was the German Carl Reinhold August Wunderlich (1815–1877). Wunderlich observed the temperatures of a massive 25,000 patients, mostly while he was in Leipzig, and he and his assistants were said to have made several million observations. He used a mercury thermometer some 22 cm long which was placed under the armpit for 15 minutes. He detailed his observations in a book published in 1868 called (in translation) *The Course of Temperature in Diseases.*[330] He gave the average normal temperature as 37°C (98.6°F) but showed that it varied during 24 hours, being higher in the day and lower at night. He defined fever as a temperature of 38°C (100.4°F) or greater, values which are more or less accepted today. But perhaps Wunderlich's most important contribution was his clear demonstration that

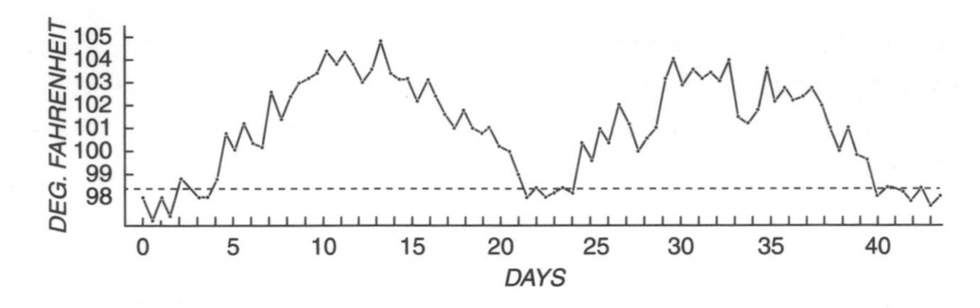

Figure 157. Temperature chart of a patient with typhoid fever showing a continuous fever. There is an apparent recovery after three weeks but then there is a recurrence. Temperature is measured in degrees Fahrenheit.

fever was a sign rather than a disease in itself. Furthermore, the pattern of fever was often of diagnostic value. This is indeed the case in typhoid fever which classically has a 'continuous pattern' in which there are small daily variations but these variations are constantly well above the normal baseline (Figure 157). Fever is often accompanied by other signs such as sweating, prostration, and wasting.

Having cleared all this out of the way, let's go back to the beginning. Typhoid fever has probably been around for millennia. Unfortunately, it is often not easy to differentiate typhoid fever from other infections that cause fever, especially early in the course of an illness. It is often said that Alexander the Great (356–323 BC) died from typhoid fever. He might well have done, but plenty of other things could have killed him, not the least being falciparum malaria. Hippocrates during the course of two successive autumns met with many cases of fever of a continuous type which could have been typhoid fever. These patients often had diarrhoea (which is not all that frequent in typhoid fever) together with vomiting, swollen abdomen and abdominal pain, red skin rashes, delirium sometimes leading to coma, and increasing emaciation, all of which are consistent with typhoid fever.

Unfortunately, there are various illnesses characterized by a fever over which there was confusion. Attempts were made at first to differentiate them on clinical grounds. The Englishman Thomas Willis (p.205) wrote a book in 1659 in Latin *De febribus,* which was published in English (*On Fevers*) in 1685.[322] In this book he tried to differentiate between diseases that we now call typhus, typhoid, the plague, and malaria, amongst others. Trying to work out which diseases he was talking about is mind-bogglingly complex and his explanations of what was going on are as though from another planet. For example, he talks about the sulphur of the blood burning, or 'expulsion of the adjust matter remaining after the deflagration of the Blood'.[322] Nevertheless, he does discuss a fever, which he calls a 'putrid fever', in a way that is very reminiscent of the clinical course of untreated typhoid fever resembling the temperature pattern shown in Figure 157:

> In this Fever four states of time are to be observed, by which, by so many Stages its course is performed; and they are these, the Beginning, the Increase, the Height and the declining state. These are wont to be pass'd over in some sooner, in some slower, and in a longer time.[322]

In the 1730s, a number of writers described typhoid fever as a 'slow nervous fever' characterized by its long duration, diarrhoea which was often bloody, abdominal pain, nosebleeds, and sweating which gave no relief. This was compared with the 'malignant continued fever' or 'putrid, malignant, petechial fever' of typhus. Confused? Well, the clinicians of those days were too.

These clinical descriptions were just stabs in the dark. A more solid distinction was made when the bodies of patients who had died from these diseases were examined. It appears that Adriaan van den Spiegel (Latin name Spigelius, 1578–1625), a Flemish physician who practised mostly in Padua, Italy, was the first to describe pathological lesions of typhoid fever when he called attention to inflammation with ulceration, and sometimes gangrene, of the small and large intestines. These observations were made more definite by Thomas Willis in 1682 when he described ulcers in the small intestine accompanied by swelling of the mesenteric lymph nodes in what he believed was putrid (typhoid) fever. The mesenteric lymph nodes are lymph glands that are located in the folds of tissue called the mesentery that bind the small intestine to the back of the abdominal wall and which drain fluid and other material (including bacteria) from the intestines. These observations were supported in 1696 by George Baglivi (Giorgio Baglio, 1668–1707), who also found intestinal inflammation and enlarged mesenteric glands. The Italian physician and anatomist Giovanni Lancisi (p.74) in 1718 first described perforation of the intestine in typhoid fever, although he mistakenly attributed this to *Ascaris lumbricoides*; these worms had merely wandered through the hole caused by the typhoid fever.

In the early nineteenth century, attention on typhoid fever switched to France. In 1826, Armand Trousseau (1801–1867), a pupil of Pierre Bretonneau (p.292), described Bretonneau's observations:

> Since 1813 he (Bretonneau) has collected a large number of cases... He has been led to distinguish a disease, the seat of which appears to be exclusively

> in the glands of Peyer and Brunner, which one finds in the jejunum, ileum and large intestine. He has given this affliction the name of dothinenteritis [from the Greek words for pustule and intestine] ... He has indicated the relationships, traced the symptoms and described with precision, the appearance of the disease, which changes on successive days.[53]

Peyer's glands, or as we call them 'Peyer's patches', are organized collections of lymphoid tissue (follicles) made up of a particular form of white blood cell (lymphocytes) located in the ileum, which is the lowermost part of the small intestine, about 2–4 metres long. Brunner's glands are in the duodenum, the first part of the small intestine, about a foot long, which produce intestinal secretions and are not these days thought to be a site particularly involved in typhoid fever. Trousseau went on to describe Bretonneau's findings. For example, on the fifth day, the earliest day on which an autopsy had been performed:

> The glands of Peyer, especially those which border on the ileocaecal valve [at the junction of the small intestine with the large intestine] are markedly swollen.[53]

On the sixth day:

> Most marked tumefaction [swelling] of the glands of Peyer ... The size of the mesenteric gland is still increased.[53]

On the ninth day:

> The glands of Peyer are now larger, more rounded ... The mesenteric glands have achieved considerable size.[53]

On the tenth day:

> One of two things, either the inflammation proceeds to resolution [healing], or it continues to go through various changes.[53]

In those who don't get better, on the 16th day:

> The core [a swollen Peyer's patch] is entirely detached ... leaving in its place a deep excavation ... the base of the ulcer rests upon the muscular coat [the outer layer of the intestine] upon the peritoneum which they perforate so frequently.[53]

Bretonneau had clearly and correctly focused on where typhoid begins—near the end of the ileum which we now know is invaded by bacteria that multiply in the Peyer's patches. But his name of dothinenteritis did not catch on.

Similar findings were described in 1828 by Pierre Charles Alexandre Louis (Figure 158), who was working in the Hôpital de la Charité in Paris, when he published a book on the subject.[195] It was Louis who was to invent the name 'typhoid', meaning 'like typhus'. Nevertheless, like his predecessors, Louis did not clearly differentiate between typhoid and typhus fevers. The problem was that no-one had the opportunity to observe both of these diseases at close hand at the same time. At that time, typhus prevailed predominantly in England and the English pathologists rarely found ulcerations of the small intestine, and when they did regarded them as a complication of typhus. Even so, the French were inclined to consider typhus in England as identical with the French typhoid.

Figure 158. Pierre Louis (1787–1872).

The first to clearly differentiate them was the American, William Wood Gerhard (Figure 159). Gerhard had gone to study with Louis in Paris and became well acquainted with the pathological lesions in typhoid fever. Before returning to the USA, he familiarized himself with the typhus which was then raging in the British Isles. When he returned to Philadelphia, he saw many cases of typhoid fever and found the same pathological lesions that he had seen in France. In 1836, typhus fever invaded Philadelphia and Gerhard was able to compare the two diseases. He was able to convincingly demonstrate that the two diseases were different:

Figure 159. William Gerhard (1809–1872).

> The anatomical characters of these varieties of fever are peculiar to themselves, and it is impossible to substitute the lesions of the follicles of the small intestine observed in the typhoid fever for the pathological phenomena of

typhus, as it is by treatment or other means to transform the eruptions of measles into the pustules of smallpox.[114]

Also, Gerhard insisted that in typhoid fever there were rose-coloured spots on the skin, whereas in typhus fever, there were plentiful petechiae (haemorrhages in the skin which would not blanch under pressure). Even so, there were still plenty of people who thought that typhus and typhoid were one and the same thing.

In 1850, an Englishman, William Jenner (1815–1898), physician to Queen Victoria, was himself afflicted with one of these two diseases, and was goaded into writing an article on whether or not typhoid and typhus were the same or different diseases. He asked whether British physicians had confounded two distinct diseases. Then he went on to marshal the facts and convincingly demonstrated that they were quite distinct on the basis of clinical features and post-mortem examination. The two conditions must therefore have different specific causes.

What were these causes? As late as 1873, the English physician Charles Murchison (1830–1879) was expounding that typhoid originated from decomposition of organic matter which fouled the air. This 'pythogenic theory' was held despite the cogent observations that had been made several decades earlier by William Budd (Figure 160). Budd was originally apprenticed to his father, a doctor, then spent four years in Paris where he came under the influence of Pierre Louis and was aware of Bretonneau's work on typhoid fever. Next, he went to Edinburgh where he graduated in medicine following which he served very briefly at the Royal Naval Hospital HMS *Dreadnought* where he suffered from a near fatal attack of either typhus or typhoid fever. He then became the medical practitioner in the village of North Tawton in Devon, where he had been born, and was thus well prepared when an outbreak of typhoid fever occurred in 1839. Within the space of four months, some 80 of the inhabitants had gone down with typhoid

Figure 160. William Budd (1811–1880).

fever, with many of them dying. Budd later wrote about this epidemic that it showed a 'strong tendency, when . . . introduced into a family, to spread through the household'.[61] What is more, a number of people who had become infected in North Tawton returned to their homes or visited others several miles away and infected other members of those households. Budd thought that this illness must be communicable from one person to another and in 1839 wrote an essay to that effect which was unsuccessful in a medical competition.

He then moved to Bristol in south-west England. In 1847, Budd realized that there was a minor epidemic in Richmond Terrace with 13 of the 34 households having at least one member with typhoid fever. These 13 households used a common well whereas the other 21 houses had a different water source. He thought that the disease must be transmitted by agency of the water. In 1863, typhoid fever broke out at the Convent of the Good Shepherd in Arno's Court, near Bristol. An inmate of the convent, after several months in a town 20 miles away, had returned to the convent while sick with the typhoid fever. Six weeks after her return, the first case of typhoid fever appeared in the convent and within a few months, more than 50 inmates of the convent had been stricken, with several of them dying. It seemed clear to Budd that this outbreak in an establishment that had never had a single case of typhoid, indicated further evidence that the condition must be contagious, even if it was a lot less contagious than typhus. He believed the poison must lie in the faecal discharges and he became an enthusiastic proponent of disinfection with chloride of lime. Budd published two articles on his views in *The Lancet* in 1856. Then, in a manner somewhat reminiscent of that of John Snow on cholera, Budd eventually gathered all his ideas in 1873 in a book. These are his prophetic words written before the bacterial cause of typhoid had been discovered: 'The living human body, therefore, is the soil in which this specific poison breeds and multiplies.'[61]

If it was contagious (or better, infectious), then what was being transferred? Enlightenment began in 1880. Who discovered the bacterial cause of typhoid fever? There is no simple answer. The difficulty lies in the fact that the organism that causes typhoid fever looks like many similar bacteria and when the discoveries were first made, we cannot really be sure of who found what. Frederick Gay, professor of pathology in the University of California, wrote a book on typhoid fever in 1918 and had this to say:

> The first observations of Bacillus typhosus [the then current name for the organism causing typhoid fever] in the tissues may be attributed to three independent investigators, namely Eberth, Klebs and Koch. Credit of priority in

> discovery of the organism is generally given to Eberth, but it may well be that injustice has been done in the general acceptance of this priority.[113]

What exactly happened? Edwin Klebs (p.296) published his first contribution on 22 April 1880. He described his findings in 24 fatal cases of typhoid fever. He found short bacterial rods in various tissues including Peyer's patches, mesenteric lymph nodes, blood vessels, and meninges (membranes around the brain), and they were absent from the intestinal wall of non-typhoid cases. The bacteria were associated with inflammation (infiltration of white blood cells) and death of the tissues. The trouble was that he also described seeing on some occasions unbranching, bacterial filaments some 50–80 μm in length.

In July 1880, Karl Joseph Eberth (1835–1926) published his observations on typhoid fever. He had been born and educated in Würzburg, Germany, and at the time of his publication was professor of pathology in Zurich, Switzerland. He concentrated mostly on histological sections of the mesenteric lymph nodes and spleen. In 18 of 40 cases of typhoid fever, he found short bacterial rods whereas he found none in 20 patients dying from other diseases. He thought that the numbers of bacteria decreased the longer the patient lived before dying. The problem with Eberth's report is that he said that the rods at times possessed spores.

In 1881, Robert Koch (p.211) published photomicrographs of the bacteria seen in various infectious diseases, including typhoid fever. In the latter, the causative organism appears with convincing clarity. Both Eberth and Klebs later agreed that the bacterial forms that they had seen were identical. Klebs has been dismissed by later commentators because he described filamentous forms even though (1) he primarily focused on the short rods and (2) it has subsequently been shown that filamentous forms do sometimes occur, especially in culture or (3) he might have been describing a secondary invader. Likewise, Eberth can be criticized for describing spores that are simply not found in the organism that causes typhoid fever. Gay concluded his analysis: 'Klebs unquestionably should be credited with suggesting the accepted name (*B. typhosus*) for the etiological agent in typhoid fever.'[113] Gay wrote relatively close to the time these events took place and we, nearly 100 years later, are not going to do any better.

The problem lies in the fact that only the first of Koch's postulates had been met. The organism needed to be grown in pure culture on artificial media. Here again there is some controversy. Klebs in his second communication in 1881 described inoculating liquid gelatine media with mesenteric lymph nodes from a case of typhoid fever and then finding rods in the turbid culture medium. He then inoculated rabbits with this culture and found haemorrhages but no ulcers in the caecum (the first part of the large intestine) of the rabbits. All this was dismissed by Gaffky, who is generally given the credit for being the first to cultivate the typhoid bacillus. Georg Theodor August Gaffky (1850–1918) worked in Robert Koch's Institute. First, he saw rods in sections of the spleen, liver, kidneys, and lymph nodes of patients with typhoid fever. He then took a piece of spleen and streaked it onto a nutrient gelatine plate. Forty-eight hours later, he had a pure culture of typhoid bacilli. Gaffky dismissed Klebs's findings on the grounds that Klebs described filamentous forms in his patients and the fact that he, Gaffky, could not produce any disease in rabbits with his own cultures. Gaffky in 1884 was certainly the first, though, to grow typhoid bacilli on solid culture media (gelatine-coated glass slides) and prove their purity. He had fulfilled Koch's second postulate.

Richard Pfeiffer (1853–1925) in Germany in 1885 reported that he had found the typhoid bacillus in faeces. The organism was observed in urine by Ferdinand Hueppe (1852–1936) in 1886 in 1 out of 18 cases. The typhoid bacillus was found in the blood of a cadaver by Eugen Fraenkel (p.253) and Morris Simmonds (1855–1925) in 1886, in skin lesions (rose spots) by Neuhaus in 9 of 15 cases in 1886, in the blood from a living patient by Dr AI Vilchur in St Petersburg in 1887 (in only one out of 37 attempts), and from the gall bladder in cases of cholecystitis in typhoid fever by Gilbert and Chirode in 1890. All of this made it clear that typhoid fever was not just an intestinal disease but an infection that affected the whole body.

But all was not entirely plain sailing. It was increasingly recognized that there were rather similar bacteria, which were all Gram-negative (stained pink on the Gram stain), that had to be differentiated from the typhoid bacillus. They occurred normally in faeces and often caused urinary tract infections. *Klebsiella pneumoniae*, which despite its name is normally found in the intestinal tract and can infect any part of the body, had been discovered by Friedländer (p.253) in 1883, *Escherichia coli* was discovered by Escherich (p.339) in 1885, and many more were to follow. Various tests had to be developed to aid in the identification of different organisms. The most important early test was the demonstration by the French physician Georges-Fernand-Isidor Widal (1862–1929) in 1896 that when serum from animals that had been inoculated with typhoid bacilli was mixed with typhoid bacilli, the bacteria clumped together (a phenomenon called agglutination, Figure 161). This enabled easy diagnosis in the laboratory. It was not until 1902 that Karl Wilhem von Drigalski (1871–1950) and Heinrich

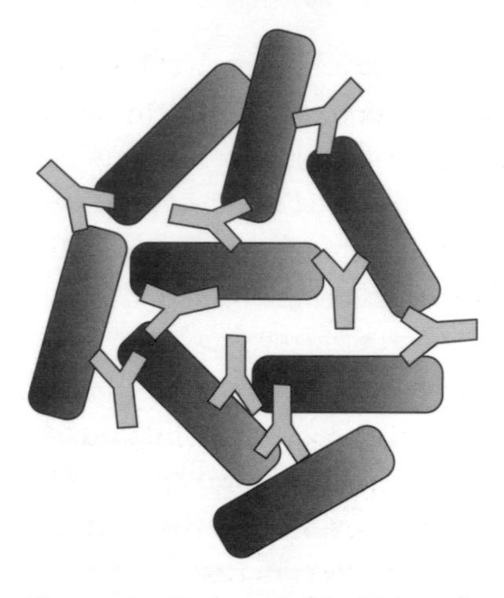

Figure 161. Each arm of the Y-shaped antibody attaches to a different bacterium causing them to clump together.

Conradi (p.354), working in Koch's Institute, developed the first culture medium that permitted selective culture (i.e. encouraged growth of typhoid bacilli but not other bacteria) and aided specific identification of the typhoid bacillus.

Shortly thereafter, an epidemic of typhoid fever occurred in Trier on the Mosel River in Germany. Koch sent out Paul Frosch (1860–1928) and co-workers to investigate. Using this new medium, they were able to trace the organism through the infected population and show its connection with sewage and contaminated drinking water. Furthermore, Koch carefully analysed the appearance of typhoid fever in a small village near Trier and realized that some infections were derived not from water or sewage but from other people who were themselves perfectly healthy. He called these people 'carriers' and said in a speech in Berlin in 1902:

> Our studies have shown that all cases of typhoid of this type have arisen by contact, that is, carried directly from one person to another. There was not a trace of a connection to drinking water.[165]

In fact, von Drigalski in 1904 showed that typhoid bacilli could sometimes be found in the faeces of people who had never had typhoid fever—so-called 'healthy carriers'. This brings us to the most famous carrier of all: 'Typhoid Mary'.

Mary Mallon (1869–1938) was the first person in the USA to be identified as a healthy carrier of typhoid fever. Mary was born in Ireland and migrated to the USA in 1884. Unfortunately, her occupation was as a cook. Large numbers of people in the households in which she worked in New York between 1900 and 1907 became infected with typhoid fever. When it was suggested to her she might be carrying the infection, she refused to be tested. Eventually she was forcibly tested, found to be positive, and in 1907 put in isolation in a hospital on North Brother Island in the East River of New York where she remained for three years. She was eventually released when she agreed that she would not work again as a cook and would be scrupulous in her personal hygiene. She did not keep her part of the bargain. She called herself Mary Brown and in 1915 infected 25 people, one of whom died, while working as a cook at the Sloane Hospital for Women in New York. She was returned to quarantine in North Brother Island where she remained for the rest of her life. Over the course of her career, she is known to have infected 53 people, 23 of whom died from the disease, but the true numbers are almost certainly much greater. When she died at the age of 69, she was still infectious and typhoid bacilli were found in her gall bladder at post-mortem examination.

You may well have realized that Koch's third postulate (that animals or humans be infected from a pure culture and produce disease) has not yet been fulfilled, although the epidemiological information (measurements of the spread of typhoid fever in communities) is pretty convincing. The reason for this is that typhoid fever in nature occurs only in humans and attempts to infect laboratory animals were disappointing. Fraenkel and Simmonds in 1885 and 1887 gave large amounts of typhoid bacilli to mice and produced ulceration of the small and large intestine but did not regard their results as similar to typhoid fever in humans. In fact, it was later shown that their findings were due to toxins (endotoxins) released by the bacteria rather than invasion of the tissues by bacteria. A similar thing happened when rabbits were used. In 1911, two Russians working in Paris, Elie (Ilya) Metchnikoff (1845–1916) and Alexandre Mikhailovich Besredka (1870–1940), after several failed attempts, finally managed to induce a disease similar to human typhoid fever in chimpanzees. Koch's postulates were fulfilled.

But what about experimental infection of humans? Surely no-one would deliberately infect humans with typhoid fever? But it was done in circumstances that you will find difficult to believe. The excuse for doing so was to assess the effectiveness of typhoid vaccines. In 1959, investigators from the University of Maryland, Baltimore in the USA embarked on a series of studies which were to span a couple of decades. First, they wanted to find out how many typhoid bacilli you had to give to someone in order to produce typhoid fever. None of those fed 1,000 organisms developed typhoid fever, 50 per cent did if you gave them 10 million bacteria, while 95 per cent of those who ingested 1 billion typhoid bacilli went down with typhoid fever. In this and all the subsequent experiments, who were these indomitable 'volunteers'? Answer: inmates of the House of Correction (i.e. prison) in Jessup, Maryland. In 1966, two of the investigators, Richard B Hornick (1929–2011) and Theodore E Woodward (1914–2005) wrote that the studies were undertaken under the sponsorship of the US Armed Forces Epidemiological Board and that 'the strictest of standards which apply to the use of human subjects as volunteers were adhered to'.[142] In case anyone worried about this, they facilely added:

> The authors gratefully acknowledge the willing participation and cooperation of the inmates of the Maryland House of Correction who have made these studies possible. Their eagerness to participate in this and other studies attests to the acceptability of such investigations. Officials of the Penal System deserve special credit for their support and interest.[142]

When he reviewed all of these studies in 1979, Woodward wrote:

> The design of the studies... was not to induce serious illness for the purpose of following its course but rather to evaluate the protection afforded by the experimental vaccine in question. If illness occurred, it was treated promptly and thoroughly. Over 16 years, 1,886 volunteers were vaccinated and 762 were challenged without a single life-threatening complication or death.[327]

Was this an adequate justification? A lot of people suffered. They could have died, even if none did. It is one thing for people to volunteer freely for experimentation in the hope of achieving a greater good. But can prisoners volunteer freely? Most people would think not. All sorts of pressures, subtle and not so subtle, can be brought to bear to induce prisoners to submit. Where do you draw the line between these studies and the experiments of Nazi doctors in concentration camps? I very much doubt a proposal to carry out similar investigations would meet the approval of ethics committees today. It is true, though, that views and perceptions change. In 1973 I studied the immune systems of some villagers high up in the Papua New Guinea Highlands, then compared them with the immune responses of the inmates of a nearby low-security prison (a barbed-wired fence) who had a much more nutritious diet. Neither I nor anyone else thought twice about it. I did not give them anything nasty. Indeed, it could be argued that I did them a favour because I immunized them against tetanus and typhoid. Whatever you may think of all this, Koch's third postulate was convincingly met in the case of typhoid fever.

We now need to spend a little time delving into the morass which is the attempts to name the typhoid bacillus and the various bacteria that were subsequently found that were closely related to it. After it was first discovered, the organism was generally known as *Bacterium typhosum* or *Bacillus typhosus* in Europe and Britain, but there was a vogue for calling it *Eberthella typhi* (the name of course in honour of Eberth) in the USA. In 1896, Emile Charles Achard (1860–1944) and Raoul Bensaude (1866–1938) in France isolated an organism from a parotid abscess, in 1897 Norman Gwyn in the USA isolated an organism from blood, and in 1919 the Pole Ludwig Hirschfeld (Ludwik Hirszfeld, 1884–1954) in Serbia reported the isolation three years earlier of another organism from the blood of patients with a clinical diagnosis of typhoid fever. These all turned out to be different in their biochemical and immunological reactions from typhoid bacilli and from each other, and came to be called *B. paratyphosus* B, *B. paratyphosus* A and *B. paratyphosus* C, respectively. Collectively, they were said to cause paratyphoid fever. In 1894, Theobald Smith (1859–1934) discovered a related organism in pigs called *Bacillus choleraesuis* (meaning that it caused diarrhoea in pigs). Smith was a research assistant to Daniel Elmer Salmon (1850–1914) who for many years was chief of the Bureau of Animal Industry of the US Department of Agriculture. Eventually it was realized that these bacteria needed to be in a new genus and the genus *Salmonella* was erected in 1900 by the French bacteriologist Joseph Léon Marcel Ligniéres (1868–1933) in honour of Salmon.

Things from there went from bad to worse and are too tortuous and tedious to follow. Suffice it to say that about 2,000 different salmonellae are now recognized on the basis of their antigenic type (immunological structures on their surface). Now, after various permutations, they have all been lumped together in the one species, *Salmonella enterica*, and subclassified as serological varieties which are given a name. A serological variety is determined by the shapes of structures on the surface of the bacterium. These induce proteins in the blood called antibodies which have complementary structures so the two fit together like a lock and key. All the different serovars are like different sets of locks and keys. Thus, the typhoid bacillus is now officially known as *Salmonella enterica* serovar Typhi. Many people don't bother with such niceties and simply refer to it as *Salmonella typhi*.

What do we know about the course of typhoid fever now? In fact, we could call it enteric fever which includes the paratyphoid salmonellae which follow a similar pattern but produce less severe disease. After ingestion, the bacteria penetrate the lining of the small intestine and enter the Peyer's patches. Here they multiply over 7–14 days. They then spread through the bloodstream and are taken up by certain cells (phagocytes) in the liver, spleen, bone marrow, and lymph nodes. It is at this point that the infected person becomes sick with fever, prostration, enlarged spleen and liver, constipation or diarrhoea, skin rashes called rose spots, and conjunctivitis. The organisms can in fact lodge anywhere and cause a wide variety of symptoms and signs. The pathogenic bacteria can be found in the blood of most patients, in their stools in about half, and in the urine in about a quarter of patients. These excreta are the source of contamination of water supplies. Moreover, faecal soiling of hands and fingers if unwashed leads to contamination of foodstuffs in which the bacteria multiply before they are ingested. Untreated, the course of illness lasts about three to six weeks, and in some patients, the symptoms return days or weeks after apparent cure (a relapse).

What about all the other serovars of *Salmonella enterica*? As we have mentioned, the old paratyphoid bacteria cause a typhoid-like illness. The vast majority of serovars cause gastroenteritis or 'food poisoning', manifesting as diarrhoea, usually without invading the tissues unless the person's immune system is impaired. These salmonellae are major causes of diarrhoea around the world in both developed and undeveloped countries.

In times past, how many developed typhoid fever and which people tended to be infected? In 1900 in the USA, there were estimated to be 350,000 cases (35.9 cases per 100,000 population) with 35,000 deaths. About half of all infections occurred in people aged between 15 and 25 years. Similar figures were noted in Europe and those other parts of the world in which they were measured. With improvements in sanitation, effective disposal of sewage, and the provision of safe water supplies, typhoid fever has retreated from countries in the First World that were once infected. Now, it is mostly a problem in tropical countries with poor sanitation, and is as frequent there as it was in the USA and Europe at the beginning of the twentieth century.

30

Escherichia coli

SCIENTIFIC NAME: *Escherichia coli*
COMMON NAME: *E. coli*
ORGANISM'S APPEARANCE: Gram-negative rod
DISEASE NAME: infections with *Escherichia coli*
DISTRIBUTION: everywhere
TRANSMISSION: ingestion of contaminated water or food
LOCATION IN A PERSON: normal or pathogenic inhabitant of the bowel
CLINICAL FEATURES: a. invasion—urinary tract infection, peritonitis, septicaemia, meningitis, etc.; b. gastroenteritis
DIAGNOSIS: isolation from urine, blood, or tissues in the case of invasive infections, and from faeces in gastroenteritis
TREATMENT: various antibiotics, depending upon the results of testing in the laboratory
PREVENTION: usually impractical

If you are a woman, there is a fifty-fifty chance that at some time in your life you will develop a urinary tract infection. And if you do so, there is a fifty-fifty chance that the infection will be due to a bacterium called *Escherichia coli*. What is the origin of this unusual-sounding name, where do these bacteria live, and what do they do?

Theodor Escherich (Figure 162) was born in Ansbach, Germany. After graduating in medicine, he decided to become a paediatrician. The year 1884 found him in Munich, Germany. He began to study the bacteria that are normally found in the stools of babies and how the stool changes immediately after birth. Meconium is the substance that fills the lumen of the intestinal tract of a fetus while in the womb and which makes up the first faecal discharges of newborn babies. The first thing that Escherich did was to show that meconium was sterile. He then found that bacteria gradually colonized the intestinal tract of

Figure 162. Theodor Escherich (1857–1911).

newborn babies, beginning immediately after birth. These bacteria came in the milk from those normally found on the mother's nipple and surrounding skin, and any other object that might find its way into the baby's mouth. The first person who had looked in babies' stools was Julius Uffelman in 1881, who had found only two types of bacteria. Escherich, on the other hand, found 19. There were a number of reasons he was able to do this. First, he used the newly described Gram stain (p.192) to separate similar-looking organisms into two groups—Gram-negative and Gram-positive. Secondly, he grew bacteria in the presence of oxygen and in its absence (anaerobically), and found that some required oxygen, some needed its absence, and some would grow whether or not oxygen was present. Thirdly, he examined whether the bacteria broke down sugars and what gases they produced. These various combinations allowed him to describe multiple species of bacteria.

He described one species in detail and named it *Bacterium coli commune*, meaning the common colon bacillus. In July 1885, he presented his findings on this organism to the Society for Morphology and Physiology in Munich. In the following year, he published a thesis entitled (in translation) 'The intestinal bacteria of the infant and their relation to the physiology of digestion', in which he showed that the then current views that bacteria were required for digestion were wrong. He died of a stroke at the age of 53 and one obituary wrote of him:

> His energy for work was tremendous, and his disposition strenuous and masterful. He is described as impulsive, uncommonly strict, strong-willed, faithful, severe with himself but kindly towards others. That the children, his patients, loved him is evidence that he loved them and, therefore, had a good heart.[288]

The first problem that needed to be solved was to differentiate the colon bacillus from the typhoid bacillus which had been cultured by Gaffky (p.335) the year before in 1884. They were both rods which looked similar, could move, and stained pink with the Gram stain (Gram-negative). They could both be found in the stools. Bacteriologists began to look at the biochemical reactions of the two bacteria. When the typhoid bacillus was grown in a broth that contained glucose, this sugar was broken down into acids (such as lactic acid) but no gas was produced whereas the colon bacillus produced both acids and gas. On the other hand, if the broth contained the sugar, lactose, nothing at all happened with the typhoid bacillus whereas the colon bacillus produced both acid and gas. Eventually Conradi and Digralski would describe a medium which encouraged the growth of the typhoid bacillus at the expense of the colon bacillus.

It gradually became clear that *Bacterium coli commune* was sufficiently different to need to be placed in a genus of its own. In 1919, Aldo Castellani (p.142) and Albert Chalmers (1870–1920) proposed that it be called *Escherichia coli* in honour of Escherich and to indicate its presence in the colon (large intestine). This name was finally accepted officially in 1958. Curiously, one textbook written as late as 1944 had this to say: 'the colon bacillus ... is almost non-pathogenic, as its normal habitat is the gastrointestinal tracts of man and animals'.[138] It is true that this organism is the most common facultative anaerobe (grows in the presence or absence of oxygen) in human faeces. It is also true that its presence there is normal and it generally does not seem to cause much trouble. But on occasion it can. And it can be deadly.

E. coli causes trouble in two circumstances. The first is when it gets into a normally sterile part of the body. The most common of these is when it infects the urinary tract. It is particularly liable to affect women because their anatomical defences are less effective because they have a shorter urethra than do men, and bacteria might ascend through this tube, particularly after sexual activity. Similarly, both men and women who have a urinary catheter passed through the urethra are prone to urinary infections. This might cause further complications by moving further up to infect the kidneys (pyelonephritis) or enter the bloodstream (septicaemia). Naturally enough, if the bowel wall is perforated (broken right through), such as in a severe case of appendicitis, *E. coli* along with other bacteria enter the peritoneal (abdominal) cavity and cause peritonitis. Again, in these days of modern medicine when chemotherapy is used to treat cancer, the lining of the bowel sometimes breaks down and allows *E. coli* (as well as other bowel organisms) to enter the bloodstream and cause septicaemia. Who first described these various infections of urine, blood, and so on, and when, is lost in the mists of time.

As the decades passed, it became apparent that although *E. coli* is a normal inhabitant of the bowel, there are some varieties of *E. coli* that caused diarrhoea, or worse, if they were ingested. In 1923, Dr A Adam in Germany identified several types of *B. coli* as a cause of severe infantile gastroenteritis by finding them on the lining of the small intestinal wall of babies who had died from the illness. He tried to differentiate them from non-pathogenic types by using biochemical reactions, especially the breakdown of sorbitol, and called the pathogenic variety *Bacterium coli* 'dyspepsiekoli'. This work was confirmed by R Goldschmidt who used serological (antibody) methods similar to those described below to identify strains from babies with diarrhoea. Unfortunately, these workers were ahead of their time. Dr A Holzel, a commentator familiar with Adam and his research, wrote in 1973:

> Adam died some years ago; a disappointed man that his studies had gone unnoticed and to a large extent deliberately ignored, because at the time the hypothesis of the food-induced severe forms of gastroenteritis was the official attitude of paediatricians in Germany.[139]

The same fate was not to await Dr John Storey Barwick Bray (1902–1985). In late 1939 at the outbreak of World War II, he was posted to the Hillingdon Hospital on the western outskirts of London, England, as an Emergency Medical Services pathologist. He expected to have to perform autopsies on victims of the bombings. Fortunately, not many of those materialized but he found himself doing post-mortem examinations on large numbers of bottle-fed babies who had died from what was then called 'summer diarrhoea' or 'cholera infantum'. There was little to find on examination of the intestines and Bray looked for known bacterial pathogens such as salmonellae and shigellae and found nothing. As he later wrote, 'all that was found was the normal *Bacterium coli* that appear on every culture plate of a faecal specimen'.[51]

After a year of this, something extraordinary happened. Bray's colleague, Dr Thomas ED Beavan, the paediatrician at the hospital, looked in on him and remarked that he could always spot a case of infantile gastroenteritis before he went into a baby's cubicle. The reason he gave was that there was always a funny smell that reminded him of semen. Bray returned to the laboratory and picked up a culture plate from a case of gastroenteritis which apparently showed nothing but normal flora. When he took the lid off the Petri dish and smelt the plate, he had an evanescent whiff of a semen-like smell. Thereafter, Bray and his technician, Mr J Stevenson, smelled plate after plate, marking them as positive or negative for the smell. Let Bray say what happened next:

> At about this time I took down to the children's wards two culture plates, one of *Proteus*, a common intestinal bacterium which has a strong smell, and one of the *Bact. coli* 'smellers', as we used to call them. I handed them to the ward sister, one after the other, and asked her to smell them. With the *Proteus* she said, 'Yes, a nasty smell, like glue', but when she smelt the seminal smelling culture she exclaimed, 'Why, that smells just like Baby Wickens.' I said, 'Sister, it is Wickens; it is the culture from Wickens.' It was at that moment that I knew the problem had been solved.[51]

Well, if not solved, Bray at least knew where to look. We have mentioned earlier that when *B. coli* (as it was then mostly called) was grown in a broth containing glucose or lactose, these sugars were broken down into acids, the presence of which was determined by a pH-sensitive dye changing colour. *B. coli* did not ferment (break down) another sugar, maltose, into acids. One day, Bray left a rack of tubes containing the suspect *B. coli* on a bench overnight. He noticed next day, that compared to normal *B. coli*, those bacteria obtained from patients with diarrhoea and which had this smell, were slowly fermenting maltose. Here was his second clue.

Bray then took home a culture of *B. coli* from Baby Wickens (who had unfortunately died) and injected the organisms into a white rabbit that he kept. The rabbit did not turn a hair, but it did make antibodies against the injected bacteria. Over the next four years, Bray used this antiserum to test *B. coli* from babies with gastroenteritis and those from normal babies to see whether they would stick to each other (i.e. agglutinate). He had many vicissitudes over this period but was able in the end to show that some 95 per cent of cases were due to the 'smeller' organism which agglutinated with the specific antiserum from the rabbit. Bray then found he could sometimes isolate these bacteria from squashed houseflies, so here was one way in which they could be spread.

But what about Koch's postulates? Bray tried to fulfil the third by infecting baby mice, rabbits, and kittens, but nothing happened. He eventually published his observations on 50 cases with 20 deaths in the *Journal of Pathology and Bacteriology* in 1945, naming the organism *Bact. coli neapolitanum*. What happened then? In a review written in 1973, Bray wrote:

> I regret to say that no one, with the exception of the editor of the journal, believed that my findings were of any importance whatever. I remember him

> coming up to me at one of the meetings of the Society of Pathology and Bacteriology at Oxford, saying, 'I'm sure that you've found the answer, but we can't say so.'[51]

Bray's regrets were not to last. Other investigators around the world replicated his findings after several years. Among them was Rolf ten Seldam. It is always a thrill to find a personal connection, however tenuous, with the story one is telling. Bray went on to write:

> I must mention, too, Dr ten Seldam, of Eindhoven, now Professor of Pathology at Perth University, Western Australia (*sic*) who got similar results in Holland, who came to see me in London, and was kind enough to name the organism 'Escherichia coli Bray'.[51]

After I moved to Perth in 1978, I had occasion to visit the department of pathology and well remember a large oil portrait of Professor ten Seldam hanging on the wall.

The antigenic structures (the patterns of the bacteria which stimulate the production of specific antibodies) of the *E. coli* in various outbreaks in Britain, Europe, and the USA were analysed in detail by the Danish bacteriologist, Fritz Kauffmann (1899–1978), beginning in 1944, and were progressively refined over the years with various co-workers. This scheme used various parts of the bacterium. The most useful were the antigens in the cell wall of the bacterium called O antigens. Bray's organism, for example, was labelled *Escherichia coli* O111.B4. Related serological varieties of these *E. coli* which caused infantile gastroenteritis were grouped together and called 'Enteropathogenic *Escherichia coli*', or EPEC for short, by the German-American, Erwin Neter (1909–1983) and his colleagues in 1955. This was used as a descriptor for strains of *E. coli* that cause disease in the intestine when compared with those strains whose normal habitat is the bowel. How do these EPEC organisms cause disease? Microvilli are multiple tiny finger-like projections into the bowel lumen from the surface of the epithelial cells which line the intestinal tract. They greatly increase the effective surface area and control the movements of salts and fluid across the bowel wall from the bowel lumen to the rest of the body. When electron microscopy (which greatly increases magnification) came to be applied to studying this disease, it was found that EPEC stick to the surface of the epithelial cells and damage the microvilli, a process called 'attachment/effacement'. This meant that the microvilli could not work properly and absorb fluid and electrolytes; these therefore continued their merry way down the intestine to cause watery diarrhoea.

But let us return to Koch's postulates. Failure to demonstrate pathogenicity when given to animals meant that this could only be proven by using humans as experimental subjects. As Bray remarked in his review in 1973: 'Obviously, one could not prove the case by trying the effect of feeding the germ to a healthy baby.'[51] Perhaps the operative words here are 'healthy baby' because Neter and Clare Shumway in 1950 gave 100 million *Escherichia coli* O111.B4 in formula milk to a two-month-old infant with what they merely described as 'multiple congenital defects, including the brain'. Within 24 hours the baby had diarrhoea and began to lose weight. After 48 hours, large numbers of these organisms were found in the throat and stools. The baby was treated with tetracycline antibiotic and recovered, with the bacteria disappearing within two days. Sixty years later, it is difficult to comment on the ethics of this experiment. Probably the congenital defects were considered so severe that the baby was not expected to live. I doubt that it would pass an ethics committee today. Whatever you may think, Koch's third postulate that a culture preparation must induce disease in a recipient and then be recovered from that person had been fulfilled for enteropathogenic *E. coli*.

What about the pathogenicity of these bacteria for adults? In 1950, Alan Kirby and colleagues in Liverpool, England, reported a small study of nine adult volunteers who were given 2 billion *Escherichia coli* O111.B4. Five became ill whereas control subjects given *E. coli* not associated with diarrhoea remained well. A much larger study was reported in 1952 by WW Ferguson and RC June, who studied a total of 114 males aged from 15 to 48 years. They were given three strains of *Escherichia coli* O111.B4 in milk. This means they were given bacteria with this particular serotype but which were isolated from three different individuals. When they ingested 530 million organisms or less, they either remained well or developed only a very mild illness. When they were given 3 billion of each of the three strains at once (making a total of 9 billion) or, in a second experiment, a total of 6.5 billion bacteria, everyone became sick, developing fever, nausea, cramps, diarrhoea, and sometimes vomiting 10 hours or so after ingestion. The illness lasted 24–96 hours. The enteropathogenic *E. coli* were excreted in the stools within 48 hours and remained positive for several days. Control subjects given un-inoculated milk remained normal. These studies convincingly proved Koch's postulates. Everything is hunky-dory. Or is it? Who were the volunteers? They were inmates of the State Prison of Southern Michigan near Jackson in the USA. The authors thanked the Corrections Commissioner who they said provided facilities and personnel and cooperated with them in every way. They don't say anything at all about what the prisoners thought about this; they apparently did not even give their 'informed consent'. Once again, we have the thorny issue of performing experiments on prisoners which we have already discussed in the context of typhoid fever.

All of this work with the various strains of enteropathogenic *E. coli* was by no means the end of the story. There are now five known groups of pathogenic *E. coli*, called 'pathotypes', generally with distinctive O serogroups, which cause diarrhoea and for which Koch's postulates have been fulfilled. Briefly, the other four are:

- enterotoxigenic *E. coli*, which produces toxins that prevent the bowel from absorbing fluid and electrolytes and cause watery diarrhoea;
- enteroinvasive *E. coli*, which causes inflammation of the bowel and causes diarrhoea with blood and mucus;
- enterohaemorrhagic *E. coli*, which produces a toxin called Shiga toxin (also found in *Shigella*), which causes inflammation of the bowel with diarrhoea with blood and mucus and sometimes kidney failure;
- enteroaggregative *E. coli*, which prevents the bowel wall from working properly and causes diarrhoea.

Today new potential pathotypes are being investigated. Finding them is like looking for multiple needles in the proverbial haystack.

All of these infections are more common in children and in adult travellers to tropical countries (where they cause 'travellers' diarrhoea). Enterohaemorrhagic *E. coli* is particularly nasty. It was first found in 1983 by LW Riley and colleagues from the US Centers for Disease Control who investigated two outbreaks of bloody diarrhoea following the consumption of undercooked hamburgers at a fast-food chain. In the same year, MA Karmali and colleagues in Canada found that some patients with this infection developed severe kidney disease as a result of the breakdown of red blood cells (known as the haemolytic–uraemic syndrome).

These pathotypes of *E. coli* are widespread and can potentially contaminate the food we eat and the water we drink. Pity the manufacturers of sausages such as salami and mettwurst who may be attacked by these invisible enemies. One such company, Garibaldi, in my home town of Adelaide, had enterohaemorrhagic *E. coli* in its garlic mettwurst in 1995 and a four-year-old girl died. The lawyers had a field day and the company was bankrupted. Diarrhoea due to *E. coli* will always be with us, but you can reduce your chances of contracting it by never eating salami and mettwurst, etc., that have not been heat-treated.

Unfortunately, it is not just meat products. When I was in Britain in May 2011, there was an outbreak of this illness centred on Hamburg, Germany. Nearly 4,000 people were infected, almost 800 of them developing kidney disease, and with 50 deaths. At first this outbreak was blamed on Spanish cucumbers, causing millions of them to be thrown away. Then the German authorities extended health warnings to all raw vegetables sparking a consumer panic across Europe that devastated the agriculture industry costing about €200 million each week. Finally, it turned out that the culprit was bean sprouts grown at an organic farm in Germany. Organically grown foodstuffs are not always what they are cracked up to be. You never know where this awful bacterium is going to turn up next.

31

Bacillary dysentery (shigellosis)

SCIENTIFIC NAMES:	*Shigella dysenteriae, S. flexneri, S. boydii, S. sonnei*
ORGANISM'S APPEARANCE:	Gram-negative rod
DISEASE NAME:	bacillary dysentery or shigellosis
DISTRIBUTION:	worldwide but worse in the tropics
TRANSMISSION:	person to person or ingestion of contaminated food or water
LOCATION IN A PERSON:	large intestine
CLINICAL FEATURES:	variable diarrhoea to dysentery (blood + mucus) in the stools, fever
DIAGNOSIS:	isolation of *Shigella* from the stools
TREATMENT:	primary—fluid replacement; antibiotics in severe cases depending upon the results of laboratory testing to find which antibiotics are effective
PREVENTION:	washing hands and other hygienic measures especially with small children

Some things remain imprinted in the memory. It was 1963 and I was a raw, 19-year-old third-year medical student attending one of my first pathological demonstrations. Stretched out on the bench before us, slit from one end to the other, was the large intestine of a four-year-old girl who had died the day before from bacillary dysentery. I still remember feeling the intimations of mortality and the unfair, unjust, and unpredictable nature of death.

We have discussed dysentery before where we defined it as loose, fluid stools containing blood and mucus, and compared it with amoebic dysentery, which lasts months to years. Bacillary dysentery, on the other hand, lasts for days to several weeks. It was well known to Hippocrates (p.116), who wrote in his *Aphorisms*:

> Dysentery, if it commences with black bile, is mortal...If in a person ill of dysentery, substances resembling flesh be discharged from the bowels, it is a mortal symptom.[137]

Indeed, the word 'dysentery' was coined by the Greeks from the Greek words *dys* and *entera*, meaning 'bad bowels'.

Although bacillary dysentery primarily kills the young it can affect anybody at any time. Indeed, throughout history, along with typhoid, typhus, plague, and smallpox, this disease has bedevilled armies and changed the course of military campaigns. Two examples will suffice. King Edward I of England (the Hammer of the Scots) reputedly died from dysentery in 1307 while campaigning against Robert the Bruce, King of Scotland. Another was King Henry V of England, who died at the age of 35 from bacillary dysentery following the siege of Meaux in France in 1422.

How does severe bacillary dysentery affect a typical patient? Here is a description by Joseph Felsen:

> The patient is in abject misery, a slave to his bedpan, vainly seeking relief from the torment of almost continuous abdominal cramps by futile or abortive attempts to move his bowels. One evacuation succeeds another until as many as fifty to one hundred daily render existence almost unbearable. He is repeatedly awakened from sleep by severe abdominal pain. Tenesmus (pain on having a bowel action)...drives the patient to distraction...The patient may be bathed in drenching sweat...After a period varying from seven to ten days the symptoms and signs gradually abate.[97]

In 1897, Kiyoshi Shiga (Figure 163) was a young medical graduate who had recently started to work at the Institute of Infectious Diseases in Tokyo directed by Shibasaburo Kitasato (p.362). Shiga had been born in Sendai in northern Japan to Shin and Chiyo Sato but was raised by his mother's relatives and later adopted her maiden name. In 1897, there was a large epidemic of dysentery in Japan with over 90,000 cases of whom almost a quarter died. Shiga was directed to investigate this disease and studied 36 patients with dysentery in the Institute. To find a bacterial cause in faeces full of bacteria was a

Figure 163. Kiyoshi Shiga (1871–1957).

daunting task. He found a particular bacillus in the stools of 34 patients and in scrapings of the bowel wall of two patients who had died (Figure 164). He did not find any bacteria in the internal organs of these two patients. He grew this bacterium on agar, gelatine plates, and potatoes. He described it thus:

> The dysentery bacillus is a short rod with rounded ends much like the typhoid bacillus and most types of coli. The organism most commonly occurs singly and occasionally in pairs . . . They are decolorised by the method of Gram [i.e. Gram-negative]. They are moderately motile . . . No spores are formed . . . Alkaline media are recommended for cultivation . . . nearly pure cultures were obtained during the height of the diarrhoea.[287]

The key to his discovery was provided by a novel approach using immunology rather than biochemistry. In August 1897, at Kitasato's suggestion, he used an agglutination technique that had been employed by Widal in typhoid fever in 1896 (p.333). Shiga mixed cultures of the organism that he grew with serum (the fluid extruded from blood after it clots) from 25 patients. He found that the bacilli stuck to each other (agglutinated) when they were mixed with serum from these patients but not when they were added to serum from normal people. Conversely, other bacteria isolated from these patients did not react with their sera. Shiga wrote about the first time he found a positive reaction:

> On August 8, 1897 the serum of one patient was found to agglutinate one kind of organism from agar cultures. It was then that I named this organism *Bacillus dysenteriae*.[287]

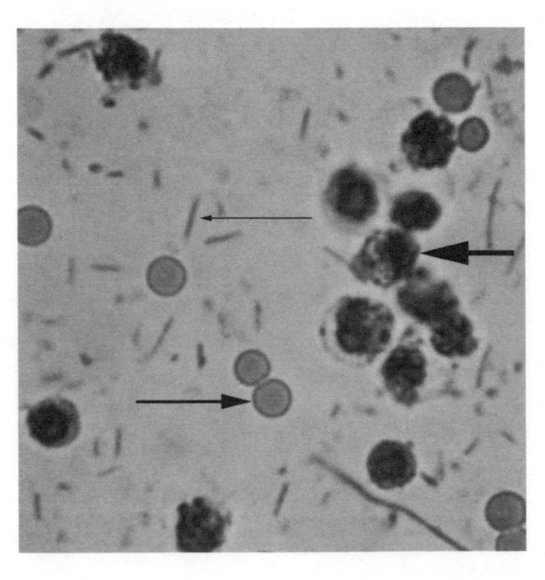

Figure 164. Dysenteric stool in a patient with shigellosis. The thin arrow points to *Shigella* bacteria; the medium arrow points to red blood cells; the thick arrow points to white blood cells.

The species name, of course, was given to indicate that the bacterium caused dysentery. Once the patients had recovered from their illness, he could no longer find the bacilli.

Being a student of Kitasato, who had in turn worked with Robert Koch, Shiga was acutely aware of the necessity to prove that this organism was the cause of dysentery by fulfilling Koch's postulates. He fed cultures of the pure organism to puppies and kittens but succeeded only in producing slimy faeces. He then gave a huge number of bacilli to a dog, which developed diarrhoea (although there was no blood) and died after five days. When he tried injecting bacilli intravenously into small animals they would die, but not from a dysenteric illness. Kitasato reported Shiga's work to a medical meeting in November 1897 and it was published in *Saikingaku Zasshi* (Japanese for *Journal of Bacteriology*) in the next month. Shiga published news of this discovery in a German medical journal in 1898. The failure to convincingly fulfil Koch's postulates caused Shiga to suggest an alternative one:

> If the serum of a person stricken with a certain disease (or having had the disease before) is able to agglutinate a distinct bacterium, then this bacterium has a close relation to the disease.[287]

In the years following Shiga's discovery, three other groups of similar organisms were found. After the Spanish–American war of 1898, Spain ceded the Philippines to the USA, which then occupied the country. In 1899, a commission led by Simon Flexner (1863–1946) from Johns Hopkins University in Baltimore, Maryland, went to the Philippines as the health of the American troops was of concern. En route, they visited Kitasato's Institute of Infectious Diseases, met Shiga, and learnt of his discovery. Once in Manila, Flexner began to examine the faeces and bowel specimens obtained at autopsy from patients with dysentery. When cultured on agar plates, they found two types of Gram-negative rods. One did not ferment mannitol (a sugar alcohol) and appeared to be the same as Shiga's bacillus. The other did break down mannitol. Each isolate was agglutinated by the patient's own serum. On the other hand, when they looked for cross-reactivity among patients, they found that serum from patients with bacilli that fermented mannitol would only agglutinate bacilli which fermented mannitol. Conversely, serum from patients with bacteria that did not break mannitol would only agglutinate mannitol non-fermenting bacteria. Thus, one type was identical to Shiga's bacillus (later called *Shigella dysenteriae*) while the other would eventually be called *Shigella flexneri*. When either type was given as live or dead cultures to small animals or monkeys, dysentery did not follow. However, a laboratory assistant accidentally ingested what was called the 'type 1 bacillus' and developed severe dysentery which lasted for a couple of weeks. Thus Flexner had found two similar but separable bacteria.

Flexner was not the only American working on this problem in the Philippines in 1899. Two US Army medical officers, Richard Pearson Strong (1872–1938) and William Everett Musgrave (1869–1927), began investigating dysentery as it was common among American troops and was impeding their fight against the Filipino resistance. They isolated a bacillus from 98 cases of dysentery, 19 of which were from autopsies, and, like other workers, showed that each patient's serum reacted with the bacilli. They described their findings in 1900 in the *Journal of the American Medical Association* but made no mention of Flexner, who, perhaps somewhat surprisingly, reported his findings in that same year in the *British Medical Journal*. Was Strong and Musgrave's bacillus the same as one of Flexner's types? It is unclear.

Meanwhile, research had started in Europe. Stimulated by Shiga's report in the German literature, Walther Kruse (1864–1943) of the Institute of Hygiene in Bonn investigated a large outbreak of dysentery just north of that city in 1899 and 1900. He found Shiga's organism in 24 patients, all of whom had high levels of antibody in their serum which agglutinated this bacterium but not other bacteria. Nor did serum from normal people agglutinate these bacteria. In 1901, he showed that the same organism was the cause of an outbreak of dysentery in a mental asylum. Thus for a long time, this organism became known colloquially as the 'Shiga–Kruse bacillus'.

Once European investigators became familiar with Shiga's bacillus, they found there was a surprise in store for them. Way back in 1888, André Chantenemesse (1851–1919) and Fernand Widal at the Institut Pasteur in Paris had obtained a certain bacillus at autopsy of five soldiers returned from Tonkin in the French colony of Indochina (now Vietnam). The description they gave of this organism was not particularly detailed or accurate (they claimed it produced spores and caused dysentery in guinea pigs). Nevertheless, the cultures were kept in the Institute and years later Louis Vaillard (1850–1935) and Charles HA Dopter (p.288) were able to show that these bacteria were identical with Shiga's bacillus.

A third serologically distinct species (i.e. with unique antigens on its surface) was found by Charles Warren Duval (1876–1950) in the USA in 1904, Castellani (p.142) in Ceylon (now Sri Lanka) in 1905–7, and by Kruse in Germany in 1907, which Kruse originally called *Bacillus pseudodysenteriae* type E. These were probably all the same organism and it was renamed *B. paradysenteriae*. Extensive studies of it were made by the Danish bacteriologist Carl Olaf Sonne (1882–1948) from 1914 onwards so the name was modified to *B. paradysenteriae* Sonne, which eventually was shortened to *Shigella sonnei* although some called it the Sonne–Duval bacillus.

A fourth serologically distinct species was described by the British Army medical officer, Major John Smith Knox Boyd (1891–1981), in Bangalore, India between 1929 and 1931. This would eventually become known as *Shigella boydii*.

All of this seems fairly straightforward but by the 1930s, it was really a pig's breakfast. There was considerable confusion over whether or not these organisms were motile; whether or not they fermented (broke down) certain sugars such as lactose and sucrose or sugar alcohols like mannitol and sorbitol quickly, slowly, or not at all; how specific certain antibody reactions were (that is, whether reactions occurred with that bacterium and only that bacterium); and whether or not they caused disease in experimental animals. As a result there were dozens of different isolates with names like Shiga–Kruse, Schmitz, Sonne–Duval, Flexner V, W, X, Y, and Z, dispar, alkalescens, Newcastle, 103, P119, D17, etc., etc. Various classifications were proposed which evolved over time, but have now settled down into four species determined by antigenic grouping. Each group has been subdivided into various serotypes. The current state of play is:

- *Shigella dysenteriae* (10 serotypes)
- *Shigella flexneri* (6 serotypes)
- *Shigella sonnei* (1 serotype)
- *Shigella boydii* (15 serotypes).

You will note that these organisms now have the genus name *Shigella*. This genus was erected by Castellani and Chalmers (p.341) in 1919 in honour of Shiga. But what are the distinctive characteristics that define this genus? Erwin Neter (p.344) wrote in 1942 that it was difficult to give a definition of the genus *Shigella* that was sufficiently distinctive to differentiate it from others and yet broad enough to include all its members. If it was difficult then, it is even worse now. If you have read the preceding chapter, it may well have occurred to you that enteroinvasive *Escherichia coli* and enterohaemorrhagic *Escherichia coli* behave in a manner very similar to *Shigella*.

This brings us to a consideration of the toxins which allow *Shigella* to cause such devastating disease. The ability of Shiga's bacillus to produce a potent toxin was first demonstrated by Heinrich Conradi (1876–1943) in 1903 while working at the Koch Institute in Berlin. He had been born Heinrich Wilhelm Cohn but was baptised as a Protestant in 1892 and took the surname of Conradi. He eventually moved to Dresden. where he lost his job because of his Jewish ancestry in 1938 and died in that city in 1943 while in the custody of the Gestapo. Conradi suspended the Shiga bacillus in saline and incubated the specimen for 24 hours, then centrifuged and removed the bacteria-free fluid from the top of the container. This was injected intravenously into rabbits or into the abdominal cavity of guinea pigs; they all collapsed with paralysis and died within 48 hours. Clearly a toxin had been released from the bacteria into the surrounding fluid. On the other hand, the Flexner and Sonne bacilli only proved fatal if given in extremely large doses.

Because the Shiga toxin caused paralysis, it was first called a neurotoxin. In 1972, however, Gerald T Keusch and his colleagues in the USA showed that this toxin also affected the bowel—it was an enterotoxin. When it was introduced into the bowel of rabbits, it caused the small intestine to secrete fluid and there was inflammation of the large intestine resembling that seen in human dysentery. The gene for this Shiga toxin was eventually determined by Nancy Stockbine and her colleagues in Bethesda, Maryland, USA, in 1988 but it was found to be present only in *S. dysenteriae* serotype 1 (which generally produces the most severe disease). Clearly other factors must be at work in other types of shigellae. On the other hand, the Shiga toxin gene (or at least, closely related genes) was found in certain enterohaemorrhagic *E. coli*, underlying the similarities of the diseases produced by these two bacteria. It is the Shiga toxin, whatever its source, that causes the haemolytic-uraemic syndrome in which red blood cells break down and the kidneys fail.

Perhaps just as important, shigellae have the ability to invade the epithelial cells which line the surface of the intestine. In them, they multiply and spread directly from cell to cell, killing them in the process and cause ulceration (Figure 165) of the mucous membrane (which is the part of the bowel wall made up of the epithelial (or lining) cells and the immediately underlying tissue).

What about the fulfilment of Koch's postulates in humans? We have mentioned the accidental infection of Flexner's assistant. In 1900, Strong and Musgrave in the Philippines reported to the Surgeon General of the United States Army that they had succeeded in producing dysentery in a condemned criminal. They gave the Shiga bacillus which had been grown in a culture of broth for 48 hours after a dose of sodium bicarbonate to neutralize gastric

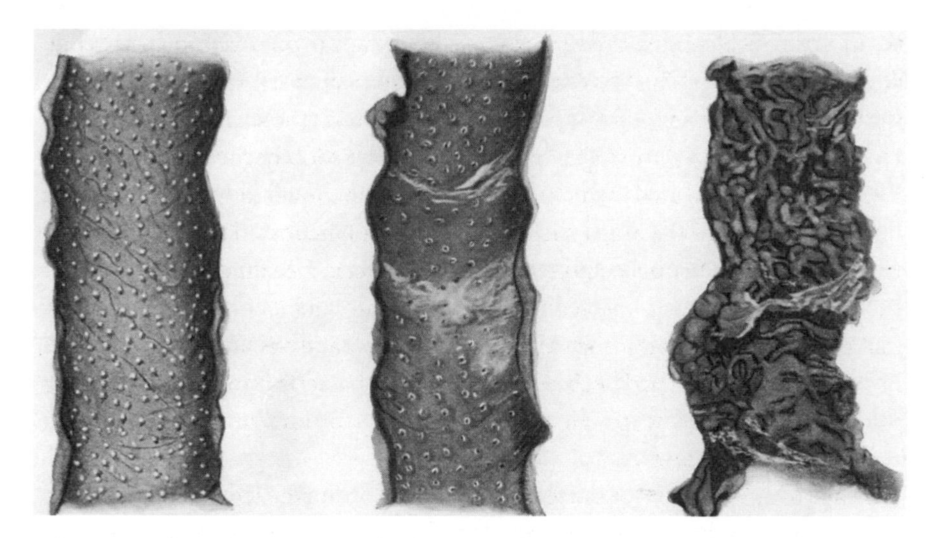

Figure 165. Drawings by Joseph Felsen in 1945 of the evolution of changes in the large intestine in bacillary dysentery. You are looking at the inside of bowel that has been cut open. Left: swellings caused by enlargement of lymphoid tissue in the mucous membrane. Centre: ulceration over those swellings. Right: spreading ulceration.

acidity. After an incubation period of 36 hours, the man developed the typical symptoms of dysentery and the organism was recovered from his stools which were full of blood and mucus. I will leave you to make your own judgement about the ethics of that experiment.

Following World War II, more detailed human studies were undertaken in the USA which further confirmed Koch's postulates. Howard J Shaughnessy (who was not a doctor but a scientist) and four colleagues, two of whom were doctors, gave shigellae to prisoners in the Joliet Penitentiary in Illinois, USA. We have discussed the ethics of medical studies using prisoners earlier (p.336). We are simply told in this study that the prisoners were 'volunteers' and the authors expressed appreciation for the aid and encouragement by, amongst others, the warden of the Illinois State Penitentiary and his staff who provided not only facilities and personnel but also their wholehearted cooperation. Whether or not the prisoners were quite so wholehearted is another matter. The strain of *Shigella* with which they were infected was called *S. paradysenteriae* Flexner W. Initially, men were given between 100 million and 80 billion bacteria in milk but only a few of them developed mild diarrhoea. The authors did not give up. They gave different strains of shigella with or without faeces in gelatine capsules and were more successful with some subjects developing severe cramps, headache, fever, vomiting, and diarrhoea with some stools becoming dysenteric. Once they had sorted all this out, they examined the effectiveness of two vaccines when compared with a control group in a total of 83 prisoners. The vaccines did not work. Finally, the subjects were treated with sulpha drugs (antibacterials) to eradicate the organisms. This study, which was very involved, had not really shed much light on the issues. The organisms they used turned out not to be very pathogenic for humans, even though they were for mice.

Some thought it was worth trying again. The long-suffering inmates of the House of Correction in Jessup, Maryland, were not only infected with typhoid fever; they were also used as guinea pigs for shigellosis. This time we are merely told they were 'well-informed volunteers'. In 1969, Herbert Dupont and his colleagues from the University of Maryland gave *Shigella flexneri* (strain 2a) in milk to a total of 43 prisoners in order to find out what dose was necessary to produce disease. One of the four men given 10,000 organisms developed dysentery, while approximately three-quarters of those given 100,000 to 100 million organisms became ill with the disease after several days, averaging eight (one person had 28) bowel actions per day. Bacteria were recovered from the stools of most patients. This organism was clearly more pathogenic than the one used in the previous study. It should be noted that *S. flexneri* generally causes less severe disease than *S. dysenteriae* but a more severe illness than *S. sonnei* or *S. boydii*.

So where does all this leave us today with bacillary dysentery which is sometimes called shigellosis? As countries have become more affluent, the numbers of cases have fallen and there has been a shift towards infections with bacteria such as *S. sonnei* that cause a mild form of the disease. Infections only occur naturally in humans and sometimes infection can follow the ingestion of as few as 10 bacteria. Infection may be acquired from contaminated food and water but it is often transferred from one person to another where disposal of faeces and hand-washing are poor, especially when people are crowded together. In poor tropical countries, it has been estimated (although the figures are rubbery) that half a million people, mostly children, die from bacillary dysentery (especially that caused by *S. dysenteriae*) each year.

32

Tetanus (lockjaw)

SCIENTIFIC NAME:	*Clostridium tetani*
ORGANISM'S APPEARANCE:	Gram-positive rod
DISEASE NAME:	tetanus or lockjaw
DISTRIBUTION:	worldwide
TRANSMISSION:	contamination of a wound with dirt
LOCATION IN A PERSON:	wound
CLINICAL FEATURES:	sneer, difficulty opening mouth, generalized tightness and spasms
DIAGNOSIS:	clinical
TREATMENT:	penicillin or metronidazole plus tetanus antitoxin
PREVENTION:	an effective vaccine is available

When I was a junior resident medical officer at the Royal Adelaide Hospital in 1967, there was a room at one end of the ward to which I was assigned that was under the stairs, dark, and lined with green velvet. It was the 'tetanus room'. The idea was that reduction of light and noise would decrease the spasms which are characteristic of tetanus. Fortunately, I never saw it in use. Within a year it was decommissioned when an Intensive Care Unit was established. Treatment with sedatives, anaesthetics, and muscle relaxants has revolutionized the outlook for those unfortunate enough to contract tetanus.

What is tetanus? It is a disease characterized by a particular set of symptoms and signs that has been recognized since antiquity. An Egyptian papyrus called the *Papyrus Ebers*, so-called because it was purchased by Georg Ebers in 1873, is one of the two oldest medical documents. It was written about 1550 BC and is 20 metres long. The seventh case concerns a patient who suffered a penetrating injury of the skull and then developed spasms of the jaw muscles so that he could not open his mouth (trismus = lockjaw) and stiffness of the neck. This is the earliest recorded case of tetanus. A striking disease of spasms, violent contractions, and death was well known to Hippocrates (p.116), who saw the disease in wounded Greeks. In his *Aphorisms* he wrote: 'Such persons as are seized with tetanus die within four days, of if they pass these, they recover.'[137] Likewise, in the first century AD, Aretaeus the Cappadocian (p.203) observed that:

> Tetanus, in all its varieties, is a spasm of an exceedingly painful nature, very swift to prove fatal...They are affections of the muscles and tendons about the jaws, but the illness is communicated to the whole frame...There are three forms of convulsion, namely in a straight line, backwards and forwards. Tetanus is in a direct line...for that backwards we call Opisthotonos and that variety we call Emprosthotonos in which the patient is bent forwards...In all these varieties...there is a pain and tension of the tendons and spine and of the muscles connected with the jaws and cheek; for they fasten the lower jaw to the upper, so that they could not easily be separated even with a lever or a wedge...Opisthotonos bends a patient backwards like a bow...An inhuman calamity! an unseemly sight! a spectacle painful even to the beholder! an incurable malady![27]

He went on to say that patients died from suffocation.

As far as the cause was concerned, Aretaeus realized that tetanus was associated with wounds of the skin and with attempts to procure an abortion:

> The causes of these complaints are many; for some are apt to supervene on the wound of a membrane, or of muscles, or of punctured nerves, when for the most part, the patients die...And women also suffer from this spasm after abortion...and they seldom recover...Children are frequently affected.[27]

A century later, Galen (p.203) noted that when a nerve was deliberately cut in a patient with tetanus, the spasms stopped but of course the muscles supplied by that nerve were paralysed.

In the fourteenth century, John of Arderne (p.260) reported a case of trismus in which he described the patient as being taken with a cramp on his cheeks that began 11 days after a gardening injury. Tetanus was also common in soldiers after battle and in women and their newborn babies when delivery was conducted in literally dirty conditions. Newborn babies were particularly at risk when they were laid on the earth and dirt contaminated the stump of their umbilical cord.

The English neurologist William Gowers (1845–1915) wrote about tetanus this way:

> Tetanus is a disease of the nervous system characterized by persistent tonic [continuous] spasm, with violent brief exacerbations [increases in severity]. The spasm almost always commences in the muscles of the neck and jaw, causing closure of the jaws [trismus or lockjaw] and involves the muscles of the trunk more than the limbs. It is always acute [sudden] in onset, and a very large proportion of those who are attacked die.[121]

Indeed, before the days of modern treatment, the majority of patients died from exhaustion, asphyxia, dehydration, and starvation. You can understand this when you look at the painting in 1809 by the Scottish surgeon and artist, Charles Bell, of a soldier with tetanus (Figure 166). In 1856, James Young Simpson (the discoverer of chloroform anaesthesia) recognized the similarities of the symptoms and signs of tetanus with those produced by poisoning with strychnine (a chemical obtained from the seeds of the tree, *Strychnos nux-vomica,* a native of South-east Asia).

The *Shorter Oxford English Dictionary* gives us quite an adequate definition:

> an acute infectious disease characterized by tonic rigidity and violent spasms of some or all of the voluntary muscles, caused... usually by contamination of a wound.

The name is derived from the Greek word *tetanos* meaning 'muscular spasm' which is in turn derived from *teinein* indicating 'stretch'. Until the later part of the nineteenth century, it was not known that it was an infectious disease. In that century, most people thought that tetanus was the result of a nerve injury and sometimes had difficulty in distinguishing it from the convulsions of epilepsy.

Figure 166. Painting by Charles Bell in 1809 showing opisthotonos (arched back) in tetanus.

Light on the true cause of tetanus began to be shed in 1884. Two Italians in Turin, Antonio Carle (1855–1927), a surgeon, and Giorgio Rattone (1857–1929), a pathologist, made use of a patient who had died of typical tetanus two days after scratching a pustule on his skin. They collected matter from the wound within two hours of his death, suspended it in saline, then injected the material into 12 rabbits. In four rabbits it was put in the sheath surrounding the sciatic nerve; the sciatic nerves are the large nerves that run from the buttock down each leg. It was thought at the time that tetanus was due to a lesion of nerves so it was logical to find out what happened when the pus was injected near a nerve. In another six rabbits, the material was injected more generally into the back muscles. In the final two rabbits, the material was inoculated directly into the spinal canal inside the vertebral column. All except one of the rabbits died of a tetanus-like disease. Blood was taken from the dead animals and injected into two more rabbits; they remained well. However, when another two animals were inoculated with an emulsion of sciatic nerve from a dead animal that had been injected at that site, they too died.

This was the first recorded transfer of tetanus. Why did they do this experiment? Tetanus had long been associated with wounds and ulcers in the skin which discharged pus. In the two or three years immediately preceding their experiment, Pasteur and Ogston had shown that streptococci and staphylococci in the pus of abscesses could cause abscesses and bloodstream infections when injected into other animals. It seemed reasonable to see if whatever it was that caused tetanus could be transmitted in a like manner. Although Carle and Rattone had shown that tetanus could be transmitted from a person to an animal, their experiment did not help them identify the nature of the noxious agent.

Figure 167. Arthur Nicolaier (1862–1942).

Major progress in this regard was made in the same year of 1884 by Arthur Nicolaier (Figure 167). Soon after his graduation in medicine, Nicolaier was working at Flügge's Hygienic Institute in Göttingen, Germany, when he made his pivotal discovery. He injected garden soil under the skin at the root of the tail of mice, guinea pigs, rabbits, and dogs. With the exception of the dogs, they all developed signs that closely mimicked human tetanus. He did this experiment because he was aware that tetanus was more common when skin lesions were contaminated with soil. By examining the effects of 18 different soil samples, he found that only the upper layers of the soil could produce tetanus. Nicolaier then proved that it was something living in soil that caused the disease. He heated soil to a temperature of 190°C for one hour and showed that tetanus did not result when this was injected. Having shown that soil could lead to tetanus, he went looking for the agent that was the actual cause. When he examined pus from the sites of injection, he found a variety of cocci and bacilli. On the other hand, when he examined histological sections of tissue at the injection site, he saw long, thin bacterial rods. He tried to grow these bacteria but was only partly successful in growing them in the depths of a blood–serum medium (where there is no oxygen); they were always contaminated with other organisms. Nevertheless, when this mixed culture was injected into experimental animals, they too developed tetanus.

Since Nicolaier could grow these bacilli but could not separate them from other bacteria, he could not prove that they were the cause. In his thesis of 1885, Nicolaier reported that the suspect bacilli were not present elsewhere in the body and suggested that tetanus was due to a strychnine-like poison. Unfortunately, Nicolaier met a sad and grisly end. He fell victim to Hitler's anti-Jewish pogroms. As he was a Jew, he was transported to Theresienstadt concentration camp, where he committed suicide by taking an overdose of morphine.

In 1886–7, another surgeon in Göttingen, Julius Rosenbach (p.234), inoculated tissue taken from a patient with frostbite into guinea pigs. They developed tetanus and he successfully transmitted tetanus serially through many generations of guinea pigs and mice. He thought the disease was due to rods with the appearance of a 'drumstick', but like Nicolaier was unable to grow them in pure culture. In 1886, Carl Flügge (1847–1923) described a terminal spore in these rods (Figure 168). We now know that spores form when the sugars which the bacteria break down are no longer present in their environment. This is a survival mechanism which allows the bacteria to persist in a dormant state until environmental conditions improve when the spores germinate and once more become multiplying rods. Flügge named Nicolaier's bacillus *Bacillus tetani*, the species name obviously reflecting its relationship with tetanus. In 1923, Bergey (p.217) and his colleagues raised the genus *Clostridium* and transferred the organism into it as *Clostridium*

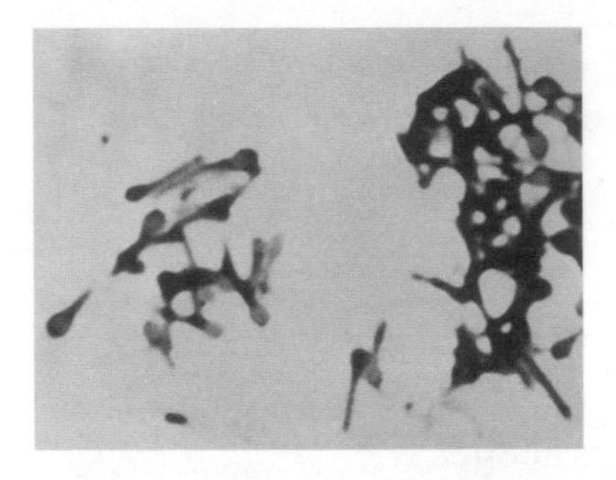

Figure 168. Gram stain of rods of *Clostridium tetani*. Note the spore at the end of each rod making it look like a drumstick or tennis racquet. If you could see this in colour, the rods would stain purple.

tetani. The genus name is derived from the Greek word *kloster* meaning 'spindle' and the suffix *idion* meaning 'small'.

The person who solved the problem of obtaining a pure culture of the tetanus bacillus was Shibasaburo Kitasato (Figure 169), who had been born on the island of Kyushu in Japan. Two years after his graduation in medicine, Kitasato was sent in 1885 by the Japanese government to study bacteriology under Robert Koch in Berlin, Germany. He was eventually assigned to work on the problem of tetanus. A soldier in the Berlin garrison died of tetanus. In 1889, Kitasato examined the pus and wrote: 'Microscopic examination of the wound revealed Nicolaier's club-shaped spore-bearing bacilli in addition to other organisms.'[158] These bacteria stained purple with the Gram stain (Gram-positive). Kitasato injected the original pus into mice. They died after several days and Kitasato noted that: 'At the site of inoculation, but not elsewhere, were club-shaped spore-bearing bacilli in addition to other organisms.'[158]

Figure 169. Shibasaburo Kitasato (1853–1931).

Kitasato isolated the spore-forming bacilli in pure form in a cunning manner. He streaked the pus on agar or coagulated serum, then incubated the media for 48 hours at 37°C. All the different types of organisms grew but club-shaped bacilli (spores) appeared. He then placed the cultures in a water bath at a temperature of 80°C. This killed all the bacteria except the spores. When these spores were injected into mice, they died of tetanus. Kitasato realized that the spore-forming bacilli only grew in the depths of cultures so presumably could not tolerate oxygen. He therefore put some culture fluid containing spores on nutrient gelatine into glass vessels which were filled with hydrogen, then the vessels were sealed and incubated at 20°C. What happened?

> After one week, colonies were seen in the vessels supplied with hydrogen, while there was no growth on the conventional dishes. After ten days, the sealed hydrogenated culture vessels were opened and the growth was examined in smear preparations. There were rods...arranged singly and in threads.[158]

The spores had germinated! Pure cultures obtained in this manner were injected into mice which died from tetanus within two to three days. All of Koch's postulates had been met. Kitasato concluded:

1. Tetanus is an infectious disease caused by a specific bacillus.
2. The causative agent...is the bacillus described by Nicolaier.
3. The bacillus is present in the pus of man and experimental animals suffering from tetanus. The bacilli form spores in the pus.
4. The bacilli can be cultured from cases of tetanus and such pure cultures can produce disease in experimental animals.[158]

In addition, Kitasato noted that spasms first appeared in experimental animals in muscles near the site of inoculation. Since no bacteria could be seen in their tissues and no bacteria grew when these tissues were cultured, like Nicolaier, he wondered whether the tetanus bacilli produced a toxin. In support of this idea, Kitasato showed that when he injected bacteria into the base of the tail of mice, then removed the injected tissue after half an hour, one hour or at hourly intervals thereafter, those in which the tissue was removed at half an hour remained alive.

That tetanus was indeed mediated by a toxin produced by tetanus bacilli was proven independently in 1890 by Knud Faber (1862–1956) in Denmark and by G Tizzani and G Gatani in Italy. They showed that the signs of tetanus were induced in experimental animals by injecting filtrates of tetanus cultures, i.e. the fluid in which they grew but with the bacteria removed. This toxin was shown in 1892 by the Italian physician, Alessandro Bruschettini (1868–1932), to be located in the base of the brain and spinal cord as well as in the peripheral nerves.

Meanwhile, Kitasato had joined forces with Emil Behring (1854–1917) to look for an antidote to the toxin. They were successful in 1890 when they inoculated rabbits with small doses of toxin that had been treated with iodine. This process immunized the animals so that they were able to neutralize the toxin. Moreover, when their serum was injected into other animals, they too were resistant to the toxin. The antibody treatment of tetanus had been born.

We now know that the toxin, called tetanospasmin, most commonly enters the nerves where they supply the muscles (called the neuromuscular junction), then travels up these motor nerves (called axons) to the spinal cord and brainstem (parts of the central nervous system). Here they interfere with the manner in which nerves communicate with each other by blocking release of the normal chemicals which transmit information between nerves. Most importantly, those nerves which stop muscles becoming overactive are blocked, so that the muscles become rigid and go into spasm.

Four major clinical syndromes are recognized:

- *generalized tetanus* is most common and is the condition described by Aretaeus. It begins with a 'sneering grin' called *risus sardonicus* and rigidity of the jaw muscles (trismus) so that the mouth cannot open ('lockjaw'). After several hours or a day or two, this is followed by generalized muscle spasms which progress to involve the whole body to cause difficulty in breathing. In addition, the autonomic (subconscious) nervous system is damaged causing a fast heart rate and fever. Even with effective treatment, recovery takes a month or two.
- *localized tetanus* is associated with rigidity of the muscles near the site of injury where bacteria are multiplying and producing toxin.
- *cephalic tetanus* is a variant of localized tetanus in which the nerves supplying the face, throat, and neck are affected.
- *neonatal tetanus* usually occurs in the second week of life with an inability to suck followed by generalized muscle spasms.

Today, tetanus has largely disappeared from developed countries thanks to widespread immunization. Unfortunately, it is all too common in poor, developing countries where few are immunized. Make sure your children are immunized. If you have a gardening injury with contamination of the wound with soil and it is ten years or more since you last had a 'tetanus shot', then go to your doctor for a booster.

33

Plague (the Black Death)

SCIENTIFIC NAME:	*Yersinia pestis*
ORGANISM'S APPEARANCE:	Gram-negative short rod
DISEASE NAME:	plague
DISTRIBUTION:	Africa, Asia, the Americas
RESERVOIR:	rodents and related animals
TRANSMISSION:	a. bubonic plague—the Oriental rat flea, *Xenopsylla cheopis*, b. pneumonic plague: aerosol from coughing
LOCATION IN A PERSON:	buboes (infected lymph nodes), then blood and any internal organs
CLINICAL FEATURES:	fever, chills, headaches, buboes, haemorrhages
DIAGNOSIS:	isolation of the organisms from a bubo, blood, or other clinical specimen
TREATMENT:	gentamicin or doxycycline
PREVENTION:	avoid and control rats

Most people when they think of the plague bring to mind the Black Death, which swept Europe in the fourteenth century. In fact, there have been three great outbreaks (also called pandemics because they spread very widely) in the last 2,000 years.

The first plague pandemic is often called the 'Plague of Justinian' because it occurred during the reign (AD 527–565) of the Eastern Roman emperor Justinian, who was based in Constantinople (now Istanbul). This pestilence found its way into the eastern Mediterranean from either north-eastern Africa or India in 541. The historian Procopius of Caesarea (AD 500–*c.* 565), recorded the awful devastation:

> It embraced the whole world and blighted the lives of all men...great swellings appeared in the groin and armpit...There was delirium, frantic restlessness and some became comatose...most died in five days. It raged in the city

> of Constantinople . . . for months and so numerous were the corpses that they were cast into hollow towers of some incomplete fortifications.[256]

Thereafter, plague recurred every three or four years for decades until well into the seventh century. In the process it carried off many millions of people and ushered in the European Dark Ages.

The second pandemic, the 'Great Pestilence' or 'Black Death', is thought to have killed one-quarter to one-third of the population of Europe. It swept out of Central Asia through the Crimea to the Mediterranean basin in 1347, ravaged its way north-west to Britain (1348), then north to Norway (1349), after which it swung east to Poland (1351) then Russia (1352). Michele da Piazza, a Franciscan friar, described the arrival of the plague in Messina, Sicily in October 1347:

> Twelve Genoese galleys, fleeing our Lord's wrath which came down upon them for their misdeeds, put in at the port of the city of Messina. They brought with them a plague that they carried down to the very marrow of their bones, so that if anyone so much as spoke to them, he was infected with a mortal sickness which brought on an immediate death in which he could no way avoid. The signs of death . . . were these: . . . nearly the entire body succumbed to the woeful disease . . . there arose certain pustules the size of a lentil on the legs or arms. The plague thus affected and penetrated the body so that its victims violently spat out blood and this . . . continued for three days until they expired.[250]

Two years later in his work *The Decameron*, the Italian writer Giovanni Boccaccio (1313–1375) described the illness the plague caused in the Italian city of Florence:

> Its earliest symptom . . . was the appearance of certain swellings in the groin or in the armpit, some of which were egg-shaped while others were roughly the size of the common apple . . . From the two areas already mentioned, this deadly swelling would begin to spread, and within a short time, it would appear at random all over the body. Later on, the symptoms of the disease changed, and many people began to find dark blotches and bruises on their arms, thighs and other parts of the body, sometimes large and few in number; at others tiny and closely spaced. These . . . were . . . (an) infallible sign that he would die.[46]

We now call the swellings that Boccaccio described 'buboes'. Hence the illness is known as 'bubonic plague'. 'Bubo' is a late Latin word derived from the Greek *boubon* meaning 'the groin' or 'swelling in the groin'. The groin is the area between the upper thigh and the abdomen. The term 'plague' is derived from the Latin 'plaga', which means a 'blow' or 'affliction'. Those living in medieval times called this devastation either the 'plague' or 'pestilence', the latter being derived from the Latin 'pestis', which means 'epidemic, destruction or curse'. The term 'Black Death' was popularized in the nineteenth century and reflects the dark blotches and bruises described by Boccaccio. He then went on to recount how the disease spread:

> Whenever those suffering from it mixed with people who were still unaffected, it would rush upon them with the speed of a fire racing through dry or oily substances... it also seemed to transfer the sickness to anyone touching the clothes or other objects which had been handled or used by its victims.[46]

In an attempt to circumvent the contagion, the most extraordinary forms of protective clothing were devised. Doctors completely covered themselves with gowns and garments and wore a beak-like mask containing perfumes that were supposed to block the odours emanating from the victims.

The plague did not quite disappear from Europe and for nearly four centuries smaller epidemics recurred at varying intervals. The Great Plague of London of 1665 killed some 70,000 out of a population of 450,000. The Great Fire of the following year ended the epidemic by incinerating the rats but no-one understood the connection at the time. Following a major epidemic in southern France in 1720–2, in which almost half of the 90,000 people living in Marseille were carried off, the pandemic petered out in Europe.

The third plague pandemic began in China in the 1860s. In 1894, it reached Hong Kong, then spread to involve the whole south China coast. Because of Hong Kong's extensive trade links, the plague was then carried rapidly by shipping around the globe. This time, armed with the new bacteriological knowledge and methods, medical scientists were ready to tackle this awful scourge and seek its cause. Several commissions were set up. Two, one from France and the other from Japan, would find the answer in Hong Kong. But like all good stories, who discovered precisely what has been a matter of some controversy for over a century. The two key players were Kitasato and Yersin.

Shibasaburo Kitasato (p.362), accompanied by the pathologist, Professor Tanemichi Aoyama (1859–1917), several medical students, and assistants, was sent by the Japanese government to investigate the cause of plague. The mission arrived with much fanfare on 12 June 1894 and was treated royally by the British, especially by Dr James Lowson, the Scottish medical superintendent of the Government Civil Hospital, who gave them good facilities in the Kennedy Town Hospital (a glassworks factory which had been transformed into an emergency plague hospital). The Japanese team had access to bodies of plague victims to examine as well as sailors to guard them and prevent them being attacked by any Chinese who might be upset with autopsies being undertaken.

They made great progress. Aoyama conducted the first post-mortem examination on 14 June. The internal organs, heart blood, and perhaps buboes (the last is a matter of some dispute) were examined. In them Kitasato found many bacilli. He then injected blood taken from the fingertips of living patients into two mice; they died two days later and the same bacilli were found in them. On 20 June, the local *China Mail* published an obsequious interview with Kitasato and Aoyama under the headline 'Discovery of the Plague Bacillus' which recounted their discoveries. Lowson telegraphed the British medical journal, *The Lancet*, in London, about all this activity and on 23 June, just eight days after the arrival of the Japanese, *The Lancet* reported in an editorial on 'The plague at Hong Kong':

> The telegram...informs us that Professor Kitasato of Tokio, late assistant in Professor Robert Koch's laboratory in Berlin, has succeeded in discovering the bacillus of the plague. Whether that be the case or not, we have as yet no means at our disposal of forming a judgement, and it is certainly premature to assume that the bacillus in question is the actual cause of this terrible disease.[18]

More information was soon forthcoming. On 11 August 1894, *The Lancet* published notes and illustrations of some preparations of the plague bacillus which Lowson said were partly prepared by himself and partly by Kitasato (Figure 170). In describing the organism, Lowson wrote:

> the ends are somewhat more rounded – when stained lightly (it) appears almost like an encapsulated diplococcus, but when more deeply stained has the appearance of an ovoid bacillus, with a somewhat lighter centre.[197]

His notes went on to say that the bacteria were seen in very large numbers in the buboes, to a lesser extent in the spleen, and occasionally in the blood. Lowson ended by remarking that he had been extremely busy with clinical and administrative duties but that he hoped soon to describe the appearance of the bacteria when cultivated outside of the body. Lowson gave the impression that he had been a major player in all of the research. It seems as though he was trying to muscle in on Kitasato's work and take credit for himself. Mind you, Kitasato could also play that game.

On 25 August 1894, an article by Kitasato himself appeared in *The Lancet*. He mentioned the pathological work of Aoyoma but did not see fit to make him a

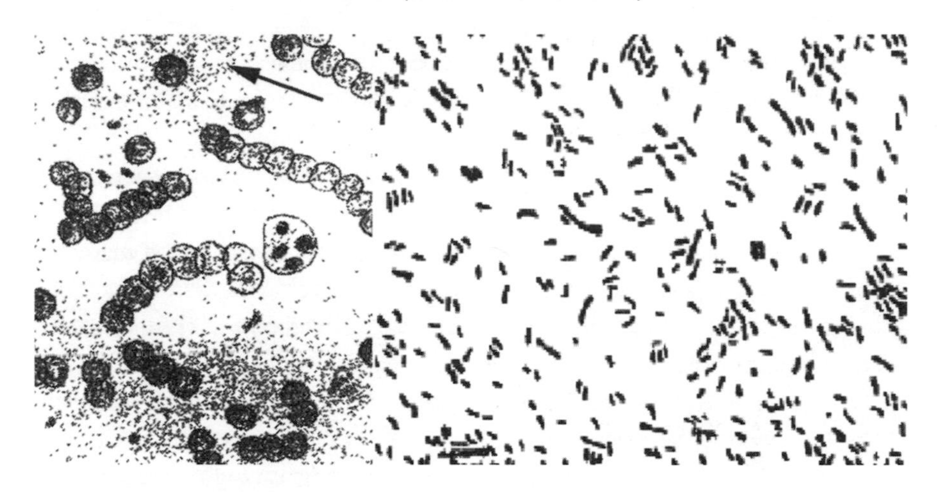

Figure 170. Kitasato's illustrations. Left: masses of bacilli (arrow) in the blood of a patient with plague. It is hard to know whether these are really *Y. pestis*. It would be unusual to see so many bacteria in the blood. Right: cultures of plague bacilli.

co-author as 'the bacteriological part [of the researches] received the care of the writer'. Kitasato was not fluent in English but the article was written in excellent English. The fourth sentence says 'Dr Lowson, Acting Superintendent of the Government Civil Hospital, put everything needful at our disposal in the most friendly spirit.'[159] I think we can guess who wrote the paper! It went on to describe the researches, some of which have been alluded to above. Altogether, 15 autopsies were done and the bacteria were always found. He described their appearance:

> The bacilli are rods with rounded ends, which are readily stained by the aniline dyes, the poles being stained darker than the middle part...and presenting a capsule, sometimes well marked, sometimes indistinct...I am at present unable to say whether or no 'Gram's double staining method' can be employed.[159]

When grown on agar, the colonies were whitish-grey in colour and had large dense centres. Kitasato reported that mice, rats, rabbits, and guinea pigs but not pigeons were susceptible to infection. What is more, he found similar bacilli in dead mice in the streets. Kitasato produced a variation of Koch's postulates:

> (1) This bacillus occurs in the blood, in buboes, and in the internal organs of the plague-stricken only; (2) This bacillus is not to be found in any other infectious disease; (3) With this bacillus it is possible to produce in animals the identical symptoms which the disease presents in human beings. From this

> evidence we must come to the conclusion that this bacillus is the cause of the disease known as the bubonic plague; therefore, the bubonic plague is an infectious disease produced by a specific bacillus.[159]

The article, which was placed and dated 'Hong Kong, 7th July 1894', ended with 'heartiest thanks to Dr Lowson for all his assistance'.

Kitasato and his team left Hong Kong on 20 July to receive a hero's welcome on their return to Japan. Once there, Kitasato launched into a series of lectures describing the findings of the team. It seemed that Kitasato was the discoverer of the plague bacillus.

Or was he? There was another player. Alexandre Yersin (Figure 171) had been born in Aubonne, Switzerland, in which country he graduated in medicine. Inordinately shy, he decided to specialize in pathology as contact with people would be minimized. After graduation he worked at the Hôtel-Dieu de Paris (a hospital, not a hotel). He accidentally cut himself in 1886 while performing an autopsy on a patient who had died from rabies. He was rescued from possible death by the rabies vaccine that had been produced in Pasteur's laboratory. So impressed was he by Pasteur and his staff that he decided to join Pasteur's team after a short stint in Robert Koch's laboratory in Berlin. In 1888, with Emile Roux, Yersin discovered diphtheria toxin. In 1890, he tossed in bacteriology and became a ship's doctor in the Far East. In late 1891, he joined the colonial medical service in French Indochina (now Vietnam, Laos, and Cambodia) but spent much of the next two and half years exploring the dangerous hinterland. In 1894, after much behind the scenes prodding by Yersin himself, he was sent by the French Colonial Office to Hong Kong to investigate the plague. He arrived unheralded and unmet in Hong Kong on 15 June 1894 together with two untrained assistants, one of whom soon absconded with much of his money.

Figure 171. Alexandre Yersin (1863–1943).

Yersin spoke little English so he solicited the aid of Father Vigano, an Italian missionary, to interpret for him and smooth his entry into officialdom. Shortly after his arrival he received a less than lukewarm reception from Lowson, who told Yersin that Kitasato had already found the bacterial cause of plague and that he, Lowson, had wired *The Lancet* with news of the discovery. A somewhat disconsolate Yersin paid a visit to Kitasato but the meeting was a great disappointment. There is also a story, which may be apocryphal, that they laughed among themselves and then, to a man, turned their backs on him. There was clearly not going to be any collaboration. Although Kitasato later claimed that he found bacteria in a bubo of the first patient, Yersin did not gain that impression and believed that Kitasato and his team were concentrating on the blood and internal organs. Yersin noted in his diary: 'I shall avoid their society; if one of us goes on a false track, the other will not imitate or follow him.'[332]

Something strange was happening. Yersin was receiving blood specimens, possibly from the same patients as Kitasato but could find no bacilli. Yersin had finally been given a small area by Lowson to use as a makeshift laboratory, but bodies of people who had died from the plague were reserved for the Japanese. A justly upset Yersin wrote to the Pasteur Institut: 'there is obviously preferential treatment in this'.[332] He therefore devised a simple plan. In broken English, he befriended the English sailors who were sentries at the mortuary, then after a couple of days offered a bribe. He was in. Yersin entered the mortuary and removed a bubo from a corpse, then hurried it back to his laboratory. He smeared some of the pus from the bubo on a glass slide and, in contrast to blood which he had looked at thus far and in which he had seen nothing, Yersin this time saw 'une véritable purée de microbes'—truly a soup of bacteria. He injected some bubonic material into mice and guinea pigs then returned to the mortuary to gather two more buboes. He saw the same bacilli. Yersin knew he was on to something.

Yersin needed more space, both to work in and to live in. He obtained permission from the authorities to construct a bamboo, timber, and grass hut. Within 24 hours or so it was built and he started using it on 21 June. Yersin realized that he could not continue clandestine operations so he used the good offices of the French consul to request the Governor of the Colony to allow him access to bodies. Within two days permission was granted. The obstruction had come from lower down the totem pole—from Lowson. But Lowson was even more devious. He had visited Yersin's laboratory, looked at bacilli in smears from the buboes, then scurried back to the Japanese with the news. On 4 August, *The Lancet* derisively dismissed Versin (as they misspelt his name) as having:

> discovered another bacillus, which he, too, claims to be the essential cause of the disease; and others, equally anxious to discover something, have swelled the list, so that, as our correspondent says 'the varieties of plague bacilli now outnumber the leaves in Vallombrosa'.[19]

Are you wondering about the leaves in Vallombrosa? John Milton in his epic poem *Paradise Lost* penned the following line: 'autumnal leaves that strow the brooks, in Vallombrosa'; Vallombrosa is a Benedictine Abbey near Florence in Italy. There were in fact no others claiming to have found the cause of plague except in Lowson's scurrilous jibes.

To Kitasato, ensconced in relatively palatial surroundings, Yersin's activities in a grass hut must have seemed pointless. But Yersin persisted. Within a week of the Governor's approval, he had studied 25 plague cadavers, isolated the plague bacteria, cultured them, and injected them into laboratory animals. Some two weeks after Kitasato's departure, Yersin too left Hong Kong for Vietnam. He departed as he had arrived—alone.

Yersin published a detailed account of his researches later in 1894. First, he described the disease he saw:

> The incubation is rapid with an incubation period of 4.5 to 6 days. The patient is prostrated. Abruptly a high fever sets in, often accompanied by delirium. On the very first day, a discrete bubo usually appears. In 75% of the cases it is located in the inguinal region [groin], in 10% of cases in the axillary region [armpit], and occasionally at the back of the neck and other regions. The nodule rapidly reaches the size of an egg. Death occurs after 48 hours or sooner... In a few cases, the bubo does not have time to form, and one will note in such cases haemorrhages on the mucous membranes or petechial spots [tiny haemorrhages] in the skin. Mortality is high; about 95% in hospitals.[331]

He then went on to describe the bacteria:

> The pulp of the buboes contain short, stubby bacilli which are rather easy to stain with aniline dyes and are not stained by the method of Gram [i.e. they are Gram-negative]. The ends of the bacilli are coloured more strongly than the centre... One can find them in large numbers in the buboes and the lymph nodes... They are seen in the blood from time to time... and only in very serious and rapidly fatal cases.[331]

Yersin then described culture of the organism:

> The pulp of buboes seeded on agar gives rise to transparent, white colonies, with margins that are iridescent when examined with reflected light... The

> most favourable medium is an alkaline solution of 2% peptone to which 1–2% gelatine has been added. Microscopical examination of the cultures reveals true chains of short bacilli.[331]

When these cultures were inoculated in mice, rats, or guinea pigs they died, pathological lesions typical of plague were seen, and the characteristic bacilli were recovered when they were autopsied. Once again, Koch's postulates had been thoroughly fulfilled.

It seemed clear that both Kitasato and Yersin had independently discovered the cause of the plague. Kitasato held pride of place because of the priority of his discovery. Then things quickly began to fall apart. The first shot across the bow was lobbed by Kitasato's colleague, Professor Aoyama, in 1895. Aoyama examined the tissues from corpses of plague victims that he had taken back to Japan from Hong Kong very carefully and showed that the bacteria in these tissues were pink on the Gram stain (Gram-negative). On the other hand, the organism recovered from the blood by Kitasato was Gram-positive. He concluded that Kitasato's organism was merely a secondary (i.e. extra infection) with a streptococcus, perhaps the pneumococcus, because secondary infections were quite common in plague. Was Aoyama cheesed off that Kitasato was receiving all the adulation and glory, or was he trying to set the record straight? Perhaps a bit of both! Then in July 1896, Yaowaka Murakami in Formosa (now Taiwan) sent a pure culture obtained from the bubo of a plague victim to Professor Okada in Tokyo. He concluded that this organism was identical with the bacillus which Yersin had found in buboes. Professor Ogata from the Tokyo Imperial University was sent to Formosa to investigate and he concurred with Okada's opinion.

Kitasato was in some trouble. He had to try to find a way out that would not cause him to lose face, and perhaps even put at risk the funding of his research institute. One of Kitasato's assistants in Hong Kong had been Tohiu Ishigami. He published a book on the plague in which he put the problem this way:

> Since the bacillus that exists in the buboes was considered to be the pathogenic element of the disease by several investigators, it naturally became our duty and responsibility to fully pursue our former incomplete investigations, so as to determine one way or other its pathogenesis.[150]

According to Ishigami, the upshot was that Kitasato's investigations of an outbreak of plague in Japan in 1899:

> proved no doubt [i.e. beyond doubt] that the bacillus of bubonic plague is identical with the Yersin bacillus which exists in the buboes, and which we

> had considered to be involution forms during our former investigations. The so-called plague bacillus... assumes, as it were, a second nature during the septicaemic stage of the disease. It first invades the blood as the Yersin bacillus. I believe Professor Kitasato will later on make public the result of this investigation.[150]

Ishigami's statement can be found in an English translation of Ishigami's book by Donald MacDonald published in 1905. MacDonald was medical superintendent of the Adelaide Government Hospital (later the Royal Adelaide Hospital), where I was to spend so much of my undergraduate and postgraduate training. While in Hong Kong, Kitasato had been obsessed with finding bacteria in the blood as a diagnostic tool and had not paid much attention to bacteria in the buboes. The key to his argument was that while in Hong Kong he had mistakenly dismissed these bubonic bacteria as 'involution forms', i.e. bacteria that changed their shape according to the environment in which they were located. Now, a few years later, he was claiming that these so-called involution forms in the buboes transformed themselves into a bacterium with a different shape and became Gram-positive. Kitasato could not bring himself to publicly admit that he had made a mistake.

Both Yersin and Kitasato had sent specimens of their cultures to the Institut Pasteur in Paris and to Koch in Berlin, respectively, from where they were disseminated for further study. No-one ever seemed to have a problem with Yersin's specimens, but Kitasato's cultures gave inconsistent results. What had happened? Commentators have spent decades arguing but it seems likely that both Kitasato, at least at first, and Yersin independently did find and culture the plague bacillus, but somewhere along the way, Kitasato's specimens became contaminated with another organism, probably the pneumococcus.

This controversy is reflected in the name of the organism that causes the plague. In 1896, Lehmann and Neumann (p.216) called it *Bacillus pestis*, then in 1900 Walter Migula (1863–1938) renamed it *Bacterium pestis*. Thus it remained until 1923 when Bergey and colleagues (p.217) erected the genus *Pasteurella* (in honour of Pasteur) and transferred the plague organism into it as *Pasteurella pestis*. In 1944, Johannes Jacobus van Loghem (1878–1968) created the genus *Yersinia*, in honour of Yersin. In 1980, the plague bacillus officially found its home as *Yersinia pestis*. Perhaps after all these tortuous convolutions, justice has finally been done to Yersin.

By the late 1890s, it was clear that Yersin's bacillus was the cause of the plague, but how did a person become infected? There had been no shortage of suggestions over the years prior to its discovery. During the scourge of the Black Death in the Middle Ages, few doubted that it was the wrath of God in response to the sinfulness of mankind. It became increasingly clear that the plague was contagious, perhaps due to the inhalation of miasmas (foul-smelling air) or the drinking of polluted water. One way of dealing with the problem when there was an outbreak of plague was to go somewhere else. So, for example, Pope Paul III in 1547 moved the Council of Trent to Bologna. A second way was to try to prevent outsiders from entering a town by putting up roadblocks to create a *cordon sanitaire*. A third measure was to stop plague-infested ships from docking at port. Thus, in 1348 the Republic of Venice excluded suspected ships for a period of 40 days—*quaranto giorni*—to allow time for the disease to declare itself aboard ship. It is from these two Italian words that we derive our word 'quarantine'.

In June 1894, when the outbreak in Hong Kong was at its peak, *The Lancet* observed that studies of the older chronicles on the plague:

> indicate that all emanations of the victims, the breath, the exhalations of the skin, and the excreta were capable of propagating the disease. 'Did anyone speak to them he was directly struck with a mortal sickness from which there was no escape.'[18]

The editorial went on to remark: 'There is no doubt that this disease is like typhus fever in its mode of origin, highly contagious and spreads wherever overcrowding and filth abounds.'[18]

When he was in Hong Kong in 1894, Yersin observed large numbers of dead rats in the streets and houses. Others had made similar observations for centuries but he wrote these prescient words:

> Dead rats . . . always harbour large quantities of the microbe in their organs. Many have real buboes. I have placed healthy mice and inoculated mice in the same cage. The inoculated ones died first, but within a few days all the others die from invasive plague bacillus. Plague is therefore a contagious and transmissible disease. It is probable that rats are the major vector in its propagation.[331]

But how exactly was the organism transmitted? The most popular views were that infection was spread in the excreta of humans and rats and that the plague bacillus contaminated dust which was absorbed through inhalation, ingestion, or skin wounds. There were, however, other ideas. Masanori Ogata, whom we have already mentioned, suggested on the basis of the distribution and spread of plague that the disease might be conveyed by insects that suck blood such as mosquitoes and fleas. In 1897, he crushed some fleas taken from rats which had died of plague and injected them into two mice; one died from the plague three days later. But he carried the idea no further.

It was at this point that Paul-Louis Simond (1858–1947) entered the scene. Simond had been born in France, studied medicine at Bordeaux, then joined the French Naval Medical Corps. In 1895, he joined the Institut Pasteur. In 1897 he was posted to Bombay (Mumbai) then Cutch-Mandvi (Mandvi) in British India to replace Yersin who had been investigating the value of an antiserum during the plague outbreak there. Simond noted that many patients with the plague had a small skin blister at the beginning of their illness. This enlarged and often developed into a bubo. This made him wonder whether the patient had been bitten by an insect that was carrying the plague bacillus. He considered and dismissed cockroaches and then concentrated on fleas infesting rats. He picked up fleas with soapy water from rats with plague and examined them under the microscope; they were full of bacteria resembling Yersin's bacillus in their digestive tract.

After a quick trip to Saigon (Ho Chi Minh City, Vietnam) he returned to India in 1898 to Kurrachee (now Karachi in Pakistan) where he undertook the following experiment in the Hotel Reynolds (the British authorities would not allow him access to the hospital). He put sand in the bottom of a large glass bottle, then put in a plague-ridden rat which had several fleas on it. For good measure, he added some unidentified fleas collected from a cat and covered the bottle with a wire mesh. Next day, when the rat was very sick, he introduced a cage which contained a perfectly healthy rat and suspended it several centimetres above the sand. This cage had strong mesh walls and the rat inside it could not have any direct contact with the sick rat, the wall of the glass bottle or the sand. The next day the original rat died. When this was examined, its blood and organs contained an abundance of plague bacilli. Five days later, the rat in the cage became ill and died the following day. It had buboes which were full of *B. pestis* as were its blood and internal organs. Simond wrote:

> That day, 2 June 1898, I felt an emotion that was inexpressible in the face of the thought that I had uncovered a secret that had tortured man since the appearance of plague in the world. The mechanism of the propagation of plague includes the transporting of the microbe by rat and man, its transmission from rat to rat, from human to human, from rat to human and from human to rat by parasites [i.e. fleas]. Prophylactic measures, therefore, ought to be directed

> against each of these three factors: rats, humans and parasites. I subsequently repeated the same experiment with similar results.[289]

Simond was convinced but most people were sceptical because initial attempts to repeat the experiment in Sydney, Australia, and Kronstadt near St Petersburg, Russia, failed. Furthermore, this experiment would not rule out transmission by the respiratory route with an infected rat coughing on a normal rat. Eventually in 1903, JC Gauthier and A Raybaud in Marseille, France, confirmed Simond's observations. They used a cage divided in the middle by two grills which were 2 cm apart which was placed in a glass jar. When fleas were present, plague was transmitted. When fleas were absent, then there was no transmission of plague. Unfortunately, the fleas involved in their experiment were not identified.

This was all very well for transmission from rats to rats but it was not possible to do an experiment to prove transmission from rats to humans. Some objected that rat fleas did not bite humans. While this was true for rat fleas most commonly found in Western Europe, it was not true for the Oriental rat flea, then known as *Pulex cheopis* (also known as *P. pallidus*; now called *Xenopsylla cheopis*), which was found to actively bite humans in places ranging from Sydney to Bombay. The matter was soon put beyond doubt by the Plague Commission set up in 1905 by the British Secretary of State for India which carried out extensive studies in and around Bombay. The Commission confirmed that an outbreak of plague in rats (called an epizootic) always preceded an outbreak of plague in humans (called an epidemic) and that the flea involved was *P. cheopis*.

The question remained as to how a flea actually transmits the bacteria. In 1914, Arthur William Bacot (1866–1922) and Charles James Martin (1866–1955) at the Lister Institute in London showed that after a flea ingests bacteria during a blood meal, the organisms multiply in the flea's foregut. So much so, that they form an obstructive mass (Figure 172). When the flea takes another meal, the ingested blood is unable to pass beyond the obstruction and the blood, together with bacilli, is regurgitated back under the host's skin. In this manner another animal or human becomes infected. The predominant mode of transmission of plague was now understood.

I have blithely mentioned that this investigator did this or another did that. What you have perhaps not entirely appreciated is that it was highly dangerous work. They did not know how the plague was transmitted and, in retrospect, there was every chance that an infected flea might have bitten them. There may

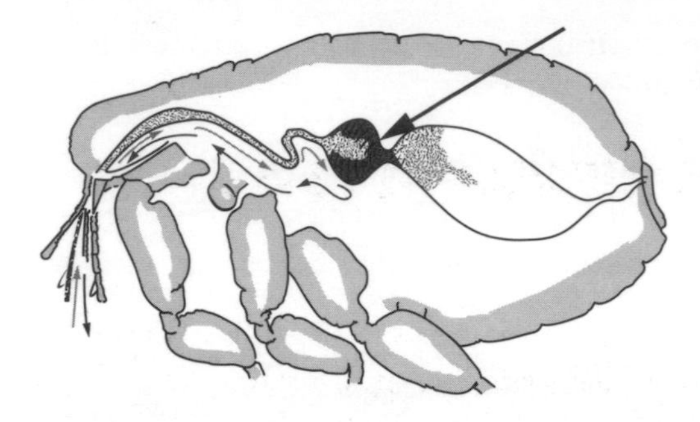

Figure 172. Diagram of a flea's gut showing the anterior portion blocked by masses of plague bacilli (arrow) so that ingested blood regurgitates. These fleas can jump 40 cm.

well have been some investigators who died from plague during their researches but I have not come across them. What I can say, though, is that the aforementioned Arthur Bacot may have avoided the plague but he died from typhus while studying that disease in Cairo, Egypt.

What we have described so far is the bubonic plague, transmitted by certain fleas, which begins with extremely painful, swollen lymph nodes (buboes). The organism then spreads through the bloodstream and in about 5 per cent of cases involves the lungs causing pneumonia. The disease can then be spread to another person directly by coughing; this is called pneumonic plague. Plague has not gone away; it is still present in the wildlife in Africa, Asia, and North and South America and from time to time jumps to humans. You might be surprised to learn that 10–20 people catch the plague in the USA every year.

34

Brucellosis (undulant fever)

SCIENTIFIC NAMES:	*Brucella melitensis, B. abortus, B. suis*
ORGANISM'S APPEARANCE:	Gram-negative cocco-bacillus
DISEASE NAME:	brucellosis, undulant fever, Malta fever, etc.
DISTRIBUTION:	many parts of the world
RESERVOIR:	*B. melitensis*—goats, sheep; *B. abortus*—cattle; *B. suis*—pigs
TRANSMISSION:	ingestion of dairy products, inhalation, skin abrasions
LOCATION IN A PERSON:	anywhere
CLINICAL FEATURES:	fever, highly variable symptoms—almost anything
DIAGNOSIS:	isolation of the organisms from blood, body fluid, or tissues;
LESS DEFINITIVE:	high level of antibodies against *Brucella* in the blood
TREATMENT:	gentamicin plus doxycycline
PREVENTION:	avoid unpasteurized milk and dairy products

In about 1970, I saw a man in his forties with fever, lassitude, and a swollen tender liver. A biopsy of his liver provided an unexpected result. His liver was infiltrated with granulomas (clumps of specialized, inflammatory white blood cells). He was a dairy farmer from the Yorke Peninsula of South Australia so this suggested a possible diagnosis—brucellosis. This was soon confirmed with a blood test. He had what is called a zoonosis. We have not used this term before: it means an illness in a human with an infection acquired from an animal. But let us go back to the beginning.

The Mediterranean region has always been a hotbed of fevers. Some have been easily recognizable such as the intermittent fevers of malaria. Others had associated symptoms which suggested the correct diagnosis as in pulmonary

Figure 173. David Bruce (1855–1931).

tuberculosis. But there were plenty of others which were very difficult to distinguish. One of them, typhoid fever, which was very common, could be diagnosed with confidence once Gaffky learnt to culture the causative bacterium in 1884.

It was in this setting that David Bruce (Figure 173) arrived in Malta in 1884. We have met him and his wife before in connection with their later researches into sleeping sickness. Bruce had been born near Melbourne, Victoria, Australia, then moved to Scotland with his family when aged five. After graduation, he joined the Army Medical Corps and was posted to Malta. This was an island in the central Mediterranean Sea of strategic naval importance and was ceded to the British Empire in 1814 as part of the Treaty of Paris. Occupation meant British troops and among them was Surgeon Captain Bruce. In Malta, he encountered an illness which was very similar to typhoid fever and was variously called Malta fever, Neapolitan fever (fever of Naples), Mediterranean fever, and so on. During the five years he was stationed in Valetta, he encountered some 400 cases and set about differentiating them from typhoid fever and looking for the cause.

In 1887 he struck gold. A number of soldiers afflicted with this condition died. At post-mortem examination, Bruce could find no evidence of ulceration over Peyer's patches, so characteristic of typhoid fever. On the other hand, when he examined a fragment of spleen stained with Gram's stain, he observed 'enormous numbers of single micrococci . . . scattered through the tissues'.[56] With the aid of Dr Guiseppe Caruana-Scicluna (1853–1921), the government analyst, Bruce prepared some tubes of nutrient agar. He had no luck in culturing any organisms from the blood of patients with Malta fever but eventually succeeded with specimens of spleens taken from patients who had died. He inoculated tubes of agar by stabbing them with platinum needles covered with splenic tissue, then incubated them at 37°C. Several days later he saw minute pearly white colonies growing along the needle tracks (Figure 174). When examined under the microscope, these colonies disclosed innumerable Gram-negative cocci (Figure 175).

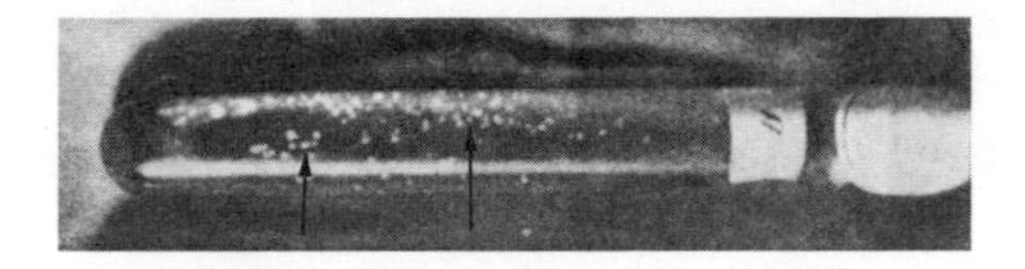

Figure 174. Agar tube with white colonies (arrow) growing along the needle tracks.

Bruce had no doubt that he had conclusively shown that a definite micro-organism could be found in the spleen in patients with Malta fever and that it could be grown on nutrient agar. Although this was very suggestive, he had to confirm the last of Koch's postulates. He therefore took a pure culture of the organism growing on agar and injected it into a monkey. The animal developed a fever and an enlarged spleen and died three weeks later. From its liver and spleen, Bruce again isolated the causative organism. Subsequently seven more monkeys were injected, all became ill, and four died. There was no longer any doubt—this organism was the cause of Malta fever.

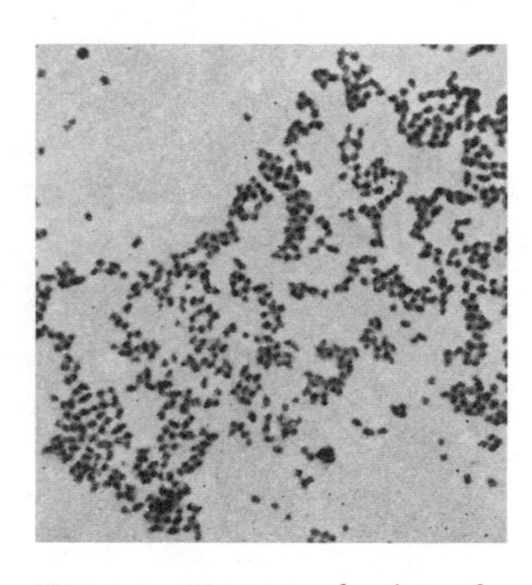

Figure 175. Gram stain of a culture of *B. melitensis*—they would look pink if you could see them in colour. Published by Hughes in his book in 1897.

Bruce described his researches in a journal called the *Practitioner* in 1887 and 1888,[56] then recapitulated his findings in the *British Medical Journal* in 1899:

> I have examined in all ten fatal cases of Malta fever for bacteria. I have also examined several fatal cases of enteric fever from the same hospital wards. The examinations by means of cultivation experiments, etc., were conducted in precisely the same way. In all the cases of Malta fever there was found a minute berry-shaped bacterium, whereas in all the typhoid cases there was found a much larger rod-shaped bacterium [Gaffky's bacillus]. The micrococcus of Malta fever planted on agar-agar, and kept at a temperature of 35°C showed no signs of growth for at least four days. On the other hand, the typhoid bacillus under the same circumstances showed marked growth at the end of twenty-four hours. Further, the micrococcus of Malta fever, inoculated into a monkey, caused death after twenty-one days, and from the spleen and liver the same micro-organism was recovered.[57]

Dr Caruana-Scicluna may well have been an important participant in these researches including culturing the bacterium from spleen samples. If so, Bruce never acknowledged it. It is likely that Caruana-Scicluna has never received the proper recognition he deserves.

Bruce, as you will have noticed, indicated that the bacterium that caused Malta fever was a micrococcus. In 1893, Surgeon Captain Matthew Louis Hughes (1867–1899: he was killed in the Boer War), who was in charge of the fever wards in the hospital in Malta, published a paper in which he named the organism *Streptococcus miletensis*.[144] His spelling was in error and later that year Bruce himself published an article in the *Annales de l'Institut Pasteur* in which he renamed it *Micrococcus melitensis*. The species name is Latin for 'of Malta' (the Romans called the island 'Melita' because of the excellence of the honey produced there (Latin 'melita' = honey)). In 1897, Hughes suggested a new name for the disease when he published a monograph on the disease.[145] He called it 'undulant fever' to indicate that the fever of this disease sometimes waxed and waned (Figure 176). This found great favour with the Maltese, who did not like their island being belittled by the name Malta fever.

So if this organism caused Malta fever, how did people acquire it? Following the urgings of Bruce, the British government in 1904 set up the Mediterranean Fever Commission to seek the answer. The team was led by Major William Horrocks of the Royal Army Medical Corps and included Royal Navy Surgeon Ernest Shaw and a Maltese bacteriologist, Dr Themistocles Zammit (1855–1931). What exactly happened has been a matter of considerable controversy.

It seems that in July 1904, Shaw injected two goats with *M. melitensis*. He found that the goats made antibodies in their blood against the bacteria. But since the animals remained well, Shaw lost interest in them. Zammit, who had at first thought that the bacteria might be transmitted by mosquitoes, tried feeding bacteria to a different batch of goats and observed that they too developed antibodies against the bacteria. It appeared that goats might be susceptible to experimental

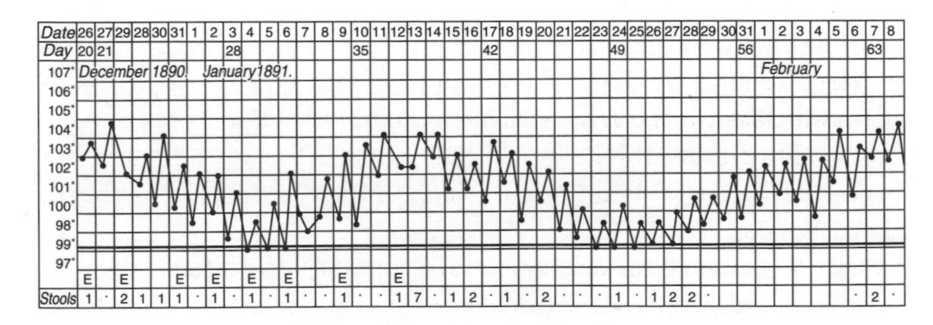

Figure 176. Temperature chart of a patient with undulant fever published in 1897 by Hughes. It is not all that different to typhoid fever. Temperatures are in degrees Fahrenheit.

infection. Thereupon Zammit purchased six healthy goats and was surprised and excited to find on testing them that they already had antibodies in their blood.

Caruana-Scicluna suggested that the milk of these goats should be examined for *M. melitensis.* On 21 June 1905, Horrocks recovered the organism from the milk of one goat, then four days later Zammit found the bacteria in blood from the same goat. Subsequent studies by members of the Commission found that 50 per cent of goats from various herds had antibodies in their blood (i.e. they were infected and perhaps infectious) and that 10 per cent were excreting the organism in their milk.

What followed was rather unsavoury. Shaw tried to muscle in and claim the credit for himself. He submitted a paper for publication in the *Journal of the Royal Army Medical Corps,* which was edited by Bruce. He made no mention of Horrocks or Zammit (remember, they were all part of the same Commission). Surprised by this, Bruce solicited the views of Horrocks, who came close to accusing Shaw of scientific fraud. Worse was to follow. Bruce, in a series of editorials and lectures, carefully left out the roles of the key players and eventually it became generally accepted that Bruce himself had discovered that these bacteria were transmitted by goat's milk. So it was that in Bruce's obituary in *The Times* in 1931, the following statement was made: 'Bruce... discovered the Micrococcus Melitensis ... and that it was transmitted by the milk of goats.'[24]

So what really happened? Zammit found that goats were naturally infected with the organism that caused Malta fever. Caruana-Scicluna appears to have first suggested that the infection might be transmitted by consumption of goat's milk and Horrocks confirmed that the organisms could indeed be found in their milk.

The role of goats was soon confirmed by two events. When Maltese goats were introduced into Rhodesia (now Zimbabwe), an outbreak of Mediterranean fever occurred. What is more, after the ship *Joshua Nelson* arrived in Malta to ship milking goats to the USA in the second half of 1905, everyone aboard who drank goat's milk, which was most of the crew including the captain, went down with Malta fever before the ship reached Antwerp. The consumption of goat's milk was banned by the military authorities and Malta fever virtually disappeared from the garrison. The same, unfortunately, could not be applied to the civilian population of Malta.

In 1908–9, an outbreak of abortions in goats occurred in southern France. Soon afterwards, a severe illness with fever appeared in humans. This outbreak was investigated by Paul Cantaloube (1880–1929), the local general practitioner, and blood tests carried out in Marseille proved that the illness was due to brucellosis. It seemed likely that farmers could contract the disease in the course of looking after sick goats as well as by drinking their milk.

We must now step back in time to 1895. The Dane Bernhard Lauritz Frederik Bang (Figure 177) qualified in both medicine and veterinary surgery. He was a professor at the Royal Veterinary College in Copenhagen when he began to investigate contagious abortions in cattle. In December of that year, Bang and his assistant, Valdemar Stribolt, examined a cow that was aborting. They removed the calf and sacrificed the cow. The only abnormality they found was a copious, gelatinous exudate between the uterine wall and the membranes. On microscopic examination, it was found to be packed with intracellular organisms. Bang was able to grow the bacteria on a medium made of serum and gelatine agar when the oxygen concentration was slightly less than its concentration in air. Bang described the organism he saw as a rod-shaped bacillus which stained negative on the Gram stain (it stained pink). Testing of more aborting cows gave the same results. Bang then fulfilled Koch's postulates by infecting pregnant heifers with cultures via the vaginal route and reproducing the disease. Not unnaturally, he wondered whether bulls had a role to play in transmission. Bang called the bacterium he had found the 'Bacillus of abortion'. This was renamed *Bacterium abortus* by Johannes Schmidt (1877–1903) in 1901 but it was popularly known as 'Bang's bacillus'. In 1911, Ernest Charles Schroeder (1865–1928) and William Edwin Cotton (1866–1969) at the Bureau of Animal Industry in Maryland, USA, showed that *B. abortus* was also excreted in cow's milk.

Figure 177. Bernhard Bang (1848–1932).

Bang did not relate the organism he found to Bruce's *Micrococcus melitensis,* nor did anyone else, presumably because one was labelled a coccus and the other was described as a bacillus. The same thing happened when the veterinarian Jacob Traum (1892–1966) discovered a similar bacterium which he said resembled *Bacillus abortus* in the liver and kidneys of an aborted swine fetus in the USA in 1914. In 1929, Irvin Forest Huddleson

(1893–1965) named this organism *Brucella suis,* the species name meaning in Latin 'of pigs'.

The connection was finally made by Alice Catherine Evans (1881–1975) of the Bureau of Animal Industry in the USA. After a discussion with Dr Adolph Eichhorn (1875–1956) in 1917, she compared Bang's *Bacillus abortus* with Bruce's *Micrococcus melitensis* because both organisms were excreted in the milk of apparently healthy animals. She found that both organisms were very pleomorphic, i.e. had variable shapes which accounted for one being called a coccus and the other a bacillus. This is what Evans had to say when she reviewed the story in 1947:

> With amazement (I) noted that one result after another pointed to a close relationship between the strains of supposedly different genera. They were alike in morphology, in staining reactions, and in biochemical reactions. They were alike culturally... They appeared alike when tested (with) an antiserum.[94]

In fact the only test which could distinguish between the bacteria from goats and that from cattle was a specialized antibody test called an 'agglutinin-absorption test'. She published her results in 1918. In 1922, Karl Friedrich Meyer (1884–1974) and Edward Byer Shaw (1895–1987) confirmed the close relationship between strains of bacteria obtained from goats and cattle and suggested that they be placed in a new genus which they named *Brucella* in honour of David Bruce. This also provided a new, non-contentious name for the disease: brucellosis.

In 1922, however, Evans was sent an isolate obtained from a patient at Johns Hopkins Hospital who had had no contact with goats or goat products, with the request to determine whether it was *Micrococcus melitensis* or *Brucella abortus.* The agglutinin absorption test showed unquestionably that it was not *Micrococcus melitensis.* In fact, subsequent studies showed that it was the organism now called *Brucella suis*; this was the first proven human infection with this organism. Meanwhile in Southern Rhodesia (now Zimbabwe), an English veterinarian named Llewellyn EW Bevan (1878–1957) used antibody tests in 1921 to describe the first known human infection with *Bacterium abortus.*

In her 1947 review, Evans thought that the differences between the three organisms were not sufficient to justify three different species. Thus was set in train a debate which was to last for many decades. A concerted attempt was eventually made to name them all *Brucella melitensis.* In 2003, the subcommittee on the taxonomy of *Brucella* concluded that in view of irreconcilable differences, a bacteriologist could use one name or different names according to his or her scientific opinion.

Koch's postulates for the three species of *Brucella* had been met in experimental animals. Surely no-one would attempt to infect humans experimentally with this potentially lethal disease, especially in the pre-antibiotic era? Extraordinarily (and perhaps today it would be viewed as criminal), in 1917–18 in California, Emanuel Charles Fleischner (1882–1926) and Karl Friedrich Meyer (1884–1974) fed milk containing large numbers of *B. abortus* to 275 children aged 0–5 years. In their defence, they thought (incorrectly as it would eventually be shown) that since it appeared to be non-pathogenic, perhaps *B. abortus* could be used to immunize against the very pathogenic *B. melitensis.* Only two children made antibodies to *Brucella* and they were said to have tuberculosis of the bone—perhaps they had brucellosis of the bone!

There were a number of other miscreants. I shall mention only one. Pablo Morales Otero (1896–1971) in Puerto Rico exposed 40 'volunteers' to one of *B. melitensis, B. abortus,* or *B. suis* by feeding or by applications to normal or abraded skin. Nine of them developed undulant fever. He found that infection was more easily transmitted through breaks in the skin than by ingestion of contaminated milk, and that symptoms appeared four to five weeks after exposure. In his experiments, *B. abortus* was pathogenic when large doses of the bacteria were administered.

Brucellosis in humans may present in an extraordinary variety of ways. Infections may be acute or chronic, and affect one organ system or many. It can be treated with antibiotics but such therapy is often not easy. Far better to avoid the infection if you live in an endemic area by not consuming unpasteurized milk, cheese, or dairy products or eating raw liver (yes, some people do!). There are thought to be over half a million cases around the world each year. At the beginning of this chapter, I described a patient with brucellosis. I could not make such a diagnosis in an Australian now, unless he or she had acquired the infection overseas, because brucellosis was eradicated from Australia in 1986. Let us hope that other countries will soon follow suit.

35

Legionnaires' disease

SCIENTIFIC NAMES:	*Legionella pneumophila, L. longbeachae*
ORGANISM'S APPEARANCE:	Gram-negative bacillus
DISEASE NAME:	Legionnaires' disease, legionellosis
DISTRIBUTION:	many parts of the world
RESERVOIR:	water and soil
LOCATION IN A PERSON:	lungs and elsewhere
CLINICAL FEATURES:	fever, muscle aches, pneumonia
DIAGNOSIS:	isolation of the organisms from sputum
FAILING THAT:	high level of antibodies against *Legionella* in the blood
TREATMENT:	erythromycin-type antibiotics
PREVENTION:	monitoring and disinfection of air-conditioning cooling towers and commercial spas; be careful with mulch

In 1976, I was working in Cleveland, Ohio, in the USA. In June of that year, I sent off an abstract of a paper I wished to present to the American Society of Tropical Medicine and Hygiene. They were to hold their annual meeting in the following November at a hotel in Philadelphia, Pennsylvania. In late July, a dreadful outbreak of a novel form of pneumonia with many deaths occurred in Philadelphia. I was somewhat nervous and hoped that it would disappear by the time I arrived in that city (it did). What happened?

The American Legion is an organization of veterans—ex-military personnel. The Pennsylvania Chapter of the Legion held a convention from 21 to 24 July at the Bellevue-Stratford Hotel on Broad Street in Philadelphia. It was attended by 3,683 delegates and companions. Over the next week or so, 149 attendees plus another 33 people who did not attend the Convention developed pneumonia; 29 of those afflicted died. What was particularly unusual about this outbreak was the cause was not apparent and patients did not respond to the penicillin-type antibiotics that were used most commonly in the treatment of pneumonia at that time.

The USA had an organization ready to deal with such a circumstance as this—the Center for Disease Control (CDC) as it was then called. So now we reach a new chapter in the search for the causes of infectious diseases. Until this point, investigators usually worked alone or in pairs or occasionally a small group. This time a large team of clinicians, pathologists, epidemiologists, statisticians, and microbiologists attacked the problem. This is best exemplified by the 1 December 1977 issue of the *New England Journal of Medicine* in which the first detailed report from CDC was published.[102] The description of the outbreak had 12 primary authors plus a further 23 members of the Field Investigation Team. The paper describing the identification of the causative organism had six primary authors plus 35 members of the Laboratory Investigation Team. Furthermore, the Pennsylvania Department of Health undertook its own investigations. Let's begin with the epidemiologists. These are the people who investigate those who became ill, where they had been, and when they developed the disease.

The first thing that became clear was that about 80 per cent of patients had either slept in the Bellevue-Stratford Hotel or attended the Legionnaires' Convention in the hotel. Another 39 patients were classified as having 'Broad Street pneumonia'. They had simply walked in the street past the hotel (23 patients) or had been at least within one block of the hotel. In contrast to all this, only one of the 400 employees in the hotel (an air conditioner repairman) developed pneumonia. Among the delegates to the Convention, the older you were, the more likely you were to become ill, with 3.7 per cent of those aged under 40 and 12.3 per cent of those 70 years or older contracting the illness. The incubation period (the time between exposure and falling ill) ranged from 2 to 10 days. Detailed studies suggested that exposure to the infection probably occurred in or around the lobby of the hotel or in the area immediately surrounding the hotel. It seemed likely that the mode of transmission was airborne, i.e. the infectious agent was present in the air. There did not seem to be any evidence of transmission of the disease from one infected person to another. But what was the infecting agent and where did it come from? The answers were far from clear but the relative absence of the illness in hotel employees suggested they were immune and that the agent had been present, perhaps intermittently, for some time.

What were the symptoms and signs of this disease? It began with lethargy and muscle aches, then after a day or so progressed to a rapidly rising fever and shaking chills. This was associated with a cough (without much phlegm), chest pain, and shortness of breath. X-rays of the chest showed pneumonia that was not distinctive from that caused by other infections. In the majority who survived, the fever began to settle after a few days and they began coughing up phlegm. Those who were going to die became unconscious before they passed away, usually a week or so after the onset of the illness.

Contemporaneous with all of this was feverish activity trying to find the cause of the illness. Sputum samples were examined as was lung tissue from patients who had died. No bacteria were seen on the Gram stain of any samples and no bacteria grew on the usual bacterial culture media. It seemed likely that the agent would turn out to be a virus, most likely a new influenza virus, particularly as Americans had been warned that year to expect an outbreak of swine flu. All attempts to find a virus were of no avail. Nor was there any evidence of fungi, chlamydiae, rickettsiae, or a toxin.

Eventually there was a suggestive clue. Electron microscopical examination of thin sections of lungs showed thin-walled bacteria in 7 of 10 patients. But the key breakthrough was made by the CDC scientist, Joseph Edward McDade (Figure 178). McDade was asked to rule out the rickettsia, *Coxiella burneti*, the cause of Q fever, as the agent of legionnaires' disease. Using a standard technique for isolating a rickettsia he injected guinea pigs with lung tissue from patients who had died. The guinea pigs became ill and died. Numerous bacilli were found in their organs. McDade then injected some spleen tissue from these guinea pigs together with antibiotics (which would kill many contaminating bacteria but hopefully not rickettsiae) into embryonated eggs. Embryonated eggs are eggs that have been fertilized and have a developing chick within them; this is a method used in studying rickettsiae and viruses. Nothing happened. In late December 1976, McDade had another look at his guinea pig preparations and this time saw a cluster of bacilli engulfed by a white cell. A new batch of embryonated eggs were inoculated, this time without

Figure 178. Joseph McDade.

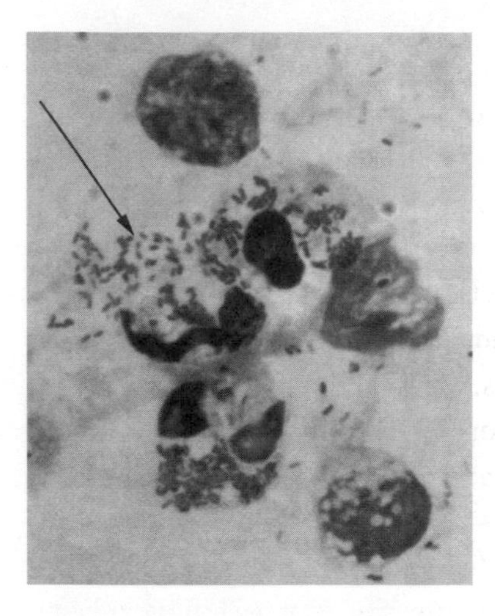

Figure 179. *Legionella* bacteria (arrow) in and around white cells in the peritoneal cells of a guinea pig. Stained with the Giménez stain.

antibiotics. Bacteria were seen in yolk sacs of the eggs (p.411), then extracts were injected into guinea pigs. The guinea pigs became ill with the same symptoms as before and the bacteria were seen once more (Figure 179). Serum from patients who had survived legionnaires' disease but not serum from other people reacted with the yolk sac bacteria. Koch's postulates had been fulfilled and the bacterium that caused legionnaires' disease had been found.

But much remained to be done. The bacteria in the yolk sacs were found to be weakly Gram-negative, i.e. they stained pink but only just. They had been too difficult to see when looking at the original lung tissues under the microscope. Most bacteria were about 2–3 µm long and 0.3 µm wide but some were much longer. Concerted attempts were then made to find an artificial medium on which the bacteria would grow. It was eventually found that if haemoglobin and a commercial mixture called Isovitalex (which contains various vitamins, amino acids, and other chemicals) were added to a commonly used medium called Mueller-Hinton medium, the bacteria would grow in the presence of 5 per cent CO_2. Since then, improved media have been developed and the colonies look as though they are made of cut glass.

The bacterium was not given a name at first, but in 1979, DJ Brenner and his colleagues erected a new genus, *Legionella,* and placed this bacterium in it as *Legionella pneumophila.* The genus name obviously reflects its discovery as a result of the outbreak of pneumonia in the American Legion. The species name means 'lung-loving', being derived from the Greek words *pneumon* and *philos* meaning 'lung' and 'loving', respectively.

Once the bacterium had been identified and characterized, it was then possible to use it to investigate unexplained outbreaks in the past from which serum from ill patients had been stored in a freezer. In 1965, there had been an outbreak in St Elizabeth's psychiatric hospital in Washington, DC, that was very similar to that seen in Philadelphia. There had been 81 cases with 12 deaths. Serum from these patients reacted with *Legionella* bacteria. In 1968, there had been an outbreak of a fever with muscle aches and minor respiratory symptoms in the county health department in Pontiac, Michigan. This 'Pontiac fever' too was found to have been caused by a *Legionella*.

What happened to the Bellevue-Stratford Hotel? It could not withstand the public hysteria and notoriety that surrounded the outbreak of legionnaires' disease. In November 1976 it closed its doors, was sold in 1978, then was refurbished and reopened in 1979 as the Fairmont Philadelphia. It went through more vicissitudes including another closure and partial conversion into offices. It is now the 'Hyatt at the Bellevue'.

The cause of legionnaires' disease had been found but where did the bacteria come from? What was their source? Environmental surveys in the USA and elsewhere showed that legionellae were widespread in nature in mud, rivers, lakes, and dams. By early 1979, legionellae had been isolated from air conditioning systems at the sites of four epidemics. It gradually became clear that evaporative air-conditioning systems were a major source of infection. Cooling towers or evaporative condensers are designed to cool water and dissipate unwanted heat to the atmosphere. They are usually located on or near the roofs of buildings such as hotels, hospitals, and office blocks. Warm water from the building's heat exchanger is dropped down from the top of the tower over splash bars and air is blown through. As the droplets pass down, they partially evaporate and cool the remaining water which collects in a pond at the base of the tower. It is in this water that legionellae are found to grow. The cool water is piped to the heat-exchanger to cool refrigerant in a different system of pipes which in turn is used to cool air entering the air-conditioning ducts. The water that is thus warmed in the heat-exchanger is returned to the top of the cooling tower. Some droplets of this contaminated water are blown out of the cooling tower and may enter the air-entry duct of the air-conditioning system if it is close by. Legionellae are thus delivered to the nice, cool area in which you are working, visiting, or living! This is not the only way you may come in contact with legionellae; they may be in spas, water misters, and water fountains as well as in water taps and shower heads. Commercial cooling towers are now routinely tested for the presence of legionellae; if they are found, the tanks are cleaned and large amounts of chlorine are added to the water in an attempt to eradicate the pathogens.

As time passed, further studies showed that not all isolates of *L. pneumophila* were equally pathogenic (capable of causing disease). They have been divided into 16 serogroups according to the antigens (shape of structures) on the surface of the bacterial walls. The Philadelphia outbreak and about 80 per cent of cases since are due to serogroup 1.

I live in South Australia, where only about half of the cases of legionellosis (as legionnaires' disease may also be called) are due to *L. pneumophila*. The rest are due to a completely different species, *L. longbeachae*. This organism was described in 1979 by Roger McKinney and 11 colleagues. It was found to cause pneumonia in four patients and was clinically indistinguishable from that caused by *L. pneumophila*. Since the first patient was in a hospital in Long Beach in California, they named the organism *L. longbeachae*.

Where does *L. longbeachae* live and how do people acquire the infection? In 1988, there was an outbreak of infection with *L. longbeachae* in South Australia with 23 cases in three months. Epidemiological investigations carried out by the South Australian Communicable Diseases Control Unit suggested that gardening was a major risk factor. Trevor W Steele (Figure 180) and his colleagues Janice Lanser and Norma Sangster from the Institute of Medical and Veterinary Science in Adelaide took up the challenge. They did not find any *L. longbeachae* in water but recovered it from a number of samples of potting mix and from soil surrounding plants in pots collected from the homes of four of the patients. I remember entering Steele's office around that time and every available space was crammed with samples of potting mix. The upshot was the finding that legionellae could persist for at least seven months in potting mixes stored at room temperature. No wonder then that *The Lancet*, when reviewing these discoveries in 1990, entitled its editorial 'Trouble in the potting shed'. It is also why the bag of potting mix in my garden shed has the following warning on it, now prescribed by law:

Figure 180. Trevor Steele (1934–). By courtesy and permission of Trevor Steele.

HEALTH WARNING

- This product contains micro-organisms
- Avoid breathing dust or mists
- Wear particulate mask if dusty
- Wear gloves and keep product moist when handling
- Wash hands immediately after use

The price of an enchanting garden or a cool room when living in a warm climate is eternal vigilance.

36

Helicobacter pylori and peptic ulcers

SCIENTIFIC NAME:	*Helicobacter pylori*
ORGANISM'S APPEARANCE:	Gram-negative spiral or curved bacillus
DISEASE NAME:	chronic, active gastritis, duodenal ulcer, gastric ulcer
DISTRIBUTION:	many parts of the world
RESERVOIR:	humans
LOCATION IN A PERSON:	stomach and duodenum
CLINICAL FEATURES:	nil or abdominal fullness or discomfort, flatulence, indigestion, soft stools
DIAGNOSIS:	isolation of the organisms from a gastric biopsy; presence of antibodies in the blood; positive breath test for the presence of gastric urease which is produced by the organism
TREATMENT:	anti-hydrochloric acid drugs + antibiotics ± bismuth
PREVENTION:	no practical measures are available

It was about 3 o'clock in the morning in May or June 1967. I was a junior resident medical officer in an operating theatre at the Royal Adelaide Hospital assisting, bleary-eyed, the senior surgical registrar repair a perforated duodenal ulcer. Bleary-eyed because I had started at 9 o'clock the previous morning and knew I would not finish until around 5 o'clock of the coming afternoon. I had seen the patient a few hours earlier. He was a man in his 50s with a long history of indigestion waking him up around 2 o'clock in the morning. which he treated with milk and antacids. This time, the pain was much more severe and was unrelieved by antacids. On examination, his abdomen was extremely tender and there was board-like rigidity of the muscles in his right upper quadrant. A chest

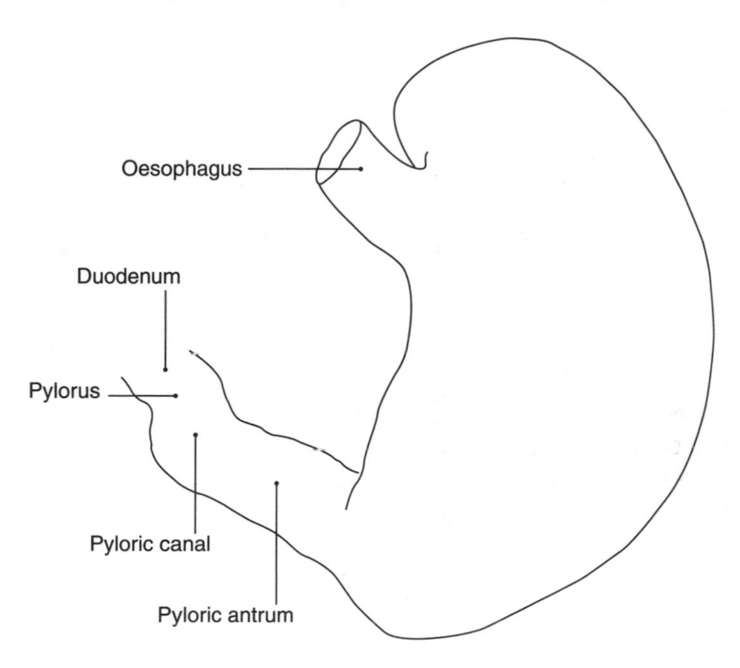

Figure 181. Anatomy of the stomach. The duodenum is the outlet to the left.

X-ray showed gas under the diaphragm, thus confirming my provisional diagnosis of a perforated duodenal ulcer. Worse, from my point of view, was that my boss decided to prolong the operation for about an hour by also performing a selective vagotomy and pyloroplasty. This involved finding the branches of the vagus nerve which supply the stomach and cutting them, then making an incision in the pyloric sphincter (valve) at the base of the stomach to relax it (Figure 181). The vagotomy impairs the production of hydrochloric acid by the stomach thus reducing the chances of a recurrence of the ulcer. Unfortunately, this operation also reduces the motility of the stomach so a pyloroplasty is necessary to assist food in leaving the stomach for the small intestine.

Peptic ulcers are generally either stomach (gastric) ulcers or duodenal ulcers (the duodenum is the first part of the small intestine, closest to the stomach) and are chronic (long-lasting). An ulcer occurs when the lining (the mucous membrane) of the stomach or duodenum is disrupted. Hydrochloric acid produced by the stomach then eats away at the wall and the hole becomes deeper and deeper and might eventually eat all the way through (a perforation). When I was a medical student, the cause of peptic ulcers was often said to be 'psychosomatic', i.e. caused by stress. This was completely wrong. No longer do people

Figure 182. Barry Marshall (1951–).

have to suffer the discomforts of indigestion due to peptic ulcers or have draconian operations such as the one I have described above. Why?

Fast forward about 35 years. Barry Marshall was sitting in my office at The Queen Elizabeth Hospital in Adelaide. I said to him 'you should win the Nobel Prize'. In 2005, Barry James Marshall (Figure 182) and John Robin Warren (1937–) did just that. How did all this come about?

In 1983, I was loitering in the University of Western Australia's department of medicine's library at the Queen Elizabeth II Medical Centre in Perth idly flicking through the 4 June issue of *The Lancet* when two sequential letters, but under the same heading, caught my eye. They emanated from the Royal Perth Hospital just a few kilometres away. The first was by Robin Warren and the second was by Barry Marshall. Warren was a consultant pathologist and Marshall was a gastroenterology registrar (a postgraduate trainee). I wondered why they had not combined the two letters to make a 'proper paper'. Perhaps they could not agree on who would be the first author? Or perhaps they thought the subject matter was so divergent that it warranted separate publications? Either way, this was the real beginning of the *Helicobacter* story.

Warren wrote that in the past three years he had examined 135 gastric biopsy specimens in which he had seen small curved or S-shaped bacilli next to the surface epithelium (lining). He noted that they were rarely present in normal stomachs. On the other hand, they were almost always present, often in large numbers, when there was inflammation of the stomach called active, chronic gastritis in which there is infiltration of white blood cells of two kinds, one called polymorphs (or granulocytes) and the other being lymphocytes. Figure 183 shows chronic active gastritis in a biopsy of my stomach taken in 1994. Warren remarked that the bacteria were difficult to see with the haematoxylin and eosin stain (which is routinely used by histopathologists to stain tissues). Actually, when you look under a very high power (which

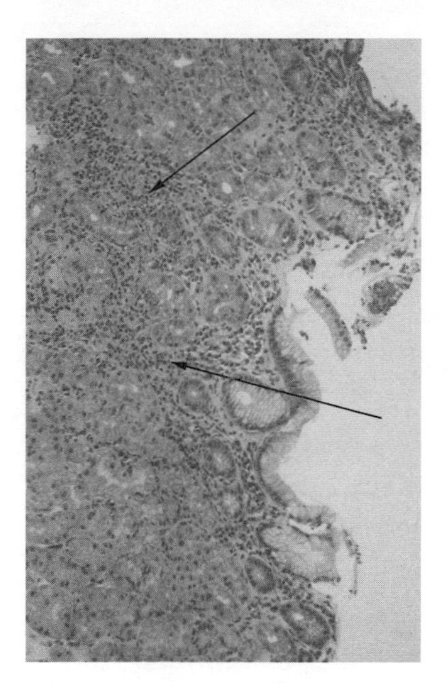

Figure 183. Biopsy of my stomach. The epithelium is on the right and the empty space is the lumen of the stomach. The arrows point to the infiltration of white cells indicating chronic, active gastritis.

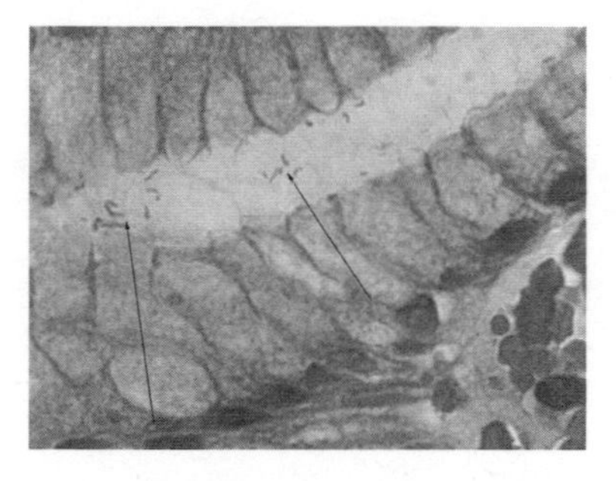

Figure 184. Biopsy of my stomach showing small curved *H. pylori* in the crypts of Lieberkühn, which open into the lumen of the stomach. They stain pink with haematoxylin and eosin.

histopathologists do not usually do) and when you know what to look for, they are quite easy to see (Figure 184). He went on to write:

> The extraordinary features of these bacteria are that they are almost unknown to clinicians and pathologists alike, that they are closely associated with granulocyte infiltration, and that they are present in about half of our routine gastric biopsy specimens in numbers large enough to see on routine histology.[316]

Warren remarked that although these bacteria resembled an organism called *Campylobacter jejuni*, it could not yet be classified. Furthermore, he hedged his bets, concluding that 'the identification and clinical significance of this bacterium remain uncertain'.[316]

The first step required to answer these questions was to isolate and grow the organism in pure culture. It was this that Marshall set out to achieve. In his quest, he was assisted by Helen Royce of the hospital's microbiology department. They cultured gastric biopsies on blood and chocolate agar and incubated them in an atmosphere that had reduced oxygen and increased CO_2 concentrations (because that is how campylobacters, which they resembled, were known to grow best). After two days in the incubator, nothing was seen and the culture plates were discarded, as is routine in microbiology laboratories, unless there is a particularly good reason for keeping them longer. Then serendipity and luck struck.

In Australia, there is a four-day break over Easter with Good Friday and Easter Monday being public holidays. The public service in Western Australia in those days was even more civilized. Easter Tuesday was also a holiday. Marshall brought in some specimens on the Thursday before Easter in 1982 and incubation was begun. Because of the holidays, the plates were not examined for six days until the following Wednesday. Tiny, faint, transparent colonies were seen which on microscopical examination were found to be made up of organisms that resembled the bacteria seen in the gastric biopsies. As Marshall later wrote:

> At first plates were discarded after 2 days but when the first positive plate was noted after it had been left in the incubator for 6 days during the Easter holiday, cultures were done for 4 days.[215]

Marshall was on his way! He described the bacteria thus:

> They are about 0.5 μm in diameter and 2.5 μm in length, appearing as short spirals with one or two wavelengths ... The bacteria have smooth coats with up to five sheathed flagellae [*sic*] arising from one end.[211]

Marshall ended his letter in a rather more upbeat fashion than Warren:

> The pathogenicity of these bacteria remains unproven but their association with polymorphonuclear infiltration in the human antrum [distal part of the stomach] is highly suspicious. If these bacteria are truly associated with antral gastritis, as described by Warren, they may have a part to play in other poorly understood, gastritis associated diseases [i.e. peptic ulcer and gastric cancer].[211]

In this he was truly prophetic.

A year later, Marshall and Warren published a joint, follow-up paper in *The Lancet*. They had now studied 100 consecutive patients who had undergone a gastroscopy. A gastroscopy is a procedure in which the stomach is looked at with a flexible optical fibrescope and which also allows small pieces of stomach wall to be collected. Flexible gastroscopy had only become routine in the previous 15 years or so. This is one explanation of why it took so long for the important discovery to be made. They saw spiral or curved bacilli lining the stomach walls of 58 patients and the organisms were grown from 11 of them. The bacteria were present in almost all patients with active chronic gastritis, duodenal ulcer, or gastric ulcer. Gastroenterologists can be a sceptical lot and when Marshall presented his findings to the Australian Gastroenterology

Association, they were met by a marked lack of enthusiasm. Microbiologists at the Second International Workshop on Campylobacter Infections in Brussels in 1983 were rather more receptive and the name 'campylobacter pyloridis' was bandied about.

Nevertheless, with the publication of the 1984 *Lancet* paper, people were going to have to take the subject seriously. There were more questions to be answered and it was to these that Marshall next applied himself, as did others whose fancy had been taken by this enigmatic organism. These included a proper taxonomic classification of the bacterium, why was it not destroyed by gastric juice, could Koch's postulates be fulfilled, did the organism cause peptic ulcers as well as chronic active gastritis, and what would happen to these diseases if the bacteria could be eradicated?

This organism was clearly important and Charles Stewart Goodwin (*c.* 1933–), the director of microbiology at Royal Perth Hospital, became interested and began to study it. In 1984, Marshall, Goodwin, the aforementioned Helen Royce, Warren, and others named the bacterium *Campylobacter pyloridis*, the species name reflecting the fact that it was most commonly isolated from the pylorus of the stomach. Their designation was rather unfortunate and in 1987 Marshall and Goodwin had to publish in the *International Journal of Systematic Bacteriology*:

> The specific epithet for *Campylobacter pyloridis* was in violation of Principle 3 and Rule 12c of the *International Code of Nomenclature of Bacteria*... Therefore, the name of this organism is revised to *Campylobacter pylori* [genitive of the noun pylorus].[214]

They had used Greek instead of Latin! The code required that Latin was used. But *Campylobacter pylori* was not going to survive either. Detailed studies showed that the bacterium was sufficiently distinctive from campylobacters to require that it be placed in its own genus. In 1989, Goodwin and seven colleagues raised a new genus, *Helicobacter*, and transferred *C. pylori* into it as *Helicobacter pylori*. There it has remained. Let Goodwin and colleagues explain why they chose this name:

> The name *Helicobacter* refers to the morphology of the organisms, which are helical in vivo but often rodlike in vitro; also, the name *Helicobacter* is euphonious and has the advantage that it is similar to the name *Campylobacter*.[118]

But did this organism really cause disease? What about Koch's postulates? In 1983, Marshall moved to Fremantle Hospital to work with David McGechie, head of the microbiology department. In 1985, Marshall, McGechie, and others published two sequential papers in the *Medical Journal of Australia*. The second described their experience at Fremantle Hospital in 1983 and 1984. Gastric biopsies were performed on 267 patients with dyspepsia (disturbed digestion); the characteristic organisms were seen on microscopy in 114 patients and the organism was grown from 88 per cent of those patients tested in 1984. The bacterium was found in 90 per cent of patients with duodenal ulcer and 68 per cent of patients with gastric ulcer. Marshall had replicated his earlier findings in a different hospital and in a different microbiology laboratory.

The first paper is even more interesting. Marshall courageously attempted to fulfil Koch's postulates. He underwent an endoscopy and showed that he had a normal stomach with no bacteria. One month later he swallowed a suspension of 1 billion cultured 'pyloric campylobacters'. Writing in the third person, he described what happened:

> On the seventh day after ingestion, he had a feeling of fullness in the epigastrium [the upper midline part of the abdomen] after taking his evening meal and felt hungry on waking early in the next morning... He had no fever but had a headache on four occasions during the second week... His faeces softened slightly... The subject was irritable and appeared ill during the second week... several colleagues observed that he had developed 'putrid breath'.[212]

Marshall had another endoscopy 10 days later and biopsy showed acute inflammation and pyloric campylobacters were seen lining the surface of the epithelial cells. He had a further endoscopy four days later which showed that the inflammation was resolving. He then took some tinidazole tablets (which had been shown in the laboratory to be effective) and 24 hours later he felt well.

This was all suggestive but not proof-positive. In an accompanying editorial, Douglas Piper, professor of medicine at the Royal North Shore Hospital in Sydney, discussed what else needed to be done:

> If the hypotheses of Marshal et al withstand the challenges forthcoming when these additional data are obtained, then their work will remain forever a landmark in our knowledge of ulcer disease.[251]

How convinced was Piper really? I know the gastroenterologists in my own department in a different hospital in Perth were rather dismissive. Did Piper encourage a then junior member of his department, Nicholas Talley, to write a somewhat disparaging letter to the same journal? Marshall, frustrated and excitable (perhaps he was still irritable from his self-infection) could contain himself no longer:

> Finally, I thank Dr Talley for his review of the epidemiology studies of gastritis. However, none of the papers he cites mentions *C. pyloridis*. This means that either God created the bacterium in Western Australia in 1982, or that the histological studies prior to that time were superficial.[213]

In 1987, Arthur Morris from Auckland in New Zealand repeated Marshall's experiment by swallowing *H. pylori* and he too developed gastritis. He made an interesting observation; the pH of the stomach changed from being very acid (1–2) to slightly alkaline (up to 7.9). This highlighted the ability of these bacteria to produce urease, an enzyme which breaks down urea to ammonium and bicarbonate thus raising the pH. This explained why gastric acid did not destroy the bacteria that were living in a neutral or alkaline micro-environment. It also provided a means for rapidly making a diagnosis. In 1987, Marshall and his colleagues showed that when gastric biopsies containing *H. pylori* were incubated in a solution of urea and a pH indicator, it was easy to show their presence by the indicator changing colour.

Nevertheless, it was never going to be possible to silence the sceptics and prove that this bacterium caused peptic ulcers by recourse to Koch's postulates because many people would have to be infected and observed for many years, it being expected that only a small proportion would develop peptic ulceration although many more would get chronic gastritis. There was a better way. Treat patients suffering from peptic ulcers with antibacterial agents. This novel idea would have seemed heretical to gastroenterologists a couple of years earlier (and it still did to many at the time).

So this is just what Marshall, who was now back at Royal Perth Hospital, and his colleagues did. Between 1985 and 1987, they investigated 100 patients who had a duodenal ulcer in a double-blind, controlled trial. This means that neither the patients nor the treating doctors knew which treatment was given. There were four treatment groups: cimetine (CIM, a drug which prevents hydrochloric acid being secreted) + tinidazole (TIN, an antibiotic), CIM + placebo (an identical-looking inert substance), bismuth (which had been shown to have antibacterial activity) + TIN, and bismuth + placebo. They reported their results in *The Lancet* in the last week of 1988. *C. pylori* was eradicated in 70 per cent of patients given bismuth + tinidazole but in hardly any of those who did not receive bismuth. What is more, after 12 months, only 20 per cent of patients in whom *C. pylori* had been eradicated had an ulcer, whereas it was still present in most patients who continued to harbour these bacteria.

The way was now open. Investigators all around the world jumped onto the bandwagon. It was soon found that even better results were achieved when combinations of antibiotics were given (this either prevented the bacteria becoming resistant or enhanced effectiveness if the organisms were already resistant to one or more antibiotics). There is now even a journal called *Helicobacter* devoted solely to the study of these bacteria. Peptic ulcers were conquered.

The twelfth-century scholar Bernard of Chartres (d.1124) in France wrote:

> We are like dwarfs standing upon the shoulders of giants and so are able to see more and see farther than the ancients.[44]

It is true that a number of observers in previous decades had seen bacteria in the stomach but knew not what to make of them. But surely the scene was set by Wye-Poh Fung, a cheery gastroenterologist I once knew, John Papadimitriou, an electron microscopist with whom I collaborated on some research papers, and Len Matz who taught me pathology as a medical student in Adelaide. In 1979, from the Royal Perth Hospital, they published a paper entitled 'Endoscopic, histological and ultrastructural correlations in chronic gastritis' in which they illustrated with an electron photomicrograph the organism we have come to know as *Helicobacter pylori*. Fung is doubtless right when he claims that this was the trigger that led Warren, who worked in the same institution, to take an interest in this beast. Yet few would have anticipated 30 years ago that a new bacterial pathogen of humans would be discovered, let alone that it causes peptic ulcers. It is to Barry Marshall's everlasting credit that he seized the idea and propelled it to fruition.

37

Typhus

SCIENTIFIC NAME:	*Rickettsia prowazekii*
ORGANISM'S APPEARANCE:	Gram-negative bacillus
DISEASE NAME:	typhus fever, gaol distemper, camp fever
DISTRIBUTION:	whenever it is introduced into an area of poverty, overcrowding, famine, or disaster
RESERVOIR:	primarily humans
TRANSMISSION:	the body louse, *Pediculus humanus corporis*
LOCATION IN A PERSON:	endothelial (inner lining) cells of blood vessels
CLINICAL FEATURES:	off-colour for a couple of days followed by sudden onset of fever and headache, stupor (impaired state of mind), coma, often a rash, cough, nausea,
DIAGNOSIS:	appearance of antibodies against *R. prowazekii* in the blood (but usually after the illness has resolved); detection of the rickettsia with the PCR test
TREATMENT:	tetracycline antibiotic, e.g. doxycycline
PREVENTION:	control lice with insecticides

Afflictions with fever, prostration, rash, and death have been with mankind for millennia. As we mentioned with typhoid fever, the Englishman Thomas Willis (p.205), in his book on fevers first published in Latin in 1659, tried unsuccessfully to differentiate the diseases we now call typhus, typhoid, plague, and malaria.[322] The first condition in this list was in fact the last for which the cause was identified, and that was not until almost 100 years ago (Figure 185).

This has not stopped many commentators from trying to ascribe the disease now properly called 'epidemic louse-borne typhus fever' as the basis for various epidemics in the past. Many of these outbreaks have been associated with armies, overcrowding, and poverty. Some have suggested that it was the cause

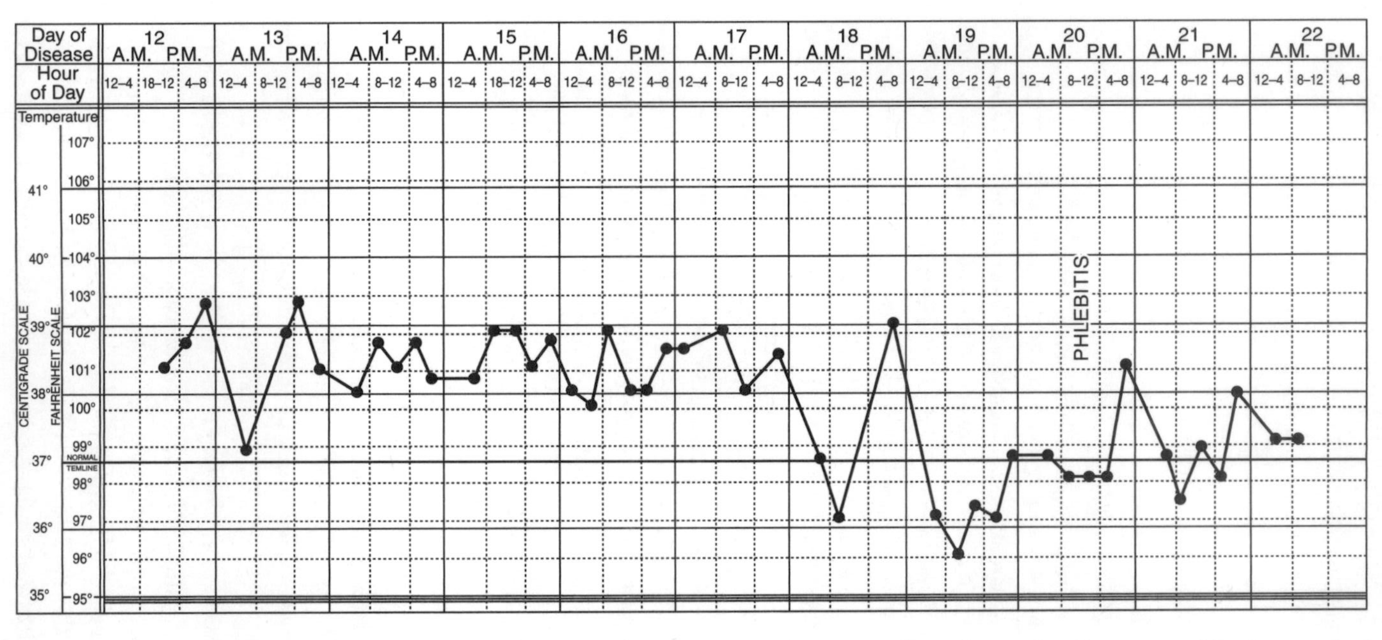

Figure 185. A typical temperature chart in typhus fever. This was of a man with severe typhus who recovered as reported by the 1920 Red Cross Commission in Poland. Temperatures are in both degrees Celsius (L) and Fahrenheit (R).

of the 'plague of Athens' in 430 BC during the Peloponnesian War but that could equally well have been due to typhoid, the plague, or something else. In smaller outbreaks, typhus has often been confused with what we now know to be meningococcal infection. Others believe that typhus was exported from the Americas after European discovery and colonization of that continent. One of the best early descriptions was by Fracastorius (p.174) in 1546:

> This fever...is contagious...but only by handling of the sick...Soon the signs of a malignant fever are revealed...(there is) weakness in the whole body and lassitude...the head becomes heavy, senses are drowsy and the mind...after the fourth or seventh day not clear, the eyes redden...the urine...afterwards red and cloudy. About the fourth and seventh day red spots appear on the arms, back and breast and often they break out like the bites of fleas...The tongue grows filthy; somnolence comes to some; wakefulness to others...Certain signs precede dying just as recovery. It is bad if the patient immediately feels himself deficient in strength...We have seen...(a patient from whom) three pints of blood burst forth from the nose and shortly afterwards died. Moreover it is bad if the urine be suppressed.[101]

Epidemics of apparent typhus recurred all over Europe in the ensuing centuries. It particularly afflicted armies including those of the Holy Roman Emperors Charles V (1552) and Maximilian II (1566), and those involved in the Thirty Years War of the seventeenth century. Napoleon's *grand armée* is said to have lost 80,000 troops to typhus and dysentery in 1812 during the Russian campaign. It was not just armies that were affected. Gaols were another focus. In England there were epidemics of typhus fever during the trials of lice-infested prisoners so the condition became known as gaol fever or gaol distemper. Typhus was abruptly reawakened during World War I (1914–18) starting in Serbia and then spreading throughout Central and Eastern Europe. It has been estimated that during 1917–25 in Russia, 25 million people contracted typhus, of whom some 3 million died. Outbreaks were likely wherever there was poverty, overcrowding, and squalor. Thus the refrain—War! Famine! Typhus! Death!

We have mentioned earlier how there was often confusion as to whether doctors were dealing with typhus or typhoid fever. We have seen that William Gerhard was the first to clearly differentiate between the two. Of prime importance in his pathological studies was the existence of lesions in the small intestine in typhoid whereas they were absent in typhus, which was characterized instead by plentiful petechiae (haemorrhages in the skin which would not whiten under pressure). The name typhus is derived from the Greek word *typhos*, which means 'hazy' or 'stupor', that is, an impaired state of mind which is one of the major features of typhus fever. *Typhos* became typhus in late Latin and the term was first used clinically by the French botanist and physician François Boissier de Sauvages de la Croix (1706–1767) in 1759.

Whatever it was called, the cause of this condition and how to control it were pressing puzzles. The Scottish physician John Pringle (1707–1782), believing that it was due to putrid odours and miasmas, urged improved ventilation. Another Scotsman, the naval surgeon James Lind (1716–1794: famous for discovering the ability of citrus fruits to prevent scurvy), thought this was a waste of time and proposed stripping and bathing together with heating of the clothing. A key observation had already been made by Tobias Cober (1570–*c.* 1625), an army surgeon from Görlitz (now the eastern-most town in Germany) in 1606 who noted a relationship between lice infestation and typhus (which he called the *Morbus Hungaricus*) in the Hungarian army during its war with Turkey. Unfortunately, this idea was not pursued.

Another 300 years were to pass before the possible role of arthropods in the genesis of typhus fever was seriously considered. In 1907, Matthew Hay (1855–1932), Scottish medical officer of health for the city of Aberdeen, put forward the hypothesis that typhus might be conveyed by vermin such as fleas. He had investigated an outbreak of 131 cases with typhus in which there were 22 deaths. He based his hypothesis on the observations that every patient had flea bites, every patient had been in contact with others with typhus who were flea-ridden, the disease did not spread in families of clean habits, and the outbreak stopped once anti-flea measures were put in place. As we shall see, Hay was on the right track but he had the wrong arthropod carrier.

But first we need to review attempts to transmit the infection more directly. In a brave but foolhardy experiment, the Russian physician Osip Osipovich Mochutkovski (1845–1903) in 1878 inoculated himself with the blood of infected patients in order to look for a bacterium. Eventually he managed to give himself clinical typhus on the eighth attempt, developing the disease after an incubation period of 17 days. In a totally reprehensible experiment, Yersin (p.370) and the French army doctor Joseph Marguerite Jean Vassal (1867–1957) in 1908 succeeded in conveying typhus fever to two Indo-Chinese subjects by subcutaneous injection of 5 ml of blood taken from patients in the second or fifth day of the disease. The incubation period in the first case was 14 and in the second 21 days. Of course, with all these experiments, we cannot be absolutely sure that the disease was really typhus because the causative agent had not yet been discovered.

The pivotal discovery was made in 1909 by Charles Nicolle (p.156) and his colleagues Charles Comte (1869–1943) and Ernest Conseil (1879–1930). Nicolle, a French physician, arrived in Tunisia (then a French colony) in 1903 and was to win the Nobel Prize in 1928 for his work on typhus. This disease was a major problem in Tunisia, recurring every winter, and Nicolle determined to investigate it. His researches did not have an auspicious start. He arranged to meet the prison doctor at a gaol 80 km south of Tunis where typhus was raging. The evening before he was due to go, Nicolle coughed up blood and aborted his visit. Nevertheless, his colleague Dr Motheau (probably René Motheau) and his servant both went. They spent one night there, contracted typhus, and promptly died. Typhus was deadly. Most of the French doctors in Tunis caught typhus at one time or another and a third of them died. It is extraordinary how many doctors and nurses looking after patients with typhus or those researching this illness succumbed to this malevolent, malign, and maleficent malady.

Nicolle made a signal observation at the local hospital in Tunis. While awaiting admission, patients with typhus often lay prostrate on the floor at the entrance. They infected their families, the doctors, administrators, and the staff responsible for taking and laundering their clothes and linen. But once they were admitted to the general wards, washed, shaved, and in fresh clothes, they did not infect the doctors and nurses. Nicolle concluded from all this:

> The contagious agent was therefore something attached to his skin and clothing, something which soap and water could remove. It could only be the louse.[230]

Nicolle, together with Comte and Conseil, then set out to prove his hypothesis by experiment. In June 1909, a chimpanzee was inoculated with blood from a patient with typhus. The animal became ill and its blood was then injected into a macaque monkey which also became ill. Two macaques were then used for the definitive experiments. They were exposed to 14 or 15 lice collected from a man with typhus fever. Exposure to the same lice was continued daily for six consecutive days in the case of one monkey and for 12 days in the other. The first monkey became ill with a fever on the 22nd day; its condition progressively worsened until its death on the 44th day. The second monkey became febrile on the 40th day, peaked on the 46th day when a rash appeared, then recovered. A post-mortem examination of the first monkey revealed only some ulceration of the caecum (the first part of the large intestine). They concluded their paper with the sage words:

> The application of this finding to the cause and the prevention of the malady in man should be made. The measures against the ravages of typhus must aim at the destruction of the parasites, especially those on the body, the body garments, the clothes and the bedding of the patients.[229]

The parasite Nicolle and his colleagues were referring to here was the ecto-parasite, the body louse, *Pediculus vestimenti* (now called *P. humanus corporis*) (p.64). Thus it was clear that not only could typhus be transmitted directly by experimental injection, but more importantly, lice transmitted the disease in nature. But they had no idea what this infectious agent was.

A clue to what it might be was also published in 1909. In the USA there is an illness called Rocky Mountain Spotted Fever (RMSF), which in many ways resembles typhus fever. In 1906, Howard Taylor Ricketts (Figure 186), based in Chicago, had travelled to Montana and shown that RMSF could be transmitted to guinea pigs by infected ticks. In 1909, he published a paper which related a new organism to RMSF. Ricketts had frequently seen bacteria-like bodies in the blood of guinea pigs and monkeys infected with RMSF as well as in some patients. He was loath to claim them as the causative agent because he could not grow them on artificial media and wondered whether they were breakdown products of white blood cells. In 1907, he had shown that ticks (identified as *Dermacentor* species) could not only infect experimental animals directly, but so could the offspring of infected female ticks. The infective agent was infecting her eggs, a phenomenon now called trans-ovarian transmission. In such eggs, Ricketts made the definitive observation when he looked at eggs just laid by a tick that had produced fatal infections in two guinea pigs:

Figure 186. Howard Ricketts (1871–1920).

> A number of eggs…were stained with Giemsa's stain. Each egg was found to be

> laden with astonishing numbers of an organism which appears typically as a bipolar staining bacillus of minute size approximating that of the influenza bacillus [*Haemophilus influenzae*]...Various forms are seen depending upon the stage of development.[270]

Ricketts then made a suspension of these organisms by crushing tick eggs (some of which contained thousands of bacteria). He found that when serum from guinea pigs that had recovered from RMSF or from infected humans was added, the bacteria clumped together, whereas they did not with serum from uninfected guinea pigs. Not unreasonably, he concluded:

> The evidence pointing to this organism as the causative agent in spotted fever, though not complete, is of a striking character...I have devised no formal name for the organism discussed but it may be referred to tentatively as the bacillus of Rocky Mountain spotted fever.[270]

Ricketts had seen the first example of an organism belonging to the group of bacteria now called rickettsiae. He went off to Mexico to study typhus there (known locally as tabardillo) and continue the work of John F Anderson and Joseph Goldberger (1874–1929). These latter researchers had just shown that monkeys and guinea pigs could be infected by inoculation with blood from patients with the disease; they abandoned their research when Goldberger went down with typhus. Ricketts and his assistant Russell Morse Wilder (1885–1952) confirmed their observations and extended them by showing that serum which had been passed through a bacterial filter was not infectious. This meant that the disease was not due to a virus. Incidentally, and confusingly, workers at this time often referred to rickettsiae as viruses; however, this was merely a synonym for a 'germ' and did not reflect the modern meaning of the word. They convincingly showed that tabardillo was completely different to RMSF. Rickets and Wilder then found bacteria that stained with Giemsa stain in the blood of patients between the seventh and twelfth days of the illness. It was 'a short bacillus which has roughly the morphology of those which belong to the "hemorrhagic septicemic group" [i.e. RMSF]'.[271] Moreover, they resembled the bacteria seen in body lice that were fed on infected patients. Ricketts and Wilder knew that this was not proof that this organism was the cause of typhus but they thought it very likely. Tragically, Ricketts died on 3 May 1910 in Mexico City from the disease he was studying.

Ricketts's observations needed to be confirmed and the scene now moves to Europe where typhus was thriving in a continent wracked by war. Stanislaus von Prowazek (p.415) and the Brazilian Henrique da Rocha-Lima (1879–1956) were sent in 1915 by the German War Ministry to investigate typhus in a prisoner-of-war camp for Russian soldiers at Cottbus in Germany. Things began well and they found organisms similar to those described by Ricketts in almost unbelievable numbers in lice in the camp, whereas they were completely absent in lice obtained from people in Hamburg who did not have typhus. Then von Prowazek died from typhus. Rocha-Lima, too, fell ill but recovered. He returned to the chase in Wloclawek (now in Poland) where he fed uninfected lice on patients with typhus. The infected lice became sick, with the abdomen swollen and red with blood, and died after a week or so. Rocha-Lima prepared histological sections of the lice and stained them with Giemsa's stain. He found that the organisms multiplied in the epithelial (lining) cells of the louse's stomach and intestine if the temperature was kept above 22°C. The cells infected with the organisms swelled and burst releasing bacteria into the gut contents. Rocha-Lima was convinced that these organisms must be the cause of typhus fever but was uncertain as to their precise nature:

> The exterior form of this micro-organism is suggestive of a bacterium but the characteristic difficulty in staining [with the usual bacterial stains], the resistance to culturing and the tendency to gather together in sharply defined parts of the protoplasm in the same manner as chlamydozoa [i.e. chlamydia] suggest rather a strongyloplasm or chlamydozoa . . . The only identifying characteristic . . . known at present is its ability to penetrate the digestive cells of the louse and there to multiply rapidly.[273]

'Strongyloplasm' is a now defunct term which was used to describe small organisms that would not stain properly with the usual bacterial stains. Rocha-Lima went on to suggest a name:

> I should like, in honour of the great researchers who have fallen as victims of typhus fever – Prowazek and Ricketts, to suggest the name *Rickettsia prowazeki*.[273]

Somewhere along the way, *prowazeki* became *prowazekii*. The latter has won the fashion stakes by a ratio of 5:1 on Google and appears to now be the official name.

There were still many unanswered questions, not the least of which were 'how does infection with this organism produce disease in man?' and 'can this organism be grown in pure culture?'. In an attempt to answer the first of these in particular, the League of Red Cross Societies sent a Medical Commission to Poland in 1920. This reborn country was being ravaged by typhus as was much of the rest of Eastern Europe. The Commission's report was published by Simeon Burt Wohlbach (1880–1954), John Lancelot Todd (1876–1949), and Francis Winslow

Palfrey (1876–1953) in 1922. They confirmed the sorts of observations we have already discussed. In particular, they affirmed the discovery of Ernst (Ernest) Maurice Fraenkel (?–1948) in 1914 that the basic pathology in typhus was a disorder of blood vessels which caused a rash very similar to that seen in meningococcal septicaemia. When examining skin lesions, he noted that they became clotted and accumulated white cells around them (later called Fraenkel's nodules). An artery that is clotted, of course, can no longer deliver oxygen and nourishment to the tissue it supplies so that tissue dies in consequence; technically, this is called necrosis and can occur anywhere in the body, but is especially dangerous in the brain. Wohlbach and his colleagues showed that this damage to the blood vessels was due to the presence of organisms in the endothelial cells (which line the inside of the blood vessels) which caused them to swell and rupture.

Then there was the question of how the lice actually transmitted infection. It was found that the lice did not transmit *R. prowazekii* directly by injection while feeding. Rather, rickettsiae excreted in a louse's faeces or released by crushing lice entered the skin through the louse bites or other abrasions or through mucous membranes such as the conjunctiva of the eyes. In fact, rickettsiae can live for up to 100 days in a dormant state in lice faeces.

What about Koch's postulates? Because these organisms could not be grown in artificial culture media, some felt the case was not proven. Wohlbach rejected such ideas, writing:

> I wish to emphasize and to insist on the importance of methods which may be employed in the face of failure to cultivate insect-borne microorganisms in artificial mediums. Properly conducted experiments in which the insect vector serves as culture tube, after natural or artificial introduction of the 'virus,' have yielded evidence fully as reliable and in my opinion less open to misconstruction than in vitro cultivation. I feel it to be a duty to challenge skepticism based on rigid adherence to Koch's postulates when dealing with insect-borne diseases.[325]

Despite significant advances being made over the years, the holy grail of growing the organisms in pure culture had not been attained. Indeed, it has still not been reached outside of cells. The reason is that somewhere along the way these organisms lost some key genetic material and rely on cells of the host to provide certain essential nutrients that allow them to grow and divide. Living organisms are grown in animals such as guinea pigs. Wohlbach and Monroe J Schlesinger (1892–1995) showed in 1923 that *R. prowazekii* could be kept alive in brain tissue cultures bathed in plasma for up to four weeks. Over the decades, improved tissue culture techniques were developed. In 1931, Alice Woodruff and Ernest Goodpasture showed that fowlpox virus grew in the chorio-allantoic membrane of chick embryos in eggs (p.411). Samuel Zia in 1934 used this system to grow *R. prowazekii* (Figure 187).

A peculiar attribute of typhus was described by Nathan Edwin Brill (1860–1925) in New York between 1898 and 1915. In many ways, the illness resembled louse-borne typhus but the mortality was exceptionally small, being less than 1 per cent. This last feature convinced Brill that it was not louse-borne typhus. In the 1930s, however, Hans Zinsser (1878–1940) showed that most patients with 'Brill's disease' had emigrated from Europe, especially from Russia, where typhus was so common. For example, Zinsser found that in Boston the disease was 15 times as common in Russian Jews as in those born in the USA. This led Zinsser to suggest, quite correctly, that these cases were due to a re-emergence of the rickettsiae which had been acquired in Europe and had remained dormant for many years. Not surprisingly, this condition became known as Brill–Zinsser disease.

For many years it was considered that typhus was exclusively a disease of humans and lice. In the 1960s and 1970s, however, the surprising finding was made that *R. prowazekii* infects the flying squirrel *Glaucomys volans* in North

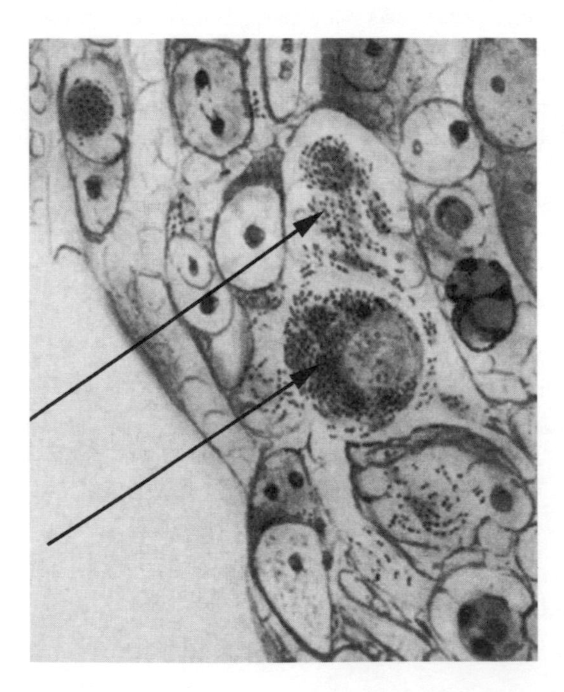

Figure 187. Illustration by Zia in 1934 of *R. prowazekii* (arrows) in the cytoplasm of cells of the chorio-allantoic membrane of a chick embryo.

America. Thus, typhus may well be a zoonosis and this discovery perhaps lends weight to the idea that the illness originated in the Americas.

So what are rickettsiae? They are prokaryotes like other bacteria, possessing both DNA and RNA. They are small Gram-negative rods. Where they differ from most other bacteria is that they are incapable of living independently in the environment. They only grow inside cells and hence are transmitted to humans (and other animals) directly by arthropods such as lice.

In the past few years, the application of modern genetic technology with the polymerase chain reaction has revolutionized the diagnosis of rickettsial infections and the identification of rickettsial species and allowed reclassification. We now recognize two species, *R. prowazekii* and *R. typhi,* which cause epidemic louse-borne typhus and endemic flea-borne typhus (murine typhus), respectively; *Orientia tsutsugamushi* causes scrub typhus in Asia and there are 10 species of *Rickettsia* which cause various types of spotted fever.

But louse-borne typhus due to *R. prowazekii* remains the grandaddy of them all. This fell disease is an ever-present threat whenever disasters in the form of earthquake, flood, famine, or man's inhumanity to man permit.

38

Chlamydia, trachoma, and urethritis

SCIENTIFIC NAME:	*Chlamydia trachomatis*
ORGANISM'S APPEARANCE:	inclusions in the cytoplasm of cells
DISEASE NAME:	a. trachoma, inclusion conjunctivitis, b. chlamydial urethritis, cervicitis, pelvic inflammatory disease; lymphogranuloma venereum
DISTRIBUTION:	worldwide
RESERVOIR:	humans
TRANSMISSION:	a. trachoma—fingers, flies, coughing, and sneezing, b. neonatal eye infections—from the birth canal during birth, c. genital infections—sexual intercourse
LOCATION IN A PERSON:	eye, genital tract
CLINICAL FEATURES:	a. trachoma—conjunctivitis, corneal scarring, b. genital—urethral discharge, abdominal pain, and infertility in women; enlarged lymph nodes in the groin (LGV)
DIAGNOSIS:	positive polymerase chain reaction test
TREATMENT:	antibiotics of the tetracycline or erythromycin groups
PREVENTION:	facial cleanliness for trachoma; avoid unprotected or promiscuous sex for genital chlamydial disease

Trachoma has probably been man's companion for much of his evolutionary history. It rose to prominence when humans first settled together in places like the fertile crescent of Mesopotamia (modern-day Iraq to Egypt) five to ten millennia ago. Experts have concluded that trachoma was described in early Egyptian, Chinese, and Indian writings in the centuries before Christ.

But what is trachoma? The name is a transliteration of *trachoma,* the Greek word for 'roughness'. Trachoma is one cause of ophthalmia, which means inflammation of the eye, and there is often uncertainty as to whether older writers were describing ophthalmia in general or trachoma more specifically. European medical practitioners were first confronted with this condition in a major way and began to differentiate it from other forms of ophthalmia during the Napoleonic Wars (1798–1815). When Napoleon invaded Egypt in 1798, many of his troops were afflicted with trachoma. The British managed to dislodge the French in 1801 and suffered in their turn. Many soldiers developed ophthalmia and 5 per cent or so became blind.

What was the cause of this condition? The French blamed noxious night vapours in the Nile delta or perhaps hot winds blowing from the south. The British, on the other hand, thought it likely that ophthalmia was contagious. Before this problem could be sorted out, it was first necessary to delineate the clinical features. Over the course of the nineteenth century, it gradually became clear that trachoma had two major phases. The first stage was characterized by active inflammation, while the hallmark of the second was scarring and its complications, although both stages sometimes could be found in the same patient. In the initial stages, the eyes become red and discharge mucus and pus as a result of conjunctivitis. Conjunctivitis is inflammation of the conjunctiva, the membrane that covers the white of the eye (the sclera) and the inside of the eyelids. But the typical feature of trachoma is the appearance of white lumps in the under surface of the upper eyelid. These are enlarged accumulations of white cells called conjunctival follicles. It is these structures that give a gritty, granular, or roughened feel and account for the name 'trachoma'. Eventually scarring develops in the inflamed eyelids and may cause distortion and buckling of the eyelids. Consequently eyelashes may rub the eye. This is particularly bad news if the lashes rub the cornea (the clear part of the eye) as this then develops opacities. Worse, the cornea may become scarred and invaded by blood vessels, a condition called pannus formation, which ultimately leads to blindness.

It was to take 50 years from the time of the discovery of the first clue as to the cause of this condition until the culprit was finally delineated. The initial breakthrough came when Ludwig Halberstaedter (1876–1949) and Stanislaus von Prowazek (Figure 188) went to Batavia (now Jakarta) in the Dutch East Indies (Indonesia) in 1907 to study experimental syphilis in monkeys with Albert Neisser (p.224). Halberstaedter, a German dermatologist and radiotherapist, was born in Silesia (now in Poland); being Jewish, he was fortunate enough to

Figure 188. Stanislaus von Prowazek (1875–1915).

be able to leave Germany in 1933 and settled in Palestine. Prowazek (after whom *Rickettsia prowazekii* was named, p.415), a microbiologist, came from Moravia, then part of the Austro-Hungarian Empire (now Czech Republic) and died in 1915 while studying typhus.

Trachoma was common in Batavia. Halberstaedter and von Prowazek became distracted from their studies on syphilis and turned their attention to trachoma. They took conjunctival secretions from patients with trachoma and infected an eye in each of a number of orang-utans. The animals developed a severe conjunctivitis but no swelling of follicles or scarring was observed. Importantly, though, the condition spread to the other eye. Furthermore, the secretions of affected orang-utans could be used to produce the disease in other orang-utans. Clearly, trachoma was infectious.

But what was the nature of the infection? Halberstaedter and von Prowazek stained scrapings of orang-utan conjunctivae with the Giemsa stain. They saw two types of bodies in the cytoplasm of epithelial cells. The first were large blue 'inclusion bodies', while the second were tiny red structures which they called 'elementary bodies' (Figure 189). They wrote:

> In the Giemsa-stained preparations, dark blue, non-homogenous irregular inclusions were visible within the epithelial cells...near the nucleus (first observed by von Prowazek). These initially small round or oval deposits gradually grew larger and assumed a mulberry-like form...(then)...underwent an increasing disaggregation which started in the centre...Within these inclusions red-stained, discrete, very fine particles appear which multiply rapidly, displacing the blue-stained masses and gradually causing their disappearance. Finally, they occupy the greater part of the cytoplasm whereas the blue-stained substances are evident only as small islands between them. In the smears one can also observe free particles beside the cells.[129]

They then found similar bodies in patients with active trachoma but not in those with old trachoma. To even see these red bodies with the old microscopes they were using was a triumph of observation as they were only about one-third of a micrometre in size. The two investigators thought that the blue bodies were made up of 'plastin' that provided a mantle or cloak surrounding the red bodies. This led them to name the cytoplasmic bodies *Chlamydozoa* from the Greek words *khlamus* and *zoon* meaning 'mantle' or 'cloak' and 'animal', respectively. Others eventually called them Halberstaedter–Prowazek (HP) bodies.

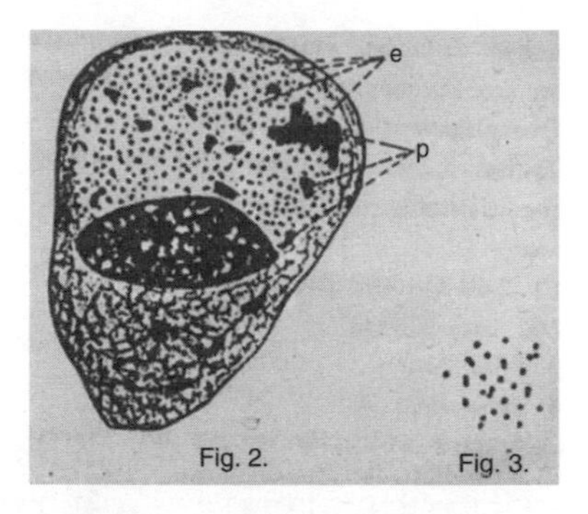

Figure 189. Halberstaedter and von Prowazek's drawings of an inclusion body which occupies the upper half of the cell. It is made up of red elementary bodies (e) and blue bodies which they called plastin. Their Fig. 3 is an illustration of free elementary bodies.

Later in 1907, Carl Richard Greeff (1862–1938), H Frosch, and Wilhelm Clausen (1878–1961) of the Prussian Commission for the Investigation of the Origin and Development of Trachoma published their observations. They had independently found minute, Gram-negative bacteria which they later concluded were identical with Halberstaedter and Prowazek's inclusions; they labelled them 'trachoma bodies'. In 1910, the Viennese ophthalmologist Karl Lindner (1883–1961) showed that plastin was really a mass of large coccoid bodies.

But the precise nature of the infectious agent was unclear. Some thought that the blue masses were infectious protozoa. Others, including von Prowazek, believed that the small red bodies were infectious protozoa. Others again, such as Greeff, believed that these bodies were bacteria; he discounted the failure to culture them as simply being due to the right culture medium having not yet been found. Even so, in reviewing the evidence in 1911, HC Solomon, a zoologist in California, wrote, 'I think that the evidence now at hand greatly favors the contention that the body is a protozoon'.[293]

It soon became apparent that these cytoplasmic bodies were not just found in trachoma but were also common in newborn babies with ophthalmia that was not due to gonococci. This was discovered in 1909 independently by Lindner, Karl Stargardt (1875–1927), and Bruno Heymann (1871–1943) in Germany as well as Halberstaedter and von Prowazek. Why newborn babies should get this 'inclusion conjunctivitis' is a subject to which we will return.

Another clue as to the nature of the cause of trachoma was provided by the observations of Charles Nicolle (p.156) and his colleagues in Tunis. In 1913, they proved that the infectious agent was very small. They passed a suspension of trachomatous material from a human through a fine Berkefeld V filter. This filtrate produced disease when inoculated into a chimpanzee and Barbary ape. Naturally, this eventually led to the agent being called a virus. Unfortunately, further attempts by various workers over the next two decades to determine whether the cause of trachoma was a filterable agent gave contradictory and inconclusive results.

Consequently, many people did not believe that HP bodies had anything to do with trachoma and continued to search for a clear-cut bacterial cause. A major red herring was promoted by Hideyo Noguchi (p.282) in 1927. He studied American Indians (now called Native Americans) with trachoma and isolated a minute bacillus that he called *Bacillus granulosis* which he used to infect monkeys. Noguchi's findings were acclaimed at first but over the next few years others could not replicate them and the idea was discarded.

In the 1930s in the USA, Phillips Thygeson (1903–2002) and his colleagues undertook what have been described as a series of hair-raising experiments in which material from patients with trachoma or inclusion conjunctivitis was inoculated into humans. All of these studies were reported in the *American Journal of Ophthalmology* and unfortunately I have not been able to view any of them to read the details. One of the studies was reported in 1935. Conjunctival scrapings were collected from 10 American Indian children with trachoma. These were pooled and passed through a filter which had pores averaging 0.6 μm in diameter. This would stop staphylococci, for example, which are 1.0 μm in diameter. The filtrate was found to be sterile when attempts were made to culture it on standard bacteriological media. Microscopical examination, however, showed that elementary bodies (whatever they were) were present in the filtrate. This was inoculated into a volunteer and it produced trachoma that progressed to scarring and pannus formation. Who was this volunteer? He was a one-eyed man named Brown who was inoculated in his only seeing eye! He was thought to be dying from a tumour in the other eye and was given $1,000 in compensation for being experimented upon. One year later, he and his wife were killed in a car crash! One commentator remarked that these studies reflected quite different ethics in research at that time. In the event, Koch's postulates were getting close to being met.

What was needed to sort all this out was to grow the organism. All kinds of tissues were tried including the chorioallantoic membrane of fertilized hens' eggs which had been used successfully for rickettsiae and viruses. They all failed. The answer was to come when the yolk sac rather than the chorioallantoic membrane of these developing chickens was used. This may have first been achieved in 1943 by the Chilean Atilio Macchiavello, who injected the yolk sac with material from a 17-year-old student and found cytoplasmic inclusions and free elementary bodies in the infected eggs, despite the frequent occurrence of contamination with bacteria. Infected material from the initial egg was then passed into another egg repeatedly. He then did the unthinkable. After nine of these passages, he experimentally infected a six-year-old child with egg yolk material. The child developed acute conjunctivitis after six days and by three weeks blood vessels were starting to grow on the cornea. The child was treated with sulphonamide antibacterial agents but took a year to become completely cured. Was it his own child? Whosever it was, it was a dreadful, unethical experiment. Macchiavello's report was written in Spanish and was published in an obscure South American journal in wartime; his observations received little recognition. Some in fact believe that he was not dealing with trachoma at all but an adenovirus (a virus that usually causes respiratory infections including conjunctivitis). We shall never know.

That recognition was duly accorded to T'ang Feifan (1897–1958) and his colleagues Chang (Zhang) Xiaolou, Huang Yuangtong, and Wang Keqian working in Beijing, China. In 1957 they successfully isolated the organism (which they thought was a virus) on three occasions from 98 samples which were injected into the yolk sacs of developing chickens together with penicillin and streptomycin (which killed contaminating bacteria but which, it was to transpire, had no effect on these organisms). They were then able to produce inclusion conjunctivitis in monkeys when the organisms were inoculated into the eyes of monkeys.

T'ang and his colleagues sent their material to colleagues in London and there was an explosion of investigations all around the world. Their results were repeated in many countries. In 1958 and 1960, Leslie H Collier (1921–2011), Stewart Duke Elder (1898–1978), and Barrie R Jones (1921–2009) reported the results of their studies with a strain which they had isolated in embryonated yolk sacs from a patient with trachoma in the Gambia, Africa. Four blind patients were inoculated and were observed for a number of months. Two developed unequivocal trachoma and the characteristic organisms were recovered once more. Koch's postulates had been fulfilled.

What was this beast? Collier and his colleagues as well as many others were still calling it a virus. It was clearly not a virus because it was soon shown that sulphonamides and tetracycline antibiotics, which have no effect on viruses, were active against these organisms. It should be a bacterium. In 1966, Israel Sarov (1934–1989), Yechiel Becker, and HP Bernkoff in Israel showed that these organisms had both DNA and RNA. As we shall see in the next chapter, this ruled out them being viruses. They must be unusual bacteria which can only multiply inside cells. They are now classified as 'obligate intracellular bacteria'. They are Gram-negative but unlike other Gram-negative bacteria do not have any peptidoglycan in their walls, as was shown when various workers in the 1950s published electron photomicrographs of these bodies (Figure 190).

We now understand their unique life cycle (Figure 191). They have an extracellular infectious form called the elementary body (EB) and an intracellular form which divides and multiplies called the reticulate body (RB). An EB, about 0.35 μm in diameter, attaches to a cell wall and is then incorporated into the cytoplasm, being surrounded by a broken-off sphere of the cell membrane in

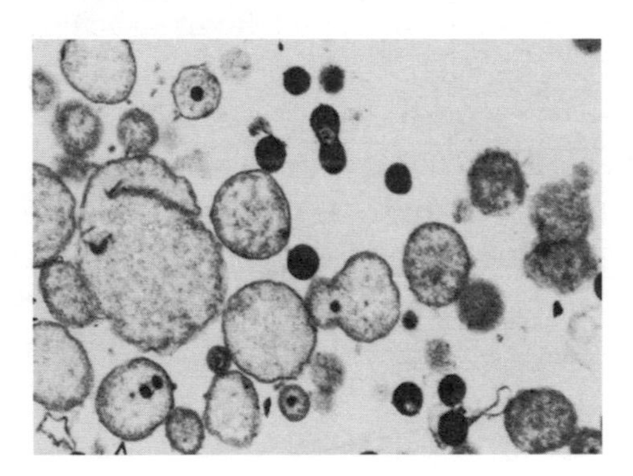

Figure 190. Electron photomicrograph of *C. trachomatis* inside an inclusion body within an epithelial cell. The large, open forms are reticulate bodies, while the small, dense forms are elementary bodies. The organisms with a central dark spot looking like an archer's target are intermediate forms.

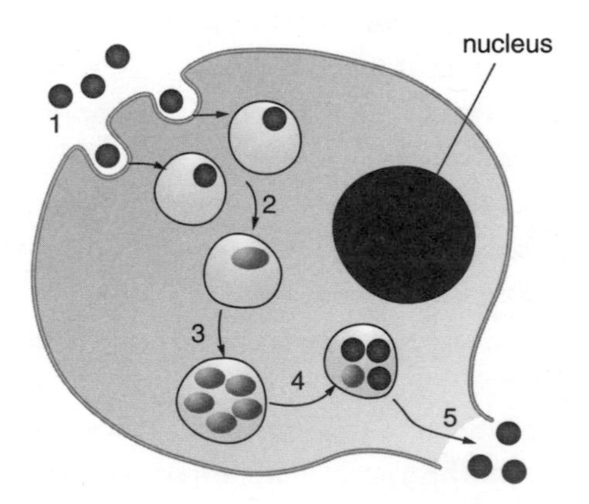

Figure 191. The life cycle of *C. trachomatis*. Elementary bodies (EBs) (1) adhere to the cell wall and are then taken into the cell and become located in vacuoles (spherical containers) in the cytoplasm. EBs change into open, large reticulate bodies (RBs) (2) which multiply (3). These then condense into EBs (4) which are released from the epithelial cell (5).

the process. This then changes into a larger RB which is about 1 μm in diameter. This divides repeatedly to make the inclusion bigger and bigger until it almost fills the cytoplasm of the cell. Clumps of these RBs are probably what Halberstaedter and von Prowazek described as blue plastin inclusion bodies. Eventually each RB condenses into multiple EBs which are released into the environment ready to infect another cell.

And what were these organisms to be called? In 1930, three different groups of workers in different countries had found similar cytoplasmic inclusions in the lung cells of people who had contracted pneumonia from parrots (psittacosis). The organisms found in psittacosis, trachoma, and inclusion conjunctivitis were eventually considered to be members of a group of atypical viruses. They were classified for a while as belonging to a new genus *Bedsonia*. This did not last and the organisms causing eye disease were often called TRIC agents (for Trachoma Inclusion Conjunctivitis). In 1935, Archimede Busacca (1893–1971) named them *Rickettsia trachomae*, which was changed two years later to *Rickettsia trachomatis*. In 1945, SD Moshkovski, following the original name of Halberstaedter and von Prowazek, called them *Chlamydozoon trachomatis*. In 1955, the International Committee on Zoological Nomenclature decided, for a variety of reasons, that this name was no good. In 1957, Geoffrey Rake (p.288) transferred Busacca's *R. trachomatis* into the genus *Chlamydia* as *C. trachomatis*.

We have already mentioned that it was found early on that newborn babies with conjunctivitis often had inclusions in the cytoplasm of their conjunctival cells which were the same as or similar to those seen in trachoma. The logical thought was that they probably acquired them from their mothers' genital tract during birth. In 1909, the aforementioned Karl Lindner took some secretions from the vagina of a mother whose baby had inclusion conjunctivitis and applied them to the eyes of a macaque monkey and baboon. They too developed conjunctivitis and in their conjunctival cells he found numerous inclusions. In the following year, Halberstaedter and von Prowazek found cytoplasmic inclusions in the cells taken with a swab from the genitalia of a woman.

If women had these bodies in their genital tract, where did they come from? Probably men. In that same year of 1910, Lindner found them in epithelial cells from the urethra of men who had urethritis (discharge of pus from the urethra) not due to gonococci. In fact in 1910, Bruno Heymann (whom we mentioned earlier), found inclusions in cells from the conjunctivae of two twins with inclusion conjunctivitis, from their mother's cervix, and from the urethra of the father. Later that year, H Fritsch, A Hoffstaedter, and Lindner produced conjunctivitis with follicles typical of trachoma in monkeys from urethral discharges from men, from genital secretions of mothers with infected babies, and from infants with inclusion conjunctivitis. It was all coming together. But more than half a century would pass until precisely what was going on was understood.

During this half-century, when a doctor was confronted with a man with a discharge from the penis (urethritis) or a woman with pelvic inflammatory disease, he or she usually thought of gonorrhoea because when a pathogen was found it was almost always the gonococcus. On those occasions when gonococci were not found, this was generally put down to a failure on the part of the laboratory, especially as it was known that these bacteria were very difficult to grow. When penicillin became available in the 1940s, almost all gonococci at that time were susceptible and penicillin cured those patients in whom gonococci had been found. It quickly became clear, however, that there were many men with urethritis in whom gonococci could not be found and whose symptoms were not cured with penicillin. Logically enough, the Americans called this 'non-gonococcal urethritis', while the British rather more vaguely named it 'non-specific urethritis' because a specific cause could not be found.

The story of the answer to this enigma really begins when Eric Dunlop from the Whitechapel Clinic (for Venereal Diseases) of the London Hospital, together with Barrie Jones and Khalaf Al-Hussaini from the Institute of Ophthalmology in London, published three papers in 1964. They took specimens from the urethras of fathers and the cervices of women who had babies with inclusion conjunctivitis. In some of them they found typical inclusions and in a lesser number they were able to grow the organism which they called a TRIC virus. They concluded:

> It is clear that in a selected group of cases so-called 'non-specific' genital infection is due to TRIC virus. But what relation this group has to the whole field of non-specific genital infection remains to be established.[90]

The following year, they reported their studies of nine men with urethritis. In three cases they found TRIC agent. Four of their female sexual partners were examined. In one woman, the organism was found in her cervix.

By the middle of the 1970s, it was clear that non-specific urethritis was the commonest sexually transmitted disease in Britain. In 1977, MD Alani and colleagues from St Bartholomew's Hospital and the Institute of Ophthalmology, both in London, reported their extensive investigations of 509 heterosexual men with urethritis, 108 of their female contacts, as well as 61 men who did not have urethritis who acted as a control group for comparison. *Chlamydia* was found in two-thirds of men with a marked discharge of pus from the urethra, in the cervix of one-quarter of their female partners, but in only 3 per cent of men without a history of urethritis. Many investigators replicated their findings. You will not be surprised to know that Koch's postulates have never been fulfilled and doubtless never will be. Who is going to volunteer to contract urethritis or pelvic inflammatory disease from a swab taken from a chicken embryo growing chlamydiae?

We now know that there are 15 serovars (antigen groups) of *C. trachomatis*. Serovars A, B, and C cause trachoma. Serovars D to K are associated with sexually transmitted genital diseases and inclusion conjunctivitis. The 15th serovar, L, causes a specific clinical syndrome called lymphogranuloma venereum. Enormous progress has been made in many parts of the world in eradicating or at least controlling trachoma, which is often transmitted by fingers, flies, coughing, and sneezing. Less success has been achieved with chlamydial genital infections. Even though these organisms are sensitive to the tetracycline group of antibiotics, chlamydia remains the commonest sexually transmitted disease in the developed world. In 2011, over 1.4 million cases were reported in the USA. Since many patients are not tested for the infection and others are not aware of any symptoms, it is believed that the true number of cases in that country is double that in people between the ages of 14 and 49 years. It just goes to show that finding the cause of a disease and the development of effective treatments do not always lead to a dramatic control of that infection.

PART VII

VIRUSES

What is a virus? Colloquially, we often refer to viruses as germs or say 'I've got the bug', especially when there is a respiratory epidemic going the rounds. On the other hand, how often have you heard the media report on the 'meningococcus virus' or the 'anthrax virus'? Clearly these reporters don't know what a virus is because those organisms are bacteria.

Nevertheless, one cannot really blame the media for getting it wrong. It took the medical profession many years to understand what a virus is. 'Virus' is a Latin word which meant 'poison' or 'venom'. For six centuries or more, European physicians and surgeons have sometimes used the word to refer to pus discharging from a wound. This gradually evolved into the idea that a virus was a substance produced within the body by a disease, especially when the disease was contagious. Note that in this concept, the diseased body produced the virus rather than vice versa. When the idea of spontaneous generation was finally dispelled with the discovery of bacteria and the development of the germ theory of disease, the word 'virus' often became applied to any infectious agent. Indeed, in 1890, Louis Pasteur wrote 'Tout virus est un microbe' (every virus is a microbe), meaning that the words virus and microbe were synonymous. As we saw on page 1, 'microbe' is a word that had been coined as a means to describe an extremely small living organism and to get around the problem of whether they were animals or plants. The term is now considered to include viruses rather than being a synonym for a virus.

As we shall see in the next chapter, it was found shortly afterwards that some infectious agents could not be seen under the microscope; they were 'ultramicroscopic'. Furthermore, they were so small that they passed through filters that were capable of blocking the passage of bacteria which are about 1 μm or more in size. Over the next 50 years, the term 'virus' was progressively restricted to these 'filterable viruses'. We now know that viruses are infectious organisms which are not capable of independent existence. They must propagate themselves inside animal or plant cells using the machinery of those cells to replicate nucleic acids and make proteins and other chemicals. Unlike all the infectious organisms we have discussed thus far, they do not have DNA and

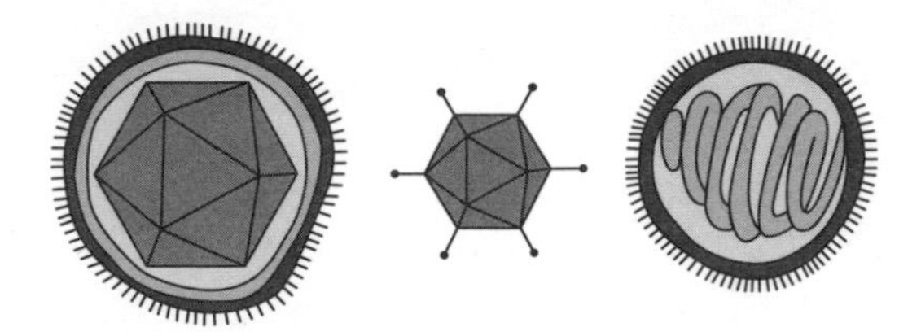

Figure 192. Schematic diagrams of viruses *Top*: chickenpox virus—enveloped, icosahedral. *Middle*: Adenovirus—naked, icosahedral. *Bottom*: Influenza virus—enveloped, helical.

RNA; they only have one or the other. This provides one way of classifying viruses—into DNA viruses or RNA viruses.

Viral nucleic acid is surrounded by a protective protein shell called a capsid. It is made up of spherical subunits called capsomeres. The shape of the capsid provides a second means of classifying viruses (Figure 192). Some viruses have a capsid made up of 20 identical, equilateral triangles joined side by side together. Geometrically this makes an icosahedron (Greek *eikosi* = 'twenty' and *hedron* = 'seat') so these viruses are said to be icosahedral. The other major group of viruses is composed of helical viruses. They are filamentous or variably shaped with a helical array of capsid proteins wrapped like a coil around a helical filament of nucleic acid. The DNA or RNA plus the protein shell is called a nucleocapsid. Unlike all other organisms, viruses do not have a nucleus or cytoplasm.

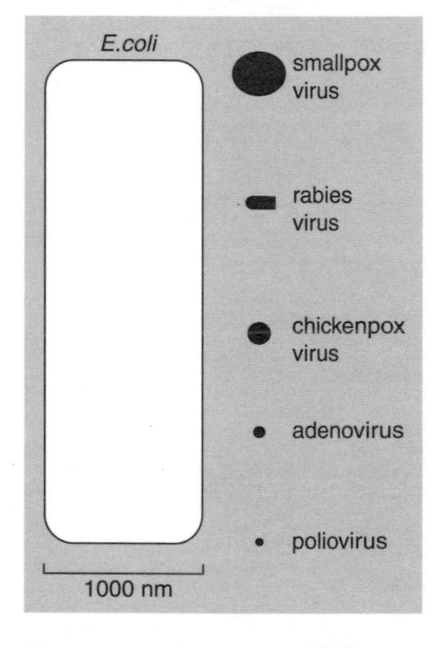

Figure 193. Relative sizes of different viruses.

A third way of classifying viruses is by whether or not they have a fatty envelope surrounding the capsid. All human helical viruses have an envelope but icosahedral viruses may be either enveloped or naked. This has practical value because viruses with an envelope, such as influenza virus, are relatively easily destroyed by alcohol disinfectants which dissolve the envelope. On the other hand, viruses without an envelope, such as poliovirus, are much more resistant to destruction by alcoholic disinfectants.

The fourth major feature useful in classifying viruses is their size (Figure 193). Poliovirus, for example, is one of the smallest viruses infecting humans, being about 27 nanometres (0.027 micrometres = μm) in diameter (a micrometre is one-thousandth of a millimetre). Near the other end of the range is the smallpox virus which is about 220 nm in size.

There are a number of other features that virologists use to classify viruses. There is in fact no agreed system of classification of viruses and several methods have been proposed, the details of which need not concern us. Let us now look at how viruses were found and then consider some of the viral infections that have been so important in human history.

39

The discovery of viruses and determination of their nature

When bacteria were discovered and the germ theory of disease took hold in the second half of the nineteenth century, investigators naturally looked for specific bacteria that might be causing each of the various infectious diseases. Sometimes they were successful. In other illnesses, no matter how hard they tried, they could not find any bacteria. Yet many of these illnesses clearly were infectious. What was the nature of these mysterious infectious agents?

Over the course of the last 100 years or more, a number of technical advances progressively allowed the definition and identification of viruses that were shown to be the causes of various clinical syndromes. But the appearance of these techniques was preceded by the seminal observations by William Jenner on smallpox and by Louis Pasteur on rabies.

Vaccination (or immunization)

As we shall see in the next chapter, William Jenner showed in 1798 that he could use fluid taken from cowpox lesions to prevent subsequent infection with the deadly smallpox. There was something in cowpox fluid that caused the human body to become resistant to smallpox. Yet it was later found that no bacteria were present in either cowpox or smallpox lesions. Similarly, Louis Pasteur in 1885 was able to take an extract of spinal cord tissue from a dog with rabies and inject it into rabbits. As we shall find, he was able to modify whatever was happening in the rabbit spinal cords and use this material to prevent infection in humans bitten by rabid dogs. Again no bacteria were found.

Neither Jenner nor Pasteur had any idea of the nature of the agents with which they were dealing. Nevertheless, there was clearly something or some things that could not be seen which could be used to generate resistance to these human infections.

Filtration

Around 1870, Edwin Klebs (p.296), then working in Switzerland, had an idea which would later have momentous consequences in the hands of others. Klebs was keen to find a method for separating bacteria from the fluids in which they were suspended. He asked his assistants Friedrich Wilhelm Zahn (1845–1904) and Ernst Tiegel to see if they could make porous filters out of white, unbaked clay and try them out. When they were connected to an air pump to hasten the movement of fluid through the filters, the system worked.

In 1877, Louis Pasteur and the physicist Jules Joubert (p.187) adopted the same principles but used a plaster of Paris filter instead of clay and connected it to a vacuum pump. By 1884, Pasteur's colleague Charles Chamberland (p.187) had developed a quite sophisticated system based on porcelain; porcelain is made from kaolin, but if it is heated insufficiently it retains some porosity.

In 1891, Hermann Nordtmeyer introduced some filters made from diatomaceous earth obtained from the Berkefeld mines in Germany, where the groundwater had been observed to be especially clear. Diatomaceous earth is derived from the fossilized remains of diatoms, microscopic algae that have shells made of silicon dioxide. These Berkefeld filters became commercially available in different pore sizes so they could filter out bacteria or other organisms of various dimensions. In 1907, Samuel Lawrence Bigelow (1870–1947) and Adelaide Gemberling (1880–1970) made filters out of collodion (a syrupy solution of nitrocellulose in ether and alcohol). In the same year, Heinrich Bechhold (1866–1937) showed that he could make a range of collodion filters with graded permeabilities which would allow the relative sizes of pathogens to be determined.

The filters made by Chamberland and Nordtmeyer were shaped like columns so they were called filter candles. Contrary to all expectations, when these filters were used to remove bacteria, it was found that some infectious agents were able to pass through the pores of the filters. As it happened, this phenomenon was seen first, not in a human or animal disease, but in a plant disease. We need to sidetrack to this story.

Tobacco plants were brought from the Americas to Europe in the sixteenth century and grown as crops. In the middle of the nineteenth century, a rapidly spreading disease appeared on these tobacco plants, especially in the Netherlands and Russia. The leaves developed mottled patterns of light and dark green and often became wrinkled. Adolf Eduard Mayer (1843–1942), a German working in Holland, began to investigate the problem in 1879. He showed it was infectious by collecting sap from diseased plants in capillary tubes and then pricking the leaves and stems of normal plants with it. They developed the same disease and Mayer in 1880 coined the term 'tobacco mosaic disease'. But he could not find any bacteria.

In 1884, Dimitri Iosifovich Ivanovski (Figure 194) in Russia took sap from diseased plants and passed it through Chamberland filters and found that the filtrate was still infectious. He did not publish his observations until 1892, but when he did he failed to interpret them correctly and believed that the disease was due to either a toxin or a bacterium that he had been unable to isolate and culture:

Figure 194. Dimitri Ivanovski (1864–1920).

> It seems to me that this can be explained most simply by the presence of a toxin elaborated by the bacteria and dissolved in the filtered sap. Of course there is still another equally acceptable explanation, namely, that the bacteria of the tobacco plant passed through the pores of the Chamberland filter candle.[151]

In 1898, a Dutch scientist and colleague of Mayer, Martinus Willem Beijerinck (1851–1931), reported his studies on the disease. Like Ivanovski, Beijerinck showed that the infectious agent was filterable and passed through porcelain filters. He then showed that it multiplied in infected tissues so it could not be a toxin (which would become progressively diluted). What is more, this mysterious agent would only multiply in living, preferably growing tissues. Beijerinck concluded that it was not a microbe (a *contagium fixum*) but a non-cellular entity which he called a *contagium vivum fluidum* (soluble living germ) and which he also called a virus. He believed that this virus was incorporated into the metabolism of the tobacco plant cells:

> The virus, even though not able to grow by itself, is carried in the growth of dividing cells where it multiplies enormously, without losing any of its individuality.[41]

Beijerinck was remarkably prescient and went on to say: 'I consider it highly probable that many other non-parasitic diseases of unknown cause may be ascribed to *contagium fluidum*.'[41]

At almost the same time, support for the concept of a filterable infectious agent was to come from investigation of an animal disease. In 1897 the German government appointed a commission to study outbreaks of foot and mouth disease that were causing terrible afflictions in the country's cattle. The illness was characterized by fever with large blisters in the mouth and on the hoofs and mammary glands. Many animals died from dehydration and starvation because drinking and eating were so painful. The commissioners were Friedrich Loeffler (p.194) and Paul Frosch (p.334). They failed to find bacteria in blister fluid but when they inoculated calves with blister fluid, the animals developed the disease. They then passed the fluid through a filter and inoculated more animals; the same thing happened. By a series of experiments they showed that this was unlikely to be due to a toxin because it could not be diluted out. Something had passed through the filters which had the ability to multiply in the infected animals. They reported:

> The agents must be so small that they are able to pass through the pores of a filter which will certainly retain even the smallest bacteria. The smallest bacteria hitherto known (have)... a length (of) 0.5–1.0μ. If the supposed agents of foot and mouth disease were only one-tenth of this size... then they would be unrecognisable even by the best of modern oil-immersion systems [i.e. light microscopes].[192]

So as the twentieth century dawned, it was realized that some infections were caused by a group of agents that were referred to as 'invisible' or 'filterable viruses'. Nevertheless, Beijerinck and Loeffler and Frosch had widely divergent ideas of the nature of these agents. Beijerinck thought the agent of tobacco mosaic disease was a non-cellular molecule that dissolved in the sap of an infected plant. The two Germans believed that the agent of foot and mouth disease was simply a bacterium that was extremely small. Decades would pass before the issue would be resolved.

Cell culture and chick embryo culture

The problem with these filterable viruses was that, unlike bacteria, they could not be seen and they could not be cultured on artificial media. Attention then turned to trying to grow them in living cells. During the first three decades of the twentieth century, attempts were made to grow poliomyelitis and rabies viruses in nerve cells, and smallpox and cowpox viruses in rabbit corneas and testes and in pulped chicken embryos. Nothing seemed to work terribly well.

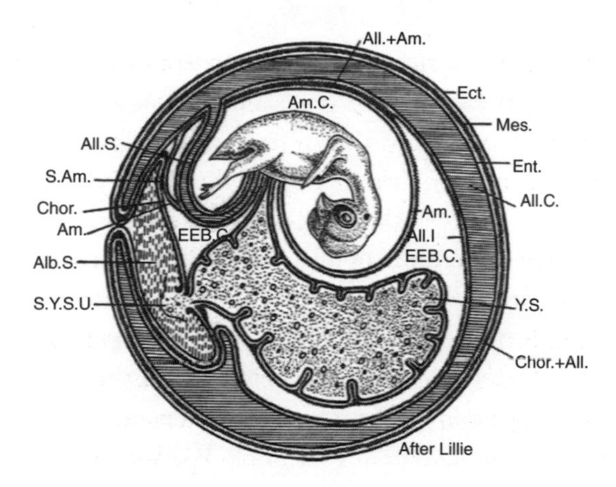

Figure 195. Woodruff and Goodpasture's reproduction of the drawing of a chicken embryo first published in 1908 in Frank Lillie's *The Development of the Chick*. 'Chor + All' (bottom right) points to the chorioallantoic membrane. YS = yolk sac. In *The Development of the Chick*, FR Lillie, Henry Holt & Co., New York, and George Bell & Sons, London, 1908.

Then in 1931 came a breakthrough. Alice Miles Woodruff and Ernest William Goodpasture (1886–1960) in Tennessee in the USA reported that fowlpox virus would grow in sheets of uniform cells on the chorioallantoic membrane of the whole developing chick embryo inside the fertilized egg shell (Figure 195) much more effectively than they would in pulped chick embryo cells. This provided a cheap and efficient experimental system and was used for studying various viruses as well as certain bacteria over the next few decades.

Cytopathic effect and plaque assays

But tissue cultures were not forgotten. Eventually a technique was devised which produced single layers of cells (monolayers). This allowed the effects of virus on these cells to be observed under the microscope. Various changes including alteration in the shape of cells, the appearance of abnormal bodies (inclusion bodies) within the cells, fusion of cells to each other, or complete breakdown of cells were seen. This phenomenon, called cytopathic, or cytopathogenic, effect provided a diagnostic system which permitted the detection of viruses in specimens.

Furthermore, it provided a means for counting the number of viruses present. In 1952 Renato Dulbecco (1914–2012), an Italian working in California, introduced his plaque technique. He used a monolayer of cells in tissue culture and exposed them to material containing virus. Wherever a virus was present, the cells developed a plaque or clear spot where the virus had killed the cells. The number of plaques could be counted and thus the numbers of virus particles determined.

Immunology

But suppose you could demonstrate the presence of a virus by cytopathic effect in a plaque assay. How could you be sure which virus you were dealing with? This is where immunology came into its own. When an animal or a human suffered a particular virus infection, the immune system responded. In many instances an antibody (protein in the blood) was made which reacted specifically with that virus and no other. If this antibody were injected into another animal prior to exposure, it might be protected from infection. Or, if you add serum containing antibody directed against that virus to a plaque assay, the antibody binds to the surface of the virus and may prevent it from infecting host cells; consequently, no plaques appear. Because it neutralized the effects of each virus, these antibodies were called *neutralizing antibodies.*

In modern laboratories, antibodies of known identity are often used to identify viruses in automated machines. Conversely, viral particles are widely used to identify antibodies in the patient's blood, thus making a diagnosis of a specific viral infection.

Electron microscopy

It is one thing to be able to infer the presence of a virus by filtration experiments or even by observing cytopathic effects or measuring the immune response. It would be another matter altogether to be able to see them. Ordinary light microscopes were of no use. Something more powerful was needed. In 1931, the German physicists and engineers Ernst Ruska (1906–1988) and Max Knoll (1887–1969) built a prototype electron microscope. In 1933 they constructed one which had a greater magnifying power than that of a light microscope. An electron microscope accelerates electrons in a vacuum, then the beams of these fast-moving particles are focused on the material being examined. The electrons are then absorbed or scattered and form an image on a photographic plate. Microscopes were continually improved and in 1938, Bodo von Borries (1905–1956) and his brothers-in-law, Ernst and Helmut Ruska (1908–1973), published photographs of the mousepox virus. This was the first virus to be visualized. Pandora's box was about to open.

Biochemistry

While all these advances were being made, discussion was continuing on the nature of viruses. Were they soluble molecules as Beijerinck thought? Were they living very small particles as Loeffler and Frosch supposed? The first clue was provided by Wendell Meredith Stanley (1904–1971) in Princeton, New Jersey, USA. He announced in 1935 that he had been able to crystallize the tobacco mosaic virus using techniques used to crystallize proteins. He found that the crystallized virus was much more infectious than diseased tobacco leaves. Since the virus was crystallized, Stanley thought that the virus must be a protein and wrote:

> Tobacco mosaic virus is regarded as an autocatalytic [i.e. stimulates itself] protein, which for the present, may be assumed to require the presence of living cells for multiplication.[294]

Stanley's report posed a big problem: how could proteins be alive? But Stanley was to turn out to be only partly right.

In 1937, Frederick Charles Bawden (1908–1972) and Norman Wingate Pirie (1907–1997) in Cambridge, England, drew different conclusions. They thought that tobacco mosaic viruses were crystalline substances which contained not only proteins but also sulphur, phosphorus, and carbohydrate indicating the presence of nucleic acids: 'The purified virus nucleic acid resembles yeast nucleic acid closely; it contains a pentose and does not give the reactions... of a desoxy pentose.'[40] In other words, the virus contained RNA—ribonucleic acid—in addition to protein; they were nucleoproteins. In 1943, Gerhard Schramm (1910–1969) in Germany demonstrated that these viruses lost their infectivity if the nucleic acid was removed.

It was eventually shown that viruses infecting humans contained either deoxyribonucleic acid (DNA) or ribonucleic acid (RNA). Of course in those early days, no-one was exactly clear what RNA and DNA did or what their structures looked like.

X-ray crystallography

This technique had been discovered by a father and son team, William Henry Bragg (1862–1942) and William Lawrence Bragg (1890–1971; born and educated in Adelaide), working in Cambridge, England. The younger Bragg found that when X-rays were passed through a crystal, they bent and created a pattern on radiographic film that showed the atomic structure of a crystal. Bragg the elder perfected a piece of equipment called an X-ray spectrometer which they could use to conduct their X-ray crystallography experiments. They were jointly awarded the Nobel Prize in 1915.

In the early 1950s, Rosalind Franklin (1920–1958) in Britain studied the appearances of DNA using X-ray crystallography. James Dewey Watson (1928–) and Francis Harry Crick (1916–2004), working in Cambridge, England, used one of her photographs to deduce a structure for DNA. In 1953, without giving proper recognition to Franklin, they proposed that DNA was made up of two complementary strands of deoxyribonucleic acid that were joined to each other and together were wound in the form of a double helix or spiral staircase. RNA turned out to be basically the same (Figure 196). The key components of viruses were now understood. They contained the building blocks of life—DNA or RNA—so this settled the question of whether they were living or were just chemicals. They were living, but could only live and reproduce inside another living cell.

Molecular biology

Many of the techniques we have discussed were cumbersome and time-consuming. In the past quarter-century, molecular biology has revolutionized virology. Molecular biology is a discipline that is chiefly concerned with the way in which cells work. In particular, it studies the way in which DNA, RNA, and proteins interact with each other and are synthesized. Perhaps the most useful molecular biological technique for virology is the phenomenon called the polymerase chain reaction, commonly called a 'PCR'.

In 1957, Arthur Kornberg (1918–2007) in the USA found an enzyme in *Escherichia coli* which stimulated the copying of DNA and replicating it. This was called DNA polymerase. But for this to work, other proteins were required to open the DNA helix described by Watson and Crick, keep it open, start the copying process (these were called primers), to remove the primers when synthesis was completed, and finally to tie the pieces together. In the 1960s, Har Gobind Khorana (1922–2011), an Indian working in the USA, pioneered a technique to make small runs or sequences of nucleic acids called oligonucleotides. Oligonucleotides with specific sequences could be used as primers for DNA polymerase. Kornberg was awarded the Nobel Prize in 1959 and Khorana received it in 1968. In the following year, Thomas Dale Brock (1926–) isolated a new bacterium called *Thermus aquaticus* from a hot spring in Yellowstone National Park in the USA. This contained a DNA polymerase enzyme that could withstand much higher temperatures than that obtained from *E. coli*. This

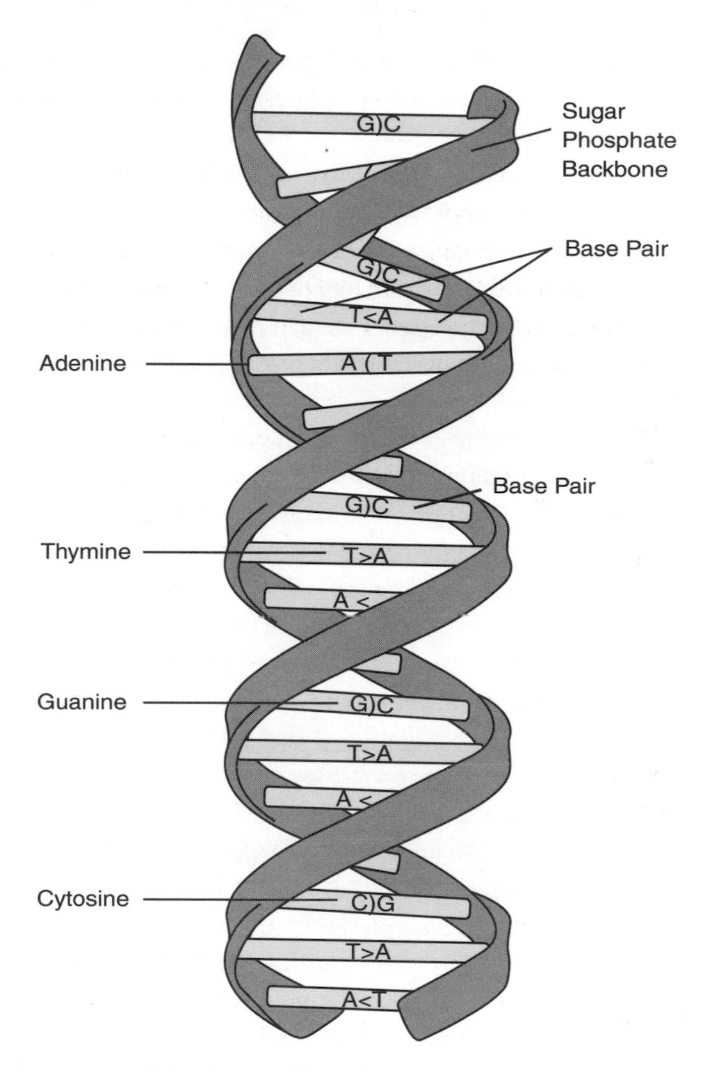

Figure 196. Diagram of the DNA spiral staircase. A (adenine) always joins to T (thymine), while C (cytosine) always joins to G (guanine). RNA is similar except that thymine is replaced by uracil and deoxyribose is replaced by ribose.

meant that the chemical reactions could be undertaken at higher temperatures, which in turn meant that the reactions would go much faster. In 1977, Frederick Sanger (1918–) in England described a method for determining the sequence of nucleic acids. For the second time, he was awarded the Nobel Prize in 1980.

Most of the building blocks were now in place to develop a method for repeatedly copying strings of DNA. Kary Banks Mullis (1944–) and his colleagues at Cetus Corporation in the USA in 1985 described the 'polymerase chain reaction'. Mullis conceived of a way to start and stop a polymerase's action at specific points along a single strand of DNA. He then realized that this target DNA could be replicated (or amplified) exponentially by repeatedly heating and cooling so that one copy became 2, which became 4, 8, 16, 32, 64, 128, 256, 512, 1024, 2048, 4096, 8192, and so on until there were millions of copies of the DNA strand in several hours. This produced plenty of material which then could be analysed and the nucleic acid sequence determined. Mullis received the Nobel Prize for his work in 1993. This avalanche of Nobel Prizes illustrates how important all this work was.

The DNA polymerase chain reaction took the virological world by storm. As early as 1987, donated blood was being screened for the presence of the human immunodeficiency virus. This technique, or variations of it, provided the tools that had been missing for identifying and understanding viruses. Their nucleic acid sequences (called the genome), whether they were DNA or RNA, could be fingerprinted exactly. So powerful was this tool that by 1989, hepatitis C virus, which had never been grown in cell culture or seen by electron microscopy, was discovered and identified by this molecular biological method. Furthermore, these techniques allowed the production of viral probes. These are stretches of DNA or RNA that are able to latch on to a sequence of DNA or RNA in a virus that is unique for that virus. If the probes are labelled with something such as fluorescein (a dye which shines under an ultraviolet microscope) or a radioisotope (which emits radioactivity which can be picked up by a photographic plate), then a virus can be shown to be present in a sample and identified.

And of course it was not just virology that was affected. We have all watched TV crime dramas where DNA evidence found the criminal; this is all courtesy of the polymerase chain reaction. Pathogenic viruses and criminals have something else in common: we don't want them.

40

Smallpox (variola)

SCIENTIFIC NAME:	variola virus, a member of the genus *Orthopoxvirus*
DISEASE NAME:	smallpox = variola major or variola minor
DISTRIBUTION:	now eradicated
TRANSMISSION:	direct contact with entry of virus probably via the respiratory tract
CLINICAL FEATURES:	fever, pustular rash with a peripheral distribution
DIAGNOSIS:	clinical recognition
TREATMENT:	no specific treatment; fluid replacement
PREVENTION:	a vaccine is available but is no longer necessary

I made a mistake. At Easter 1964, I was planning to go canoeing on the Glenelg River in Victoria. Instead I spent the break lying semi-delirious on my bed. My mistake? I had a smallpox vaccination 10 days earlier. I was one of those who had a severe reaction. As we shall see, smallpox vaccination really means immunization with vaccinia virus. I had this vaccination, of course, to reduce my chances of contracting smallpox, a devastating illness to which any medical student at the time could unexpectedly have been exposed. In a wonderful turn of events, the disease (although not the virus) has vanished from the face of the earth and very few are immunized any more. How did this happen?

Smallpox only affects humans. Where and when it began is shrouded in mystery but it seems to have afflicted the human race for two or three millennia. At the beginning of the Christian era, smallpox was probably endemic in the Nile and Ganges river valleys. There is considerable uncertainty but one view has it that smallpox spread to China in the second century and reached Japan in the tenth. Smallpox followed the march of Islam into the Mediterranean littoral in the eighth century, then spread northwards into Europe. From the fifteenth century onwards, European sailors, explorers, and colonists took this malevolent illness with them to the uttermost parts of the earth.

The term 'smallpox' is derived from the Latin 'pocca' meaning a 'bag' or 'pouch'. This became pockes in Anglo-Saxon and later 'the pox', referring either to the pustules (blisters containing pus) typical of the disease or to ulcers. With the appearance of syphilis in Europe in the late fifteenth century, it became necessary to call it smallpox to distinguish it from the great pox (or ulcer) of syphilis, first appearing in print as 'small pokkes' in a letter by the cleric and diplomat Richard Pace to Cardinal Wolsey in 1518.

What causes smallpox? No-one knew. Rhazes (865–925), a Persian physician in Baghdad in the early tenth century, thought that smallpox arose when the blood putrefied and fermented so that excess vapours were expelled from it. It would take nearly a millennium to find out what was really going on. Until the nineteenth century, smallpox was regarded as a uniformly severe disease with about one in three or four patients dying. The illness began about 12 days after exposure, with the afflicted person developing a high fever and complaining of a splitting headache and severe lassitude while many vomited. Within two or three days, red spots rather like measles appeared on the skin, which rapidly developed into fluid-filled blisters (vesicles), which then turned into pus (pustules). More and more lesions developed and large areas of skin became involved. Physicians recognized that the lesions of smallpox were mostly on the face and the arms and legs, whereas the similar lesions of chickenpox were much more prevalent on the trunk. Over the next week or so the pustules dried into scabs which then fell off, often leaving scars. Other patients, though, developed haemorrhages, became delirious, and died.

It was eventually realized that the people who had survived smallpox never got smallpox again. Then it was noticed that people who were accidentally scratched on skin contaminated with pustular fluid usually developed a less severe form of the disease. Two and two were put together so that about 1,000 years ago in India, smallpox was induced by deliberately scratching pustular fluid or the dried scabs from patients into the skin. Around about the same time, the same process was evolved in China except that the material was introduced into the nostrils. The Indian technique, at first called inoculation and later variolation, spread to the Middle East in the thirteenth century and then to Turkey.

Lady Montague (1689–1742), wife of the British ambassador to Turkey, came to hear of this procedure. She had been severely pockmarked by an attack of smallpox in 1715 and was anxious her children should not suffer the same fate. Her daughter in 1721 became the first person to be professionally inoculated in

England. She persuaded the Princess of Wales to authorize an experiment in which six inmates of Newgate Prison who were condemned to death would be pardoned if they agreed to be inoculated. They did and they survived—later on, one even slept in the bed of a smallpox victim with no ill effects. The idea caught on and a memorial to Lady Montague was erected in Lichfield Cathedral in England in 1789:

> Sacred to the memory of the right honourable Lady Mary Wortley Montague, who happily introduced from Turkey into this country, the salutary art of inoculating the smallpox. Convinced of its efficacy, she first tried it with success on her own children and then recommended the practice to her fellow citizens. Thus, by her advice, we have softened the virulence and escaped the danger of this malignant disease.[10]

Hyperbole perhaps, as she by no means discovered the procedure, but nevertheless she did popularize it in her own country to great benefit.

There were disadvantages, however. One per cent or so of those inoculated became so ill that they died. Furthermore, these patients could infect other people who developed normal smallpox with its usual high mortality. In the late eighteenth century, a belief arose among certain English country folk that milkmaids did not get smallpox because they had developed pustules on their fingers while milking cows that had a blistering rash on their teats (cowpox). The person who took hold of this idea, investigated it, and brought it to fruition was Edward Jenner (Figure 197), a general practitioner in Gloucestershire, England. On 14 May 1796, Jenner took fluid from a lesion on a finger of a milkmaid, Sarah Nelmes, and inserted 'a little Vaccine Virus' into the arm of an eight-year-old lad named James Phipps, the son of Jenner's gardener. He induced a lesion on the boy's arm similar to that seen with smallpox inoculation but the child suffered barely any general side effects. Seven weeks later, Jenner challenged the boy with smallpox inoculation to see what would happen. He wrote to his friend Gardner:

Figure 197. Edward Jenner (1743–1829).

> But now listen to the most delightful part of my story. The boy has since been inoculated with smallpox, which, as I ventured to predict, produced no effect. I shall now pursue my experiments with redoubled ardour.[153]

Jenner then went on to repeat the experiment on a number of children with similar success. After some initial scepticism, he was able to persuade his colleagues to copy the procedure. Hundreds did so and came to the same conclusion. With superb confidence Jenner wrote in 1801:

> It now becomes too manifest to admit of controversy, that the annihilation of the smallpox, the most dreadful scourge of the human species, must be the final result of this practice.[154]

We might think this prediction is over the top. The British House of Commons did not: they voted him 10,000 pounds. In 1978 Jenner would be proven right.

This is not to say that things necessarily went smoothly. After Jenner's publication, Benjamin Jesty, a farmer of Dorset, claimed that he had inoculated his family with cowpox in 1774. But he never challenged them with smallpox, never published his findings, and his claims were much disputed.

Furthermore, cowpox vaccine was on occasion contaminated with smallpox, as happened when William Woodville (1752–1805) at the London Smallpox and Inoculation Hospital first tried to repeat Jenner's technique. At that time, there were often shortages of cowpox vaccine and sometimes horsepox vaccine was substituted. What is more, a specimen was often taken from a lesion on a recently vaccinated person and another person immediately vaccinated with it. The physicians of the day did not realize that this provided an excellent means of transferring hepatitis, syphilis, and other infectious diseases. Moreover, theological (it was interfering with the will of God) and philosophical objections were sometimes raised and the introduction of compulsory vaccination to Britain in the mid-nineteenth century produced an outcry. Nevertheless, the benefits were undeniable.

In his first publication in 1798, Jenner coined the term 'variolae vaccinae' meaning 'smallpox of the cow' (Latin: 'vacca' = 'cow') and talked of inoculating the 'vaccine virus'. In 1881, Pasteur generalized the use of the term 'vaccination' to include prevention with all kinds of infectious agents. When it became apparent in the second half of the twentieth century that these processes were working by stimulating the immune system, the term 'vaccination' gradually fell out of favour and was replaced with 'immunization'.

Neither Jenner nor Pasteur nor anyone else had any idea what this 'vaccine virus' was. When bacteria were recognized to cause disease, cowpox lesions and smallpox lesions were examined assiduously for bacteria but none could definitely be seen and certainly not cultivated. However, smallpox virus is among the biggest of the viruses (Figure 193); at about 250 nm in size, it is a quarter that of a staphylococcus. This is right at the limit of resolution of an optical light microscope.

In 1887, John Buist (1846–1915), an Edinburgh physician, described minute bodies in vaccine fluid which had been stained with gentian violet and thought these must be the causative agent. In 1905, von Prowazek (p.415) found that the particles could be stained with Giemsa stain and called them 'elementary bodies', a view with which Enrique Paschen (1860–1936) in Germany in 1906 so ardently concurred that they were often called 'Paschen bodies'. In that same year, Adelchi Negri (p.447) in Italy passed vaccine fluid through a filter that held back bacteria and found that the fluid was still infective; they were indeed 'filterable viruses'.

All of these studies were undertaken with material used for vaccination. What exactly was the virus in them? Jenner's original 'variolae vaccinae' was cowpox virus. In 1938, Allan Watt Downie (1901–1988) in Britain discovered that the vaccine then in use consisted not of cowpox virus but of a very similar virus now called vaccinia virus. The origins of this latter virus are obscure but for practical purposes it did not matter much since both cowpox virus and vaccinia virus induce immunity in humans to smallpox.

In 1932, Goodpasture and his colleagues in the USA grew the virus we now call vaccinia on the chorioallantoic membrane of chick embryos. In 1944, EA North and his colleagues in Australia did the same thing for smallpox virus. In 1940, Charles Hoagland and his colleagues in New York showed that vaccinia virus contained DNA but not RNA, while in 1948, Frederick P Nagler and Geoffrey Rake (p.288) used the electron microscope to observe the smallpox virus (Figure 198). An understanding of the nature of the virus that caused smallpox was gradually being compiled.

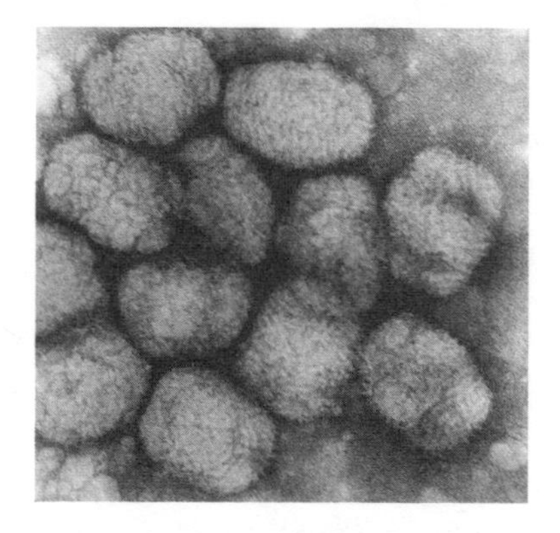

Figure 198. Electron microscopic appearance of variola virus.

You will have noticed that most of these studies were undertaken with vaccinia virus rather than smallpox virus because the latter is much more dangerous. Let's stop and recapitulate on these three viruses—cowpox, vaccinia, and smallpox. They are all closely related but nevertheless have important differences. Coxpox virus and vaccinia virus both infect a broad range of animals in nature; the former was found predominantly in Europe but the latter occurred worldwide. Smallpox virus, on the other hand, only infects humans in nature. Its technical name is variola virus. That is because an alternate medical name for smallpox is 'variola'. 'Variola' is a medieval Latin word meaning 'pustule' and is in turn derived from the Latin 'varius' meaning 'speckled' or, less likely, from 'varus' indicating a 'pimple'.

One more thing. In 1824, the Scottish physician John Thomson published an account of an epidemic of a mild form of smallpox which he called varioloid. Similar outbreaks were described by Izzett Anderson who named it varioloid varicella in Jamaica in 1865 and WE de Korté in South Africa in 1904 called it maasa or kaffir milk-pox. Yet others called it alastrim (derived from a Portuguese word), and later still it became known as variola minor to contrast it with variola major, the severe form of smallpox. This form of smallpox had a mortality rate of only about 1 per cent.

Around 1960, the World Health Organization (WHO) began, initially in a very desultory way, a campaign to eradicate smallpox based on isolation of cases and immunization of people who came in contact with them. Despite widespread scepticism that this would work, it was surprisingly and dramatically successful, with one country after another becoming free of smallpox. By late 1977, Somalia was the only country left still infected. On 22 October 1977 in Merca, Somalia, Ali Maow Maalin went down with smallpox. He was isolated and all those with whom he came in contact were quarantined and vaccinated. There were no more cases. On 17 April 1978, the WHO field office in Kenya sent a telegram to Geneva:

> Search complete. No cases discovered. Ali Maow Maalin is the world's last known smallpox case.[25]

He was the world's last naturally acquired case. There would be one more—a laboratory worker became infected and tragically died. Stocks of variola virus were then moved to just two centres—one in Moscow, Russia, and the other in Atlanta, USA. Medical science and human cooperation had triumphed. For the first and so far the only time in history, one of humanity's most dreadful scourges has been eradicated.

41

Rabies (hydrophobia)

SCIENTIFIC NAME:	rabies virus, a member of the genus *Lyssavirus*
DISEASE NAME:	rabies
COMMON DISEASE NAME:	hydrophobia
DISTRIBUTION:	many parts of the world, especially Asia and Africa
TRANSMISSION:	bite of an infected animal, usually a dog but also other mammals including monkeys and bats
CLINICAL FEATURES:	a. furious rabies (involving the brain)—muscle spasms, terror often provoked by attempts to drink water (hydrophobia), hallucinations, coma, and death b. paralytic rabies (involving the spinal cord)—paralysis, beginning in the bitten limb
DIAGNOSIS:	clinical suspicion; biopsy of the brain
TREATMENT:	no specific treatment; sedation, pain relief, and intravenous fluids replacement
PREVENTION:	avoid rabid animals, especially dogs; immunization before or immediately after a bite

I have a photograph taken years ago of my eldest son when he was 15 months old in front of the Taj Mahal in India communing with a monkey about his own size. I thought nothing of it at the time. Now I shudder whenever I think of it. In 1987, a 10-year-old boy was the first person to die from rabies in Australia. He had been bitten eight months earlier by a monkey in India. It was a salutary reminder that this foul disease is not just caught from the bites of rabid dogs.

Why did I call rabies a foul disease? I have had a number of young Indian doctors working with me over the years. One of them had worked in a hospital for the poor in India and he told me how they managed a person with rabies there.

They put him or her in a room, shut the door, and locked it. I was shocked. Yet, on reflection, I can see the logic. Once a person has the signs of rabies, it is a death sentence. What else can you do with such a patient when there is no effective treatment, money is scarce, and the patients are a danger to those around them because in their agitation or delirium they might bite someone with their infectious saliva? What would we do in a Western hospital? Probably put sufferers in an intensive care unit, anaesthetize them, relax their muscles, and ventilate them with a machine until they died!

Rabies in humans has been known for millennia as has its relationship with sick dogs. The first written record comes from the twenty-third century BC. A tablet found at Eshunna in ancient Sumeria (now Tell Asmar in Iraq) was inscribed with cuneiform text which has been translated as:

> If a dog is mad and the authorities have brought the fact to the knowledge of its owner; if he does not keep it in, and it bites a man and causes his death, then the owner shall pay two-thirds of a mine of silver.[152]

Of course, dog bites can cause bacterial infections which were sometimes fatal within days in the pre-antibiotic era. But the operative words here are 'a dog is mad' meaning it was behaving in an irrational and dangerous manner. Indeed, the modern French word for rabies is 'la rage', or as we would say in English 'rage', to reflect this behaviour.

In the fourth century BC, Aristotle in Greece recognized that mad dogs could transfer this state to other animals:

> Rabies drives the animal mad, and any animal whatever, excepting man, will take the disease if bitten by a mad dog so afflicted; the disease is fatal to the dog itself and to any animal it may bite, man excepted.[28]

Strangely, Aristotle thought that humans were unaffected by the disease. Galen (p.203) in the second century AD did not agree for he said that humours of mad dogs became so corrupted that if their saliva merely fell upon a man, that person would become rabid.

The ancient Greek playwright Homer (*c.* twelfth century BC) may have had the sufferings of rabies in mind when describing the torment of Tantalus in Hades, who was unable to drink water. This brings us to another name for the disease, 'hydrophobia', derived from the Greek words *hydros* and *phobia* meaning 'water' and 'horror'. The awfulness of this disease is described in graphic detail by the *New York Times* of 8 March 1872 in an article headed 'The Horrors of Hydrophobia—Six Children Bitten by a Rabid Dog in Brooklyn'.

> On the 9th of February Julia Connolly eleven years of age... while returning home... was bitten on the cheek and upper lip by a rabid dog... It was hoped the girl would escape an attack of hydrophobia but on Monday last she was taken ill, and physicians... informed (her parents)... that their daughter was suffering from rabies. Since that time she has rejected all feed and drink... She continually froths at the mouth and suffers great agony. She seems to be conscious, however, and expressed a constant desire to go to her mother... Julia was yesterday placed in an arm chair and so violent were her writhings that it required several persons to hold her down. The physicians say she cannot recover and death may occur at any moment.[14]

Even worse was a farmer in Indiana, USA in 1891:

> He became irrational, frothed at the mouth and became so violent that it was found necessary to bind him to the bed with ropes, but this did not prevent the madman from throwing his head wildly and snapping his teeth with savage vigour that made even strong men in the room shudder. All day Sunday and that night these spells continued until at seven o'clock Monday morning death removed the unfortunate victim from more suffering.[17]

He had been bitten by his pet spaniel.

All this brings us to the origin of the word 'rabies'. It is derived from the Latin word 'rabere', which means to rage, rave, be mad, or be savage. It well describes the disease in both people and dogs.

Although it was long recognized that rabies appeared to be transmitted by the bites of mad dogs, there was less certainty about precisely what was being transmitted. One widespread myth was that rabies was caused by a small 'worm' at the base of the tongue and its removal would cure the disease. In 1804, Georg Gottfried Zinke (?–1813) in Jena, Germany, formally demonstrated that rabies was transmissible. He brushed saliva from an infected dog into incisions made on the foreleg of a young daschund. A week later it became sick, refused food and water, and crawled into its cage. By the tenth day, the signs of rabies were obvious. In 1813, the pioneering neurologist François Magendie (1783–1855) and the anatomist Gilbert Breschet (1784–1845) in France performed an experiment which was the reverse of the usual state of affairs. They inoculated a dog with saliva taken from a human with rabies and induced rabies in the dog.

There were no visible worms in the saliva and so the nature of whatever was causing the disease was unclear. When it was realized that bacteria were the causes of some diseases, Louis Pasteur (p.182) in France naturally looked for a bacterial cause of rabies. He examined the saliva of a boy with rabies and observed pairs of coccoid bacteria (probably pneumococci). When he injected the saliva into rabbits, they died and Pasteur recovered bacteria with the same appearance from the rabbit's blood. These were in turn injected into other rabbits to once more cause disease, but the disease was nothing like rabies. The mystery remained.

There were other questions. Where did this mysterious agent in saliva come from and where did it go after a bite. During the nineteenth century, a number of investigators showed that rabies could be transmitted by injecting saliva into veins, the eyes, the nasal passages, and the nerves. In 1879, the veterinarian Pierre Victor Galtier (1846–1908) in Lyon, France, found that rabbits injected with saliva from a rabid dog became ill within a week but that blood from these rabbits was not infectious when injected into other rabbits. It must have gone somewhere else. These experiments with dogs, as you can imagine, were very dangerous. Rabbits, on the other hand, lacking ferocious teeth, were much less so, especially as rabid rabbits became languid rather than furious.

Galtier's work aroused the interest of Louis Pasteur. Pasteur transferred rabies repeatedly from rabbit to rabbit (technically called 'passage') and found that rabies was invariably induced after one week. In fact, he did this 90 times. He then noticed that if he took the spinal cords from rabbits that died from rabies and then hung them up to dry for a couple of weeks, the spinal cord material would not induce disease. This process is called attenuation. Even more importantly, if an animal which had been injected with this altered spinal cord preparation was then challenged several weeks later with fresh rabies material, it was protected from developing rabies. Using this technique, Pasteur successfully immunized 50 dogs. Pasteur had been able to induce a change which made the active agent (whatever it was) behave like Jenner's cowpox in protecting against smallpox.

Then came the celebrated case of Joseph Meister. Pasteur's laboratory studies had convinced him that he had found a way of preventing death from rabies in both animals and humans. Meister was a nine-year-old boy from Meissengott in Alsace who had been bitten repeatedly by a rabid dog on 4 July 1885. In desperation, his mother brought him to Pasteur two days later. Pasteur was not a medical practitioner and consulted Drs Alfred Vulpian (p.300) and Jacques-Joseph Grancher (1843–1907) of the Académie de Médecine. With their encouragement, the boy was injected with some emulsion from the spinal cord of a rabbit that had died of rabies on 21 June and had been kept in dry air for 15 days. He was then given 14 inoculations over 10 days with portions of the cord that were progressively fresher (and thus more virulent which caused Pasteur some trepidation). Four months later he remained well and was pronounced cured.

In October, Jean-Baptiste Jupille, a 15-year-old shepherd, was also successfully treated. By the end of 1886, 350 patients from all over Europe, Russia, and America had been treated. Rabies was preventable if treatment was started early enough! A statue commemorating Jupille and the dog that bit him remains at the original Institut Pasteur (now Musée Pasteur) in Paris to this day.

Even so, the nature of the rabies agent was still obscure. In 1903, Paul Ambroise Remlinger, a Frenchman working in Turkey, showed that the agent of rabies could be passed through a filter candle that retained bacteria—it was indeed a filterable virus. In the same year, the Italian pathologist Aldechi Negri (1876–1912) in Pavia discovered inclusions in brain cells of humans and animals with rabies (Figure 199); these were later called 'Negri bodies'. Negri thought they were protozoa. Remlinger thought they were filterable viruses. They could not both be right and argument went on for some time.

In 1936, Leslie Tillotson Webster (1894–1943) and Anna D Clow in the USA grew the agent of rabies in tissue cultures made from the brains of embryonic mice and chickens. However, the major advances in understanding the nature of the rabies agent came in the 1960s. In 1963, Seiichi Matsumoto, a Japanese working in Baltimore, used the electron microscope to find the characteristic bullet shape of rabies viruses in nerve cells in infected mouse brains. In 1965, Kaneatsu Miyamoto and Matsumoto in Japan showed that Negri bodies consisted of masses of excess material produced during the synthesis of new rabies

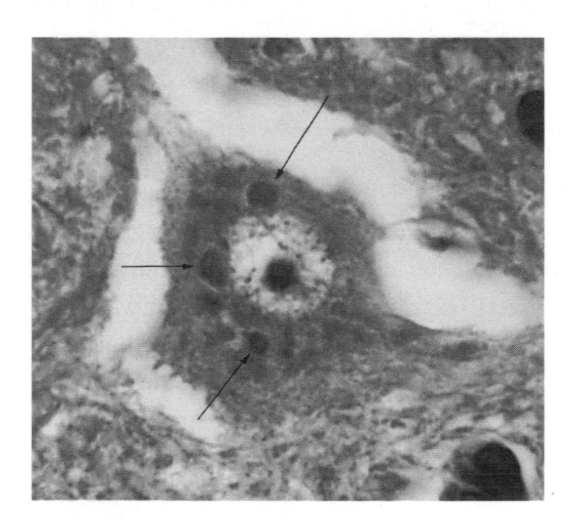

Figure 199. Negri bodies (arrows) in the cytoplasm of a nerve cell in the brain.

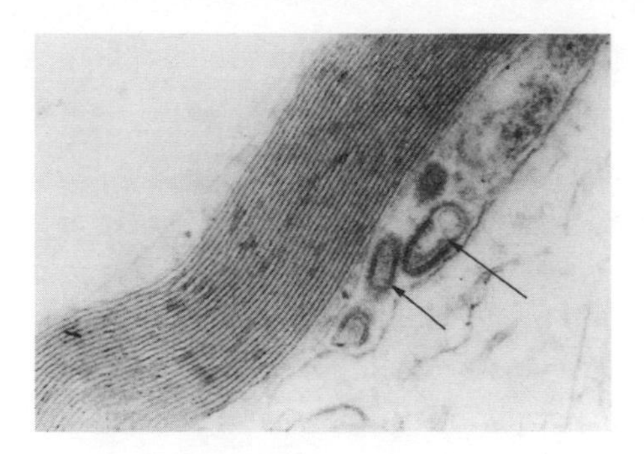

Figure 200. Bullet-shaped rabies viruses (arrows) travelling along a nerve fibre.

viruses. In 1968, Frantisek Sokol and colleagues showed that purified rabies virus contained RNA.

There remained an important question: how does the virus get from the site of the bite to the brain and spinal cord? As early as 1881, Roux had suggested that the rabies agent travelled from the entry wound to the brain and spinal cord via the nerves themselves rather than in the bloodstream. Nearly 100 years were to pass before he was shown to be correct by experiments in which limbs were amputated and nerves were cut in experimentally infected animals. What is more, rabies virus was observed travelling in nerve fibres (Figure 200).

We now know that all warm-blooded animals are susceptible to rabies and that bites of humans by any infected animal may transmit the disease. Of these, by far the most important are dogs, followed by monkeys and bats. Dogs usually become irritable and aggressive; this is known as 'furious rabies' and is very dangerous. Sometimes they become listless and apathetic, known as 'dumb rabies'.

We began with India. Let us finish there. It has been estimated that a person is bitten by an animal in that country every two seconds, just over 90 per cent of those bites being from dogs. Someone dies from rabies in India every 30 minutes. If you do the maths, it means that about 1 in 1,000 bites are from rabid animals and that roughly 20,000 people die from rabies every year. And of course rabies is a major problem in many other countries besides India. There is a long way to go to control, let alone eradicate, this disease.

42

Yellow fever

SCIENTIFIC NAME:	yellow fever virus, a member of the genus *Flavivirus*
DISEASE NAME:	yellow fever
DISTRIBUTION:	parts of northern South America and tropical Africa
TRANSMISSION:	a. urban—bites of humans by infected mosquitoes, most commonly *Aedes aegypti* b. jungle (sylvatic)—bites of monkeys by infected mosquitoes of the genus *Aedes*
CLINICAL FEATURES:	fever, muscle aches, yellow eyes and skin, haemorrhages, and kidney failure
DIAGNOSIS:	clinical sometimes supported by finding antibodies to yellow fever virus in the blood
TREATMENT:	no specific treatment; fluid replacement
PREVENTION:	an effective vaccine is available; avoid mosquitoes

In 1793, Philadelphia was capital of the fledgling, newly independent United States of America. In June, it had a population of nearly 40,000. Three months later, 4,000 or so of its inhabitants were dead and many of the remainder including President George Washington and his cabinet had fled. What had happened?

In July, large numbers of refugees arrived in Philadelphia from Santo Domingo in the Caribbean fleeing revolution and pestilence. They brought with them the dreaded yellow fever. Benjamin Rush (1746–1813), a physician in Philadelphia, had seen this condition before. In his account of the illness, which he called 'bilious remitting yellow fever', he recounted the description he had recorded while an apprentice during the epidemic of 1762:

> The patients were generally seized with rigors, which were succeeded with a violent fever, and pains in the head and back. The pulse was full and sometimes irregular. The eyes were inflamed and had a yellowish cast and vomiting almost always attended. The 3rd, 5th, and 7th days were mostly critical, and the disease generally terminated on one of them, in life and death. An eruption (of the skin) on the 3rd or 7th day over the body proved salutary. An excessive heat and burning about the region of the liver, with cold extremities, portended death to be at hand.[279]

To this, in 1793, he emphasized sallow skin, yellow sweat, and highly coloured or turbid urine. All of this we would now recognize as symptoms of hepatitis or inflammation of the liver. Rush was convinced he knew the cause—foul miasmas in the streets emanating from rotting coffee dumped on the wharf.

It seems that this disease was brought to the New World along with slaves captured and transported from the West African coast in the seventeenth century, perhaps being first recorded in Yucatan (Mexico) and Havana (Cuba) in 1642. Over the next 150 years, outbreaks reared their ugly heads not only in the Americas but also in Spain, France, and Britain. At the Greenwich Seaman's Hospital in England, patients were segregated and dressed in jackets marked with a yellow patch to forewarn others, and a yellow flag, nicknamed the 'Yellow Jack', flew over the quarantined area. Hence the disease often became known colloquially as 'yellow jack'.

Yellow fever did not seem to be transmitted from person to person nor did contaminated food or water appear to be the cause. During the later part of the nineteenth century, attempts were made to find bacteria in yellow fever. Giuseppi Sanarelli (1864–1940), an Italian working at the University of Montevideo, gave a lecture in that city, translated in the *British Medical Journal* in 1897, in which he claimed to have found the cause of yellow fever. He believed it was due to a toxin-producing bacterium which grew on agar, often in association with many other bacteria, and which he named *Bacteroides icterus*. He reported that he had produced disease in a variety of animals and also in humans by injection of toxins prepared from filtering cultures of bacteria to remove the organisms themselves. He clearly thought he was dealing with an illness analogous to that produced by the toxin of diphtheria and wondered whether the bacterium was transmitted through the air or in water. We shall return to *B. icteroides* later.

A novel means of transmission of the infecting organism, whatever it was, through the air had been put forward in 1881. Carlos Juan Finlay (Figure 201), the son of a Scottish doctor and a Frenchwoman, had been born in Cuba. He received his early education mainly in France, graduated in medicine in Philadelphia, and then practised in Havana, Cuba. In 1879, the USA sent a Commission, led by Stanford Chaillé, a pathologist from New Orleans, to Cuba to study yellow fever. Chaillé's microscopical observations indicated that damage to the blood vessels by the infectious agent allowed red blood cells to leak out and produce the haemorrhages commonly found in yellow fever. This suggested to Finlay that there must be a vector or carrier of the infectious agent. Mosquitoes seemed to Finlay to fit the bill. In December 1880, he began to study the mosquitoes of Havana. Two were common but he singled out the mosquito known then as *Culex fasciatus* (later renamed *Stegomyia fasciata* and now called *Aedes aegypti*) because it lived in inhabited areas, bred by depositing eggs in stagnant water, and bit humans primarily in the early morning and late afternoon.

Figure 201. Carlos Finlay (1833–1915).

Finlay then embarked upon a series of extraordinary experiments which he continued over the next 20 years. I say 'extraordinary' because from today's perspective, if not that of his own time, they would have to be considered grossly dangerous, irresponsible, and unethical. Yet I have not read any twentieth- or twenty-first-century review that does not eulogize Finlay and his work. Having obtained permission from Spain (Cuba was a Spanish colony), he began to expose newly arrived Spanish soldiers to mosquitoes that had been allowed to feed on patients with yellow fever. Finlay knew that only female mosquitoes ingested blood. The first 'volunteer' was a 22-year-old named Francis Beronat, who was inoculated on 30 June 1881; nine days later he developed fever and jaundice. The next two were said to have 'abortive yellow fever' and the last two were only mildly ill. This was convincing enough for Finlay, who described his preliminary findings to the Academy in Havana on 14 August 1881:

> Three conditions will therefore be necessary in order that yellow fever may be propagated: (1) the existence of a yellow fever patient into whose capillaries

> the mosquito is able to drive its sting and to impregnate it with the virulent particles . . . (2) that the life of the mosquito may be spared after its bite upon the patient until it has a chance of biting the person in whom the disease is to be reproduced. (3) the coincidence that some of the persons whom the mosquito happens to bite thereafter shall be susceptible to contracting the disease.[98]

All this was received by his colleagues with incredulity and ridicule. Unabashed, Finlay continued his experiments. By June 1900, he had inoculated altogether 102 people, usually by allowing only a single bite from contaminated mosquitoes. Of these subjects, seven developed typical yellow fever, two had 'abortive yellow fever', and one had 'ephemeral' (transient) yellow fever. Finlay was convinced but many others were not. Perhaps the most important of the sceptics was George Sternberg (p.253), the US Army Surgeon General, who dampened any enthusiasm for Finlay's observations.

And there things stood until the USA decided to expand its empire and invaded Cuba to wrest it from Spanish control in 1898. More US troops died from malaria, typhoid fever, and yellow fever than from Spanish bullets. In May 1900, the US Army appointed a four-man 'board' of medical practitioners under the command of Major Reed to go to Cuba to investigate infectious diseases on the island, particularly yellow fever. Walter Reed (Figure 202) had been born in Virginia and graduated in medicine after a two-year course at the very young age of 18. A few years later he joined the US Army and in 1893 was posted to the Army Medical School where he was professor of bacteriology and clinical microscopy. His colleagues included James Carroll (1854–1907), who had been born in England, migrated to Canada, graduated from Baltimore, Maryland, where he also trained in bacteriology and pathology, and then joined the Army. Next was Jesse William Lazear (1866–1900), a civilian physician, who was chief of the clinical laboratories at Johns Hopkins Medical School in Baltimore. The last was Aristides

Figure 202. Walter Reed (1851–1902).

Agramonte (1868–1931), a civilian pathologist in Havana, who had been born in Cuba but had trained in medicine in New York.

The first task the board undertook on arrival was to examine Santarelli's *Bacillus icteroides*. They failed to find this organism in the blood of 18 patients with yellow fever or in the corpses of 11 patients who had died from the disease. They concluded that this bacterium had nothing to do with yellow fever. They were aware of Finlay's ideas about the transmission of yellow fever and were stimulated by the recent discoveries regarding the transmission of malaria by anopheline mosquitoes. After visiting Finlay on 1 August and discussing the situation with him, they decided to turn their attention to Finlay's *Culex fasciatus* mosquito.

The problem of course was how to tackle the question. Since no susceptible animals were known, the answer would most likely come from experiments on humans but that was clearly dangerous if the hypothesis was correct. The four members of the Commission decided that they should be among the first guinea pigs, who should all be fully informed of the risks and give their free consent. This decision having been made, Reed then returned to Washington! So Lazear opened the batting. He hatched mosquitoes from some eggs obtained from Finlay, fed them on patients with yellow fever, then almost immediately set them upon himself and eight others. Nothing happened.

A second experiment was then undertaken on 27 August 1900 in which Lazear exposed Carroll to a mosquito that had been fed on four yellow fever patients 2–12 days previously. Two days later Carroll became mildly ill, then on 1 September he was carried off to the yellow fever camp. His life was in the balance with high fever, delirium, headache, backache, swollen gums, and yellowing of his skin and eyes. But he survived. Since Carroll had been in contact with yellow fever patients for several days prior to his inoculation, it was not entirely clear that his illness was a result of mosquito bites. Consequently, an American soldier, Private William Dean, who had not been near yellow fever patients, was exposed by Lazear on 31 August to the same mosquito that had bitten Carroll, as well as three other mosquitoes that had fed on yellow fever patients. Dean developed a case of moderate yellow fever from which he recovered.

Such good fortune was not to attend Dr Lazear. On 13 September he was collecting blood from yellow fever patients when he was bitten on his hand by a *Culex* mosquito that had been flying in the ward. Since he had been bitten previously by a contaminated mosquito without ill-effect four weeks earlier, he deliberately let the insect finish feeding. Five days later he became ill. On 25

September he died from overwhelming yellow fever. Carroll later wrote, 'I shall never forget the expression of his eyes when I last saw him alive on the third or fourth day of his illness.'[66]

So what was the count after Lazear's tragic death? Three cases of yellow fever out of 12 attempts. You might not think that was any different to Finlay's experience. Nevertheless, Reed (who had now returned to Cuba but stayed for only 11 days) and his colleagues were convinced and wrote in a preliminary note which was published in at least two versions in late 1900, one saying:

> The mosquito serves as the intermediate host for the parasite of yellow fever, and it is highly probable that the disease is only propagated through the bite of this insect.[264]

This did not meet with universal acclaim. The *Washington Post* of 2 November 1900 declaimed:

> Of all the silly and nonsensical rigmarole of yellow fever that has yet found its way into print – and there has been enough of it to build a fleet – the silliest beyond compare is to be found in the arguments and theories generated by the mosquito hypothesis.[20]

There was clearly still some uncertainty. Consequently, an experimental station named Camp Lazear, with seven hospital tents and two wooden buildings, was set up in an open, well-drained field about a mile from the nearest town in late November 1900. This would allow the movements of all experimental subjects to be controlled and all sources of infection other than mosquitoes to be excluded. Three types of experiment were to be undertaken. First, volunteers would be exposed to mosquitoes that had fed upon yellow fever victims. Second, volunteers would be exposed to inanimate objects (fomites) taken from places housing yellow fever patients. Third, blood would be taken from patients with yellow fever and then injected into volunteers. Eighteen Americans, mostly military personnel, volunteered as did 13 Spanish immigrants who were offered 100 dollars in gold and a further 100 dollars if they contracted yellow fever.

The first two recruits were Privates John Kissinger and John Moran from Ohio who refused any financial compensation, with Kissinger saying that they volunteered solely in the interest of humanity and the cause of science. Braver soldiers could not be found! Within two weeks, five cases had been induced in 12 subjects by infected mosquitoes. Those who lived in a hut with fomites including clothing and bedding from yellow fever victims did not become infected. Then one building had a wire screen partition placed across its middle.

Fifteen infected mosquitoes were released on one side of the screen. The soldier who stayed in this compartment developed yellow fever whereas the two who lived on the other site of the partition remained well. There did not seem to be any doubt that mosquitoes were the carrier.

But when were patients infectious for mosquitoes? In December 1900 and January 1901, blood was taken one or two days after the onset of yellow fever in several patients and 0.5–2 ml injected into five recipients. Four of them developed yellow fever after four days or so. The experiment was repeated the following October with blood collected three or four days after the onset of the illness. All four who were given fresh blood or filtered serum became ill but yellow fever did not develop in the three subjects who were given blood that had been heated for 10 minutes at 55°C. Clearly mosquitoes were not necessary for transmission although it appeared that this was the way it occurred in nature and they could be infected by patients in at least the first four days of the illness. What was more, the infective agent could be destroyed by gentle heating.

The next question concerned how soon after biting a patient with yellow fever a mosquito became infectious to another human. A series of experiments showed that mosquitoes were not infectious to humans until 12 days after they had bitten a yellow fever sufferer and that this infectiousness was maintained for at least a further six weeks.

Twenty-two of the 31 volunteers developed yellow fever. Remarkably, even though a death rate of 20–40 per cent had been predicted at the onset of the experiments, none of them died; Lazear was the only death. Even so, their courage was incalculable, especially as no treatment was known. Would such an experiment be sanctioned by an ethics committee today? No. Did these experiments lead to a major advance in medical understanding and a significant improvement in human health? A resounding yes! Why then did the volunteers volunteer and how should we view their actions over a century later? Cuba had been in the grip of yellow fever for 140 years and many reasoned that they would get the disease anyway. For example 48 of the 102 volunteers who Finlay inoculated with mosquitoes and failed to develop yellow fever did contract the illness naturally in the years to come. The altruism and heroism of all these volunteers must be applauded. And, of course, it was not just the volunteers who were at risk. So were the medical staff who lived in Camp Lazear and attended them, some of whom, as we have seen, deliberately infected themselves.

In the event, these experiments led to extraordinarily concrete results. As a consequence of mosquito control measures introduced by Major William Crawford

Gorgas (1854–1920), yellow fever was controlled in Havana within six months. The same measures paved the way for the successful construction of the Panama Canal, previous attempts having been defeated by malaria and yellow fever.

But there was still uncertainty as to the precise nature of the infectious agent of yellow fever. No bacteria could be grown from the blood on the usual media. Moreover, in the direct inoculation experiments referred to above, two subjects were given serum that had been passed through a Berkefeld filter and was thus free of bacteria. This led Reed in 1902 to conclude:

> that these experiments appear to indicate that yellow fever, like the foot and mouth disease of cattle, is caused by a micro-organism so minute in size that it might be designated as ultra-microscopic.[265]

In other words, it was likely to be caused by what we now recognize as a virus. This was among Reed's last words on the subject. Although he escaped yellow fever, he died from appendicitis on 22 November 1902.

This was not to say that everyone paid much attention. In 1903, Finlay waxed eloquent on hypothetical sporozoites, schizonts, merozoites, and gametes as in malaria. In 1909, Harald Seidlin (1878–1932), a Dane working in Yucatan, found minute protozoan-like bodies which he called *Paraplasma* in the red blood cells of patients with yellow fever; then in 1913 JW Scott Macfie and JEL Johnston, Britons working in Nigeria, thought they had confirmed his findings. This was not to be for in 1915 Charles Wenyon and George Low (p.61) found indistinguishable bodies in normal guinea pigs.

In 1927, Adrian Stokes (1887–1927), Johannes Henrik Bauer (1890–1961), and Noel Paul Hudson(1895–1987) of the Rockefeller Foundation working in West Africa discovered that rhesus monkeys were susceptible to yellow fever. Hudson showed that the pathological changes in monkeys and humans were similar with widespread haemorrhages, jaundice, breakdown of liver cells, and degeneration of the kidneys. The infection could be transmitted in various ways including filtered serum. The infective agent must be a virus. Unfortunately, Stokes and two other members of the Rockefeller Commission, Hideyo Noguchi (p.282) and William Young (who performed the autopsy on Noguchi and died two weeks later), caught yellow fever and died.

Nevertheless, these studies paved the way for Max Theiler (1899–1972), a South African, and Hugh H Smith (1902–1995), both working for the Rockefeller Foundation in New York, to make the next advance. In 1937, they reported that the organism had been repeatedly grown in whole mouse embryos, chick

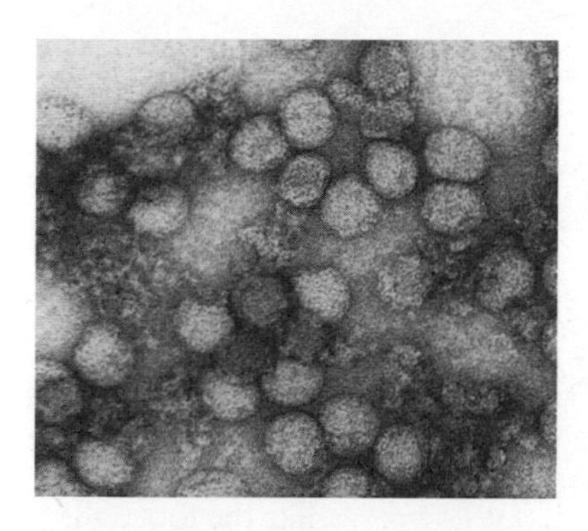

Figure 203. Electron micrograph of yellow fever virus.

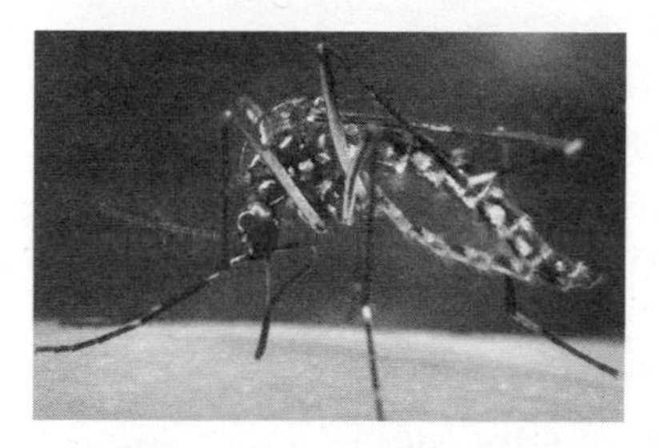

Figure 204. *Aedes aegypti.*

embryos from which the head and spinal cord had been removed, and testicular tissues of mice and guinea pigs. Organisms grown in the chick embryos lost their pathogenicity and were suitable for a vaccine. For this work, Theiler received the Nobel Prize in 1951.

In 1953, Reginald L Reagan and AL Bruechner from the University of Maryland in the USA described the appearance of yellow fever virus when viewed with the electron microscope. It was a spherical virus with a slightly irregular contour 40–50 nm in diameter (Figure 203). In 1960, WG Bearcroft reported that yellow fever virus replication was associated with ribonucleoprotein synthesis in the nucleus; then in 1963, G Neilson and J Marquardt extracted ribonucleic acid from the virus. We now know it is an RNA virus with a fatty envelope.

Today yellow fever still occurs sporadically in certain parts of tropical and subtropical Africa and South America. If there is an effective vaccine, you might be wondering why the disease has not been eradicated? Following the discovery that monkeys could be infected experimentally, it was observed that the infection occurs naturally among monkeys in the jungle, where several species of mosquitoes transmit infection. It will never be eradicated! Unfortunately, the vector of yellow fever in humans, *Aedes aegypti* (Figure 204), occurs not only in these endemic areas but also in parts of Asia and Australia. If it were to gain a foothold in these regions, the results could be catastrophic. That is why, if you live in one of these places and visit a yellow fever area, you must be immunized against yellow fever and carry the little yellow vaccination booklet certifying that fact.

43

Dengue fever (breakbone fever)

SCIENTIFIC NAME:	dengue virus, a member of the genus *Flavivirus*
DISEASE NAME:	dengue fever
COMMON DISEASE NAME:	break-bone fever
DISTRIBUTION:	many parts of the tropics and subtropics
TRANSMISSION:	bites of humans by infected mosquitoes, most commonly *Aedes aegypti*
CLINICAL FEATURES:	fever (often biphasic), severe muscle pains, rash which may peel or flake; sometimes haemorrhages and shock
DIAGNOSIS:	detection of virus or antibodies to dengue virus in the blood
TREATMENT:	no specific treatment; fluid replacement
PREVENTION:	avoid and control mosquitoes

'Few diseases have had more numerous synonyms than dengue, owing chiefly to local or regional variations in the predominance of one or other of its phenomena.'[290] So wrote William Smart, inspector-general of the Haslar Naval Hospital in Hampshire, England, in 1877. It is this variation in symptoms which has caused much controversy among medical historians who have looked back at epidemics as far afield as Batavia (Jakarta) and Cairo in 1779 to Philadelphia and Zanzibar, Calcutta, South Carolina, and the West Indies in the 1820s. From today's perspective, some of these outbreaks were probably caused by dengue virus, others by Chikungunya virus, and some perhaps by different arthropod-borne viruses such as Ross river virus.

So let us begin with Smart's description of dengue fever which is much as we would recognize it today. He remarked that the illness began with:

> heat (fever) and sweating . . . and a sense of cold in a painful degree. This stage may last (up to) 48 hours during which the whole surface is of a more or less scarlet hue, and there are more or less acute pains in the head, spine, and

> limbs, creating much alarm. It often terminates abruptly, on awaking, in apyrexia (normal temperature)...a general stiffness and soreness in motion alone remaining...Frequently, after a period of from 1–4 days, there comes a second accession of fever, characterised by a rubeoloid (German-measles-like skin) eruption which ends by desquamation (peeling of the skin), and by a return of the pains...The continuance of this stage is uncertain, lasting but three or four days, then disappearing gradually, but sometimes very slowly.[290]

This description of a biphasic temperature pattern with a fever occurring, disappearing, then recurring is typical of many cases of dengue fever. So are the severe pains which led to another common designation—break-bone fever.

The origin of the name dengue (pronounced den-gee with a hard 'g') is rather obscure. African slaves in the West Indies in 1827 called it 'dandy fever', apparently because victims held their painful limbs in weird positions. This was in turn translated into Spanish as 'dengue', which means 'fastidiousness' or 'prudery'. Others have claimed it arose from the Swahili word 'dinga' meaning 'cramp-like seizure' used to describe an epidemic in Zanzibar in 1723. Perversely, it seems quite likely in retrospect that the dengue of the West Indies in 1827 was in fact caused by Chikungunya virus but we will never know for sure.

Importation of this disease or complex of diseases into one or other location and its rapid spread implied that it was infectious. If so, what was the nature of the infectious agent? In 1886, James Wharton McLaughlin (1840–1909) in Texas reported that he had found a bacterium (a micrococcus) in the blood in dengue fever but his findings were never confirmed and were doubtless contaminant skin organisms. In 1903, Harris Graham (*c.* 1862–1922), a Canadian missionary in Beirut, Lebanon, claimed that a protozoon similar to malaria (but lacking in pigment) in red blood cells caused dengue fever. What is more, he reported that he had seen the organisms in the salivary glands of the mosquito *Culex fatigans* (now called *C. quinquefasciatus*) and had transmitted dengue fever to five of six men exposed to infected mosquitoes. He might have transmitted dengue fever but, if so, it was not due to parasites; no-one else could find them. His idea that dengue fever might be transmitted by mosquitoes had more merit but he did fix on the wrong species, his pools of *C. fatigans* probably being contaminated with *Stegomyia fasciata* (now called *Aedes aegypti*).

Thomas Lane Bancroft (p.60) in Brisbane, Australia, turned his attention to the problem in 1905. Bancroft was used to mosquitoes, having studied their role in the transmission of filariae. He was aware of the recent discovery of the transmission of yellow fever by *S. fasciata*. He thought this species was the more likely candidate for transmitting dengue fever as it was a day-biter and he was aware of victims who had only been bitten during the day. He fed stegomyiae (as he called them) on patients in the first to third days of illness, kept them for 10–17 days, then exposed five men to them. Only two became ill and Bancroft wondered whether he had kept the mosquitoes too long before using them. He found neither parasites nor bacteria and was inclined to think the organism was ultramicroscopic.

The next advance was made by Captain Percy Moreau Ashburn (1872–1940) and Lieutenant Charles Franklin Craig (1872–1950) of the US Army. The USA had dispossessed Spain of the Philippines in 1898 and taken it over as their own colony. Ashburn and Craig were instructed to look for the cause of dengue fever and its possible transmission by mosquitoes at Fort William McKinley, four miles from Manila when an epidemic occurred there in 1906. First, they examined the blood of a number of cases and found neither bacteria nor protozoa. Next, they injected unfiltered blood from patients with dengue fever into 11 US volunteer soldiers, seven of whom developed the fever. Ashburn and Craig justified these and later experiments thus:

> We were dealing with a disease which in the young and vigorous is not dangerous to life, and for this reason we felt justified in making the experiments.[29]

Then they passed blood through a filter to remove bacteria and two more men were injected. Both developed dengue fever indicating the organism was an ultramicroscopic virus. Finally, nine men were exposed to *C. fatigans* that had fed on dengue fever patients a day or so previously. Only one developed dengue fever, with the investigators blaming the eight negative cases on pre-existent immunity or perhaps a failure of mosquitoes to bite their victims. This did not stop them writing:

> We realize that the work that we have been able to do as regards mosquito transmission is very incomplete and that a very great deal remains to be done before this feature of the etiology of dengue is fully elucidated, but we believe that we have confirmed Graham's results in this respect and that we have proven experimentally that this disease can be transmitted by the mosquito, *Culex fatigans*.[29]

Are you convinced that they had proven mosquito transmission? I'm not, nor were Cleland and his colleagues who wrote in 1918 that 'we can hardly understand the importance attributed to this isolated case by most text-books'.[72]

One of the big problems faced by all these investigators was that their studies were carried out in an endemic area which meant that subjects might be infected coincidentally in a natural fashion or they could possibly be immune due to prior infection. An outbreak of dengue fever in Queensland and northern New South Wales, Australia in 1916 provided the solution for John Burton Cleland (1878–1971), Burton Bradley, and W McDonald. They travelled to two towns in northern New South Wales and took blood from infected patients and samples of *C. fatigans* and *S. fasciata* caught in the rooms of patients back to Sydney where dengue had never occurred. They were particularly suspicious of *S. fasciata* because this species had the same distribution in Australia as dengue fever, whereas *C. fatigans* was also common in the southern parts of the continent where dengue fever never occurred. Four of seven volunteers bitten by infected *S. fasciata* developed dengue fever, whereas neither of the two volunteers bitten by *C. fatigans* did. Blood was taken from three of the volunteers and injected into more subjects; they too developed dengue fever. They then carried out a complex series of investigations producing dengue fever by injecting volunteers with whole blood, serum, plasma, washed red cells, or serum passed through a Chamberland filter, this last group not providing any clear-cut answer.

What about the ethics of all these experiments? Transmission experiments with mosquitoes are perhaps one thing—an unpleasant illness might be caught but death was unlikely. But injection of blood is another thing as hepatitis viruses, the bacterium of syphilis, and other pathogens might be transmitted. But the early investigators knew little of these things. By the time of Cleland and his colleagues, the cause of syphilis was known so they did a blood test (Wasserman reaction) to exclude such donors. With today's standards, no ethics committee would approve such studies. But back then? I leave you to judge.

Time would show that Cleland and his colleagues were right—*S. fasciata* (= *Aedes aegypti*, Figure 204) is the principal mosquito that transmits dengue fever virus although sometimes other species of that genus, especially *A. albopictus*, might do so. Various investigators showed that the virus needed to be incubated in the mosquitoes for one to two weeks for them to become infected. This time was needed for the viruses to invade the cells of the insect's gut and multiply, then spread to the salivary glands from which they could be injected next time the mosquito bit someone.

During World War II, investigators in Japan, including R Kumuru and Susumu Hotta, injected blood from humans with dengue fever into mouse brains, found the agent was a filterable virus, and serially transmitted it from one mouse brain to another. Within a few months, Albert Sabin (p.468) and Robert Walter Schlesinger (*c.* 1914–2003) in Hawaii, unaware of the Japanese work, independently made the same discovery. In 1950, Schlesinger achieved sustained growth of the virus in chick embryos. In the following decades, the virus was propagated in a variety of mammalian and insect cell cultures.

In 1950, Sabin used antibody reactions to show that there were two distinct types of dengue viruses. In 1956, William M Hammon (1904–1989), working in the Philippines, found two more, now called dengue viruses 3 and 4. In 1965, TM Stevens and Schlesinger showed that dengue viruses were RNA viruses. In 1970, TJ Smith and his colleagues used the electron microscope to visualize the virus in infected mouse brains and cell cultures. It is spherical in shape being about 50 nanometres in diameter (Figure 205) and was shown to have a lipid envelope.

Dengue fever usually cures itself. Sometimes though, it is associated with widespread haemorrhages and/or shock (blood pressure collapse) which is often fatal, especially in young children. This 'dengue haemorrhagic fever/dengue shock syndrome' was possibly first described in 1897 by FE Hare, a general practitioner in Charters Towers, Queensland, Australia. However, it was not until the 1950s that this syndrome came into its own as the four different serotypes of dengue virus spread rapidly around the world. In the 1960s and 1970s, William Hammon and others showed that these complications generally occurred when a person who had previously had dengue fever was subsequently infected with one of the three other dengue virus serotypes.

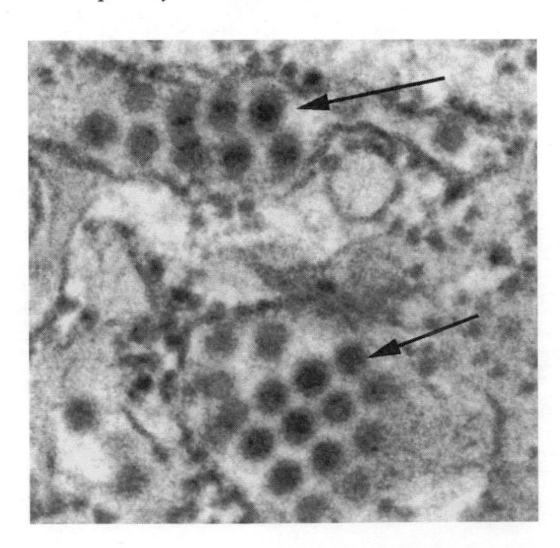

Figure 205. Electron micrograph of dengue virus (arrows) in tissue cells.

Dengue fever is a disease of humans although asymptomatic infections can occur in monkeys. The World Health Organization estimates that there are about 50 million cases of dengue fever around the globe each year. Developed countries are not necessarily immune—it is endemic in northern Queensland, Australia. Since dengue fever is for all practical purposes a disease of humans, you might think it is a prime candidate for eradication. Unfortunately, there is as yet no effective vaccine against all four serotypes of dengue virus. Minimization of cases of dengue fever is dependent upon control of *Aedes* mosquitoes—a difficult and never-ending task.

44

Poliomyelitis (infantile paralysis)

SCIENTIFIC NAME:	poliovirus, a member of the genus *Enterovirus*
DISEASE NAME:	poliomyelitis
COMMON DISEASE NAMES:	polio, infantile paralysis
DISTRIBUTION:	some parts of the Third World, especially south Asia
TRANSMISSION:	ingestion
CLINICAL FEATURES:	usually no symptoms; 1% have fever followed by flaccid paralysis of one or more limbs; if the lower brainstem is involved, breathing may be impaired
DIAGNOSIS:	isolation of the virus or a positive PCR test
TREATMENT:	no specific treatment; rarely an iron lung is needed to support breathing
PREVENTION:	effective vaccines are available

It was about 1958. My classmates and I were lined up outside the Health Department caravan standing in the middle of the high school quadrangle. One by one we mounted the steps and were jabbed in the arm with the new Salk polio vaccine. If my memory serves me correctly, the needles were sterilized by being held in the flame of a gas ring then allowed to cool for the next recipient—a total no-no these days because we now know about the risk of contamination with blood-borne viruses. But the Salk vaccine, named after Jonas Salk (1914–1995), the developer of the first effective polio vaccine, marked the beginning of the end for the poliovirus.

What is poliomyelitis, also known as polio for short? It is an illness characterized by fever followed by paralysis which may affect one or more limbs and sometimes loss of ability to breathe if the chest wall and diaphragm are involved. You may know someone who is paralysed as a result of a stroke. That form of paralysis is known as spastic paralysis because although the sufferer cannot move the limb, when another person moves it passively for them, there is some resistance because the muscle tone is increased. This contrasts with polio in which the paralysis is flaccid, that is the limb loses its tone and is floppy. After weeks or months, the muscles waste away so that the affected limb becomes not only paralysed but withered.

Figure 206. Carving on an Egyptian stele showing a man with a withered leg.

It seems that polio has been around for millennia. An ancient Egyptian wall carving from the eighteenth dynasty about 3,500 years ago shows a lame man with a withered leg and contractures around the knee and ankle joints supporting himself with a crutch; he probably had polio as a child (Figure 206). The Scottish author, Walter Scott, at the age of 18 months developed a fever which was followed several days later by paralysis of his right leg which remained smaller and weaker than his left leg for the rest of his life. Shortly thereafter in 1789, the English paediatrician, Michael Underwood (1737–1820), provided an account of what he called 'debility of the lower extremities':

> It usually attacks children previously reduced by fever; seldom those under one, or more than four or five years old. It is a chronical complaint...the first thing observed is a debility of the lower extremities which gradually become more infirm, and after a few weeks are unable to support the body.[305]

Until the early nineteenth century, this condition tended to occur sporadically with a case here or there, but in the 1830s clusters or mini-outbreaks occurred in places as far apart as England and the South Atlantic island of St Helena. Because mostly small children were affected, it came around this time to be called 'infantile paralysis' although it was also known as Heine–Medin disease after the German and Swedish physicians who described small epidemics in 1840 and 1890.

What was the pathological nature of this disease, what caused it, and how did one get it? Was it infectious? These were questions that exercised the minds of many. The answer to the first question was provided by the French neurologist Jean-Martin Charcot (1825–1893) and his assistant Alexis Joffroy (1844–1908) in 1869. They performed an autopsy on a woman who had contracted infantile paralysis as a child and examined the cord with a microscope. They found that the nerve cells in the part of the grey matter of her spinal cord called the anterior horn were destroyed. This meant that the nerve fibres which emanated from them and supplied the muscles were also destroyed (Figure 207); this accounted for both the flaccid paralysis and the wasting. This soon led to a new name being coined—poliomyelitis—derived from the Greek words *polios* and *myelos*, meaning 'grey' and 'marrow' with 'itis' indicating inflammation. This name superseded 'infantile paralysis' when it was realized that the disease could afflict adults as well as children, the most famous example being Franklin D Roosevelt, later president of the USA, who was diagnosed with this illness at the age of 39. I must say, however, that Roosevelt may not have had polio but a mysterious condition called the Guillain–Barré syndrome.

Was it infectious? It seemed that it might be because epidemics appeared in Europe, North America, and Australia in the first decade of the twentieth century. Otto Ivar Wickman (1872–1914) studied in detail a large outbreak of 1,031 cases which began in Sweden in 1905. He examined where and when these cases as well as what he called 'abortive cases' (i.e. the illness started similarly but did not reach the stage of paralysis) occurred. Often there were small clusters and Wickman thought this suggested the disease was infectious. Moreover, he realized that it was quite possible that healthy people could carry and transmit the infectious agent for there was often no direct contact between patient and patient.

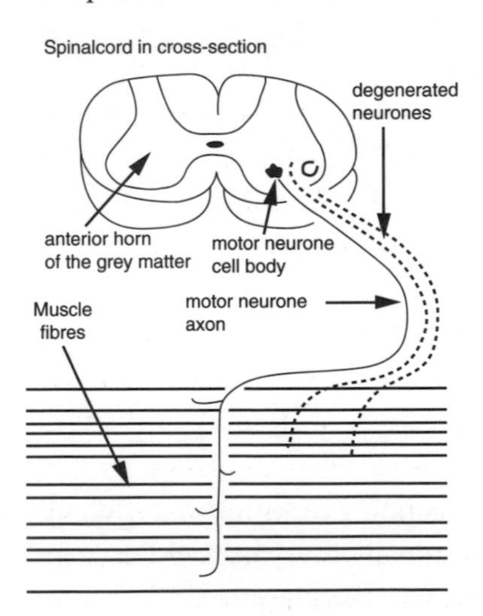

Figure 207. Diagram showing loss of motor neurones with destruction of neurone cell bodies in the anterior horn and the axons supplying the muscle fibres which atrophy.

That Heine–Medin disease (as Wickman called it) was indeed infectious was confirmed by the dramatic discovery of Landsteiner and Popper in Vienna. Karl

Landsteiner (1868–1943) would win the Nobel Prize in 1937, not for his work on polio but for his discovery of the ABO blood groups which set the stage for blood transfusions and saved countless lives. In November 1908, Landsteiner and Erwin Popper (1879–1955) obtained the spinal cord from a boy who had died from polio, ground it up, and then injected the saline suspension into the abdominal cavity of two monkeys as well as rabbits, guinea pigs, and mice. Nothing untoward happened to the latter animals but the monkeys became paralysed 6 and 17 days later. At post-mortem examination, their spinal cords showed the same kind of damage that afflicted the boy:

> the pia [membrane around the spinal cord] was infiltrated with small, round, deeply-staining cells . . . The substance of the cord showed . . . areas of inflammation, which . . . occurred chiefly in the anterior grey matter. . . . Haemorrhages were also present . . . The ganglion cells (neurones) of the anterior horns especially showed severe degeneration and were often surrounded by, or actually invaded with, mononuclear and polynuclear cells.[175]

They then injected spinal cord material from these monkeys into two further monkeys but they remained well. In the following year, Landsteiner teamed up with Constantin Levaditi (1874–1953) of the Institut Pasteur in Paris and proved that the infectious agent was a filterable virus that passed through a bacteria-retaining Berkefeld filter, the filtrate inducing disease in monkeys.

The baton then passed to the Rockefeller Institute for Medical Research in New York which had both money and monkeys and with its director, Simon Flexner (p.351), was to dominate polio research for the next 25 years or so. In 1910, Flexner and Paul Aldin Lewis (1879–1929, who died while researching yellow fever in Brazil) reported their extensive studies on 81 monkeys (mostly *Macacus rhesus*) that were infected initially by injecting some spinal cord material, obtained from a one-year-old boy who had died from polio, into the brain through a small hole in the monkeys' skulls. The infection was later transferred from monkey to monkey by injections into a variety of anatomical sites.

Unfortunately, Flexner was later to muddy the waters in two ways. First, he believed for a number of years that the infecting organisms were minute 'globoid bodies' about 200 nm in diameter. Eventually, everyone became disenchanted with this idea. Secondly, he was convinced that the portal of entry for the virus was through the nasal passages with the virus then passing to the brain via the nerves supplying that area. Such was the power of his prestige that during one epidemic in the USA, people walked around with zinc sulphate paste in their nostrils as a preventive—a complete waste of time.

There were, however, occasional suggestions that infection might be acquired by ingestion. In one experiment, common houseflies were allowed to crawl on infected material and then on bananas. When monkeys ate those bananas, they developed polio. The matter was put beyond all doubt by a mammoth study reported in 1941 by Albert Bruce Sabin (1906–1993) and Robert Ward in Cincinnati, Ohio, USA. They performed autopsies on nine patients who had died from polio and carefully collected tissues from 19 different sites including the nasal passages, tonsils, lower part of the small intestine, brain, spinal cord, and so on as well as the contents of the descending colon (effectively faeces). The separate collections were then injected into more than 200 rhesus monkeys in order to determine which ones had poliovirus as shown by the monkeys becoming ill. They demonstrated conclusively that:

> the virus is distributed predominantly in two systems: (a) certain regions of the nervous system, and (b) the alimentary tract... Poliomyelitis virus was demonstrated in the walls of the pharynx, ileum [last part of the small intestine], and only once in those [i.e. the wall] of the descending colon, while the contents of the descending colon regularly contained the virus.[281]

No virus was found in the nasal passages. The portal of entry of the virus was now clear—it was by ingestion, the virus multiplied in the lowermost parts of the small intestine, and it was excreted in the stools.

It seems that the virus multiplies in epithelial cells of the intestine and white blood cells found in Peyer's patches in the last part of the small intestine. How does it get from there to the spinal cord and brainstem? It is still not entirely clear, but it probably depends upon the virus entering the bloodstream, at least for a short time. One hypothesis has it that the virus somehow enters the central nervous system from the blood and infects nerve cells. The other suggests that when viruses reach the muscles, they enter the nerves where they join to the muscles and then pass up the axons, like rabies virus, to affect the nerve cell bodies in the anterior horn of the grey matter of the cord and sometimes the medulla oblongata at the bottom of the brainstem. I would put my money on this latter suggestion as it explains the peculiar location of the pathological changes.

What exactly is the virus? In 1946, Hubert S Loring and colleagues first described the appearance in the electron microscope of the Lansing strain of poliovirus which had been adapted to cotton rats. It was a very small, icosahedral virus, about 30 nm in diameter without an envelope (Figure 208). In 1957, J Colter and colleagues showed that it was an RNA virus.

What would the virus grow in? In 1935, Sabin and Peter Kosciusko Olitsky (1886–1964) grew it in nerve tissues from three- to four-month-old human

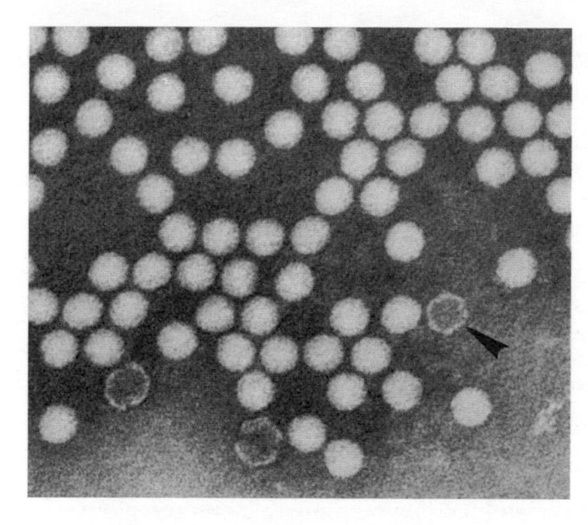

Figure 208. Electron micrograph of poliovirus. The arrow points to a virus that has lost its RNA core.

embryos but not in non-nerve tissues. In 1948, John Franklin Enders (1897–1985), Thomas Huckle Weller (1915–2008), and Frederick Chapman Robbins (1916–2003) succeeded in growing poliovirus in human embryonic cell cultures composed chiefly of skin, muscle, and connective cells from the arms and legs or bowel tissue, and in nervous tissue. Moreover, Robbins was the first to recognize differences in the appearances of infected cells, a phenomenon which Enders called cytopathic effect. For these discoveries, they were awarded the Nobel Prize in 1954.

The stage was now set to grow poliovirus so that a vaccine could be made. But first a question needed to be answered—was there just one poliovirus or more than one? In 1931, Frank Macfarlane Burnet (1899–1985) and Jean Macnamara (1899–1968) in Melbourne, Australia, showed that poliovirus in Melbourne did not induce immunity against the Rockefeller strain and vice versa. No one paid much attention 'since it came from unknown investigators on a remote continent'.[91] The question was reinvestigated in the early 1950s and three distinct strains were identified. All three serotypes were then used in the vaccines developed by Salk and Sabin.

The Salk vaccine is made from killed virus and is given by injection. The Sabin vaccine contains live viruses that have been rendered non-pathogenic (i.e. they are attenuated) and is taken by mouth. Each vaccine has its pros and cons. Salk and Sabin vigorously promoted their own. Some countries give a Salk injection first then follow it at bimonthly intervals with two doses of the Sabin vaccine; others use only three doses of the inactivated Salk vaccine.

In the 1950s, mass immunization campaigns began in many countries. In 1988, the World Health Organization started a campaign to eradicate poliomyelitis from the rest of the globe. The results have been spectacularly successful. The last case in the Americas was in Peru in 1991, while the last in Europe was in Turkey in 1998. In 2000, eastern Asia and the Pacific were certified polio-free. There are only four countries in which polio transmission has never been stopped—Afghanistan, Pakistan, India, and Nigeria. Unfortunately, the virus has been re-introduced into another dozen countries or so, often with transmission being re-established. Nevertheless, polio only infects humans in nature, all the necessary tools are available, and the world is tantalizingly close to achieving eradication of this nasty virus. But all this comes to naught with man's inhumanity to man—around 16 health workers distributing polio vaccine were killed by Taliban terrorists in Pakistan in December 2012 and January 2013 and the programme has been halted!

45

Measles (rubeola)

SCIENTIFIC NAME:	measles virus, a member of the genus *Morbillivirus*
DISEASE NAME:	measles or rubeola or morbilli
DISTRIBUTION:	many parts of the world
TRANSMISSION:	aerosol
CLINICAL FEATURES:	fever, runny nose, conjunctivitis, cough, diarrhoea, rash of slightly raised red spots called a maculo-papular rash
COMPLICATIONS:	pneumonia and middle ear infection (both usually caused by secondary bacterial infections), corneal ulcerations, and rarely inflammation of the brain
DIAGNOSIS:	clinical; measles antigen in respiratory secretions
TREATMENT:	no specific treatment; fluids
PREVENTION:	an effective vaccine is available

When I was a child I had measles. Most of my generation did. We did not particularly fear it. It was unpleasant but it did not kill in our community and rarely caused serious complications. The measles we had was similar to that described by the English physician Thomas Sydenham (1624–1689) in the seventeenth century:

> The measles generally attack children. On the first day they have chills and shivers, and are hot and cold in turns. On the second they have the fever . . . thirst, want of appetite, a white . . . tongue, slight cough, heaviness of the head and eyes, and somnolence. The nose and eyes run continually . . . To this may be added sneezing, a swelling of the eyelids a little before the eruption, vomiting and diarrhoea . . . The symptoms increase till the fourth day. Then or sometimes on the fifth there appear on the face and forehead small

> red spots, very like the bites of fleas. These increase in number, and cluster together, so as to mark the face with large red blotches. They are formed by small papulae [lumps], so slightly elevated above the skin, that their prominence can hardly be detected by the eye, but can just be felt by passing the fingers lightly along the skin. The spots take hold of the face first; from which they spread to the chest and belly, and afterwards to the legs and ankles. On these parts may be seen broad, red maculae [red spots], on, but not above, the level of the skin. About the eighth day they disappear from the face, and scarcely show on the rest of the body. On the ninth, there are none anywhere.[300]

In 1896, the New York paediatrician Henry Koplik (1858–1927) described an almost fool-proof diagnostic sign now known as 'Koplik's spots' which were seen in the mouth, on the cheeks and lips as 'small irregular spots of a bright red colour. In the centre of each spot there is... a minute bluish-white speck.'[167] Spare a thought for Reubold, Murray, Gerhardt, and Flindt, all of whom described the sign before Koplik!

Measles itself had in fact been first described by Rhazes (p.438) in the ninth century who distinguished the disease from smallpox. He said that the rash in measles was preceded by a sore throat, hoarse voice, and red cheeks after which the rash came out quickly whereas the rash of smallpox appeared more quickly. Nevertheless, measles continued to be confused with various red rashes, including smallpox, for years to come. The name 'measles' has been used with various spellings for centuries and is probably derived from Old Dutch or German words, 'mazelen' and 'maselen', which mean red spots on the skin. Alternative names include 'rubeola' derived from the Latin 'rubeus', meaning 'reddish', and 'morbilli', also from the Latin 'morbus' indicating 'disease'.

Relatively mild measles seems to occur in communities in which it has long been present and in which the population has adapted to it. Such is not the case when measles is introduced into a community that has never seen the disease. Thus, when measles was carried to Fiji from Sydney, Australia, in 1875, 20,000 people, 40 per cent of the population, died from the infection over the next four months. In the same way several centuries earlier, smallpox and measles together ravaged the Indians of the American continents. Why the virus should have these variations in behaviour in different populations is still not entirely clear.

The rapid spread of measles in households, villages, and large communities naturally suggested that it was infectious. A severe epidemic occurred in Edinburgh, Scotland, in 1755 in which 10 per cent of cases died. Francis Home (1719–1813), a local physician, wondered whether the problem could be dealt with in the same way that variolation was used in smallpox. He cut the skin where the rash was thickest and soaked the blood in cotton. Then he made an incision in the arm of his experimental subject and put the cotton in. Twelve children were inoculated; nine developed measles. He also tried inoculations from nose to nose but without success. Similar experiments were repeated by many others for 150 years up to and including Ludwig Hektoen (1863–1951). Hektoen, in Chicago, USA, in 1905 reported the results of his own investigations. He took blood from two patients and cultured it in a mixture of broth and ascites fluid (fluid taken from someone with fluid in their abdominal cavity) for 24 hours. No bacteria grew in this broth but when it was injected into two young men, one of whom was a medical student who 'readily gave his consent', measles developed almost two weeks later. Such dangerous experiments of course would never be permitted these days.

The scene was set for delineating the causative agent when John F Anderson (1873–1958) and Joseph Goldberger (p.409) in 1911 successfully transmitted measles by injecting blood, serum, and washed red blood cells as well as mouth and nasal secretions to monkeys by injections into various sites. They then passed the infection from monkey to monkey in six successive experiments. Ten years later, Francis Gilman Blake (1887–1952) and James Dowling Trask (1890–1942) showed that both unfiltered nasal secretions and secretions that had been passed through Berkefeld filters to remove bacteria caused measles in monkeys when inoculated into the respiratory tract. The causative agent was definitely a filterable virus.

In 1938, Harry Plotz (1892–1947) reported that this virus could be grown in chick embryonic tissue. Another 16 years passed before John Enders (p.469) and Thomas Chalmers Peebles (1921–2010) in Boston in the USA took blood and throat washings from a boy with measles named David Edmondston in whom the rash had just appeared and inoculated them into roller tube cultures of human and monkey kidney cells, and later human amnion cells. The cells became infected, and after four to nine days multinucleate giant cells appeared in the sheets of cells. Such giant cells can sometimes be seen in the lungs of patients who have died from measles (Figure 209).

Naturally, researchers wondered about the nature of this virus. Reginald L Reagan and his colleagues in the USA in 1952 used the electron microscope to view the virus. They found that it had a lipid envelope and was variable in size ranging from 100 to 300 nm in diameter (Figure 210). After another decade, in 1963 and 1964, several groups of workers showed that measles virus contained ribonucleic acid (RNA).

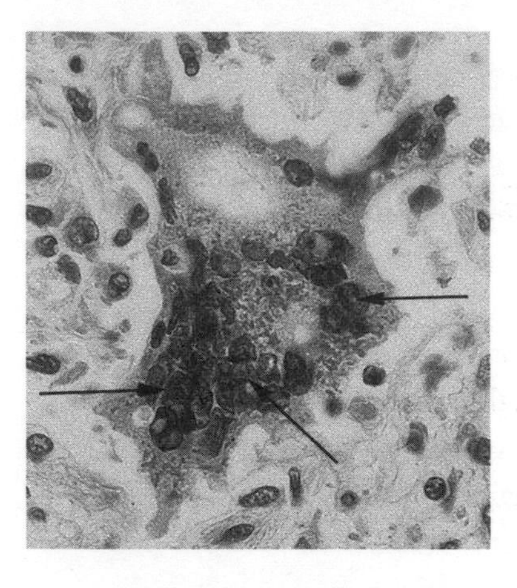

Figure 209. Giant cell with multiple nuclei (arrows) in measles pneumonia.

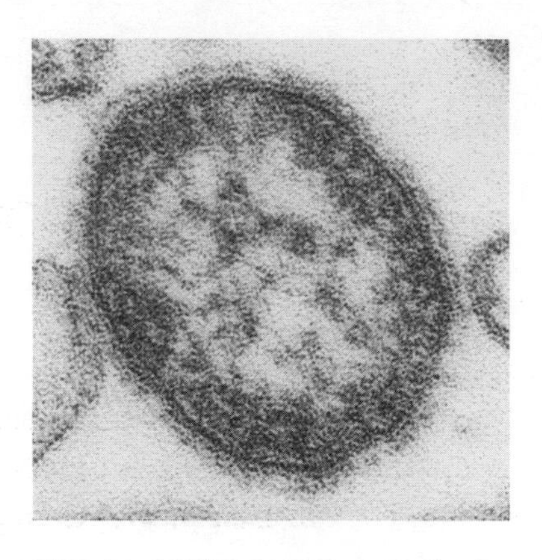

Figure 210. Electron micrograph of measles virus.

We now know that measles is acquired when the virus is breathed in. It multiplies in the lining of the nasal passages, throat, and windpipe as well as the conjunctiva. A few days later it passes via the bloodstream to the liver and spleen where the viruses replicate again. Around day 8 of the infection, the viruses are carried in the blood to the target organs which include the skin, eye, lung, and bowel, where they multiply once more. The rash which appears about two weeks after infection is due to a strong allergic reaction to the virus. In the First World, any secondary bacterial infections are treated and almost everyone recovers. In the Third World, especially where there is malnutrition or suppression of the immune system with AIDS, the illness is often severe and death may follow.

The discovery that measles virus could be grown in cell cultures allowed the production of virus in sufficient quantity to make a vaccine. The first vaccines were introduced in the 1960s, and in 1971 a combined measles, mumps, and rubella vaccine appeared and came into general use. The three viruses are alive but weakened (attenuated). Like the poliovirus, measles virus only infects humans in nature and an effective vaccine is available. Will measles ever be eradicated? There is no official policy to embark on global eradication. The challenge will be greatest in Africa where there is so much HIV infection. However, measles has been eliminated from the Americas and Australia although there are frequent re-importations from overseas so eternal vigilance is required. So what is the answer to my question? Time will tell but I rather doubt it unless *everywhere* in the world there is a simultaneous immunization campaign. Can you imagine that happening?

46

German measles (rubella)

SCIENTIFIC NAME:	rubella virus, the only member of the genus *Rubivirus*
DISEASE NAME:	German measles or rubella
DISTRIBUTION:	many parts of the world
TRANSMISSION:	aerosol
CLINICAL FEATURES:	swollen lymph nodes in the neck and back of the head, mild fever, rash of slightly raised red spots called a maculo-papular rash
CONGENITAL RUBELLA:	when the mother is infected in the first three months of pregnancy it may lead to cataracts, deafness, heart disease
DIAGNOSIS:	clinical; appearance of rubella antibodies in the blood
TREATMENT:	no specific treatment;? abortion for congenital rubella
PREVENTION:	an effective vaccine is available

In 1972 I attended a scientific meeting in Sydney. The airlines were on strike so I took the overnight sleeper from Adelaide to Melbourne and then the day train to Sydney. It was one of those trains that had a table between facing seats. I clearly remember resting my elbow on the table with my chin on my hand and feeling a large, swollen lymph gland under my jaw. Being medical, I naturally thought I had leukaemia. I stayed with a friend but hardly had the energy to lift myself off the floor. When both the meeting and the strike were over, I flew home to be greeted by mother who looked at my face and exclaimed how sunburnt I looked. I went to have a shower before going off to play cricket. I looked down my body and saw a fine red rash. What a relief—I did not have leukaemia; I had German measles. I gave up all idea of playing cricket.

Who first described German measles is a matter of some uncertainty. Bartolommeo Castelli's *Lexicon medicum graeco-latinum*, which went through many editions in the seventeenth and eighteenth centuries, ascribes the word 'rubella' to Paracelsus (1493–1541) but this does not have anything to do with German measles. Various reviewers say that several German physicians including Daniel Sennert (1572–1637) in Wittenberg in 1619, Friedrich Hoffmann (1660–1742) in 1740, De Bergen (or De Berger) in 1752, and Orlow in 1758 first described this disease. What these good doctors actually wrote I am afraid I cannot tell you, nor can I tell you any more about the doctors themselves, but the illness they described acquired the German name *rötheln*. Matters became somewhat clearer, at least for English speakers, when the English physician, William George Maton (1774–1835) in 1815 published a paper called 'Some account of a rash liable to be mistaken for scarlatina'. He saw a family in which eight members, aged from 1 to 28 years, were afflicted with an illness characterized by malaise (marked lethargy) followed by swollen lymph glands in the neck and at the back of the head and a generalized skin rash. This is recognizably German measles, just like my illness. In England, this illness was often called 'Third disease' by medical practitioners to distinguish this feverish illness with a rash from measles and scarlatina (= scarlet fever). Others preferred the designation 'German measles', perhaps reflecting its description by German clinicians, while others think the name is a corruption of 'germane' (i.e. pertaining to) measles. Henry Veale, a surgeon with the British Royal Artillery, seems not to have liked any of these names and he certainly did not like rötheln. When in 1866 he described 30 cases of this disease in schoolchildren at Mt Abu, India, he proposed naming the illness 'rubella'.

> The name of a disease is always a matter of some importance. It should be short for the sake of convenience in writing and euphonious for ease in pronunciation. It should, if possible... indicate a definite group of pathological conditions... Rotheln is harsh and foreign to our ears... I therefore propose Rubella... as a name for the disease.[308]

The word 'rubella' is the feminine form of the Latin word 'rubellus' meaning 'reddish'. As we shall see, his choice of the feminine gender was rather appropriate.

In 1914, Alfred Fabian Hess (1875–1933) in New York showed that German measles could be transmitted to young rhesus monkeys with blood taken from humans with the disease thus proving the clinical observation that it was infectious. In 1938, Y Hiro and S Tasaka transmitted the disease to children by inoculating filtered nasal secretions, thus showing that the infectious agent was a filterable virus which was present in respiratory fluids.

Nevertheless, clinicians were not particularly interested. In the northern winter of 1935–6, an epidemic of German measles occurred in Montreal, Canada, with perhaps 100,000 people being affected. Even so, when Harvey B Cushing, professor of paediatrics at McGill University in Canada, reviewed the outbreak in 1938, he was moved to write:

> German measles impresses one rather as an unmitigated nuisance than a serious menace. If it were not for the eruption, the disease would be utterly negligible, as it is much less serious in course and sequelae than the common cold.[78]

Three years later, he was shown to be comprehensively wrong.

In 1941, Norman McAlister Gregg (Figure 211), an ophthalmic surgeon in Sydney, Australia, realized that he was seeing an unusually large number of babies with congenital cataracts (clouding of the lens of the eye). Moreover, many of them had heart defects. He wondered whether these malformations could have been caused by a toxin or infection during pregnancy. Part way through his series of patients, Gregg had a brainwave:

> By a calculation from the date of the birth of the baby it was estimated that the early period of pregnancy corresponded with the period of maximum intensity of the very widespread and severe epidemic in 1940 of the so-called German measles.[123]

Thereafter, Gregg enquired of his new patients whether the mother had had German measles during pregnancy and went back and asked the same question of the mothers of his initial patients. He looked after 13 patients of his own and saw seven being treated by his colleagues. Furthermore, ophthalmologists from other hospitals up and down the east coast of Australia sent him details of their cases. Altogether, he had 78 cases of congenital cataracts. Sixty-eight mothers gave a clear-cut history of German measles during early pregnancy and in some of the remaining 10 women, the infection could

Figure 211. Norman Gregg (1892–1966).

not be ruled out. Many children had small eyes and 44 had heart murmurs. In those days, diagnosis of cardiac defects was imprecise, but in some of those who died and were examined at autopsy there was a condition called patent ductus arteriosus in which a large blood vessel fails to close after birth.

Gregg's paper was published during the darkest days of World War II and it took several years for doctors in the rest of the world to examine the issue. But it was taken up in Adelaide, Australia by Charles Spencer Swan (*c.* 1912–*c.* 1963) and his colleagues. He reversed Gregg's approach by writing to all the medical practitioners in South Australia seeking mothers who had recently had German measles and then assessing their babies. He found 49 women and infants. Fourteen of the babies had eye disease (13 with cataracts), 17 had congenital heart disease, and many had a small head. Swan also found a new association—seven of the children were deaf. Finally, he showed that almost all the congenital defects happened when the mother had German measles during the first three months of pregnancy.

These observations caused others to revisit rubella. German measles was not trivial. But what was the precise nature of the filterable agent that caused it? In 1953, Reginald Reagan and his colleagues infected rhesus monkeys with blood from an infected patient and then concentrated the monkeys' blood to observe the virus with the electron microscope. They found viral particles 90–100 nm in diameter. This size was somewhat overestimated as it was eventually shown that, including a lipid envelope, the virus measures 50–70 nm in diameter with 6 nm spikes projecting from the envelope as in Figure 212.

In 1962, two groups of investigators in the USA independently grew the virus of German measles in cell culture. Paul Douglas Parkman (1932–) and his colleagues in Washington, DC, infected kidney cells from African green monkeys with throat washings from Army recruits with rubella. At the same time in

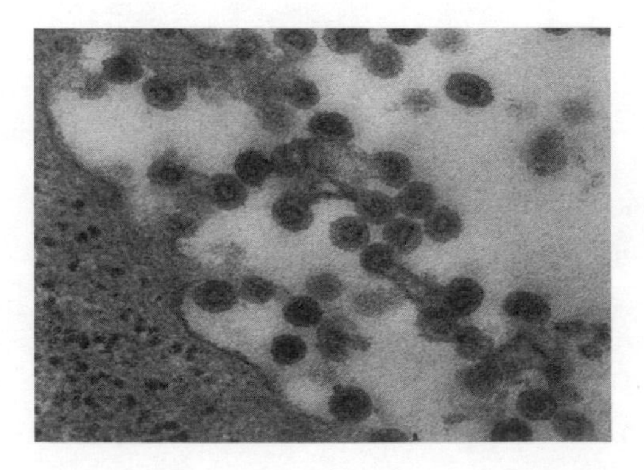

Figure 212. Electron micrograph of multiple rubella virus particles.

Boston, Massachusetts, Thomas Weller (p.169) and Franklin Allen Neva (1922–2011) found cytopathic effects in human amnion cells infected with blood and urine from patients with German measles.

Beginning in 1964, various groups of investigators produced indirect evidence to suggest that rubella virus contained RNA. This was then confirmed directly in 1969 by Ken T Wong and colleagues in Stanford, California, as well as by T Hovi and A Vaheri in Helsinki, Finland, and by W David Sedwick and Frantisek Sokol in Philadelphia, USA, in 1970.

We now know that rubella only infects humans in nature. It is spread in droplets in the air and two weeks or so later produces malaise, swollen glands, a red rash, which usually begins on the face then spreads to the rest of the body, and a mild fever.

The ability to grow rubella virus in cell culture led to the development of effective vaccines around 1970. The first vaccine to be introduced was at first given to school-age girls to prevent congenital rubella. Later, it was incorporated into a combined measles–mumps–rubella vaccine. German measles and the dreadful scourge of congenital rubella should become a thing of the past, at least in the First World. In 2011, there were hardly any cases in the USA but on a global scale, 114,000 cases, no doubt a considerable underestimate, were reported to the World Health Organization. Babies are still at risk!

47

Mumps

SCIENTIFIC NAME:	mumps virus, a member of the genus *Paramyxovirus*
DISEASE NAME:	mumps
DISTRIBUTION:	many parts of the world
TRANSMISSION:	aerosol
CLINICAL FEATURES:	swollen, painful parotid and other salivary glands, fever
COMPLICATIONS:	inflamed testicles, meningitis, encephalitis
DIAGNOSIS:	clinical; appearance of mumps antibodies in the blood
TREATMENT:	no specific treatment
PREVENTION:	an effective vaccine is available

You might have had mumps. I cannot remember whether I did or not as a child. Childhood mumps, mainly affecting the salivary glands, especially the parotid glands just in front of and below each ear, is one thing. Mumps after puberty in males is another. I had a relative who acquired the disease while a serviceman during World War II. He had inflammation of his testes (known as orchitis) which left him unable to have children, a well-recognized complication.

Mumps has been around for a long time. Hippocrates (p.116) gave an excellent description of the illness nearly 2,500 years ago, recognizing involvement of both the parotid glands and the testicles:

> Swellings appeared about the ears, in many on either side, and in the greatest number on both sides, being unaccompanied by fever so as not to confine the patient to bed; in all cases they disappeared without giving trouble... They were of a lax, large, diffused character, without inflammation or pain... They seized children, adults, and mostly those who were engaged in the exercises... Many had dry coughs without expectoration, and accompanied with

> hoarseness of voice. In some instances earlier, and in others later, inflammations with pain seized sometimes one of the testicles, and sometimes both; some of these cases were accompanied with fever and some not; the greater part of these were attended with much suffering.[137]

Mumps was brought back onto the medical map, at least as far as the English-speaking world was concerned, by Robert Hamilton (1721–1793), a physician in Norfolk, England. In 1773, he read a paper to the Philosophical Society of Edinburgh (published in 1790) which gave a very detailed account of his observations over more than a decade of probably hundreds of patients, mostly males between the ages of puberty and 30. After remarking that the Scots called this disease 'branks' and which 'the common people in England vulgarly called the mumps',[130] he suggested the name 'angina maxillaris' (swelling of the jaws). Fortunately 'mumps' won. Hamilton recognized the swelling of the parotid salivary glands and sometimes the testicles that occurred and described the course of time to resolution, noting that the testicles may sometimes be left wasted. He was also aware that occasionally after resolution of testicular inflammation 'a fresh exacerbation of fever ensues, the head is affected, delirium follows, with convulsions and other dreadful symptoms, and sometimes death closes the scene'.[130] He was describing what we now call mumps encephalitis—inflammation of the brain. He was uncertain whether it was 'contagious' because sometimes it went through a family but sometimes it did not. Hamilton himself had a severe case of the mumps but none of the six adults or four children in his family suffered from it. A year later, his six-year-old daughter had mumps but no-one else caught it. But Hamilton did remark that he had never seen anyone have a second bout.

Where did this word 'mumps' come from? It is all rather unclear. Suggestions include 'mump', an old English word meaning 'lump', the Icelandic 'mumpaskaeler', indicating a grimace, and the Dutch 'mompen', signifying 'mumbling'. One can imagine all of them fitting the symptoms and signs of mumps. Take your pick.

When bacteria were recognized to cause disease in the second half of the nineteenth century, many attempts were made to find a bacterial cause for mumps. Various cocci were suggested but nothing came of them. In 1908, Saverio Granata in Italy suggested that mumps was caused by a filterable virus. He inoculated sterile salivary filtrate from two patients into rabbits and two weeks later the parotid glands became swollen. In 1913, the Frenchmen Charles Nicolle (p.156) and Ernest Alfred Conseil (p.407) aspirated fluid ect from the parotid glands of children and then injected it into the parotids of monkeys; they all developed a fever and in one the parotid gland swelled. In the following year, Mervyn Henry Gordon (1872–1953) in Britain inoculated a sterile filtrate of saliva from patients with mumps into the brains of 10 monkeys: four died from meningitis. Another monkey, into which the fluid was injected intravenously, developed swollen parotid glands 11 days later. Gordon could not transfer the disease from one monkey to another and concluded that mumps was due to a filterable virus of comparatively low virulence. In 1916, Martha Wollstein in New York claimed to induce a disease similar to human mumps in cats, but later investigators thought this was questionable.

But then in 1925, Yves Kermorgant in France muddied the waters by claiming that mumps was caused by a spiral bacterium. Therefore Claude Johnson and Ernest Goodpasture (p.431) in Nashville, Tennessee believed that the whole subject needed to be looked at again. They collected saliva from six patients early in the disease, passed it through a Berkefeld filter, then injected the material into the parotid duct, which opens into the mouth, of six rhesus monkeys; four developed inflamed parotid glands. They then transferred the illness from monkey to monkey through seven passages. Saliva from normal people had no effect. Not unreasonably, Johnson and Goodpasture concluded in 1934 that this virus was the causative agent of mumps.

In 1945, Karl Habel of the US Public Health Service reported that he had been able to grow mumps virus obtained from infected monkey parotid glands but not from human saliva in the yolk sac, amnion, and chorioallantoic membrane of chick embryos. Within a year, William Ian Beveridge (1908–2006) and colleagues in Melbourne, Australia, succeeded in growing mumps virus from the saliva of three patients in developing chick embryos.

This all seems relatively straightforward. How then can one explain the dreadful experiments of Gertrude Henle (p.502) and her colleagues from the Children's Hospital in Philadelphia, USA, which were reported in 1948? They wanted to find out when patients became infectious after exposure, how long their saliva remained infectious, and whether virus could be recovered from subjects who were inoculated with the virus but failed to develop overt disease. They selected *institutionalized children* in good physical condition without a history of mumps, checked that this was the case by showing the absence of antibodies in the blood, and then these children 'were considered susceptible to mumps and, therefore, chosen for these experiments with the permission of parents and guardians'.[135] Fifteen children had virus sprayed into or around their mouth; seven of them developed mumps after about two weeks. Clearly, we do need ethics committees to stop out-of-control investigators.

Electron microscopical studies began with the reports of WJ Elford and colleagues and of ML Weil and colleagues in 1948. The virus is very variable in shape ranging in size from 100 to 600 nm (Figure 213). In 1955 Henle and Friedrich Deinhardt reported that they were able to isolate and propagate mumps virus in tissue culture.

We now know that mumps only infects humans in nature. It is spread in droplets in the air and two weeks or so later produces malaise, mild fever and swollen salivary glands, especially the parotid glands which produce lumps near the ears. The testes may become swollen a few days later, usually in males infected after puberty. The virus may invade the central nervous system, but fortunately symptoms are rare.

The advent of tissue culture systems led to the determination that mumps virus, like other paramyxoviruses, contained RNA. It also smoothed the path

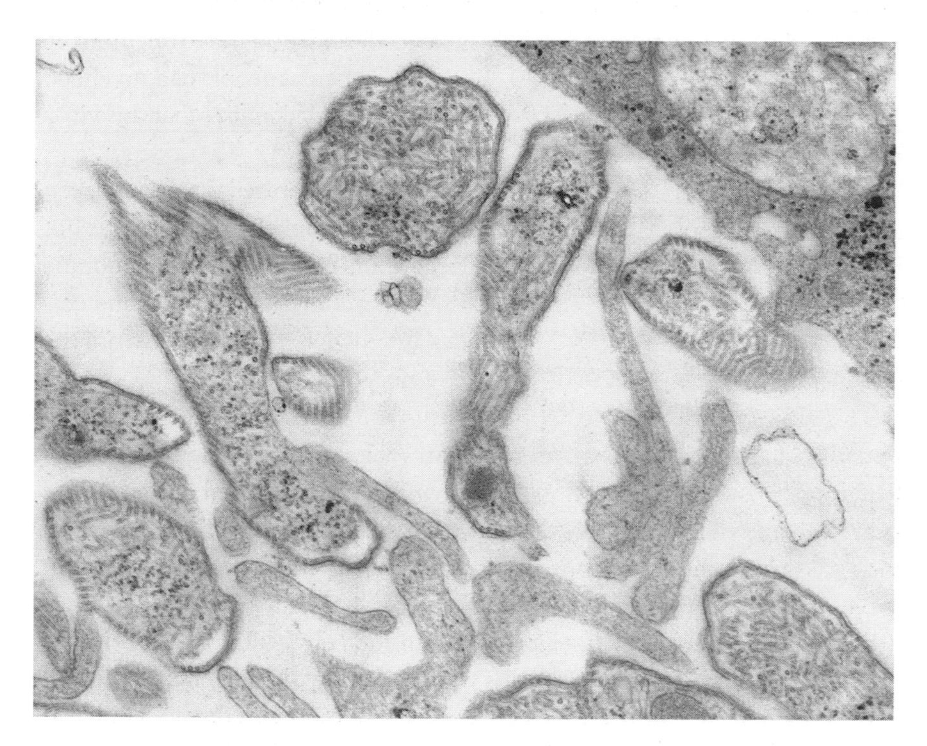

Figure 213. An electron micrograph showing the various sizes and weird shapes of mumps viruses.

for the development of vaccines. A mumps vaccine was first released in 1967 but was soon incorporated into the measles–mumps–rubella vaccine which is still in widespread use. The numbers of cases of mumps have declined dramatically in developed countries. Nevertheless, outbreaks occasionally occur among those who have not been immunized, the worst being in the United Kingdom in 2004–5 when over 56,000 people went down with the disease. Most of those affected were young adults. Over 6 per cent of them required admission to hospital; three-quarters had orchitis but significant numbers had meningitis or inflammation of the pancreas (pancreatitis). Partial immunization of a population against mumps may delay infection in the unimmunized until adult life when the consequences of infection may be more severe. Keep your children immunized against mumps!

48

Varicella (chickenpox and shingles)

SCIENTIFIC NAME:	varicella-zoster virus, a member of the genus *Varicellovirus*
DISEASE NAME:	a. generalized—chickenpox = varicella, b. localized—shingles = herpes zoster
DISTRIBUTION:	most parts of the world
TRANSMISSION:	aerosol
CLINICAL FEATURES:	a. chickenpox—centripetal rash with papules, vesicles and crusts, fever, b. shingles—pain in the distribution of a nerve root with a localized rash appearing a day or two later
COMPLICATIONS:	a. chicken pox—pneumonia, encephalitis, b. shingles—persistent pain called post-herpetic neuralgia
DIAGNOSIS:	clinical
TREATMENT:	antiviral agents such as aciclovir
PREVENTION:	an effective vaccine is available for chickenpox and a partially effective one for shingles

I was about five years old. I distinctly remember sitting up in my parents' bed covered with the itchy scabs of chickenpox which were in turn covered with calamine lotion. Depending upon which country you live in (because they introduced a vaccine at different times), you may well have had chickenpox if you were born before 1990–2000. Chickenpox usually begins with a mild fever, lethargy, and aches and pains, then after a day or two widespread blisters (vesicles) appear on the skin, become scabs, and eventually fall off after a week or so.

How long has chickenpox been around? It is hard to say but probably for millennia. The reason for this uncertainty is that doctors of olden times were faced with fevers associated with skin blisters which we now know are due to either chickenpox or smallpox, and the latter illness might be either very severe with many deaths or relatively mild with few deaths. It is not surprising that they had some trouble in differentiating them. Rhazes (p.438) in the tenth century wrote a tract describing the differences between smallpox and measles and some believe that he also recognized a 'mild or weak smallpox' (i.e. chickenpox) which did not confer protection against catching smallpox later. In 1553, the Sicilian physician Giovanni Filippo Ingrassia (1510–1580) described what appears to be chickenpox and which he called 'crystalli'. The first person to clearly differentiate between smallpox and chickenpox in the English language was the English physician William Heberden (1710–1801) in 1768. This was important, he said, because:

> though it [chickenpox] be so insignificant an illness . . . yet it is of importance on account of the small pox, with which it may otherwise be confounded and so deceived the persons who have had it into a false security which might keep them either from keeping out of the way of the smallpox or from being inoculated.[134]

He then goes on to describe what he considered the key differentiators:

1. The appearance on the second or third day from the eruption of that vesicle full of serum upon the top of the pock.
2. The crust, which covers the pocks on the fifth day, at which time the small pox are not the height of their suppuration.[134]

All was not plain sailing, however, for not everyone agreed. For example, Ferdinand von Hebra (1816–1880) in Vienna, perhaps the most famous dermatologist of the nineteenth century, and his acolytes did not accept that chickenpox and smallpox were different. But clinical differentiation gradually became clearer. In 1902, William McConnell Wanklyn (*c*.1867–1929), who had the grandiose title of 'Smallpox referee and medical superintendent of the River Ambulance Service of the Metropolitan Asylums Board' in London, England, published a useful paper. He had the opportunity of seeing 200 cases of chickenpox sent to the diagnosing station during an outbreak of some 7,000 smallpox cases. He believed that the distribution of the rash was of cardinal importance. Smallpox lesions were most abundant on the face, wrists, hands, and feet but comparatively light upon the trunk. In contrast, the rash of chickenpox was most prominent upon the trunk (Figure 214). In 1906, Ernest Edward Tyzzer (1875–1965), an American working in the Philippines, noted that most patients he saw in a prison with chickenpox had smallpox vaccination scars, thus

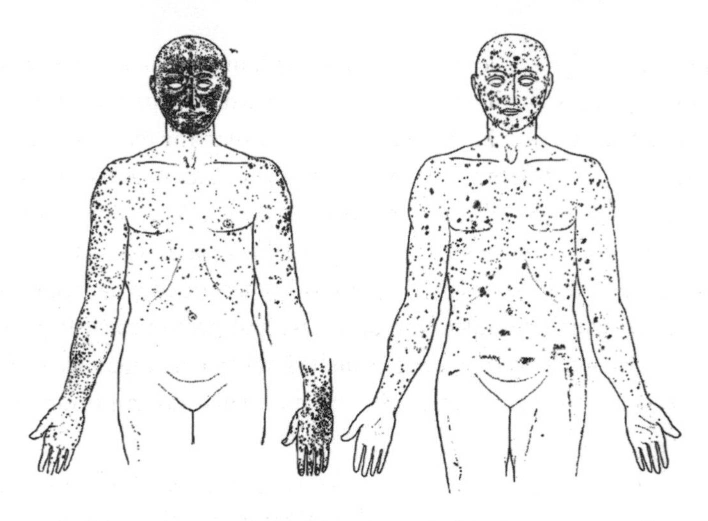

Figure 214. Wanklyn's diagrams of the distribution of the lesions of smallpox (left) and chickenpox (right).

suggesting the two diseases were different as exposure to one did not prevent the other. Furthermore, he found multinucleated giant cells in fluid from chickenpox lesions but not from smallpox vesicles. But the proof of the pudding would have to await the finding of the causes of the two diseases.

What did cause chickenpox? In 1843, Mr Erasmus Wilson, consulting surgeon to the St Pancras Infirmary, England, who was a firm believer in the clinical distinction between chickenpox and smallpox, wrote in *The Lancet*:

> The cause of varicella is an infectious and contagious poison, identical with that of smallpox, and occurring under circumstances where the poison is modified or diluted.[323]

He was on the right track although wrong in certain major respects. Could chickenpox be transferred by inoculation? The answer was not clear-cut. When Tyzzer reviewed the literature, he found that seven authors were unable to transmit infection in humans whereas three authors reported that they had. He therefore tried inoculating the skin of monkeys and the cornea of rabbits with fluid from chickenpox lesions but nothing happened. In 1923, Thomas M Rivers and William S Tillett in New York thought that they could transmit chickenpox as a filterable virus to rabbits but this turned out to be a red herring; they were transmitting a different, natural virus infection of rabbits.

Could an infectious agent be seen in chicken pox blisters? No-one could find any bacteria. In 1911, Henrique de Beaurepaire Aragão (p.158) in Brazil reported

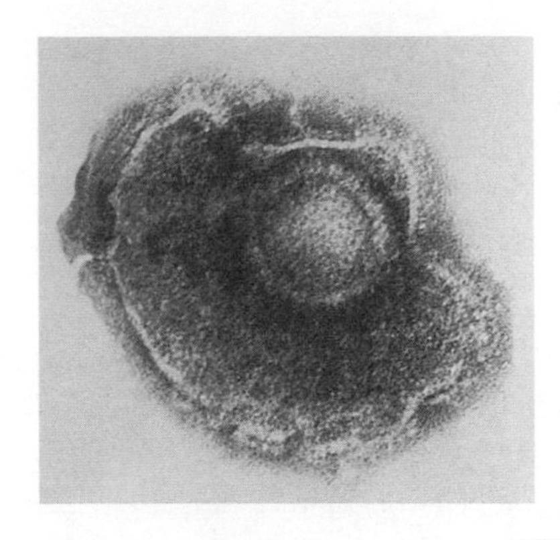

Figure 215. Electron micrograph of varicella-zoster virus.

seeing small bodies in fluid from chickenpox lesions. Then these 'elementary bodies' were seen again by Enrique Paschen (p.441) in 1917: they were similar to those he had found earlier in smallpox. In 1933, Cleeve Russell Amies (1903–1975) noted that these bodies clumped together when they were mixed with serum from a patient with chickenpox, thus suggesting they were the cause of the disease. The first drawings of what these looked like under the electron microscope were provided in 1943 during World War II by the German physician Helmut Ruska (p.432); he thought they were round with a central dot. Five years later, Frederick P Nagler and Geoffrey Rake (p.288) provided photographs of chickenpox virus, then others showed that the virus has a lipid envelope and is 150–200 nm in diameter as shown in Figure 215.

Why is this disease called 'chickenpox' (sometimes spelled as chicken pox and chicken-pox)? It is not easy to say. The term appears to have been first used by the English physician Richard Morton (1637–1698) in 1694 as 'chicken pocks' to describe what he thought was a mild form of smallpox. Various fanciful suggestions have been made as to the origin of the name. It is simplest to accept the explanation of Samuel Johnson in his famous first dictionary of the English language of 1755—'so-called from its being of no very great danger', i.e. it's really chicken-feed or no big deal when compared with smallpox. 'Varicella' is an alternative name that has been used since the middle of the eighteenth century. It is probably simply an appropriation of the French 'la varicelle' and Italian 'varicella' which are their words for chickenpox. Where did they come from? They are said to be 'irregular diminutives' (i.e. meaning something lesser) of variola (= smallpox). Today in English, chickenpox and varicella are used interchangeably. But this is by no means the end of the story.

Shingles (herpes zoster)

When one of my sons, who had chickenpox when he was 18 months old, was approaching his thirteenth birthday, he complained of a sharp pain in his shoulder and upper chest. I told him that he was trying to bowl the cricket ball too fast. Two days later, I had egg on my face. He showed me a rash on his chest which was classical shingles. This word, in one spelling or another, has been used in the English language for over 600 years and is a corruption of the medieval Latin 'cingulus', which is in turn derived from the old Latin word 'cingulum' meaning 'girdle' or 'belt'. The other name given to it is 'herpes zoster'. We shall discuss the origin of the Greek word *herpes* in the next chapter. It was used by Hippocrates (p.116) some 2,400 years ago, though whether he was referring to shingles is a matter of some conjecture. *Zoster* is the Greek word for 'girdle' and is descriptive of the girdle of rash often seen in this condition. Some authorities have concluded that the Roman, Pliny the Elder (AD 23–79), gave a recognizable description of shingles which he called zoster. The problem is that over the ensuing centuries, these words were applied indiscriminately to a variety of dermatological conditions.

There is not much doubt, however, about the account by the afore-mentioned William Heberden because he describes not only the rash but also the characteristic pain:

> The herpes, or shingles, has begun with a pain which has lasted in some for two or three days before the eruption appeared. It consists of a heap of watery bladders, itching at first, of which there are sometimes nearly so many as to surround the body, whence it has its name of shingles from *cingulum*...It will be several days before they are quite healed. But the greatest part of the misery is many times to come after they are perfectly well, and the skin has recovered its natural appearance, for I have known a most pungent, burning pain left in the part, which has teased the patient for several months, or even for two or three years.[134]

In this marvellous account, Heberden describes the pain which precedes the appearance of the rash and the sometimes unpleasant sequel of a nasty pain lasting months or even years, which we now call post-herpetic neuralgia.

When Heberden talked about the rash nearly 'surrounding the body', he probably meant only one side of the body because that is what shingles does. In 1831, Richard Bright (p.248) suggested that the rash of shingles coincided with the distribution of a nerve coming from a single level of the spinal cord. In 1862, the German dermatologist Friedrich Wilhelm Felix von Bärensprung (1822–1864) confirmed the correctness of this deduction when he performed an autopsy on a patient who died from tuberculosis 40 days after the appearance of shingles. He found that the dorsal root ganglion was damaged although he did not properly understand the anatomy of this structure. It needs some explanation. Four nerves, two on each side, come out from the spinal cord at the level of every vertebra (Figure 216). On each side, one is at the front (the ventral or

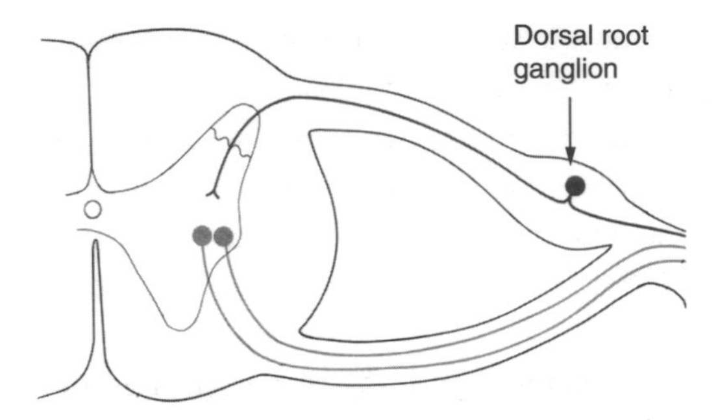

Figure 216. Diagram of the sensory nerves that pass through the dorsal root ganglion.

anterior root) and one is at the back (the dorsal or posterior root). They then join to form the nerve that travels to the structures it supplies. The ventral root carries motor fibres which cause muscles to contract. The dorsal root carries fibres of sensation (touch, pain, and temperature, etc.) from the skin. But the cell body with the nucleus of the sensory neurone is not located in the grey matter of the spinal cord but in a swelling in the dorsal root called a ganglion. Bärensprung's observation was expanded in great detail by the English neurologist Henry Head (1861–1940) and Australian pathologist Alfred Walter Campbell (1868–1937) working in London. In 1900, they reported a mammoth paper 170 pages long detailing their post-mortem studies of 21 patients with herpes zoster in which they precisely mapped sensory dermatomes. They described what they saw under the microscope:

> the universal characteristic of the changes we have described in the affected ganglion is the presence of inflammation in every degree, from scattered collections of small round cells to profound haemorrhages in the centre of inflammatory foci.[133]

Furthermore, they demonstrated:

> that when a certain ganglion is affected the eruption most frequently lies over a definite tract of skin, which may be called the normal area from which fibres enter that ganglion.[133]

It is this area supplied by each nerve root that we now call a dermatome. Head and Campbell had no idea what caused this condition:

> Of the nature of the agent which is responsible for this process we are completely ignorant. Microscopically, we have not been able to find any signs of bacterial infection... But this agent, whatever it may be, seems to have a specific attraction for the posterior root ganglion exactly as the equally unknown cause of acute anterior poliomyelitis attacks the substance of the anterior horns.[133]

Even so, there was a clue. The individual lesions of localized shingles looked very similar to the generalized lesions of chickenpox. Were they the same condition or different diseases? The next advance was made by the Hungarian paediatrician János Bókay (also spelled Janus von Bokai, 1858–1937). In 1892 and again in 1909, he reported that he had seen children who had contracted chickenpox after exposure to adults with shingles, thus suggesting that they were the same disease.

Karl Kundratitz (1889–1975) and Benjamin Lipschütz (1878–1931) tried the effect of direct experiment in 1925. They inoculated 28 children under the age of five with fluid from the lesions of zoster in the hope of immunizing them against chickenpox; 17 developed local lesions but none had chickenpox. Similar observations were made by Johan Gustav Edvin Bruusgaard (1869–1934) in Norway in 1932. He inoculated a total of 18 children who had never had chickenpox with fluid from five patients with shingles. Four children developed localized reactions at the site of inoculation and another four children went down with typical chickenpox. Then other children caught chickenpox from these cases. On the other hand, Ernst Lauda and D Stöhr in 1926 did not produce any local lesions in 54 children inoculated with material from 17 zoster cases although three developed chickenpox. Very confusing! All these investigations would not be permitted today because of the risk of transmission of serious blood-borne viruses. Perhaps because of these seemingly contradictory results it was still being taught at Harvard University in 1940 that chickenpox and shingles were distinct, unrelated clinical entities.

Clearly this dispute was not going to be answered until the agents that caused these diseases could be isolated and compared. A start was made by Ernest W Goodpasture (p.431) and Katherine Anderson (1909–1995) in 1944. They grafted fragments of human skin onto the chorioallantoic membrane of chick embryos and then inoculated fluid from shingles vesicles. The eosinophilic lesions in nuclei that were typically seen in chickenpox and zoster lesions were seen in the skin samples a week or so later. Incidentally, Ernest and Katherine married in the following year. Their observation was supported when the aforementioned Rake and his colleagues showed in 1948 that the virus found in shingles looked just like that of chickenpox when examined under the electron microscope.

In 1953, Thomas Weller (p.469) successfully inoculated cultures of human skin and muscle embryonic tissue with vesicle fluid from cases of chickenpox and from people with shingles. In both cases, there were similar inclusion bodies in the cell nuclei. Five years later, Weller and his colleagues showed by immunological studies that these viral inclusions were identical. When he reviewed these studies nearly 30 years later, Weller wrote:

> When my group's definitive papers were published in 1958...we concluded that the evidence that varicella and zoster were caused by the same virus was irrefutable; the agent was named 'VZ virus'.[318]

I could not find these precise words when I read the relevant papers but the virus of chickenpox and shingles did indeed come to be called varicella-zoster virus or VZV. Harvard University (where Weller worked) had to change its teaching!

What sort of a virus is VZV? All the viruses that we have discussed thus far are RNA viruses. In 1972, H Ludwig and colleagues broke up infected cells and obtained cell-free virus harbouring DNA.

Another important question still exercised many minds—how did chickenpox lead to shingles and vice versa? In 1943, Joseph Garland in Boston, USA, postulated that the shingles 'reaction is probably due to the activation of a latent virus by factors acting as *agents provocateurs*'.[111] Support for these ideas was provided by Robert Edgar Hope-Simpson (1908–2003) in England in 1965. He had observed 192 cases of shingles in his general practice over 16 years. Household contacts of patients with herpes zoster did not themselves develop shingles nor did people exposed to chickenpox get zoster. On the other hand, people who had never had chickenpox did develop varicella after they were exposed to shingles. He went on to show that the frequency of zoster in a population increased with advancing age. Hope-Simpson suggested that when the chickenpox virus is first caught, it invades the bloodstream and travels to the skin where it produces lesions. This fits in nicely with Tyzzer's observations way back in 1906 that in chicken pox there were inclusions in the nuclei of endothelial cells lining small blood vessels. From there Hope-Simpson hypothesized:

> virus also enters...the endings of the sensory nerves, whence it is transported up the sensory fibres until it arrives at the sensory ganglia, where it becomes established as a latent infection in the nuclei of the neurones.[141]

In 1972, Margaret Esiri and AH Tomlinson working in Oxford, England, proved he was right. They examined the skin, nerves, and trigeminal ganglion (found near the base of the skull) of a woman who had died four days after the onset of ophthalmic shingles. They found virus in skin, Schwann cells around nerve fibres, and in the nuclei of neurones in the ganglion as well as their satellite cells.

QED? Not quite. Why does this latent virus occasionally reactivate? What are Garland's *agent provocateurs*? Sometimes it happens when the immune system is impaired but in most cases, we still simply do not know why.

VZV only infects humans. Even though a vaccine has been available for 10–20 years, it is going to be around for many decades to come as all those who have had chickenpox are harbouring the virus and it may spring into life as shingles at any time. There is some bright news on the horizon. A vaccine to prevent reactivation of latent virus has recently been introduced which is claimed to halve the likelihood of an attack of herpes zoster.

49

Herpes simplex (cold sores and more)

SCIENTIFIC NAME: herpes simplex viruses 1 and 2, members of the genus *Simplexvirus*
DISEASE NAME: cold sores, herpes labialis, genital herpes
DISTRIBUTION: all parts of the world
TRANSMISSION: contact
CLINICAL FEATURES: ulcers of the lips and genitals; rarely encephalitis
DIAGNOSIS: clinical
TREATMENT: aciclovir and related antiviral agents
PREVENTION: no measures are available for cold sores but use of condoms and the avoidance of promiscuous sex may prevent genital herpes

You have probably had cold sores—blisters which may ulcerate and crust on or around your lips. The medical name for these lesions is 'herpes labialis'—herpes of the lips. The word *herpes* is a Greek word which meant 'creeping'. It was used for a plethora of dermatological conditions by the ancient Greek and Roman doctors for conditions that have nothing to do with our modern, rather inappropriate, use of the term. Even as late as the nineteenth century, medical opinion was divided as to significance that should be attached to the term 'herpes'. In 1813, the Englishman Thomas Bateman (1778–1821) published an influential book[39] in which he described six forms of herpes characterized by one or more localized crops of small superficial vesicles (blisters) on the skin which healed spontaneously within 10 to 12 days. His groups were herpes phlyctaenodes, herpes zoster, herpes circinatus, herpes labialis, herpes praeputialis, and herpes iris. Eventually views such as his prevailed. Three of Bateman's forms have survived—herpes labialis, herpes praeputialis (= genitalis), and herpes zoster.

Irrespective of the niceties of terminology, it seems that an early description of herpes labialis was given by Herodotus, a Roman physician of the first or second centuries, who noted that it heralded the termination of a fever. This association was also recognized by Richard Morton (1637–1698), who in 1694 called the lesions 'herpes febrilis' (herpes of fever). Why are these lesions called cold sores? Perhaps because an association was noted between the onset of the common cold and the appearance of these lesions.

Cold sores are an irritating nuisance but not devastating. Similar lesions on the penis or vulva are a different thing altogether. The first person to describe genital herpes, although he did not use that name, was the French physician Jean Astruc (1684–1766) in 1736. He said the lesions on the penis were like:

> watery or crystalline bladders, which are filled with lymph, which is thin or thick or diaphanous...(and)...these disorders...are common to women from the same cause...The labia...clitoris and prepuce of the clitoris...are swelled and inflated in the same manner as the prepuce or glans in men.[30]

Were cold sores and genital herpes different manifestations of the same disease or were they different conditions? The answer to that question would depend upon finding the cause or causes. In the nineteenth century, herpes labialis was often said to be due to a nervous condition. This was perhaps not as silly as it sounds because an attack often coincided with undue stress. But others wondered whether it was an infection. The first experiment to find out whether these lesions were transmissible was undertaken by Jean-Baptiste Emile Widal (1825–1893) in 1873 in France. He took fluid from an herpetic lesion and re-inoculated it into other anatomical sites in the same person (apparently himself). New lesions appeared suggesting there was indeed an infectious agent. In fact, this procedure using oral or genital material would be repeated many times in France, allegedly for therapeutic reasons, for the next 60 years or so.

We now need to go off on a tangent. In 1871, Johann Friedrich von Horner (1831–1886) in Zurich, Switzerland, described inflammation in the cornea (the clear part of the eye), now called herpetic keratitis, which sometimes accompanied pneumonia and became recurrent. Another Swiss ophthalmologist Emil Emmert (1844–1911) in 1885 described the creeping ulceration of the cornea which came to be called a dendritic ulcer, and its association with cold sores was increasingly recognized. In an attempt to find out whether this was transmissible, the German ophthalmologist Wilhelm Grüter (1882–1963) in 1912 took material from the cornea of a patient with herpetic keratitis and inoculated it on the cornea of a rabbit and reproduced the disease. Then in a perhaps understandable experiment, he transmitted the disease back from the rabbit to the normal cornea of a man who was already blind.

Grüter's study was not published until 1920, probably because of the interference of World War I. In the previous year, A Löwenstein had simulated Grüter's study except that this time he used fluid from cold sore blisters to produce the disease in rabbit corneas. Bacteria were not present in these samples; clearly the disease was being caused by a filterable virus which could cause both cold sores and lesions on the cornea. In fact, the virus could do even more than that. Also in 1920, the Swiss virologist Richard Doerr found that inoculation of fluid from cold sores onto the rabbit cornea sometimes also led to encephalitis, i.e. inflammation of the brain, in these rabbits.

It was clear that not only were these lesions caused by filterable agents, but it was looking likely that the same virus could cause these different syndromes. In fact in 1906, these various conditions had been lumped together as 'herpes simplex', 'simplex' being the Latin word for 'single'. The infectious agent was given a name, even though it had not yet been found and its presence simply inferred. It was labelled herpes simplex virus, although there was for a time a vogue for calling it herpesvirus hominis.

But what about herpes zoster which was also characterized by blisters in the skin? In 1922, Levaditi (p.467) showed that blister fluid from herpes zoster did not produce disease on rabbits' corneas whereas that from cold sores did. The two diseases were clearly due to different agents and infectivity for the cornea was the first method for showing this. Then in 1933, James Dawson in Nashville, USA, reported the successful cultivation of herpes simplex virus in chick embryos, with microscopic changes similar to those seen in human herpes lesions, particularly inclusions in the nuclei of cells.

We have mentioned herpes encephalitis induced experimentally in rabbits. The first human case was described by M Hass in 1935 in which a newborn baby died from herpes simplex virus encephalitis. Then it was found that this condition could affect adults too when Chris J Zarafonetis and Joseph E Smadel (p.539) in 1944 described fatal herpes simplex encephalitis in an adult. Disease induced by this organism could range from the common and trivial cold sore through relatively frequent, disturbing genital herpes to the rare and devastating brain inflammation.

How do infections with herpes simplex begin? When one of my sons was two years old, he developed a dreadful ulcerative inflammation of his mouth.

He would not eat and hardly drank. It was awful. He had what is now recognized as herpetic gingivostomatitis (inflammation of the gums and mouth). As we have seen, by the late 1930s, the usually recognized forms of herpes simplex infections were cold sores of the lips, genital herpes, and corneal herpes. In 1938, Katherine Dodd and colleagues in Nashville, USA, described a new syndrome. They saw 88 children with similar symptoms to those described for my son. Samples were taken from 12 children and inoculated into the corneas of rabbits. In each case, eye disease typical of herpes simplex infection developed. In the following year, Burnet (p.469) and Dora Lush (1910–1943) in Melbourne, Australia, showed that these primary infections were followed a couple of weeks later by the appearance of antibodies against herpes simplex, confirming that this disease was due to the herpes simplex virus.

What does the virus that causes these lesions look like? Lewis Coriell and colleagues in the USA in 1950 reported the results of electron microscopical examination of fluid from herpetic lesions taken from the lips, genitalia, skin, and chick embryo cultures. They found the same viruses, about 175 nm in diameter, which would turn out to look very similar to those of varicella-zoster virus (Figure 215). What type of nucleic acid did this virus contain? In 1961, M Anthony Epstein (p.502) in London demonstrated that herpes simplex viruses grown in a culture of cells called HeLa cells contained DNA.

The following year, Karl Eduard Schneweis reported that herpes simplex viruses could be separated into two groups on the basis of their immune structure. These were called herpes simplex type 1 and herpes simplex type 2. This was rapidly followed by the demonstration that type 1 usually caused infections of the mouth and face (so-called 'above the belt infections'), while genital infections were mostly due to type 2 (below the belt infections). This led to the demonstration that genital herpes was transmitted sexually whereas orolabial infections were passed on by contact and the respiratory route.

There was still another puzzle. It was common observation that both oral and genital herpes lesions frequently recurred. Yet in 1930, Andrewes (p.507) and E Arnold Carmichael had shown that patients with recurrent herpes labialis had antibodies against herpes simplex virus in their blood; indeed these antibodies were found in almost 80 per cent of the population. Clearly these antibodies had not eliminated the infection. It seemed that the virus remained quiescent and occasionally flared up. If so, where was it remaining latent?

One logical place to look was in cells at the site of the initial lesion. However, attempts to isolate the virus from the site of a previous lesion during a remission never had any success. An alternative idea was suggested by the observation of Harvey Williams Cushing (1869–1939) in 1904. He noted that treatment of trigeminal neuralgia (pain in the face) by destruction of the trigeminal ganglion precipitated herpes labialis. Could the virus be latent in a nerve ganglion? This is reminiscent of varicella-zoster virus which is latent in the dorsal root ganglia. As far back as 1925, Goodpasture (p.431) had shown that when herpes simplex virus was injected into the muscles of the jaw, it travelled up the nerves to the brain. It was eventually found that the virus remains latent in nerves, especially the trigeminal ganglia at the base of the skull following an initial herpes infection of the mouth, and in the sacral ganglia at the base of the spine following genital infection. Why latent infections should occasionally flare up remains a mystery. Herpes simplex only infects humans in nature. Almost everyone has herpes simplex type 1. It is going to be with us for a long time yet.

50

Glandular fever (infectious mononucleosis)

SCIENTIFIC NAME:	Epstein–Barr virus of the genus *Lymphocryptovirus*
DISEASE NAME:	glandular fever or infectious mononucleosis
DISTRIBUTION:	all parts of the world
TRANSMISSION:	contact, especially oral
CLINICAL FEATURES:	sore throat, generalized enlarged lymph nodes, malaise, fever, and sometimes enlarged liver and spleen
DIAGNOSIS:	atypical lymphocytes, heterophile or specific EBV antibodies in the blood
TREATMENT:	supportive: fluids and paracetamol or aspirin
PREVENTION:	no practical preventive measures

Glandular fever is another infection that has afflicted one of my offspring, this time while a teenager at university. Some people think this condition may have first been described independently by Filatov and Pfeiffer. Nil Fedorovich Filatov (1847–1909) in Moscow, Russia, in 1887 published an account of an illness in children, mostly between the ages of two and four years, in which there was fever and enlarged, tender lymph nodes under the ears and behind the jaws as well as enlargement of the liver and spleen. The fever lasted for a week and the swollen nodes for two to three weeks, then the illness resolved spontaneously. Emil Pfeiffer (1846–1921) of Wiesbaden, a paediatrician in Germany, in 1889 described two sorts of patients. The first were children five to eight years old who had the sudden onset of high fever, severe pains in the arms and legs, and numerous tender, swollen lymph nodes in the neck; after a few days everything returned to normal. Perhaps more recognizable is his description of his second type:

> of cases which last longer and the disease...can prolong itself eight or ten days....the fever continues for several days...the glands swelling up, first on one side of the neck, then on the following day, those on the other side begin to be painful and to enlarge...the throat becomes red and there is slight pain on swallowing...On the third or fourth days...the liver and spleen are definitely enlarged...and...there is pain in the lower abdomen...The chronic cases give rise to more errors, false diagnoses, and difficulties than the light cases.[249]

Because swollen lymph nodes (= glands) and fever were so prominent, he named this condition 'Drüsenfieber', which is German for 'glandsfever'. Not surprisingly, this illness became known in English as glandular fever. Pfeiffer did not have much doubt that this was an infectious disease, writing, 'clinical experience characterizes this disease...as an infectious disease, because it appears in epidemics and indeed in house epidemics'.[249]

The illness was soon recognized in other countries. However, most observers agreed that Pfeiffer was wrong when he said that lymph nodes in the groin and under the armpits did not enlarge. This was important as it indicated the illness was a generalized infection rather than being secondary to a throat infection. Whether or not these conditions were really glandular fever we will never know as neither Filatov nor Pfeiffer examined the blood.

In 1920, Thomas Peck Sprunt (1884–1955) and Frank Alexander Evans (1889–1956) in Baltimore, USA, described a series of six young adults with symptoms similar to those alluded to above in whom there was a significant increase in the numbers of a particular type of white cell known as 'mononuclear cells' in the blood. They realized that the patients had a benign, self-limited illness, not a spontaneously resolving leukaemia as some authors had thought, and coined the term 'infectious mononucleosis' (IM) to describe it. They did not relate the findings to Filatov's or Pfeiffer's glandular fever. Nevertheless, their observations stimulated Henry Letherby Tidy (1877–1960) in England in 1921 to examine the blood in 18 of 24 schoolboys in whom he had made a diagnosis of glandular fever. He found that the numbers of mononuclear cells (of a certain type called lymphocytes) were increased in glandular fever. He published his findings in 1923, considering that glandular fever and infectious mononucleosis were one and the same:

> It is consequently considered that glandular fever and infective mononucleosis are identical, and that an absolute lymphocytosis [increased number of lymphocytes] is a normal, though not invariable, occurrence in glandular fever.[303]

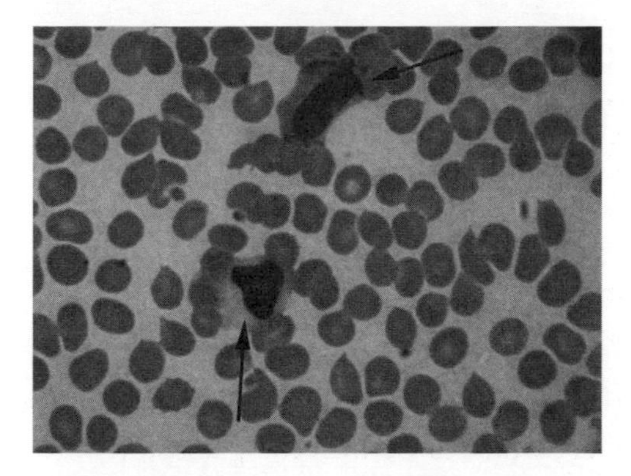

Figure 217. Reactive lymphocytes (arrows) between red blood cells.

In this same year, Hal Downey (1877–1959) and Chauncey Angus McKinlay showed that not only were these lymphocytes increased in number but their appearance was changed. These white blood cells, called reactive or atypical lymphocytes, became the hallmark for the diagnosis of infectious mononucleosis (Figure 217).

But sometimes there was uncertainty. It would be useful if another diagnostic test could be discovered, especially to help differentiate the condition from leukaemia. Serendipity was to play a part. John Rodman Paul (1893–1971) in Connecticut, USA, was examining the possible role of heterophile antibodies in rheumatic fever. Heterophile antibodies are antibodies that cross-react with antigens in a completely different species. Paul looked for antibodies in human serum that would cause sheep red blood cells to clump together. In the process he used some serum from a patient with infectious mononucleosis and this had the highest level of heterophile antibodies ever reported. Investigation by Paul and Walls W Bunnell (1902–1966) of other patients with infectious mononucleosis led in 1932 to the development of the 'Paul–Bunnell test'. In 1937, Israel Davidsohn (1895–1979) in Chicago showed that the reliability of the test could be improved by mixing the serum first with guinea pig kidney cells which removed the heterophile antibodies of infectious mononucleosis whereas they were not removed by beef red blood cells. The Paul–Bunnell test, in one form or another, is still used for the diagnosis of glandular fever today.

But what actually caused the disease? Around the turn of the nineteenth and twentieth centuries, various bacteria were suggested but none stood the test of time. And how was it acquired? Robert J Hoagland, a medical officer in the US military reported an important clue in 1964. He noted that the large majority of cases of glandular fever in cadets at West Point, an officer training school for the US Army, occurred in February, August, and September. This was a month or so after the cadets were on vacation—for the rest of the year, the cadets were generally unable to leave and their opportunities for social contact were minimal. The prominence of sore throats and enlarged lymph glands in the neck suggested the causative agent might enter via the mouth or respiratory tract. Not surprisingly, the cadets often gave a history of kissing their girlfriends during their period of leave. Thus, infectious mononucleosis became known colloquially as the 'kissing disease'. It helped explain why the peak incidence of glandular fever was in the teens and early twenties. It seemed likely that glandular fever was due to a virus, but what virus? Once more we need to go off on a tangent.

Denis Parsons Burkitt (1911–1993) was an Irish missionary surgeon who was working in Uganda when he encountered a novel form of cancer which especially affected children in the jaw. He reported his observations in 1958 and it came to be known as Burkitt's lymphoma. Michael Anthony Epstein (1921–), Bert Geoffrey Achong (1928–1996), and Yvonne M Barr (1932–) in London, England, began to investigate cells (lymphoblasts, i.e. young lymphocytes) from these tumours by electron microscopy. They reported in 1964 that in the first two specimens they had examined, they saw a virus that had a remarkable resemblance to herpes simplex virus (and varicella-zoster virus, Figure 215). This virus was famously to become known as Epstein–Barr virus (EBV). Spare a thought for Achong!

Gertrude Henle (1912–2006), her husband Werner Henle (1910–1987), and Volker Diehl (1938–) in Philadelphia, USA, became interested in EBV. They had developed a technique for growing lymphocytes (white blood cells) in which antigens (particles) of EBV could be found. So they began to investigate sera of children who had 'viral infections' to see if they had antibodies to EBV, but the results were inconclusive. Then they had a spot of luck. One of their technicians donated her blood for experiments. In January 1967, her blood cells would not grow and no antibodies for EBV were present in her serum. In August of that year, she developed glandular fever. This time her white cells grew and 2 per cent or so of them had EBV antigens in them. What is more, she had high levels of antibody against EBV in her serum. They then examined sera from 42 patients with glandular fever and 50 control sera obtained from students at Yale University. All of the glandular fever patients had high levels of antibody against EBV. Only 24 per cent of the control students had any antibody, and that was at low levels. Not unreasonably, they wrote:

> Patients with infectious mononucleosis regularly develop antibodies to the herpes-type virus (EBV)... The antibodies persist for many years and are distinct from heterophile antibodies. The epidemiology of IM and the

> seroepidemiology of EBV share many features. Thus, it appears that EBV, or a close relative of it, is the cause of IM.[136]

It did not take long for others to replicate their findings and show that antibodies against EBV appeared after an attack of glandular fever. Moreover, like other herpes viruses, EBV was a DNA virus. The cause of infectious mononucleosis had effectively been found, although as far as I know, there have been no attempts to deliberately transmit the virus to see if it caused glandular fever.

In most people, the symptoms of glandular fever resolve within a month or so. But we now also know that the virus replicates in the cells lining the mouth and causes a lifelong infection in a type of lymphocyte called B cells. There was a vogue 20 years or so ago for linking chronic EBV infection with chronic fatigue syndrome but this association has not withstood the test of time. It is likely that EBV infection and glandular fever will remain with us for the foreseeable future.

51

Influenza (the flu)

SCIENTIFIC NAME:	influenzavirus A, B, and C, each virus being a separate genus of the same name
DISEASE NAME:	influenza or the flu
DISTRIBUTION:	all parts of the world
TRANSMISSION:	aerosol infection of the respiratory tract
CLINICAL FEATURES:	fever, cough, sore throat, chills, runny nose, muscle aches, and pains
DIAGNOSIS:	clinical or finding influenza antigens in sputum
TREATMENT:	supportive—fluids and paracetamol or aspirin
PREVENTION:	isolation, wearing masks, washing hands; immunization, and influenza antivirals may be partially effective

In 1957, I was in my second year at high school. I remember looking around the class one day and noticing that half the students were missing. Over the next few weeks, almost everyone was away from school sick. I emerged unscathed. It was the 1957 Asian flu. In September 1968, I spent a month doing a locum in general practice, often peering down patients' throats, then planned to have a week's holiday before starting a new hospital job. Instead I spent the week on my bed with fever, sweats, generalized aches and pains, a hacking cough, and abject misery. I had not escaped the 1968 Hong Kong influenza pandemic.

A pandemic is an epidemic of an infectious disease that spreads across large areas of the world affecting great numbers of people. Influenza has probably been around for many centuries. Even so, it is only from the sixteenth century onwards that we can identify epidemics with a reasonable degree of certainty. One epidemic afflicted Edinburgh in 1562 and the English ambassador wrote to London referring to Mary, Queen of Scots:

> immediately upon the Quene's arivall here, she fell acquainted with a new disease that is common in this towne... which passed also through her whole courte, neither sparinge lordes, ladies nor damoysells not so much as ether Frenche or English. It ys a plague in their heads that have yt, and a sorenes in their stomackes, with great coughe, that remayneth with some longer, with others shorter tyme... The queen kept her bed six days. There was no appearance of danger, nor manie that die of the disease, excepte some olde folkes.[262]

Johann Boekel of Germany wrote of an epidemic in 1580:

> In the space of six weeks it afflicted almost all the nations of Europe, of whom hardly the twentieth person was free of the disease... Its sudden ending after a month, as if it had been prohibited was as marvellous as its sudden onset.[47]

Boekel, writing in Latin, called this disease 'catarrhum febrilem' meaning febrile or feverish catarrh; catarrh indicates running eyes and nose. The word 'influenza' entered the English language in 1743, when it was appropriated from the Italians who used it to name the epidemic raging there that year. In Italian, 'influenza' means 'influence' for some thought this influenza was a result of the influence of an unusual conjunction of the planets at the times of epidemics of coughs, colds, and fever. Thus the disease we are considering became 'influenza di catarro'. Not to be outdone, the French also made their mark with their name 'la grippe'; 'grippe' was popular for a century or so in England as an alternative name for the flu. It is derived from 'gripper' meaning 'to grip' and presumably refers to the disease gripping one in the throat.

Thomas Willis (p.205) in England was perhaps familiar with the idea of a celestial influence because in his description of an epidemic in 1658, he had this to say:

> being sent as it were by a blast from the stars, seis'd a great many; that in certain towns in a Weeks space, above a thousand Men lay ill at once. The pathognomonick symptom of this disease... was a troublesome Cough, with a copious Spitting, and a Catarrh falling on the Palate, Throat and Nostrils; there is also a feverish Distemperature... a want of Appetite, a spontaneous Lassitude, and a great Pain in the Back and Loins.[322]

Large epidemics came every decade of two. One researcher has identified 16 major outbreaks between 1700 and 1900. Bad though these were, they were nothing compared with the catastrophic 'Spanish flu' which ravaged the globe in 1918 and 1919. Where and precisely when it began is still unclear. It certainly was not Spain—it was just that it afflicted Spain one month before it hit England. The English called it the Spanish flu and the name stuck. It has been estimated that more than 20 million people died from complicating pneumonia. It was the large number of deaths in which young as well as old were affected that marked this outbreak as being different. Five hundred thousand died in the USA and 200,000 in Britain, countries which were reasonably good at keeping accurate figures. Few places escaped. Eskimo villages were wiped out. A quarter of the population of Samoa died. It even reached the New South Wales village of Queanbeyan, adjacent to where the Australian capital of Canberra was to rise. The following ditty appeared in the *Queanbeyan Age* newspaper in 1919:

> A splitting head and limbs of lead
> A burning throat and dry,
> Add to these woes a snuffling nose
> A red and streamy eye.
> Thermometer beneath my tongue
> Reads 'one-nought-four point two'
> Instead of normal 'ninety-eight.'
> I've got the Spanish flu.[22]

What caused the disease and how to prevent it were pressing questions. Prevention was difficult. The following is an extract from a notice which appeared outside a theatre in Chicago, USA:

> If you have a cold and are coughing and sneezing do not enter this theatre. GO HOME AND GO TO BED UNTIL YOU ARE WELL. Coughing, sneezing and spitting will not be permitted in the theatre.[21]

So, isolation was one preventive measure. Another was the wearing of masks. In November 1919, all the inhabitants of Chicago were ordered to wear a mask in public, failure to do so incurring a fine or being sent to jail.

As to the cause, that was quite unclear in 1918–19 although there did not seem any doubt that it was infectious. This was not for the want of trying. During an outbreak of influenza in 1889–90, Richard Pfeiffer (p.333) in Germany had isolated a bacterium in great numbers in the throats of patients with the flu. Unsurprisingly, he called this *Bacterium influenzae* (now *Haemophilus influenzae*). Not everyone was convinced. The possible role of this organism was re-examined during the Spanish flu. Drops of cultures of Pfeiffer's bacillus were instilled in the noses of volunteers; only rarely did influenza follow (and that was coincidental).

A pointer to the right direction was given by John S Koen, a veterinarian with the US Bureau of Animal Industry in Iowa. In 1918, he saw a disease in pigs which he thought was the same as the Spanish flu:

> We were confronted with a new condition, if not a new disease... The similarity of the epidemic among people and the epidemic among pigs was so close... that an outbreak in the family would be followed immediately by an outbreak among the hogs and *vice versa*... It looked like 'flu' and until proved it was not 'flu', I shall stand by this diagnosis.[166]

Koen had described swine flu. But in 1921, another false trail was followed by Peter K Olitsky and Frederick L Gates in New York. They claimed to find in the respiratory secretions of patients with influenza during the first 36 hours of the disease 'a minute bacilloid body', small enough to pass through the usual bacterial filters, which could be cultivated in kidney tissue under anaerobic conditions. They called it *Bacterium pneumosintes* and believed it was the cause of the flu. Others had difficulty in replicating their results and their proposed pathogen disappeared from view.

Koen's ideas were put to the test in 1928. Charles N McBryde and colleagues in the US Bureau of Animal Industry transmitted influenza by taking mucus from the respiratory tract of pigs and instilling it into the noses of healthy pigs. However (and somewhat confusingly), when they used material that had been passed through a bacterial filter, nothing happened. In 1928 and 1929, though, Richard Edwin Shope (1901–1966), in New Jersey, repeated the experiment and produced a mild flu in pigs using filtered material suggesting a filterable virus was indeed the cause. However, when he reported his results in 1931, Shope muddied the waters by claiming that a concurrent infection with *Bacillus influenzae* of pigs was necessary to produce the severe disease.

Nevertheless, the cause of human influenza was still pretty much of a mystery. The breakthrough came in London during an epidemic in 1933, when Wilson Smith (1897–1965), Christopher Howard Andrewes (1897–1965), and Patrick Playfair Laidlaw (1881–1940) turned their attention to the problem. They obtained throat washings from a number of patients with influenza, passed them through a bacterial filter to remove bacteria, then inoculated a variety of animal species with the filtrate. At first they had no luck. Then they heard about an outbreak of what seemed like influenza, not only in the staff but also in ferrets kept at the Wellcome Laboratories at a farm at Mill Hill in north London. They therefore turned their attention to trying to infect normal ferrets. Filtered throat washings from eight people with the flu (one of whom was Andrewes) were inoculated into ferrets; five produced influenza in the ferrets as did a filtrate prepared from the lung of someone who had died from the flu. On the other hand, filtered throat washings from four normal humans and nasal secretions from a person with a severe common cold had no effect on the ferrets. The disease was transferred from ferret to ferret by inoculating them with material prepared from the nasal passages or simply by placing an infected animal with a healthy ferret in the same cage. What is more, ferrets who recovered from the illness developed antibodies in their serum; if this was mixed with a viral filtrate, it prevented illness developing in the recipient ferrets. They then obtained samples of swine influenza virus from Shope and found that this produced the same disease in ferrets. They concluded:

> Our results with ferrets, so far as they have gone, are consistent with the view that epidemic influenza in man is caused primarily by a virus infection. It is probable that in certain cases this infection facilitates the invasion of the body by visible bacteria giving rise to various complications.[291]

In the following year, they reported that they had adapted the influenza virus to be infective for mice when injected directly into their lungs. Incidentally, Smith and his colleagues had been very lucky in being stimulated to look at ferrets; it eventually turned out that the original supposed influenza epidemic in ferrets was not the flu at all but canine distemper (a measles-like virus infection of dogs)!

So far, influenza virus had been transferred from humans to ferrets but the circle had not been completed. That was achieved by accident in 1936. A ferret was infected with influenza virus that had been passed in series through 196 ferrets during a time when there were no human influenza infections in the community. The ferret sneezed upon one of the researchers, Dr Charles Stuart-Harris, who was to become an eminent virologist. He went down with the flu.

In 1935, Wilson Smith showed that influenza virus would grow on the chorioallantoic membrane of embryonated chicken eggs; then five years later Macfarlane Burnet (p.469) showed that much better viral growth could be obtained by direct inoculation of the amniotic and allantoic cavities of chick embryos (see Figure 195).

In the late 1940s and early 1950s, various investigators began to study the structure of influenza viruses with the electron microscope. They were found to be roughly spherical in shape and 80–120 nm in diameter (Figure 218), although sometimes filamentous forms were seen. They have an internal nucleocapsid core and a surrounding lipid envelope. In 1959, three groups of investigators showed that influenza viruses contained ribonucleic acid (RNA).

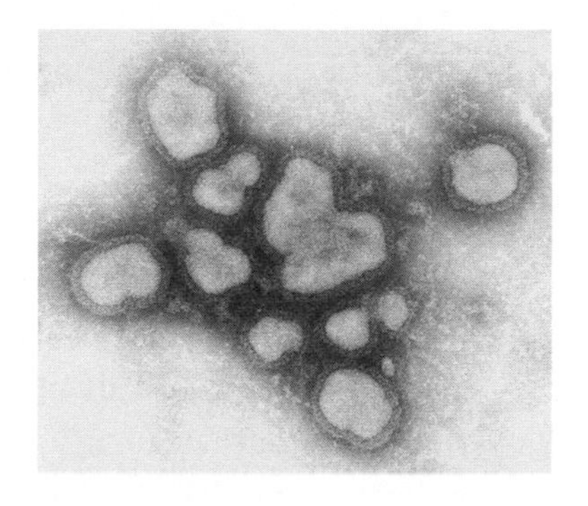
Figure 218. Electron micrograph of influenza virus.

But there was not just one sort of influenza virus. In 1940, Thomas Francis (1900–1969) and Thomas P Magill (1903–1999) in New York independently isolated a new influenza virus from patients by infecting ferrets and mice. This virus produced different antibodies because of the differences in the internal structures, later shown to be ribonucleoprotein antigens. This new virus was called influenza type B and the original one isolated in 1933 became influenza type A. In 1944, Francis and colleagues exposed 30 subjects to a spray containing influenza B virus; 27 developed the flu. Who were these volunteers? Inmates of the Ypsilanti State Hospital for the criminally insane in Michigan! This is yet another example of the need for ethics committees.

In 1949, Richard M Taylor (1887–1981) discovered an influenza virus with a third type of core antigen which is called influenza virus type C. We now know that type A infects humans, a variety of birds, and certain animals, most importantly, swine. Types B and C only infect humans.

Of even greater importance than these internal antigens, were antigens on the surface of the virus. In 1941, George K Hirst (1909–1994) discovered the important haemagglutinating antigens on the surface. He accidentally tore a blood vessel of an infected chick embryo and noticed that the escaping red blood cells agglutinated, or clumped, around influenza viruses in the infected fluid from the embryo. Hirst also noted in 1942 that if haemagglutinated clumps of red cells were warmed up, the clumps broke down. He presumed there must be an enzyme that caused the release of the virus particles from the red cell surfaces by destroying the receptors on the red cell surfaces. This enzyme was identified in 1957 by Alfred Gottschalk (1894–1973) in Melbourne, Australia as an enzyme named neuraminidase. These two antigens on the surface of the virus would turn out to be terribly important. Haemagglutinin facilitates attachment of influenza virus to the surface of a cell thus allowing its entry. Neuraminidase permits the release of newly formed viruses from an infected cell.

As the twentieth century progressed, it became apparent that most cases of the flu and all the major pandemics were due to influenza virus type A. Attention then turned to why in most years, there were relatively mild epidemics but every 10–40 years there were major pandemics of a serious illness. The importance of haemagglutinin and neuraminidase antigens in pandemic spread first became apparent in the outbreak of Asian flu which began in China in 1957. The virus that caused this pandemic was found to have novel haemagglutinin and neuraminidase antigens which were named Haemagglutinin 2 (H2) and Neuraminidase 2 (N2), while those which present on the previously circulating strain of influenza (which had caused the Spanish flu) were therefore labelled H1 and N1. Another pandemic began in Hong Kong in 1968. This had a new haemagglutinin which was named H3. More and more types of both haemagglutinin and neuraminidase were found in influenza viruses isolated from wildlife, especially birds. The next major pandemic was the 'bird flu' of 2004 which was due to influenza virus H5N1. In 2009, a pandemic of 'swine flu' due to influenza virus H1N1, the same as caused the Spanish flu, appeared. We have gone full circle back to John Koen.

Major changes in the haemagglutinin and neuraminidase which led to pandemics are called 'antigenic shift', i.e. a major change in the immunological structure. Consequently, people were not immune to such strains and the infection spread rapidly and caused severe disease. Why does a major change happen? It seems that occasionally genetic material from a bird and perhaps pig strain becomes mixed with a human virus to produce a novel human strain. But there is also what is known as 'antigenic drift' between pandemics. This means that each year the virus retains the current haemagglutinin and neuraminidase antigens but they undergo minor modifications. Pre-existent antibody from a previous infection may give some protection so these infections tend to be milder and epidemics less severe.

This all poses a major challenge to the producers of vaccines. They have to modify their vaccine every year as a new strain appears and that takes time. If we are unlucky, the haemagglutinin and/or neuraminidase antigens may change so that the virus spreads and does its damage before the manufacturers can catch up and the medical profession can administer the vaccines—a clear case of shutting the stable door after the horse has bolted.

Politicians and public health authorities like to appear to be doing something during a potential or actual influenza epidemic. They often exhort the public to have a 'flu shot'. In my humble opinion, that is often wasted effort. Drugs called neuraminidase inhibitors are not much better. They are pretty useless once a patient has the flu. They do have some effect (perhaps 25 per cent) in preventing acquisition of infection so they must be taken continually during an epidemic; they are most valuable for medical and nursing staff and others looking after patients with the flu. We need to remember the lessons of yesteryear. Stay home if you are sick—you will get better more quickly and are less likely to infect others. If you must go out, wear a mask if you are sick (to reduce droplet spread to others) or if you are well to reduce your chances of inhaling the virus. There is a very effective mask called an N95 mask but it is expensive and difficult to wear. My dictum says: any mask is better than no mask, a surgical mask is better than a paper mask, an N95 mask is better than a surgical mask. And remember to wash your hands.

52

Viral hepatitis (A, B, and C)

SCIENTIFIC NAMES: hepatitis viruses A, B, and C
DISEASE NAMES: hepatitis A, B, or C
DISTRIBUTION: all parts of the world
TRANSMISSION: a. hepatitis A: ingestion of contaminated food and drink, b. hepatitis B: from mother to child at birth, sexual intercourse, injection of contaminated blood or body fluids, c. hepatitis C: injection of contaminated blood or body fluids
CLINICAL FEATURES: acute infections—lethargy, fever, jaundice; chronic infections (hepatitis B and C)—cirrhosis of the liver
DIAGNOSIS: finding antibodies or the virus itself (HBV and HCV) in the blood
TREATMENT: a. acute infectious hepatitis—supportive with fluid and rest, b. chronic hepatitis B—interferon or lamivudine, c. chronic hepatitis C—interferon plus ribavirin
PREVENTION: effective vaccines are available for hepatitis A and B but not hepatitis C

The first textbook of medicine I bought was the 11th edition of 'Cecil and Loeb' edited by Beeson and McDermott published in 1963. I looked at what it had to say about viral hepatitis. I searched in vain in the section on infectious diseases but found it under 'Inflammatory diseases of the liver'. That is not entirely illogical because hepatitis means 'inflammation of the liver', being derived from 'hepar', the Latin word for the liver. The article said that there were two clearly defined forms of viral hepatitis as a result of studies during World War II of epidemics of 'acute catarrhal jaundice' and 'homologous serum jaundice'. It went on to say that these diseases were now called 'acute infectious hepatitis' and 'serum hepatitis'. In 2013, we have an alphabet soup of viral hepatitis caused by hepatitis viruses A–G. Hepatitis F virus has disappeared from the scene but there may be more hepatitis viruses waiting to be discovered. We shall concentrate on viruses A–C.

The cardinal clinical sign of severe hepatitis is jaundice—yellowing of the whites of the eyes (the sclerae) and, in the case of white people, of the skin in the more marked cases. Jaundice occurs when there is an increased concentration in the blood of a chemical called bilirubin. Bilirubin is formed from the breakdown of red blood cells, processed in the liver, excreted through the bile ducts to the duodenum (upper part of the small intestine), and passed out of the body in the faeces. Whenever this sequence is disturbed, bilirubin levels rise. Unfortunately, the presence of jaundice does not necessarily mean hepatitis. It is often due to obstruction of the bile duct by cancer, or gallstones, or occasionally even bacterial abscesses in the liver. This means that the recognition of viral hepatitis as a clinical entity is only relatively recent even though these infections have doubtless been around for many years. There have been countless epidemics of disease associated with jaundice over the last few centuries. Some of these may well have been due to the disease we now call hepatitis A but they could just have easily been due to other infections such as typhus, typhoid fever, brucellosis, and malaria.

The easiest place to begin with definite viral hepatitis is Bremen in Germany in 1883. In August of that year, 1,289 shipyard workers were vaccinated against smallpox with what has been described as glycerinated human lymph, i.e. the vaccine contained human body fluids. Several weeks to six months later, 191 of the men went down with hepatitis. In contrast, none of 500 new employees who had not been vaccinated became jaundiced. This outbreak was described by Dr August Lürman in 1885, who unsurprisingly wrote, 'I am not in a position to supply an explanation for this curious chain of cause and effect.'[199] Later in the same year, Dr Gottfried Jehn reported a similar outbreak among the inmates of an asylum for the insane in Merzig in Germany; 144 people out 510 who had been vaccinated against smallpox became jaundiced. Over the next half-century, various outbreaks of jaundice were noted, for example in a diabetic clinic in Sweden that used the same instrument for obtaining blood to measure glucose levels, in clinics for the treatment of syphilis in which arsenic was given with previously used and improperly sterilized syringes, and in England in 1937 following the administration of a measles vaccine that contained pooled human plasma.

The clues were there but the penny did not drop until the catastrophic outbreak of jaundice in over 28,000 US servicemen who were inoculated with a yellow fever vaccine that contained human serum in 1942; 62 died. Various experiments showed that the cause of the jaundice was not the yellow fever virus itself (yellow fever vaccine contains living yellow fever virus which has been rendered harmless). Once new preparations of vaccine that did not include human serum were made, the problem went away. This disease was called homologous serum jaundice since there seemed to be something in serum that caused jaundice and 'homologous' means serum of the same species, in this case, injection of human serum into humans.

What about the other form—acute catarrhal jaundice. The great German pathologist Rudolf Virchow (1821–1902) mistakenly ascribed the form of jaundice occurring in epidemics and often associated with poor sanitation to a plug of thick mucus and bile in the ampulla of Vater (the opening of the bile duct into the small intestine) caused by inflammation or 'catarrh' (which means a discharge) of the duodenum, hence the name 'catarrhal jaundice'.

Dr F Lainer in Vienna in 1940 was perhaps the first to attempt to transmit what was probably catarrhal jaundice, later called infectious hepatitis, and now called hepatitis A. He failed to infect people given either 300 ml of blood or 300 ml of duodenal fluid taken from 15 patients 3–10 days after the appearance of jaundice. On the other hand, H Voegt in Vienna in 1942 reported more success. He and three medical student volunteers swallowed duodenal fluid from a patient with hepatitis; within three to four weeks, they all became ill and two were jaundiced. It seems likely that 'H Voegt' was in fact Hans Voegt, who went on to conduct hepatitis transmission experiments in involuntary subjects in Nazi concentration camps. For obvious reasons, this information never came to the attention of the English-speaking world. Around the same time, JDS Cameron, a British Army doctor in Palestine, in 1941 injected serum or blood from patients with infective hepatitis into the buttocks of seven volunteer soldiers from a British cavalry regiment. Six patients could be followed up: jaundice developed in all of them after one to six months. He and his colleagues showed that bedbugs failed to transmit infection but they abandoned further experiments after some patients (not in the experimental groups) died from hepatitis.

Returning to homologous serum hepatitis, the most definitive study was that of John W Oliphant (*c.* 1902–1952) and his colleagues of the United States Public Health Service in 1943. The aforementioned yellow fever vaccine was given to 50 study patients; 24 per cent became jaundiced. A pool of serum from nine patients who had developed jaundice was then injected into another 30 subjects; eight became ill. Finally, a specimen of serum taken from a patient just before he showed jaundice caused the disease in 28 per cent of 14 subjects into whom it was injected. On the other hand, serum from the same patient taken 10 weeks after the onset of jaundice failed to infect any of 15 persons. Rather similar results in a lesser number of subjects were then reported in 1944 by MacCallum and Bauer.

What did all this mean? It seemed clear that there was an infectious agent in blood. No-one had found a bacterium so it was probably a virus, although no-one at that point seems to have demonstrated that it was a filterable virus. Frederick O MacCallum wrote in 1944:

> There are many debatable points in favour of and against the identity of infective hepatitis and homologous serum jaundice but there is no conclusive evidence one way or the other. It is essential to find a satisfactory susceptible experimental animal before much further knowledge can be obtained.[202]

In other words, he did not know if there were one or two viruses and a suitable experimental animal would need to be found in order to determine the issue. As we shall see, such animals are few and far between and the studies had to be done on humans. Two years later though, as a result of a large number of reports, MacCallum was clearer on some aspects:

> The available evidence suggests that the agent responsible for most cases of homologous serum hepatitis is not the same as that causing naturally occurring infective hepatitis. The agent will pass through the usual filters.[203]

What was this evidence? Patients recovering from infectious hepatitis were still susceptible to inoculation, accidentally or experimentally, with serum of patients with serum hepatitis and vice versa. In other words, exposure to one virus induced immunity to that virus but not to the other. And the agent was a filterable virus!

You might be looking at all these attempts to transmit hepatitis to humans with eyes somewhat askance. Were these experiments ethical? From today's perspective, we would say decidedly not. Yet there were mitigating factors. These were the years of World War II and soldiers falling ill or dying from hepatitis impeded the war effort. Something needed to be done. And no-one then had any idea that some forms of viral hepatitis could cause chronic liver disease. It would be nice to be able to say that the aforementioned Oliphant and his colleagues used volunteer soldiers or sailors who deemed dodging viruses better than German or Japanese bullets. Regrettably no. All the authors record is that 'volunteers' were obtained from an (unnamed) institution with a population of 1,700. It seems that they took their samples from the Virgin Islands to the Colony for Epileptics and Feeble-minded in Lynchburg, Virginia to continue their studies! Was this much different from Voegt in his concentration camp?

In the ensuing years, all attempts to culture these viruses in the laboratory met with frustration as did searches with electron microscopes to see the agents. However, one important advance was made. Questions had been raised as to whether you could get ill with hepatitis but not so severely as to cause jaundice. For example, Cameron in Palestine had written in 1943:

> We are of opinion that in some instances hepatitis is present without jaundice occurring at all, although the other clinical symptoms and signs characteristic of the disease are evident.[65]

Unfortunately, short of putting a needle in the liver and removing a piece for examination under the microscope, there was no way of knowing. And in 1943, that was a dangerous procedure. In 1954, however, measurement of the level of an enzyme in the blood called serum glutamic oxaloacetic transaminase (SGOT, now known as alanine transaminase or ALT) was introduced to help diagnose damage to heart muscle in heart attacks. In the following year, Felix Wróblewski and John S La Due in New York observed that there was a striking increase in the levels of this enzyme in the early stages of acute infectious hepatitis. The way was now open to understanding transmission of infection with hepatitis viruses in a population.

It was in this environment that a highly controversial series of investigations was initiated in 1956 at the Willowbrook State School in New York in the USA, where viral hepatitis had been endemic since 1949. 'State School' is rather a misnomer. Willowbrook was in fact a residential care facility for grossly intellectually retarded children run by the state of New York. The authorities of the school approached Drs Robert Ward and Saul Krugman (1911–1995) of the department of paediatrics at New York State University for advice. Ward and Krugman could perhaps have recommended spending money on buildings and staff to improve facilities and hygiene. Instead, they proposed using the children as a laboratory for experimentation to find out more about viral hepatitis. Thus began a series of experiments that were to extend for the next 14 years, largely funded by the United States Army Medical Research and Development Command and approved by various authorities.

Over the years, hundreds of children were fed or injected with preparations of faeces or serum taken from other children with hepatitis, and whether or not hepatitis was induced was determined by looking for the appearance of jaundice and measuring some 25,000 SGOT levels in serum. The information given to parents, initially at least, was scanty in the extreme and arguably ingenuous and misleading. The letter of 15 November 1958 was addressed to Dear Mrs... and said:

> We are studying the possibility of preventing epidemics of hepatitis on a new principle. Virus is introduced and gamma globulin given later to some, so that either no attack or only a mild attack of hepatitis is expected to follow. This may give the children immunity against this disease for life. We should like to give your child this new form of prevention with the hope that it will afford protection.[278]

The numbers of inmates in Willowbrook increased from 200 in 1949 to over 6,000 by 1963. Eventually a new ward was built, but only children whose parents agreed to their taking part in the study were admitted to the institution. Studies continued for a decade, and in 1967 Krugman and his colleagues were able to put it all together:

> The identification of two types of infectious hepatitis with distinctive clinical, epidemiological and immunological features provided an explanation for the recurrence of second attacks of the disease. One type resembled classical infectious hepatitis (IH); it was characterized by an incubation period of 30 to 38 days, a relatively short period of abnormal serum transaminase activity (up to 19 days)...and a high degree of contagion. The other type resembles serum hepatitis (SH); it was characterized by a longer incubation period (41–108 days), a longer period of abnormal transaminase activity (35–200 days)...(and) contrary to commonly accepted concepts, the SH type was moderately contagious. Patients with IH type were later proved to be immune to the same type. Patients with the SH type were not immune to the IH type of infection.[168]

These various studies were reported in premier US medical journals including the *New England Journal of Medicine, Journal of the American Medical Association* and the *American Journal of Medicine*. By publishing these reports, these respected journals gave their tacit support to the ethics of these studies. Were they ethical? By 1971, concerns were beginning to be raised, not in the USA, but in Britain. In April of that year, Stephen Goldby wrote to *The Lancet*:

> I believe that...the whole of Krugman's study, is quite unjustifiable, whatever the aims, and however academically or therapeutically important are the

results. I am amazed that the work was published and that it has been actively supported editorially by the *Journal of the American Medical Association* and by Ingelfinger in the 1967–68 *Year Book of Medicine*. To my knowledge only the *British Journal of Hospital Medicine* has clearly stated the ethical position on these experiments and shown that it was indefensible to give potentially dangerous infected material to children, particularly those who were mentally retarded, with or without parental consent, when no benefit to the child could conceivably result... (The) attempted defence is irrelevant to the central issue. Is it right to perform an experiment on a normal or mentally retarded child when no benefit can result to that individual? I think that the answer is no, and that the question of parental consent is irrelevant.[116]

Another Englishman, Michael Pappworth, then added:

The experiments at Willowbrook raise two important issues: What constitutes valid consent, and do ends justify means? English law definitely forbids experimentation on children, even if both parents consent, unless done specifically in the interests of each individual child. Perhaps in the U.S.A. the law is not so clear-cut... No doctor is ever justified in placing society or science first and his obligation to patients second. Any claim to act for the good of society should be regarded with distaste because it may be merely a high-flown expression to cloak outrageous acts.[238]

In 1986, Krugman wrote an impassioned and robust defence of his studies.[169] If you are interested, read it for yourself. Personally, I did not find the case he made either convincing or compelling.

Figure 219. Baruch Blumberg (1925–2011).

Hepatitis B virus

When Krugman and his colleagues defined the clinical and epidemiological characteristics of two forms of hepatitis in 1967, they were still working in the dark. Light was about to be shed from an unlikely source. Baruch Samuel Blumberg (Figure 219) was a doctor working in the USA who was interested in the variations of proteins in the blood (a process known as polymorphism). He decided to determine whether patients who received large numbers of blood transfusions (such as haemophiliacs who bleed a lot) might develop antibodies against one or more polymorphic serum proteins which they themselves had not inherited, but which the blood donors had. This could be determined by finding bands of precipitate when two different sera were allowed to diffuse on an agar gel and interact. When he examined sera from New York haemophiliacs, he found one particular serum that reacted strongly only with serum that had been obtained from an Australian Aborigine. Why was there a reaction between a New York haemophiliac and an Australian Aborigine? He then looked for evidence of this reaction, which he called 'Australia antigen', in the vast number of serum specimens stored at the National Institutes of Health. It was found in only about 1 in 1,000 US sera but in 5–15 per cent of sera from tropical populations. But among Americans, it was relatively common in patients with leukaemia (who often received blood transfusions), including patients with Down's syndrome (mongolism) who are more prone to develop leukaemia.

One of these Down's syndrome patients was named James Blair. In early 1966 he did not have Australia antigen in his serum but then it appeared. He was admitted to hospital for investigation. Since many proteins are made in the liver, tests of liver function were performed. On 28 June 1966, Alton Sutnick, one of Blumberg's colleagues, wrote in the patient's case notes: 'SGOT slightly elevated! Prothrombin time low! We may have an indication of [the reason for] his conversion to Au +.'[298] A low prothrombin time is another indicator of liver dysfunction. Sutnick's prediction proved correct. The diagnosis of hepatitis was confirmed by a liver biopsy on 20 July 1966. Naturally enough, Blumberg and his colleagues turned their attention to patients with hepatitis and found Australia antigen in many of them, although it was usually only present for a few days to weeks. They published their first tentative paper in 1966 suggesting that Australia antigen was associated with the agent responsible for causing hepatitis. Kazuo Okochi in Tokyo confirmed Blumberg's findings and showed that it could be transmitted by transfusion and some people developed hepatitis. Consequently, Blumberg and his colleagues now tested blood prepared for transfusion for the presence of this antigen and discarded it if the antigen was present. The incidence of post-transfusion hepatitis fell dramatically.

What was this antigen? Manfred E Bayer together with Blumberg and Barbara Werner examined a preparation in 1968 with the electron microscope and found particles 20 nm in diameter, some of which were elongated like a sausage, and which aggregated when mixed with antiserum to Australia antigen. Two

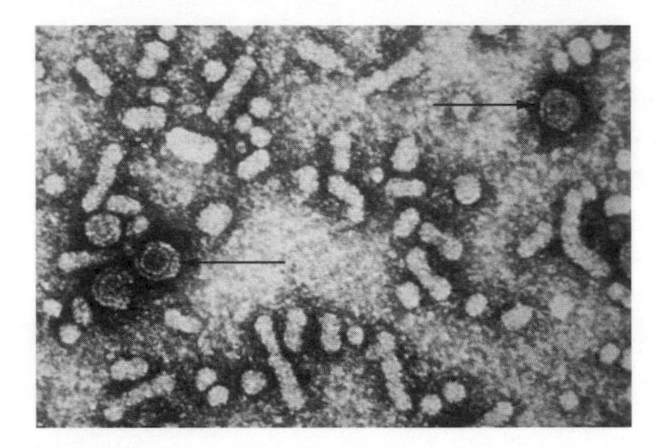

Figure 220. Electron photomicrograph of hepatitis B virus particles. The arrows point to Dane particles, the intact virion.

years later, David S Dane and colleagues in Britain found the whole virus which was a 42-nm particle (Figure 220). The search was effectively over. All of these particles had Australia antigen on their surface so Australia antigen was renamed 'hepatitis B surface antigen'. Why hepatitis B? Because this was the second form of hepatitis described earlier by Krugman and his colleagues. Indeed, Krugman and Joan Giles went back and re-examined their sera for the presence of what they called 'hepatitis associated antigen'. In 1970, they confirmed that it was present in the blood of patients with serum hepatitis (hepatitis B) but not in those with infectious hepatitis (which naturally enough is now called hepatitis A). What is more, they found that the virus persisted in the blood of one-third of the children; these patients would later be called chronic carriers of hepatitis B virus.

What sort of virus is hepatitis B? In 1973, Paul M Kaplan and colleagues in the USA found that Dane particles contained DNA polymerase, an enzyme which acts on DNA, thus suggesting that it was a DNA virus. This was confirmed in the following year by William S Robinson and colleagues who demonstrated DNA in Dane particles and observed that it was in two circular strands with one strand being longer than the other. Because it causes damage to the liver (hepar) and contains DNA, hepatitis B virus is now classified as a hepadnavirus. Unfortunately, attempts to grow hepatitis B virus in tissue culture have not met with success. In 1975, KR Berquist and colleagues found that chimpanzees could be infected but these remain the only animals susceptible to this virus. So important was the discovery of Australia antigen by Blumberg and the events that it set in train, that he was awarded the Nobel Prize in 1976.

How do people acquire this infection? Clearly one way is by exposure through the skin as in blood transfusion or use of needles contaminated with blood. Two other ways turned out to be even more important and were the main reasons for the high frequency of infection in various tropical populations. The first was the discovery in the early 1970s of transmission from an infected mother to her baby, who often remained infected for life. The second, also found around the same time, was the finding that the virus could be transmitted by sexual intercourse.

Hepatitis A virus

By the early 1970s, the major points of serum hepatitis or hepatitis B virus infection seemed to have been sorted out. What about hepatitis A virus? The logical place to look was in faeces since this form of hepatitis had been clearly shown to be transmitted by ingestion of faecally contaminated food or other materials. In 1973, Stephen M Feinstone, Albert Z Kapikian, and Robert H Purcell at the National Institutes of Health in the USA reported success with a technique called immune electron microscopy. They examined the stools of four volunteers who had been infected with infectious hepatitis several years earlier and whose specimens had been stored. Who were these volunteers? Their communication did not explicitly say but other reports indicated that they were inmates of Joliet prison in Illinois, USA. Specimens of faeces collected before and after deliberate infection were examined with the electron microscope. Nothing of interest was found in the pre-infection stools but those collected when the volunteers were acutely ill revealed small spherical virus particles 27 nm in size in two volunteers (Figure 221). These particles reacted with antibody taken from the same volunteers after they had recovered and also from patients with naturally acquired infectious hepatitis in outbreaks in Massachusetts and American Samoa. What is more, these particles did not react with antibodies in the blood of patients with hepatitis B. Not unreasonably, they concluded: 'The findings suggest that it is the etiologic agent of hepatitis A.'[96]

But the point needed to be proven. Jules Dienstag joined Feinstone and his colleagues and in 1975 they reported further details of two Joliet prisoners who had been infected in 1968. They found that the virus appeared in the stools up to five days before biochemical evidence of hepatitis with elevation of serum transaminases but then disappeared from the faeces within 6–10 days. The case was looking even more convincing and was further buttressed by another report from the same authors later in 1975. They managed

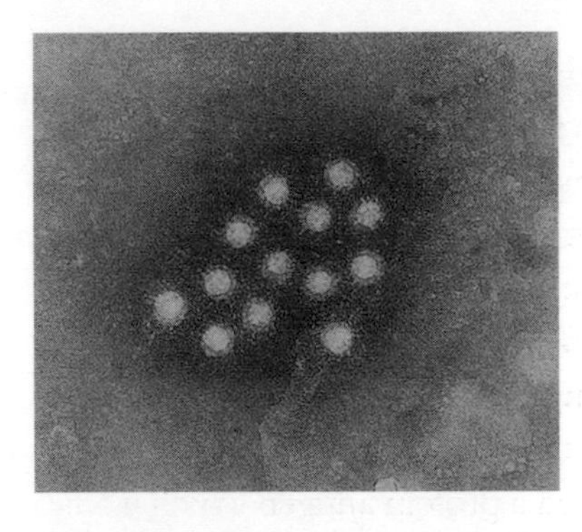

Figure 221. Electron micrograph of hepatitis A viruses.

to infect two infant chimpanzees by inoculation with a virus-rich faecal filtrate, given orally to one and intravenously to the other. The chimpanzees developed hepatitis and the virus appeared in their stools in a pattern similar to that in humans. There was no longer any doubt. Hepatitis A virus had been discovered. In 1979, Philip J Provost and Maurice R Hilleman in the USA reported that hepatitis A virus could be grown in tissue culture in marmoset monkey liver cells and in rhesus monkey kidney cells. In 1978, Günter Siegl and Gert Frösner in Switzerland and Germany showed that hepatitis A virus was a single-stranded RNA virus. Since this virus contained RNA and as it is very small, hepatitis A virus is now classified as a picornavirus (a picometre is 1 million-millionth of a metre).

Hepatitis C virus

All was now done and dusted. Both hepatitis A virus and hepatitis B virus had been found and seemed to explain the two known types of viral hepatitis. But was this the case? It became clear in the mid-1970s to several groups of investigators that sometimes hepatitis occurred seven weeks or so after transfusion in patients in whom no evidence of hepatitis B virus could be found. Nor for that matter was there any sign of hepatitis A virus. Unsurprisingly, this condition was labelled 'non-A, non-B viral hepatitis'.

Naturally everyone raced off to find this new virus. But apart from the demonstration independently in 1978 by two US teams that whatever caused non-A, non-B hepatitis could be transmitted to chimpanzees, for nearly 15 years there was a deafening silence.

Then came an extraordinary breakthrough using the new science of molecular biology in which everything was done backwards. In May 1988, Chiron Corporation of California announced at a press conference in Washington that they had identified the virus. But it took nearly a year before the details were published in the scientific literature. Qui-Lim Choo and colleagues in Michael Houghton's laboratory at Chiron Corporation took plasma from a chimpanzee which was highly infectious with non-A, non-B hepatitis, ultracentrifuged a small pellet of nucleic acids, and constructed what is known as a cDNA library. Since they did not know whether the virus was a DNA or RNA virus, they used an enzyme called reverse transcriptase to produce fragments of complementary DNA (i.e. the DNA that would produce that RNA sequence). These were then inserted into *Escherichia coli* bacteria using a bacteriophage (a virus that infects bacteria) and colonies of individual *E. coli* were grown. The bacteria then produced proteins whose nature depended upon the sequence of nucleic acids that had been inserted. Choo and colleagues then looked at about a million different bacterial colonies for one that produced a protein antigen which would react with serum presumed to contain antibodies from a patient with non-A, non-B hepatitis. They eventually found one such colony. They then identified the nucleic acid sequence that had been inserted and worked out the sequence for the whole virus. It turned out to be an RNA virus and they were able to demonstrate the presence of this RNA in the liver of infected chimpanzees.

The virus had been identified and sequenced but no-one knew what it looked like for no-one had seen it. Nevertheless, knowing the nucleic acid composition of the virus and being able to harness the antigens it produced enabled the development of a specific test to diagnose hepatitis C infection by finding antibodies in the blood of patients. It then became apparent that only about 75 per cent of people with antibodies were infectious; the remainder had somehow eliminated the virus and become immune. So a new test was necessary. The polymerase chain reaction was used to look for the presence of the virus itself in a patient's serum. If it was present, then the patient had an ongoing infection and was infectious to others. The virus has proved difficult to grow in cell culture. In 1994, M Kaito and colleagues used the electron microscope to find the virus: it is about 60 nm in diameter and has a lipid envelope. The virus has been classified as a hepacivirus. The derivation of this name is pretty obvious.

So where are we now? Hepatitis A cures itself and a vaccine is available which can prevent hepatitis A infection. Hepatitis B cures itself in most people who are infected as adults but persists in a carrier state in most people who acquire the infection around the time of birth. In carriers it may cause permanent damage with scarring of the liver called cirrhosis. A vaccine is available to prevent hepatitis B infection. Hepatitis C cures itself in about one quarter of cases but persists in the remainder, some of whom may go on to develop cirrhosis of the liver. There is as yet no vaccine available to prevent hepatitis C. Treatment of chronic carriers with hepatitis B or hepatitis C virus is available but is successful in only a proportion of patients. We still have a long way to go.

53

Human Immunodeficiency Virus and the Acquired Immune Deficiency Syndrome

SCIENTIFIC NAMES: human immunodeficiency viruses 1 and 2, members of the genus *Lentivirus*

DISEASE NAMES: HIV infection; Acquired Immunodeficiency Syndrome

DISTRIBUTION: all parts of the world

TRANSMISSION: heterosexual and homosexual intercourse; from mother to baby; contaminated syringes (mostly drug abusers), and rarely (now) in blood transfusion

CLINICAL FEATURES: acute infections—flu-like illness with a rash followed by a period of 5–10 years with no symptoms or enlarged lymph nodes

AIDS: weight loss, fever, opportunistic infections including tuberculosis, MAC infection, candidiasis, cytomegalovirus infection, toxoplasmosis, pneumocystosis, Kaposi's sarcoma, lymphoma, and dementia

DIAGNOSIS: finding antibodies and virus (using the polymerase chain reaction) in the blood

TREATMENT: a combination of antiretroviral agents (the first effective drug was zidovudine = azidothymidine = AZT)

PREVENTION: use of condoms, avoidance of promiscuous sex and intravenous drug abuse, and administration of antiviral agents to infected women during childbirth

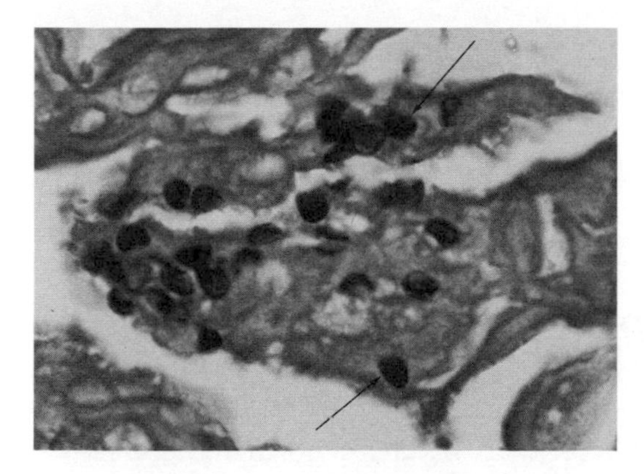

Figure 222. Lung tissue showing the dark *Pneumocystis* organisms (arrows).

As I drove down a busy street in Johannesburg, South Africa, several years ago, the salutary thought struck me that about one in six of the adults walking down that street were likely to be infected with HIV, the human immunodeficiency virus. By 2012, perhaps 30 million people have died from this dreadful affliction. All of this in 30 years! How did it happen?

In 1981, Michael Stuart Gottlieb (1947–), a physician specializing in immunology at the University of California, Los Angeles Medical Center in the USA, asked one of his colleagues if he knew of a patient with interesting features that he could use for teaching medical students. This led him to a 31-year-old young man with unexplained fevers, marked weight loss, and severe thrush (candidiasis) in the mouth. A week or two later, the patient developed pneumonia and a sample of lung tissue showed that he was infected with an unusual organism, then called *Pneumocystis carinii* (now *P. jirovecii*, once thought to be a protozoon but now recognized as a fungus (p.167); Figure 222). Over the next several months he came across several similar patients with fevers, weight loss, swollen lymph nodes, thrush, *Pneumocystis* pneumonia, and often infection with a virus called cytomegalovirus (a virus related to Epstein–Barr virus).

These unusual infections suggested that the patients' immune systems might be impaired; it is for this reason that they are called 'opportunistic infections' because they take advantage of weakened defences. One major arm of our defences is a particular form of white blood cell called lymphocytes. There are two major sorts of lymphocytes. One group is known as B lymphocytes, which turn into plasma cells, which make antibodies. The other type is called T lymphocytes; they assist the body to make an inflammatory reaction to get rid of unwanted invading organisms. Gottlieb and his colleagues soon found that T lymphocytes were depleted in these patients, in particular a subtype of T lymphocytes called 'helper cells' which could be identified by a certain marker on their surface called CD4. In the patients that Gottlieb saw, these helper cells were virtually absent.

The patients shared another feature in common—they were all men who were practising homosexuals. Gottlieb knew he was on to something. He contacted the editor of the *New England Journal of Medicine* (*NEJM*) with a view to submitting an article on his observations. Because of the need to alert public health physicians and the delay inherent in publishing in the *NEJM*, the editor advised him to submit a brief article to *Morbidity and Mortality Weekly Report* (*MMWR*), and that this would not prejudice his chances of publication subsequently in the *NEJM*. Gottlieb did just that. *MMWR* is a weekly bulletin of statistics about diseases and notes of newsworthy events published by the United States Centers for Disease Control (CDC). In the issue of 4 June 1981, the following note appeared:

> 5 young men, all active homosexuals, were treated for biopsy-confirmed *Pneumocystis carinii* pneumonia at 3 different hospitals in Los Angeles, California. Two of the patients died. All 5 patients had laboratory-confirmed previous or current cytomegalovirus (CMV) infection and candidal mucosal infection.[119]

An editorial note went on to add:

> The fact that these patients were all homosexuals suggests an association between some aspect of a homosexual lifestyle or disease acquired through sexual contact...All the above observations suggest the possibility of a cellular-immune dysfunction related to a common exposure that predisposes individuals to opportunistic infections such as pneumocystosis and candidiasis.[119]

So began the first documentation of the disease we now call HIV/AIDS.

Gottlieb's observations were indeed detailed later in 1981 in the *NEJM* in a paper which in the title included the words 'evidence of a new acquired cellular immune deficiency'.[120] Others then found similar patients and in the following year, reports appeared in *MMWR* of persistent enlargement of lymph nodes (technically called lymphadenopathy) for no apparent reason in homosexual men and the occurrence of a hitherto rare skin tumour called Kaposi's sarcoma in Haitians living in the USA. Between 1 June 1981 and September 1982, the Centers for Disease Control had received reports of 593 cases of what was now called the acquired immune deficiency syndrome (AIDS), with 243 of those patients having died.

Clearly something important was going on. Was the condition infectious? Was AIDS due to something else like a poison? The race was on to find the cause. In a remarkably short time, the cause was indeed found in separate laboratories in France and in the USA. If the cause was a virus, then the logical place to look was inside lymphocytes because these cells were being destroyed in AIDS. In order to understand what happened, we need to backtrack.

In the decades before the appearance of AIDS, it had been found that a number of animal tumours could be transmitted by filterable viruses. These included a sarcoma in chickens, breast cancer in mice, and leukaemia in cats. Some of these viruses were found to be RNA viruses and were called RNA tumour viruses. When viewed under the electron microscope, the viruses resembled a C-shaped particle so they were sometimes also called C-type or type C viruses. Then in 1970, two researchers in the USA, Howard Martin Temin (1934–1994) and David Baltimore (1938–), discovered that these viruses had a unique enzyme which reversed the normal genetic process of DNA→RNA→protein. This enzyme, which is called reverse transcriptase, caused the host cell to turn the viral RNA into viral DNA. In this way, the virus was able to take over the genetic processes of the cells and any daughter cells formed by replication of that cell. Because of this amazing characteristic, these sorts of viruses were renamed 'retroviruses' (they worked backwards). Incidentally, Temin and Baltimore won the 1975 Nobel Prize for their discovery.

If retroviruses could cause tumours (cancers) in animals, could they also do so in humans? In 1976, Robert Charles Gallo (1937-) and his colleagues devised a method to grow lymphocytes in a test tube and defined a growth factor called (interleukin-2) required to maintain their growth. Gallo used this technique to grow lymphocytes from people with diseases of lymphocytes. In 1980, he and his colleagues found a retrovirus in lymphocytes from a patient with a condition called cutaneous T cell lymphoma (also known as mycosis fungoides). This virus was named human T lymphotropic (or leukaemia) virus, or HTLV for short (later HTLV-I); 'lymphotropic' means the virus preferentially invades lymphocytes. In the following year, Gallo and his colleagues found a similar but distinct retrovirus in lymphocytes from a patient with hairy cell leukaemia (also known as Sézary T cell leukaemia, leukaemia meaning cancer of white blood cells). This virus was named HTLV-II. Incidentally, the first author of both these reports was Bernard J Poiesz but all of the work was carried out in Gallo's laboratory.

The stage was now set to look for a retrovirus in the lymphocytes of patients with AIDS. Two different groups, one in France in the laboratory of Luc Montagnier (1932–) at the Institut Pasteur, and the other in the USA, in the laboratory of Gallo at the National Institutes of Health, had success in a remarkably short time. But who discovered what and who found what first is still a matter of some debate and controversy.

Let's begin in France. In December 1982, the virologists Françoise Barré-Sinoussi (Figure 223), Jean-Claude Chermann (1939-), and Montagnier obtained some lymph node tissue from a patient named Frederick Bru... (being the first three letters of his surname). They cultured lymphocytes from this lymph node, found viral reverse transcriptase in them, took electron photomicrographs (Figure 224) of the virus, then infected healthy lymphocytes from normal people with a cell-free solution from the culture. What is more, using antibodies to HTLV-I and HTLV-II provided by Gallo, they showed that this virus appeared to be distinct from the known HTLVs. They submitted their paper to *Nature* but it was rejected. They were encouraged to submit it to *Science* by Gallo, who also encouraged its acceptance. The French researchers were circumspect in their conclusions:

Figure 223. Françoise Barré-Sinoussi (1947–).

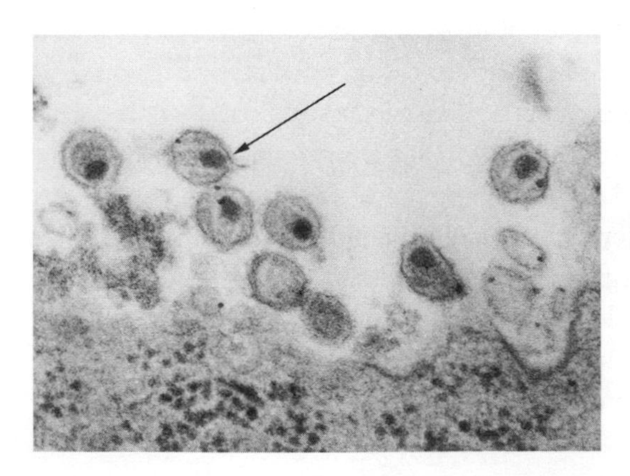

Figure 224. Electron photomicrograph of human immunodeficiency viruses. You can see that the one that the arrow points to looks a bit like the letter C.

> From these studies it is concluded that this virus as well as the previous HTLV isolates belong to a general family of T lymphotropic retroviruses that are horizontally transmitted in humans and may be involved in several pathological syndromes, including AIDS.[38]

The Montagnier group's paper (first author Barré-Sinoussi) was published in *Science* in 1983 immediately following a paper by Gallo and his group.

What had Gallo and his colleague's done? They collected blood from a patient in the USA who had AIDS and isolated a retrovirus which turned out to be HTLV-I. They did not claim that this virus caused AIDS. They also reported that they had recovered retroviruses from two other patients, both women living in France, one a Haitian and the other a Caucasian who had visited Haiti, but indicated that they had not yet been able to type these retroviruses.

At this point it was looking like the French group had isolated a virus that may well be the cause of AIDS and there was much excitement. But then confusion set in. A panel of experts thought that the electron microscopic photographs were of arenaviruses, not retroviruses, and others had doubts as to whether the virus they had isolated was really different from HTLV-I or II. Even recently, one Swedish expert in 2009 reviewed all the information and concluded that they had probably isolated HTLV-II. On the other hand, most people, including the Nobel Prize committee as we shall see, have accepted the validity of the 1983 French paper as the first description of the AIDS virus.

A more thorough description of the virus was given by Montagnier in an oral presentation at a meeting in Cold Spring Harbor in New York State, USA, in September 1983. He indicated that it had now been found in five patients with lymphadenopathy and in three patients with AIDS. Montagnier called the virus 'lymphadenopathy-associated virus' (or LAV for short) and showed that it had a conical core on electron microscopy, demonstrated that the virus selectively invaded CD4-containing lymphocytes, and noted that some patients made antibodies against specific protein components of the virus. However, Montagnier's presentation was not peer-reviewed and was not published for another year.

By that time, however, more papers had been published and there was no doubt. In April 1984, Montagnier's group (first author Etienne Vilmer) published an article in *The Lancet* describing the isolation of what they called 'a new human T lymphotropic virus' from two brothers who were suffering from haemophilia, a bleeding disorder which had necessitated blood transfusions. One brother was suffering from AIDS but the other was at that stage as well.

The authors said that this virus was different to HTLV-I but was similar to the virus they had earlier described as LAV.

In the following month, Gallo and his colleagues reported a series of four papers in *Science*. First, they reported their methods for recovering the virus and then they described how they had recovered a retrovirus from 48 patients with AIDS or with symptoms and signs preceding the onset of full-blown AIDS. They called this virus HTLV-III:

> That these new isolates are members of the HTLV family but differ from the previous isolates known as HTLV-I and HTLV-II is indicated by their morphological, biological and immunological characteristics. These results and those reported elsewhere in this issue suggest that HTLV-III may be the primary cause of AIDS.[110]

Third, they described immunological methods for distinguishing this virus, while the last paper reported that antibodies could be found in the blood of nearly 90 per cent of men with AIDS.

It became clear that LAV and HTLV-III were the same virus. What should it be called? The US CDC sat on the fence for a while and called it HTLV-III/LAV. Detailed studies of the nucleic acid sequences showed that this virus was sufficiently different from those of HTLV-I and HTLV-II to warrant a separate classification so in 1986 it was renamed Human Immunodeficiency Virus (HIV for short) in the genus *Lentivirus* of the family Retroviridae ('lentus' is Latin for 'slow' because these are 'slow viruses'). The virus and the disease it caused thus became known by the unwieldy term HIV/AIDS. To complicate matters even more, Montagnier and his colleagues in 1986 reported a similar but different virus which they called LAV-2 and which was later renamed HIV-2. The viruses originally isolated all belonged to the same type and were thus HIV-1; this virus remains the cause of the vast majority of cases of AIDS.

A rather more important question was 'how was the virus transmitted?'. By September 1982, CDC had reports of 593 cases of AIDS. Three-quarters of them were in homosexual or bisexual males, 20 per cent of whom were intravenous drug abusers. In the 20 per cent of heterosexual cases, 60 per cent were intravenous drug abusers. Of the remainder, some had had blood transfusions. It thus seemed that the virus could be transmitted by blood and body fluids, most commonly by anal intercourse among homosexual males but also by contaminated syringes or directly by blood transfusion. Initial attempts to control infection mostly focused on the behaviour of homosexual men having unprotected sex and drug users using contaminated needles and syringes. But then the true enormity of the problem became apparent when it was discovered that there were vast numbers of people infected in Africa and elsewhere in whom transmission was occurring by heterosexual intercourse.

The development of a blood test provided a means for rejecting any infected blood donated for transfusion. This was rapidly introduced into the USA and many other countries. Unfortunately, the French health authorities chose not to use this test, perhaps because they wanted to develop their own French test. Eventually the former Health Minister Edmund Herve was found guilty of allowing this to happen but not penalized, whereas several blood transfusion officials were found guilty and imprisoned. Even more extraordinary was the denialism that HIV caused AIDS in South Africa under the presidency of Thabo Mbeki and his Health Minister Manto Tshabalala-Msimang, who famously claimed that the people who were ill could be treated in a perfectly adequate manner with garlic, lemon, and beetroot. This led to drug treatment being withheld for infected South Africans and countless needless early deaths.

Unfortunately, a dispute also broke out between the two prime movers, Luc Montagnier and Robert Gallo. As one commentator remarked: 'The story is labyrinthine in its complexity and difficult for the uninitiated to follow.'[75] The French had sent Gallo a sample of their virus on the written understanding that it would not be used for commercial gain. They were understandably unimpressed when in 1984, Margaret Heckler, the US Secretary for Health and Human Services, announced at a press conference, with Gallo standing next to her, that Gallo had discovered the cause of AIDS and named the virus HTLV-III. Gallo neglected to mention the research of the French group. In fact, the French and American researchers had exchanged samples and it was eventually to turn out that the electron photomicrograph that Gallo displayed at the press conference was in fact Montagnier's LAV. When LAV and HTLV-III were analysed, they were found to be identical, whereas a high degree of variability was found among subsequent HIV isolates. This led to accusations that Gallo's specimens had become contaminated with the French LAV, or, worse, that he had appropriated the French virus to make his discovery. To make matters even more explosive, Gallo that same day applied for a US patent for a blood test which could be used to diagnose the infection in people. This enraged the French and led to a prolonged and acrimonious dispute which was not resolved until 1987 when the French Prime Minister Jacques Chirac and the US President Ronald Reagan announced that Montagnier and Gallo had agreed to share credit for the discovery of the AIDS virus.

Questions were again raised in 1990 about Gallo's conduct at the National Institutes of Health, and he was subjected to further enquiries which ultimately eventually cleared him of scientific misconduct although they were critical of his general scientific behaviour. Gallo did acknowledge in 1991 that the virus which he announced he had discovered in 1984 was in fact the same virus as the one Montagnier had sent him the year before. It seems that this virus had contaminated Gallo's cultures. As an editorial in *Nature* in 1991 put it:

> It is wrong that Gallo should be judged guilty of all the offences in the calendar simply because one strain of virus may have contaminated his own cultures... There is no evidence that Gallo stole Montagnier's virus, but he may have stolen the limelight.[26]

Perhaps all this controversy explains why the Nobel Committee waited a quarter of a century until 2008 before it awarded the prize for the discovery of the greatest infectious plague of modern times. Who were the recipients? Françoise Barré-Sinoussi, the prime mover in all of this, and Luc Montagnier in whose laboratory she worked. Gallo was omitted. Naturally enough, this decision created its own controversy. An editorial in the journal devoted to retroviruses had this to say:

> These two [Barré-Sinoussi and Montagnier] indeed deserve our congratulations. It would be churlish however not to acknowledge that the work of others, notably the Gallo lab – but also other scientists working at the time, and since – made very significant contributions without which it is likely that the unprecedented advances we have seen in our understanding and successful therapy of HIV might have been much slower in coming.[185]

This pretty much sums it up. The French workers seem to have discovered the virus we now know as HIV but they would never have done so without the prior work of Gallo on culturing lymphocytes and isolating other human retroviruses, and who was, moreover, hot on the trail as well. It seems a shame that Gallo was not a co-recipient of the prize. Instead, the Nobel committee did award the prize to a third person, Harald zur Hausen, for his work on papillomaviruses and cancer—it's a pity that choice was not put off for another year.

So what do we know about AIDS now? It is transmitted in blood and body fluids. In the developed world, the clear majority of cases still result from sexual contact between men followed by heterosexual activity and contaminated intravenous drug abuse. In the Third World, especially in Africa, however, it is a different story altogether, with most transmission occurring between men and women. Consequently husbands and wives have often become infected and died, leaving by the end of 2009 an estimated 17 million orphans, mostly in sub-Saharan Africa. To compound the disaster, about a quarter of the babies born to infected women, perhaps half a million each year, are infected, acquiring the disease from their mothers during pregnancy or at or after birth.

You might be wondering why this virus first afflicted mankind in the last quarter of the twentieth century and where it came from. HIV-1 is derived from a retrovirus called simian immunodeficiency virus, which infects chimpanzees and perhaps gorillas in the forest of western Central Africa. HIV-2 is largely confined to West Africa and came from sooty mangabey monkeys that live in that region. It is thought that HIV jumped from apes to humans sometime in the first half of the twentieth century, probably in equatorial Africa in or near Cameroon, though the precise circumstances remain shrouded in mystery. The most popular theory is that humans became infected while hunting and butchering apes for 'bush-meat', a practice that is still pursued in some African villages. Then, of course, it spread from human to human and was carried around the world.

When adults become infected, they frequently develop a flu-like illness, often associated with a transient rash a few weeks after contracting the virus. This is followed by a latent or quiescent period of 5–10 years while the person remains generally well although some people develop persistent swollen lymph nodes. Unfortunately, the virus is silently destroying helper lymphocytes and eventually they are so depleted that the immune system no longer functions effectively. The infected person loses weight and often develops fever. He or she is prone to a variety of infections for which the body no longer has any resistance. These include *Pneumocystis*, *Candida*, and cytomegalovirus already mentioned, the last sometimes affecting the eyes and causing blindness. In addition, tuberculosis is common as is infection with the related bacterium, *Mycobacterium avium-intracellulare*, and toxoplasmosis, a protozoan infection that particularly affects the brain and spinal cord. There may also be opportunistic tumours such as lymphoma and skin tumours called Kaposi's sarcoma. On top of that, the human immunodeficiency virus itself can cause dementia. Once these opportunistic conditions appear, death usually follows in untreated patients within a year or so.

Fortunately a number of drugs called antiretroviral agents are now available. Unfortunately, they do not cure the infection but can often control it when a combination of drugs is given and has to be taken for life. This is not as simple as it sounds as many of them has significant side effects, they are expensive, and they have to be taken in combination otherwise the HIV develops resistance to them. Much effort has been spent over nearly 30 years to develop a vaccine against this infection. This has all been to no avail. I suspect it will be many years, if ever, before an effective vaccine is discovered. With regard to HIV infection, never was this dictum more true: prevention is better than cure. This is the basis of the 'ABC campaign': Abstain (from sex), Be faithful (i.e. monogamous), or use a Condom, in decreasing order of effectiveness but the last much better than nothing.

PART VIII

PRIONS

Prions (pronounced 'pree-ons') are totally weird. Are they infectious, that is, able to transmit disease from one person or animal to another? Yes. Are they able to replicate? Yes. Do they contain nucleic acids like other infectious organisms? No—they harbour neither DNA nor RNA. It is almost as if the long-discredited theory of spontaneous generation is being resurrected after all these years. Prions are proteins, a particular polypeptide chain of amino acids. How then do they replicate? They do not somehow double themselves up and split down the middle. It turns out that there are normal prion proteins with the same amino composition in the cells of every mammal. The disease-forming prions have exactly the same amino acid sequence but, for unclear reasons, have an abnormal shape. The abnormally shaped disease-causing prions become attached to normal prion proteins and push them into changing their shape to that of the disease-causing prion. The amount of disease-causing prions increases—effectively, it multiplies.

And these disease-causing prions are extraordinary in another way. One of the first things I learnt about proteins when I studied biochemistry in 1962 was that proteins are easily denatured, i.e. change their shape, alter their physical characteristics, and lose their function, by gentle heating. You can see this when an egg is boiled for several minutes and the soluble clear part of the egg turns into solid white. But you can put prions in a steam autoclave at a high temperature and pressure (as for sterilizing bacteria) and they won't turn a hair. In numerical terms, human disease caused by prions is small. But their biology is so astoundingly different that we need to be aware of them. We shall consider kuru, localized to Papua New Guinea, and variant Creutzfeldt Disease, which affected mostly Britain. Who knows where future research on these enigmatic pathogens may lead?

54

Kuru, mad cows, and variant Creutzfeldt–Jacob Disease

NAME: kuru prion; variant CJD prion
DISEASE NAMES: kuru, variant Creutzfeldt–Jacob Disease
DISTRIBUTION: a. kuru: Papua New Guinea, b. variant CJD: mostly United Kingdom
TRANSMISSION: ingestion of contaminated human tissue (kuru) or beef (vCJD)
CLINICAL FEATURES: depression and anxiety leading to progressive difficulty in walking, involuntary movements, coma, and death
DIAGNOSIS: clinical features, microscopical examination of the brain (usually at autopsy) and, less reliably, biopsy of the tonsil
TREATMENT: none available
PREVENTION: don't eat human flesh (kuru); do not feed cows with animal offal (vCJD)

It was January 1966. We reached a clearing on a ridge in the jungle somewhere west of Bulolo in the then Australian Territory of Papua and New Guinea (TPNG). The patrol leader pointed north-west to a mountain range in the far distance and said 'We can't go there—that's kukukuku (cooker-cooker) country. It's uncontrolled and there are still cannibals there.' I did not think any more about cannibals until I was sitting in the common room of the Institute of Human Biology at Goroka, TPNG, in 1973. I was researching there at the time and had a bedroom adjacent to the common room. A colleague asked if he could borrow it to examine a patient. Half an hour later, I saw a young man in his twenties come staggering out. 'What was wrong with him?' I asked my colleague later. 'Kuru' he replied. I had shared my bed with a cannibal!

Why is kuru (pronounced koo-roo) so important? As we shall see, it was integral to the discovery of an astounding new concept in infectious diseases. The Australian Administration set up a police post at Okapa in the Eastern Highlands District in 1951, and a resident Australian patrol officer was stationed there from 1954. You might think it surprising that this should occur so late in the twentieth century. In fact, it was not until the 1930s that the first white explorers traversed the Highlands of TPNG and found that it held a large population. Conversely, it was not until the appearance of white patrol officers, the occasional exploring miner, or nosey anthropologist that the Fore people at Okapa had much inkling of the outside world apart perhaps from the odd Allied or Japanese aeroplane flying over in World War II. Among the first things the patrol officer and his assistants did was to discourage fighting between villagers and suppress the cannibalism which was the norm in that society. As well as enemies, the cultural practice of the Fore people was to eat the complete bodies of dead relatives in order to incorporate them into the bodies of the living. Another task of the patrols was to take a census in each village, counting each person, noting their gender, and guessing their age. It soon became apparent that many people were suffering and dying from a dreadful disease that was called kuru, meaning 'trembling' in the local language, and was thought by the locals to be due to sorcery. Thus Patrol Officer JR MacArthur recorded on 6 December 1953 that on reaching a certain village in that region:

> I observed a small girl sitting down beside a fire. She was shivering violently, and her head was jerking spasmodically from side to side. I was told that she was a victim of sorcery and would continue this shivering unable to eat until death claimed her within a few weeks.[200]

This was brought to the attention of the district medical officer, Vincent Zigas (?–1983), who had been born in Estonia and trained in Germany. In October 1956, he obtained serum specimens from 26 subjects and the brain from one patient and sent it all to Dr Stewart Gray Anderson at the Walter and Eliza Hall Institute of Medical Research (WEHI) in Melbourne, Australia. No pathogens including bacteria or viruses were found. Zigas reported the seriousness of the problem to his superiors. Dr JT Gunther, the director of public health in the country, responded by sending in an anthropologist and soliciting the help of Dr Anderson, who seems to have been none too keen because he replied to Gunther in February 1957:

> Both my wife and Sir Macfarlane [Burnet, director of the WEHI] have independently expressed concern at the possible danger of investigating 'sorcery' among the Fore people. I assume that this area is fully controlled and that there would be no appreciable risk from hostile natives singly or en masse. I would like to convey to my wife and Sir Mac your personal assurance on this point before visiting the area.[5]

Anderson's dilatoriness was rapidly overtaken in the form of an intrepid whirlwind, Daniel Carleton Gajdusek (Figure 225). Gajdusek was born in New York to Slovak parents. A Harvard medical graduate, Gajdusek spent a year or more in 1956 at the WEHI working with Burnet (who was to win the 1960 Nobel Prize for his work in immunology). Burnet (p.469) had a perceptive insight into Gajdusek, writing to a colleague:

> He has an intelligence quotient in the 180s and the emotional immaturity of a 15-year-old. He is quite manically energetic when his enthusiasm is aroused... He is completely self-centred, thick-skinned and inconsiderate, but equally won't let danger, physical difficulty or other people's feelings interfere in the least with what he wants to do. He has apparently no interest in women but an almost obsessional interest in children, none whatever in clothes and cleanliness and he can live cheerfully in a slum or a grass hut.[63]

Figure 225. Carleton Gajdusek (1923–2008).

We will hear more about his obsessional interest in children later. It seems likely that Gajdusek heard about kuru in Melbourne and decided to go himself. With the permission and encouragement of the new director of public health, Roy Scragg, he turned up on Zigas's doorstep on 13 March 1957 at Kainantu in the Eastern Highlands District. Two old women with kuru were nearby. They could not walk, shivered uncontrollably, and had slow, continuous, involuntary movements (called athetosis). Their speech was slurred and they had frequent smiles and facial grimaces. Gajdusek's interest accelerated to volcanic proportions and next day he and Zigas set off for Okapa, four hours away by four-wheel drive. In an amazingly short time, they collated clinical, epidemiological, and pathological data and published two papers late in 1957. One, a long report in the *Medical Journal of Australia*, provided detailed information on the history, geography, distribution, clinical features (which they said had some resemblance to paralysis agitans = Parkinson's disease), and laboratory investigations as well as being profusely illustrated with many photographs of patients with kuru. Altogether there were 154 patients ranging from less than 5 to over 40 years in age. But there was a most unusual gender difference. Of the 49 aged under 20 years, 18 were male and 31 were female. But of the 105 aged more than 20, only 6 were men. Kuru seemed to be a disease mostly of children of either sex or of adult women.

What caused the disease? The other paper published a week earlier in the *New England Journal of Medicine* covered some of the above information but was more speculative. Zigas had originally thought that the illness was probably infectious but no evidence for this could be found. There were no clinical signs of inflammation such as fever or white blood cells in the cerebrospinal fluid that surrounds the brain. So Gajdusek began to think that the illness might be genetic, being passed from mother to child. He and Zigas concluded this paper:

> The etiology of kuru therefore remains obscure... The peculiar age and sex distribution of cases, the high familial prevalence in an intermarried community... along with the type of picture that the disease presents, all support the suspicion that strong genetic factors are operating in the pathogenesis, probably in association with as yet undetected ethnic-environmental variables... If the degeneration of kuru is a postinfectious phenomenon the antecedent illness must be so mild or subtle as to escape detection by the natives and ourselves.[106]

Clearly specimens of diseased brain were critical. This is how Gajdusek described the first autopsy he performed:

> I did it at 2 a.m., during a howling storm, in a native hut, by lantern light, and sectioned the brain without a brain knife. But the brain in 10% Formol is off to Melbourne for neuropathology... along with pieces of all organs I sampled.[105]

Unfortunately, Gajdusek was not a trained pathologist and he mangled the brain beyond repair. A brain needs to be removed intact and immersed in a bucket of formalin for three weeks before it is cut. His boss, Joseph E Smadel

(1907–1963), back at the National Institutes of Health (NIH) in the USA sent him instructions on how to do it properly. Gajdusek then shipped a series of brains to NIH where they were processed and examined by Igor Klatzo (1916–2007), who found that the most extensive damage was in the part of the brain called the cerebellum, which is located below the cerebral hemispheres at the base and back of the skull. The cerebellum controls movement so this finding explained many of the patients' clinical signs. There were no signs of inflammation nor were there any signs of an infectious agent. Klatzo thought perhaps the problem was an environmental toxin. Eventually Klatzo noticed tiny knots of protein in the brain called 'amyloid plaques', which were common in elderly patients with dementia but which he had never seen in children. Here was a clue that both Klatzo and Gajdusek did not at the time follow up.

However, Klatzo did raise the possibility of a resemblance to a disease called Creutzfeldt–Jacob disease although he had never seen a case; only 20 cases worldwide had been reported to that time, none of them in America. In 1913, Hans Gerhard Creutzfeldt (1855–1964) in Germany observed a dreadful degenerative disease of the brain in a 23-year-old young woman who developed tremors, jerks, slurred speech, emaciation, and eventually coma and death. Because of World War I, he was not able to publish this case until 1920. In the following year, Alfons Maria Jacob (1884–1931) described four similar cases. This condition is now known as Creutzfeldt–Jacob disease (CJD). CJD usually infects older people, is rare, and is sporadic rather than being an epidemic in a localized geographical region.

Gajdusek was still favouring a genetic basis for kuru when he received a letter in 1959 from William John Hadlow (1921–), an American veterinarian working in England while researching a disease called scrapie. Scrapie had been known in England for a couple of centuries. The sheep itch so much that they rub themselves against posts, trees, and so on scraping off wool, hence the name. Eventually they get the staggers, tremors, go blind, fall down, and die. In 1936, two French veterinarians, Jean Cuillé (1872–1950) and Paul-Louis Chelle (1902–1943), had shown that if they took tissue from sheep with scrapie and injected it into healthy sheep and goats, the animals went down with the disease after a long period of up to five years. Hadlow knew that where neurones died, spaces appeared so that sections of tissue looked like a sponge (Figure 226). He found that in scrapie, this 'spongiform change', rather like looking at slices of Swiss cheese, affected not only the cerebellum but also the main part of the brain, the cerebral hemispheres. Hadlow also noticed that there was a proliferation of

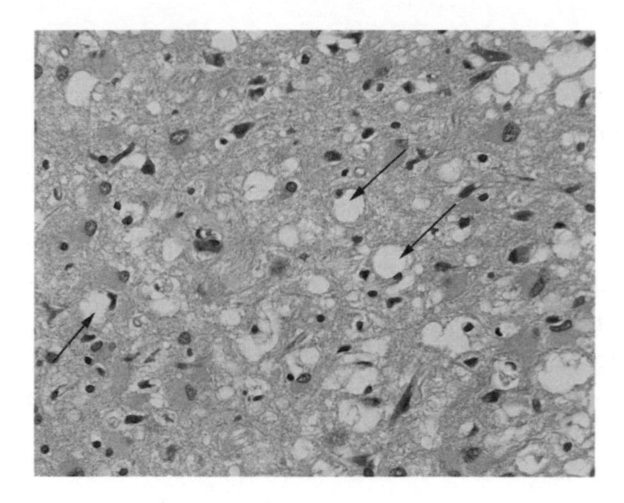

Figure 226. Spongiform change in the brain. The arrows point to holes where the neurone cell bodies had been. This photograph is of variant CJD but scrapie and kuru look the same.

cells called astroglial cells, which are normally found in smaller numbers in the brain between the neurones.

Then there was a piece of serendipity. Hadlow visited the Wellcome Medical Museum in London, which just happened to be hosting an exhibition on kuru prepared by Gajdusek. Hadlow immediately recognized an uncanny resemblance between the lesions of scrapie in sheep and those in kuru in humans as well as the clinical nature and course of the diseases. He dashed off a letter to *The Lancet* with a copy to Gajdusek in TPNG.

Hadlow's letter stunned Gajdusek because Hadlow went on to point out that scrapie was known to be transmissible and perhaps attempts should be made to transmit kuru to subhuman primates. Gajdusek still favoured the genetic hypothesis but knew he had to test the transmission theory. He also knew, on the basis of other people's experience with scrapie, that he might have to wait months or years for a result. He returned to the USA to find a place where he could carry out such experiments, preferably somewhere no-one would know exactly what he was doing. He settled on Patuxent Wildlife Research Center in Maryland. It took over two years for the NIH and the Department of the Interior's Fish and Wildlife Service to come to an agreement and for facilities to be built. Someone needed to run the programme at Patuxent and this fell to a virologist named Clarence Joseph Gibbs Jr (1924–2001).

Meanwhile, at the end of 1961, a young Australian doctor named Michael Philip Alpers (1934–) arrived with his wife and infant daughter in Okapa as the government medical officer charged with investigating kuru. Fortunately, he and Gajdusek got on well together. By early 1963, Gibbs had acquired some two-year-old chimpanzees and was ready. Alpers collected brains from dead kuru patients in TPNG and sent the specimens to the USA. Three chimps were inoculated by injection directly into the brain with brain preparations from different kuru patients, the first (Daisy) in August 1963, the second (Georgette) in September 1963, and the third (Joanne) in February 1964. Alpers then moved to Patuxent to join Gibbs and make movie films of the chimps. Georgette in May 1965 was the first to show signs of kuru and became progressively sicker until she was put down in October 1965. Daisy became ill a few days later and lasted until December of that year. Joanne became ill in August 1965 and her disease followed a similar course. Elizabeth Beck, a London neuropathologist, came to collect Georgette's brain and took it bathed in formalin back to London. Eventually she was able to examine the brain and sent a telex. Alpers remembers it as saying PATHOLOGY OF GEORGETTE INDISTINGUISHABLE FROM HUMAN KURU.[4] In their report in *Nature* in February 1966, Gajdusek, Gibbs, and Alpers concluded that their findings 'lead us to believe that kuru has been transmitted experimentally to these chimpanzees'.[107] Then in 1958, Gajdusek and his colleagues reported that they had transmitted CJD from a 59-year-old Englishman to a chimpanzee by intracerebral inoculation.

But this begged the question as to how it was being transmitted in TPNG since the Fore people were not inoculating one another's skulls. The role of cannibalism was revisited. Cannibalism had been suppressed, and 1967 was the last year in which a child under 10 years of age died from kuru. By now it was also realized that it was the women and children who ate the brains, not the men who were given the tastier meat. By stopping the practice of cannibalism, the transmission of kuru had serendipitously been interrupted. This idea was supported by the observations of a British veterinary researcher, William Gordon, who had found in the 1960s that scrapie could be transmitted to goats and sheep orally as well as by injection.

Kuru continued to decline dramatically as the years passed and the age at which infection was first seen progressively increased. The last case, after an incubation period of about 50 years, occurred in 2009. Meanwhile, Gajdusek and his team in America transmitted not only kuru and CJD but also scrapie and mink spongiform encephalopathy to an extraordinary array of chimpanzees, gibbons, a dozen or so species of monkeys, calves, ferrets, hamsters, mice, rats, and so on. There could no longer be any doubt. The spongiform encephalopathies were transmissible.

Gajdusek was lionized for his work and awarded the Nobel Prize in 1976. But as he got older, things became unstuck. Over the years, he had brought back 56 children, mostly boys from the South Pacific, to live with him. One of them, when an adult, accused Gajdusek of molesting him as a child. Gajdusek was charged with paedophilia based not only on the accusation but also on incriminating passages in his personal diaries. He pleaded guilty and was sentenced in 1997 to one year in prison. After his release, he was permitted to serve his probation in Europe. He never returned to the USA and died in Norway in 2008 at the age of 85. In the following year, the BBC aired a television programme in which it was revealed that seven men testified that Gajdusek had sex with them as boys and Gajdusek himself openly admitted to molesting boys and approving of incest. Macfarlane Burnet had been very perceptive all those years ago.

But to return to our story, if kuru and CJD could be transmitted, what exactly was the nature of the infectious agent? It was assumed that it must be a virus because no bacteria or other pathogens had been found. Since there was such a long period between inoculation and the development of disease, this 'organism', like that of the agent of scrapie, was thought to be a 'slow virus'. But it was a very strange organism. Between 1939 and 1953, David R Wilson, an Edinburgh researcher, had subjected the scrapie agent to a variety of insults including boiling for half an hour, freezing for two months, soaking in formalin or carbolic acid, or left in dried brain for two years. When injected into sheep, they became ill no matter what had been done to the preparations. So unusual were these results that Wilson hesitated to publish them and they never became widely known. This line of thought was followed by Tikvah Alper (1909–1995), DA Haig, and MC Clarke in Britain. They bombarded scrapie-infected mouse brain with electrons in a linear accelerator (which should destroy or damage an organism) and were able to calculate that if the scrapie agent was a virus, it was 1,000 times smaller than the smallest known virus. Moreover, ultraviolet light, which is able to kill viruses by damaging their nucleic acid, had no effect whatever on the scrapie agent. They wrote in 1966 that 'this suggests that the agent may be able to increase in quantity without itself containing nucleic acid'.[3] This heretical idea seemed to go against every tenet that followed from Watson and Crick's nucleic acid model put forward only 13 years previously. An infectious agent that reproduced without nucleic acid would be unique in biology.

John S Griffith, a mathematician in London, pondered the matter on logical grounds and came up with two possibilities that would fit the observed facts.

The first was that the scrapie agent could be a protein that switched on host genes that were suppressed so that they made something that caused the damage. The second idea was that the scrapie protein was an aberrant or deformed type of a normal host protein that might develop spontaneously as a result of a mutation and then this might act as a template or pattern to form more of the abnormal protein. Griffith, not a biologist, was so close to the mark.

The next piece in the puzzle was put in place by Patricia Merz, an electron microscopist working in New York. In 1978, while looking at scrapie brain, she saw 'sticks', short broken lengths of twisted thread lying in the debris between cells (Figure 227). They were always in scrapie brain and never in normal brain. She called them scrapie-associated fibrils and thought perhaps they were fibrils of amyloid, plaques of protein that occur in Alzheimer's dementia and which Klatzo had seen years before in kuru brain, but no-one would agree with her. Then she found them in the brain in CJD. What's more, she then found the same fibrils in spleen samples from animals with scrapie; subsequently they were found in many tissues. One night in 1980 while washing the dishes after the evening meal, she suddenly realized that she might well be looking at the agent or cause of spongiform encephalopathies.

At this point, the baton passed to Stanley Prusiner (Figure 228), a neurologist and biochemist in San Francisco. First, he devised a way of measuring the infectivity of scrapie in hamsters much more quickly, which speeded up experiments enormously. He then reviewed the evidence that chemicals that destroyed proteins also eliminated the infectivity of scrapie material. He published his findings in 1982:

> Because the novel properties of the scrapie agent distinguish it from viruses, plasmids and viroids, a new term 'prion' is proposed to denote a small *pro*teinaceous *in*fectious particle.[257]

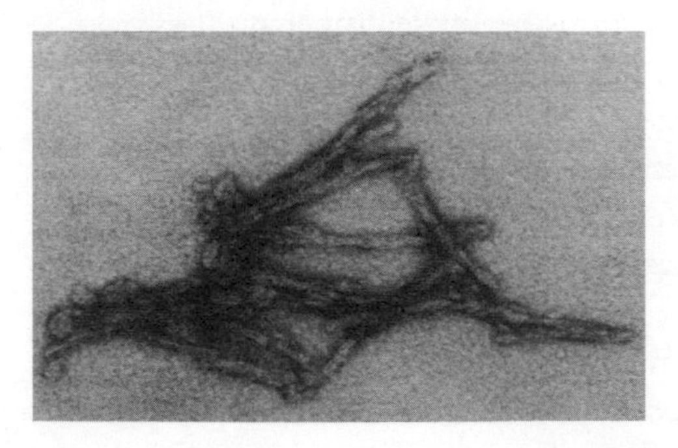

Figure 227. Electron photomicrograph of scrapie-associated fibrils.

Figure 228. Stanley Prusiner (1942–).

There was of course much argument but eventually most people agreed that the scrapie agent did not contain DNA. But since the scrapie agent was a protein, i.e. a chain of amino acids, at some point there must have been a sequence of DNA that would code for its production. How could it be found? Prusiner and his colleagues eventually purified the scrapie agent and showed that a thousand or more molecules aggregated into rod-shaped particles. These undoubtedly were the same as the scrapie-associated fibrils that Merz had seen. They then worked backwards. In the mid-1980s, Prusiner and various colleagues determined the sequence of amino acids in the prion protein and were therefore able to deduce the genetic DNA code that directed the production of the protein. To everyone's surprise, this DNA code was found in cells of infected and normal animals and humans. Everyone produced a protein that had the same amino acid sequence without causing any disease. But humans or animals with spongiform encephalopathy had an abnormal form even though the amino acid sequence was the same. This is where the terminology became confused because both the normal and abnormal proteins were called prions, the difference being designated by prion^{C} for the normal prion and prion^{Scr} for the diseased or scrapie prion. Clearly prion^{C} is not infectious—it is normal; it would have been far better to give it another name.

How could normal and diseased prion proteins have identical amino acid sequences? And how could the two forms be differentiated? A difference was found. The normal prion protein was easily, rapidly, and completely broken down by proteinase K, an enzyme which destroys proteins, whereas the diseased prion protein was only partly broken down. These enzymes work by locking on to a shape in the protein. This finding suggested that the diseased protein had changed its shape at some point after its initial, normal DNA-directed production. But why should it change its shape?

It would be years before that question was answered with any certainty. But the necessity of normal prion proteins in the generation of these encephalopathies was underscored by some remarkable experiments by Charles Weissmann (1931–) and his colleagues in Zurich in 1993. Mice had previously been shown to succumb to inoculation by the scrapie agent. The Zurich researchers therefore genetically modified some mice to remove the gene that coded for the normal prion protein so they could not make it. The mice remained perfectly well (what is the point of it then, you may well ask). And when these mice were inoculated with scrapie prions, they still remained well! What is more, when they put the hamster gene for normal prions into these mice then challenged them with hamster scrapie, they became ill. Without a doubt, normal prions were a prerequisite for development of a transmissible spongiform encephalopathy.

In 1994, a group of chemists at the Massachusetts Institute of Technology (first author: David A Kocisco) reported that they had used a cell-free system in which normal prions were mixed with scrapie proteins and they found that some of the normal proteins converted to the abnormal form in the test-tube. In the following year, two of the co-authors of that original paper, PT Lansbury and B Caughey, published a new model in a subsequent paper entitled 'The chemistry of scrapie infection: implications of the "ice 9" metaphor'. They postulated that a crystal of abnormal prion protein could act as a seed or nucleus around which normal prion protein would precipitate causing the normal prion to fold in an abnormal pattern, ultimately forming, when large enough, what are called amyloid or kuru plaques. So abnormal prions don't divide and reproduce—they somehow persuade the similar normal prion proteins to alter their shape to that of the abnormal prion proteins.

In 1997, Richard Rhodes wrote a book in which he goes to some pains to write a section which in the index is referred to as 'Prusiner, manipulation by'. He was not altogether impressed by Prusiner:

> While Prusiner was accomplishing this important work of his own, he also invaded and colonized the work of others in his apparent pursuit of the Nobel Prize... (He refused to be interviewed for this book)... Because priority – who first makes a discovery – is the accepted measure of achievement in science, he continued to try to claim priority for work other people had already done.[269]

When he wrote those words, little was he to know that in the same year that his book was published, the 1997 Nobel Prize in Physiology and Medicine would be awarded to Prusiner 'for his discovery of Prions—a new biological principle of infection'.[231] Did the Nobel committee get it wrong? Even a cursory reading of this chapter will reveal how many people were involved in the identification and elucidation of these enigmatic transmissible agents.

Meanwhile a catastrophe had occurred in Britain. In April 1985, a vet was called to a dairy farm in Kent, England, where a cow had become irritable and aggressive, uncoordinated and stumbling. Over the next 18 months, seven more cows became ill with the same signs and died. By 1986, cases were being seen in south-western England. By the middle of 1987, the UK Central Veterinary Laboratory had found spongiform changes in the cows' brains and identified scrapie-associated fibrils. The cows had bovine spongiform encephalopathy or BSE (bovine means 'of cattle'). Because the normally peaceful cows became nervous and aggressive, the media nicknamed the illness 'Mad Cow Disease'. The name stuck. By the end of 1987, there were 420 confirmed cases from all over England. Where did the infection come from?

Dairy farms generally had no sheep so it seemed unlikely that they could have got scrapie from them. It seemed more likely that the cows were being infected in their feed. Instead of just feeding cows on grass and grain, it had been the practice for many years to supplement their food (but not that of beef cattle) with 'meat and bone meal'. This was made after slaughtering dairy cows that were no longer producing milk as well as 'downer animals' (cattle and sheep that could no longer stand) by a process called 'rendering'. In this procedure the carcass is chopped, ground, cooked, and dissolved, then the fatty tallow separated from the gritty remainder which is then turned into meat and bone meal. You can see where this is heading. It seemed likely that sheep scrapie-infected meat and bone meal had entered the bovine food chain. In view of this possibility, the British government banned feeding of meat and bone meal to cows in July 1988. This of course would do nothing to prevent disease in those cattle already infected. By 1995, over 140,000 cases of BSE in cows had been reported; this number would grow to 180,000.

Was beef safe for humans to eat? The UK government appointed a committee to investigate the matter. It handed down its report in February 1989. Since people had been eating lamb and mutton, some of whom may have had scrapie, for generations without getting ill, the committee concluded that any risk to human health from the BSE epidemic was remote. Was it right?

Two British farmers died of CJD in 1993. It could be argued that this was just a chance finding because CJD does rarely develop in older people in the population. But then in 1993 a 15-year-old girl fell ill with the disease. In 1995, a 16-year-old girl and an 18-year-old boy succumbed to CJD, and by the beginning of 1996 seven other young people had either died or were gravely ill from CJD. Things were looking ominous. In that year, James Ironside, a pathologist in Edinburgh, realized that these cases had plaques of prion protein that were larger and more florid than those seen in sporadic CJD. In fact, Gajdusek would later identify them as being identical to the plaques seen in kuru. This condition was therefore named variant CJD.

On 20 March 1996, the UK Secretary of State for Health told a stunned nation that BSE had probably spread to humans who had eaten contaminated beef. Pandemonium followed. McDonalds changed to Dutch beef in its burgers. Confirmation of the Secretary's statement followed when BSE was transmitted to monkeys and they had identical changes in the brain to those seen in variant-CJD of humans. The political response was mass slaughter of 4.4 million cattle, which decimated the British farming industry and cost billions of pounds. By 2011, 170 people had died from variant-CJD. In that year though, only 17 cases of mad cow disease were reported worldwide. On 13 July 2011 the British newspaper, the *Telegraph*, felt able to declare 'The end of BSE: Good news, 20 years on, the BSE epidemic is finally over'. It was not just a British disease. Twenty-five French people contracted the illness whereas only 22 people in the rest of the world are known to have acquired it. Well may the French have said 'Perfidious Albion'.

So what do we know now? Prion protein is found in the membranes of all mammals. Its function remains unknown. But in the diseases we have been considering, its shape is modified and it becomes pathogenic, being deposited in the brain, destroying neurones, and after a long incubation period causing disease and, invariably, death. The most specific diagnostic test is by finding the abnormal prions and other characteristic features in brain but this is usually a post-mortem procedure as expensive neurosurgical instruments cannot reliably be decontaminated. Simpler but less reliable in the case of variant CJD is biopsy of a tonsil. The amino acid sequence of the prions found in the different diseases is identical but their shapes probably differ, thus accounting for the variations in the clinical features in the different spongiform encephalopathies. After oral exposure, the agent replicates in the spleen, lymph nodes, and tonsils, then enters the brain. What probably happened in kuru is that someone spontaneously developed a rare prion disease, then that person entered the human food chain because he or she lived in a cannibalistic society. There is still an awful lot to discover about prions.

PART IX

UNDE VENIS ET QUO VADIS?

This Latin tag roughly translated means 'where have we been and where are we going?' I said in the first chapter that medical scientists have made enormous progress, especially over the last 150 years or so, and that it is a tale of dogged determination, perseverance, flashes of insight, luck, serendipity, controversy, and bravery. If you have read the intervening chapters, I hope you will have come to the same conclusion.

The period from about 1860 until the end of World War I was a time when most of the important discoveries in bacteriology, parasitology, and mycology were made and immunology began to find its place. This set the scene for the advent of vaccines against some infections and the remarkable discoveries of antibacterial agents, first with the sulphonamides in the 1930s, then with antibiotics, beginning with the application of penicillin in the early 1940s.

All this produced a remarkable change in the psyche of humankind. By the time I was a medical student in the early and middle 1960s, those diseases which had been dreaded by earlier generations were no longer part of the lexicon of life. Immunization had seen a dramatic fall in the prevalence of those feared diseases of childhood, diphtheria and whooping cough. Tetanus in all ages had become rare. Antibiotics meant that tuberculosis was no longer a death sentence. There was hope for those suffering from leprosy. Death from meningococcal meningitis was not inevitable. The reappearance of plague and cholera was no longer at the forefront of people's minds. Effective antibiotic treatments were now available for staphylococcal, streptococcal, and pneumococcal infections, typhoid fever, typhus, and bacillary dysentery as well as a plethora of other infections. Even rheumatic fever was disappearing.

So you can understand the reaction of an older and supposedly wiser physician when I went to see him in 1971. I was a young physician trying to decide upon a specialty career. There was no such thing as a specialty of infectious diseases in Australia at the time, so when I told my mentor that this was what I was interested in, he looked askance at me and said 'Infectious diseases are dead. Antibiotics and vaccines have solved all that.' Disconsolate, I took his advice and embarked upon an academic career in immunology. Fortunately for me, infections produce immune responses and it was by that route over the years that I melded my way back into the study and practice of microbiology and infectious diseases.

My mentor was right in some respects, yet he was comprehensively wrong in others. How was he to know that within five years, a brand new cause of bacterial pneumonia in the form of legionnaires' disease would be discovered, or that within a dozen years or so a bacterial cause would be found for peptic ulceration, a disease which he would have confidently put down to acid, stress, and smoking. And how could he have predicted the tsunami of HIV infection with the devastating acquired immune deficiency syndrome which would sweep the globe in the 1980s and beyond. Or that this disease, because of its suppression of the immune system, would bring to the fore small print organisms like *Mycobacterium avium-intracellulare, Pneumocystis jirovecii, Toxoplasma gondii,* cytomegalovirus, or the enigmatic Microsporidia.

It can truly be said that the period since the end of World War II has seen the flowering of our understanding and the beginnings of effective treatments for viral diseases. New viruses are constantly being discovered, whether they be common and widespread ones like hepatitis A, B, or C, or less common or geographically restricted viruses from the viral haemorrhagic fevers due to Ebola, Lassa, and Marburg viruses in Africa to the bat-borne Lyssa and Hendra viruses in Australia. Then something will come out of the woodwork like SARS (severe acute respiratory syndrome), which caused widespread panic when it spread across the globe from late 2002 until the middle of 2003; it was later shown to be a coronavirus which had probably been acquired from animals in China. Then of course every year our public health authorities bemoan what it is likely to happen from the ubiquitous influenza viruses.

There are hundreds and hundreds of organisms that can infect humans. In this book, we have only been able to address those that seem most important. Are there any bacterial diseases yet to be discovered? If history is any guide, it would be a brave person who says no. Attempts were made to blame atherosclerosis (hardening of the arteries) on infection with *Chlamydia pneumoniae,* a bacterium similar to *C. trachomatis,* which we discussed in an earlier chapter, but this does not seem to have withstood the glare of intensive scrutiny. If you were to ask me where I would look, I would turn my attention to Crohn's disease and ulcerative disease. They rather remind me of tuberculoid leprosy, where there is an intense inflammatory reaction, but bacteria are few and far between. I would not be at all surprised if these are one day shown to be due to certain bacteria lurking hidden among the masses of bacteria in the bowel.

Are there any viral diseases to be discovered? Without doubt. As well as novel viruses that appear out of nowhere to afflict the human race, it may well be shown in the fullness of time that diseases as diverse as various forms of cancer, multiple sclerosis, chronic fatigue syndrome, and perhaps even conditions like rheumatoid arthritis are due to now obscure viruses. If you want to see how vast this field is, take a look at *Emerging Infectious Diseases*, a journal published by the United States Centers for Disease Control and Prevention and which is freely available over the Internet.

And there is another area that my mentor did not foresee, although it was already happening in his own time. Many bacteria have not lain back and taken a beating. They have fought back. Many bacteria are not only resistant to the first antibiotics introduced but are developing ways around the new antibiotics which chemists devised when bacterial resistance first appeared. Hence we have the 'superbugs' beloved of the media. And it is not only bacteria. Chloroquine was the wonder drug for the treatment of malaria when it was first introduced. Now it is often not much better than water. Then as new agents are introduced, the clever parasites find their way around those as well.

Should we put all our faith in vaccines? Many extremely effective vaccines are available for a number of bacteria and viruses yet others remain stubbornly obstinate. One of my favourite quotations is of a certain luminary who said in print in 1986 that within three years, his institution would have an effective vaccine against malaria. We are still waiting! And perhaps most depressingly of all, despite the enormous amount of money thrown at it over the last 30 years, there is no sign of a vaccine against HIV on the horizon.

Is there any good news? Yes. We have eradicated smallpox. We are tantalizingly close to eradicating poliomyelitis; all the tools are there if only political factors would allow their deployment. Then there is Guinea worm (*Dracunculus medinensis*), which we have not had space to discuss in any detail. In 1986 when the Carter Center (after President Jimmy Carter) began leading a campaign to eradicate this infection, there were some 3.5 million cases in Africa and Asia. When I clicked on their website at http://www.cartercenter.org/health/guinea_worm/mini_site/index.html in January 2013, it informed me that there were only 542 cases in 2012. But where were they? In war-torn Mali, South Sudan, Chad, and Ethiopia. We can only hope.

Infectious diseases are dynamic and always changing. There are plenty of challenges that we know we have already, and there will certainly be many new ones to come. We will always need microbiologists, immunologists, chemists, infectious diseases physicians, and public health personnel. They stand tall upon the legacy of the wonderful foundation left by their predecessors. It is their story that I have tried to tell.

REFERENCES

1. Abu Hassan. (10th century). Translated in (64).
2. Adler S. (1941). The transmission of *Leishmania tropica* by the bite of *Phlebotomus papatasii*. *Indian Journal of Medical Research*, **29**, 803–9.
3. Alper T, Haig DA, Clarke MC. (1966). The exceptionally small size of the scrapie agent. *Biochemical and Biophysical Research Communications*, **22**, 278–84.
4. Alpers MP. (*c*.1966). Cited in (269).
5. Anderson SG. (1957). Cited in RF Scragg, Kuru memories from 1957. *Philosophical Transactions of the Royal Society of London B Biological Science*, **363**, 3661–3, 2008.
6. Annesley J. (1828). *Sketches of the most prevalent diseases of India*. Thomas and George Underwood, London, 464 pp.
7. Anonymous. (*c*.2300 BC). Cited in J Théodoridès, *Histoire de la Rage*. Masson, Paris, 1986 and translated in (152).
8. Anonymous. (*c*.430 BC). Cited and translated in (138).
9. Anonymous. (1587). *Mirror for Magistrates*, edited by LC Campbell, Cambridge University Press, Cambridge, 1938.
10. Anonymous. (1789). Cited in F Fenner *et al.*, *Smallpox and its eradication*. World Health Organization, Geneva, 1460 pp., 1988.
11. Anonymous. (1803) *Practical rules for the management of negro slaves and medical treatment of negro slaves in the sugar colonies*. Printed by J Barfield for Vernor and Hood, London, 468 pp.
12. Anonymous. (1833). London Medical Society: Living Hydatids. *Lancet*, **i**, 719–20.
13. Anonymous. (1857). In: DP Kidder, JC Fletcher, *Brazil and the Brazilians*. Childs and Peterson, Philadelphia, 630 pp.
14. Anonymous. (1872). *New York Times*, 8 March, cited in (152).
15. Anonymous. (1878). Is the mosquito the intermediate host of the *Filaria sanguinis hominis*? *British Medical Journal*, **i**, 904.
16. Anonymous. (1884). Koch on cholera. *British Medical Journal*, **ii**, 427–8.
17. Anonymous. (1891). *New York Times*, 1 July, cited in (152).
18. Anonymous. (1894). The plague at Hong Kong. *Lancet*, **i**, 1581–2.
19. Anonymous (1894). The plague in China. *Lancet*, **ii**, 266.
20. Anonymous. (1900). *Washington Post*, 2 November, cited in (236).
21. Anonymous. (1918). Cited in (236).
22. Anonymous. (1919). *Queanbeyan Newspaper*, 7 February.
23. Anonymous. (1922). Cited in (218).
24. Anonymous. (1931). Obituary of David Bruce, *The Times*, 28 November, p.12.
25. Anonymous. (1978). Cited in JB Tucker, *Scourge: the once and future threat of smallpox*. Atlantic Press Monthly, New York, 291 pp., 2001.

26. Anonymous. (1991). Gallo vs Montagnier? *Nature*, **351**, 426.
27. Aretaeus. (AD *c*.50). In: F Adams, *Extant works of Aretaeus the Cappadocian*. The Sydenham Society, London, 510 pp., 1856.
28. Aristotle. (*c*.320 BC). In: *Historia Animalium*, translated in L Wilkinson, Rabies—two millennia of ideas and conjecture on the aetiology of a virus disease. *Medical History*, **21**, 15–31, 1977.
29. Ashburn PM, Craig CF. (1907). Experimental investigations regarding the etiology of dengue fever. *Journal of Infectious Diseases*, **4**, 440–75.
30. Astruc J. (1736). *De morbis venereis*, Paris, translated in DC Hutfield, History of herpes genitalis. *British Journal of Venereal Diseases*, **42**, 263–8, 1966.
31. Atkins J. (1742). The Navy surgeon. J Hodges, Printer, London, 378 pp.
32. Atkinson EL. Cited in GS Nelson, A milestone on the road to the discovery of the life cycles of the human schistosomes. *American Journal of Tropical Medicine and Hygiene*, **26**, 1093–110, 1977.
33. Avenzoar. (11th century). Translated in (64).
34. Baillou G de. (1578). Cited and translated in (138).
35. Bancroft J. (1877). Cited in TS Cobbold, Discovery of the adult representative of microscopic filariae. *Lancet*, **ii**: 70–1.
36. Bancroft TL. (1899). On the metamorphosis of the young form of *Filaria bancrofti*. *Journal and Proceedings of the Royal Society of New South Wales*, **33**, 48–62.
37. Barbot J. (1732). *A description of the coasts of North and South Guinea and of Ethopia inferior*. London, six volumes.
38. Barré-Sinoussi F *et al.* (ending with Montagnier L). (1983). Isolation of a T-lymphotropic retrovirus from a patient at risk for acquired immune deficiency syndrome (AIDS). *Science*, New Series, **220**, 868–71.
39. Bateman T. (1813). *A practical synopsis of cutaneous diseases*. Longman, London, 342 pp.
40. Bawden, FC, Pirie NW. (1937). The isolation and some properties of liquid crystalline substances from solanaceous plants infected with three strains of tobacco mosaic virus. *Proceedings of the Royal Society, London B*, **123**, 274–320.
41. Beijerinck MW (1898). Over een contagium vivum fluidum als oozraak van de vlekziekte der tabaksbladen. *Verslag van de Gewone Vergadering van de Afdeling Natuurskunde, Koniglijke Nederlandse Akadamie van Wetenschappen, Amsterdam*, **7**, 229–35, translated in (179).
42. Bell B. (1793). *A treatise on gonorrhoea virulenta and lues venerea*. J Watson & Co, Edinburgh, two volumes.
43. Benedictus A. (1497). *Veronensis physica historiae humani*. Cited and translated in (260).
44. Bernard of Chartres. (*c*.1100). Cited and translated in RK Merton, *On the shoulders of giants: a Shandean postscript*. Free Press, 299 pp., 1965.
45. Bilharz T, von Siebold CT. (1852). Ein Beitrage zur Helmintholgraphia humana. *Zeitschrift für wissenschaftliche Zoologie*, **4**, 53–76, translated in (157).
46. Boccaccio G. (1350). *Decameron*, translated by GH McWilliam, Penguin, London, second edition, 909 pp., 1995.

47. Boekel. J. (*c.*1580). *Europis novi morbi quem plerique medicorum catarrhum febrilem etc*, translated in C Creighton, *A history of epidemics in Britain: from the extinction of the plague until the present time*. volume 2, 1894.
48. Bonomo GC. (1687). In: Redi F, translated in (64).
49. Borovsky PF. (1909). [On Sart sore], [Military Medical Journal], Part **145**, 925–41. In Russian, translated in (157).
50. Bouillaud JB. (1837). New researches on acute articular rheumatism in general, and especially on the law of coincidence of pericarditis and endocarditis with this disease. Haswell, Barrington and Haswell, Philadelphia, 64 pp.
51. Bray J. (1973). Bray's discovery of pathogenic *Esch. coli* as a cause of infantile gastroenteritis. *Archives of Disease in Childhood*, **48**, 923–36.
52. Bretonneau PF. (1821 and 1855). In *Memoirs on diphtheria: from the writings of Bretonneau, Guersant, Trousseau, Bouchut, Empis and Daviot*, translated by RH Semple. The New Sydenham Society, London, 407 pp., 1859.
53. Bretonneau PF as reported by Trousseau A. (1826). [Concerning the disease to which M. Bretonneau, physician of the hospital of Tours, has given the name of dothinenteritis]. *Archives Générales de Médecine* **10**, 67, translated in (205).
54. Brock TD. (1961). *Milestones in microbiology*. Prentice Hall, Englewood Cliffs, New Jersey, 276 pp.
55. Brock TD. (1988). *Robert Koch, a life in medicine and bacteriology*. Science Tech Publishers, Madison, Wisconsin, 364 pp.
56. Bruce D. (1887). Note on the discovery of a microorganism in Malta fever. *Practitioner*, **39**, 161–70.
57. Bruce D. (1889). Observations on Malta fever. *British Medical Journal*, **i**, 101–1105.
58. Bruce D. (1895). *Preliminary report on the tsetse fly disease or nagana in Zululand*. Bennett and Davis, Durban, 28 pp.
59. Bruce D, Nabarro DN, Greig ED. (1903). Further report in the sleeping sickness in Uganda, 87 pp., Abstracted as, The etiology of sleeping sickness, *British Medical Journal*, **ii**, 1343–50 + 2 plates, 1903.
60. Budd W. (1867). The nature and the mode of propagation of phthisis, *Lancet*, **ii**, 451–2.
61. Budd W. (1873). Typhoid fever: its nature, mode of spreading, and prevention, extracted in *American Journal of Public Health*, **8**, 610–12, 1918.
62. Buffon, Comte D. (1749–1804). *Histoire naturelle*. Sonnini-Dupont, Paris, 44 volumes, 1749–1804. Cited in L Descours. *Pasteur and his work*, translated by AF and BH Wedd. Fisher and Unwin, London, 256 pp., 1928.
63. Burnet FM. (*c.*1960). Cited in (269).
64. Busvine JR. (1976). *Insects, hygiene and history*. Athlone Press, 262 pp.
65. Cameron JDS. (1943). Infective hepatitis. *Quarterly Journal of Medicine*, **12**, 139–55.
66. Carroll J (1900). Cited in (236).
67. Castellani A. (1903). On the discovery of a species of *Trypanosoma* in the cerebro-spinal fluid of cases of sleeping sickness. *Proceedings of the Royal Society of Medicine*, **71**, 501–8.

68. Celsus AC. (AD *c*.50). *De medicina*, translated by WG Spencer. Loeb Classical Library, Heinemann, London, 3 volumes, 1948–1953.
69. Chagas C. (1909). Nova tripanozzomize humana. *Memórias do Instituto Oswaldo Cruz*, **1**, 159–218, translated in (157).
70. Chagas C. (1922). The discovery of *Trypanosoma cruzi* and of American trypanosomiasis: historical retrospect. *Memórias do Instituto Oswaldo Cruz*, **15**, 3–11, 23–2.
71. Cheadle WB. (1889). Harveian Lectures on the various manifestations of the rheumatic state, as exemplified in childhood and early life. *Lancet*, **i**, 821–7.
72. Cleland JB, Bradley B, McDonald W. (1918). Dengue fever in Australia. Its history and clinical course, its experimental transmission by *Stegomyia fasciata* and results of inoculation and other experiments. *Journal of Hygiene*, **16**, 317–418.
73. Clifford A (Countess of Dorchester). (1603). In KO Acheson (editor), *The memoir of 1603 and diary of 1616–1619 of Ann Clifford*. Broadview Press, 2007.
74. Cohn F. (1876). Untersuchungen über Bakterien IV. Beiträge zur Biologie der Bacillen, *Beiträge zur Biologies der Pflanzen*, **2**, 249–76, translated in (55).
75. Connor S. (1992). Gallo v Montagnier. *British Medical Journal*, **304**, 1319.
76. Cumano M. (*c*.1496). Observations of Marcellus Cumanus. In: CG Gruner, *Aphrodisiacus*, Jena, 1789, translated in (260).
77. Cunningham DD. (1885). On the presence of a peculiar parasitic organism in the tissue of a specimen of Delhi boil. *Scientific Memoirs by Medical Officers of the Army of India*, **1**, 21–31.
78. Cushing HB. (1938). German measles or rubella. *Canadian Medical Association Journal*, **i**, 24–5.
79. Danielson L, Mann E. (1806–1807). Letter to the *Medical and Agricultural Register*, reprinted in *Reviews of Infectious Diseases*, **5**, 969–72, 1983.
80. Darwin C. (1835). *The voyage of the Beagle*. Everyman's Library, London, 496 pp., 1961.
81. Davaine CJ. (1863). Recherches sur les infusoires du sang dans la maladie connue sous le nom de sang de rate. *Comptes Rendus de l'Académie des Sciences*, **57**, 220–3, 351–3, 386–7, translated in KC Carter, *The rise of causal concepts of disease: case histories*. Ashgate Publishing, Aldershot, 237 pp., 2003.
82. Davis D. (1906). The bacteriology of whooping cough. *Journal of Infectious Diseases*, **3**, 1–36.
83. Demarquay J. (1863). Note sur une tumeur des bourses contenant une liquide laiteux (galactocoele de Vidal) et renferment des petites êtres vermiformes. *Gazette Médicale de Paris*, **18**, 665–7, translated in (157).
84. Descours L. (1928). *Pasteur and his work*, translated by AF and BH Wedd. Fisher and Unwin, London, 256 pp.
85. Diamond LS, Clark CG. (1993). A redescription of *Entamoeba histolytica* Schaudinn, 1903 (Emended Walker, 1911) separating it from *Entamoeba dispar* Brumpt, 1925. *Journal of Eukaryotic Microbiology*. **40**, 340–4.
86. Dick GF, Dick GH. (1923). The etiology of scarlet fever. *Journal of the American Medical Association*, **81**, 1166–7.

87. Dick G. (1924). Recent developments in scarlet fever. *American Journal of Nursing*, **24**, 939–42.
88. Dubini A. (1843). Nuovo verme intestinal umano (*Agchylostoma duodenale*) costituente un sesto genere dei nematoidea proprri dell'uomo. *Annali Universali di Medicina*, **106**, 5–13, translated in (157).
89. Duclaux E. (1896). *Pasteur: the history of a mind*, translated by Erwin F. Smith and Florence Hedges. Scarecrow Reprint Corp, Metuchen, New Jersey, 363 pp., reprinted 1973.
90. Dunlop EM, Jones BR, Al-Hussaini MK. (1964). Genital infection is association with TRIC virus infection of the eye. III. Clinical and other findings: preliminary report. *British Journal of Venereal Diseases*, **40**, 33–42.
91. Eggers HJ. (1999). Milestones in early poliomyelitis research (1849–1949). *Journal of Virology*, **73**, 4533–5.
92. Elsdon-Dew R. (1968). The epidemiology of amoebiasis. *Advances in Parasitology*, **6**, 1–62.
93. Esmarch JF, Jessen PW. (1857). Syphilis und Geistesstörung. *Allgemeine Zeitschrift für Psychiatrie*, **14**, 20, translated in M Moore, HC Solomon, Contributions of Haslam, Bayle, and Esmarch and Jessen to the history of neurosyphilis. *Archives of Neurology and Psychiatry*, **32**, 804–39, 1934.
94. Evans AC. (1947). Brucellosis in the United States. *American Journal of Public Health*, **37**, 139–51.
95. Fantham HB, Porter A. (1916). The pathogenicity of *Giardia (Lamblia) intestinalis* to men and to experimental animals. *British Medical Journal*, **ii**, 139–41.
96. Feinstone SM, Kapikian AZ, Purcell RH. (1973). Hepatitis A: detection by immune electron microscopy of a virus-like antigen association with acute illness. *Science*, **182**, 1026–8.
97. Felsen J. (1945). *Bacillary dysentery, colitis and enteritis*. WB Saunders, Philadelphia and London, 618 pp.
98. Finlay C. (1881). The mosquito hypothetically considered as the agent of transmission of yellow fever, Read to the Royal Academy of Havana on 11 August 1881, reprinted in HA Kelly, *Walter Reed and yellow fever*. Medical Standard Book Company, Baltimore, second edition, 310 pp., 1906.
99. Fournier JA. (1875). De l'ataxie locomotrice d'origine syphilique. *Annales de Dermatologie et de Syphilgraphie*, **7**, 187–97, translated in R Nitrini, The history of tabes dorsalis and the impact of observational studies in neurology. *Archives of Neurology*, **57**, 605–6, 2000.
100. Fracastorius. (1546). De contagione, contagiosis morbis et eorum curatione, cited and translated in (54).
101. Fracastorius. (1584). *Opera omnia*. Venice, translated in (205).
102. Fraser DW *et al.* (1977). Legionnaires's diseases: description of an epidemic of pneumonia. *New England Journal of Medicine*, **297**, 1189–97.
103. Friedländer R. (1886). Weitere Arbeiten über die Schizomyceten der Pneumonie und der Meningitis. *Fortschritte der Medizin*, **4**, 702–5, translated in R Austrian,

Pneumococcus and the Brooklyn connection, *American Journal of Medicine*, **107**, 2S–6S, 1999.

104. Fung WP, Papadimitriou JM, Matz LR. (1979). Endoscopic, histological and ultrastructural correlations in chronic gastritis. *American Journal of Gastroenterology*, **71**, 269–79.
105. Gajdusek DC. (1957). Cited in (269).
106. Gajdusek DA, Zigas, V. (1957). Degenerative disease of the central nervous system in New Guinea: the endemic occurrence of kuru in the native population. *New England Journal of Medicine*, **257**, 974–8.
107. Gajdusek DC, Gibbs CJ, Alpers M. (1966). Experimental transmission of a Kuru-like syndrome to chimpanzees. *Nature*, **209**, 794–6.
108. Galen CC. (AD *c*.190). Translated in AS Pease. Some remarks on the diagnosis and treatment of tuberculosis in antiquity. *Isis*, **31**, 380–93, 1940.
109. 'These experiments, as well as bringing the promise of a possible psychiatric treatment for some psycho-pathological states, have produced findings that are interesting to compare with the known facts of the natural and experimental pathology [of the condition].' Gallais P *et al.* (1953). Étude clinique, biologique, electroencephalographique, parasitologique de la trypanosomiase d'inoculation. *Médecine Tropicale*, special edition, December, pp. 807–43.
110. Gallo RC *et al.* (1984). Frequent detection and isolation of cytopathic retroviruses (HTLV-III) from patients with AIDS and at risk for AIDS. *Science*, New Series, **224**, 500–3.
111. Garland J. (1943). Varicella following exposure to herpes zoster. *New England Journal of Medicine*, **228**, 336–7.
112. Garnham PC. (1967). Reflections on Laveran, Marchiafava, Golgi, Koch and Danilewski after 60 years. *Transactions of the Royal Society of Tropical Medicine and Hygiene*, **61**, 753–64.
113. Gay FP. (1918). *Typhoid fever considered as a problem of scientific medicine*. The McMillan Co, New York, 286 pp.
114. Gerhard WW. (1837). On the typhus fever which occurred at Philadelphia in the spring and summer of 1836. *American Journal of Medical Science*, **19**, 289–92, 298–9, 302–3.
115. Goeze JAE. (1784) *Versuch einer Naturgeschichte der Eingeweidewürmer thierischer Körper*. Blankenburg, 471 pp., translated in (157).
116. Goldby S. (1971). Experiments at Willowbrook State School. *Lancet*, **i**, 749.
117. Golgi C. (1889). Sul ciclo evolution dei parassiti malrici nella febbre terzana. *Archivio per le Science Medichi*, **13**, 173–96, translated in (157).
118. Goodwin CS *et al.* (1989) Transfer of *Campylobacter pylori* and *Campylobacter mustelae to Helicobacter* gen. nov. as *Helicobacter pylori* comb. nov. and *Helicobacter mustelae* comb. nov. respectively. *International Journal of Systematic Bacteriology*, **39**, 397–405.
119. Gottlieb MS *et al.* (1981) *Pneumocystis* pneumonia—Los Angeles. *Morbidity and Mortality Weekly Report*, **30** (21): 1–3.

120. Gottlieb M *et al.* (1981). *Pneumocystis carinii* pneumonia and mucosal candidiasis in previously healthy homosexual men: evidence of a new acquired cellular immunodeficiency. *New England Journal of Medicine,* **305**, 1425–31.
121. Gowers W. (1889). *A manual of diseases of the nervous system.* Blakiston, Philadelphia, volume 1.
122. Grassi GB, Parona C, Parona E. (1878). Intorno all'Anchilostoma duodenale (Dubini). *Gazzetta Medica Italiana Lombardia,* **38**, 193–6, translated in (157).
123. Gregg NM. (1941). Congenital cataract following German measles in the mother. *Transactions of the Ophthalmological Society of Australia,* **3**, 35–46 (dated 1941 but published in 1942).
124. Griesinger W. (1854). Klinische und anatomische Beobachtungen über die Krankheiten von Egypten. *Archiv für Physiologie Heilkunde,* **13**, 528–75, translated in (157).
125. Griffith F. (1934). The serological classification of *Streptococcus pyogenes. Journal of Hygiene,* **34**, 542–84.
126. Grove DI. (1990). *A History of Human Helminthology.* CAB International, Wallingford, 848 pp.
127. Gruby D. (1841). *Comptes rendus de l'Academie des Sciences,* **13**, 72, cited and translated in (179).
128. Gruby D. (1843). *Comptes rendus de l'Academie des Sciences.* **17**, 301, cited and translated in (179).
129. Halberstaedter L, von Prowazek S. (1907). Zur Aetiologie des Trachoms. *Deutsche klinische Wochenschrift,* **33**, 1285 and *Arbeiten aus dem Kaiserlichen Gesundheitsamte,* No. 26, 1907, cited and translated in HR Taylor, *Trachoma,* Centre for Eye Research, East Melbourne, 282 pp., 2008.
130. Hamilton R. (1790). An account of a distemper, by the common people in England commonly called the mumps. *Transactions of the Royal Society of Edinburgh,* **2**, 59–72.
131. Hansen GA. (1874). Undersøgelser Angående Spedalskhedens Årsager. *Norsk Magazin for Laegervidenskaben,* **4**, 1–88, translated in SY Tan, C Graham Armauer Hansen (1841–1912): discoverer of the cause of leprosy. *Singapore Medical Journal,* **49**, 520–1, 2008.
132. Hartmannus PJ. (1688). *Anatomia glandiorum,* translated in (157).
133. Head H, Campbell AW. (1900). The pathology of herpes zoster and its bearing on sensory localisation. *Brain,* **23**, 353–523.
134. Heberden W. (1768). On the chicken pox. *Medical Transactions of the Royal College of Physicians,* **1**, 427–36. This quotation, Variolae pusillae; the chicken pox, from *Commentaries on the history and cure of diseases,* third edition (first English edition, preface to the original Latin edition dated 1782), printed for T Payne, London, 1806.
135. Henle G, Henle W, Wendell KW *et al.* (1948). Isolation of mumps virus from human beings with induced apparent or inapparent infections. *Journal of Experimental Medicine,* **88**, 223–32.
136. Henle G, Henle W, Diehl W. (1968). Relation of Burkitt's tumor-associated herpes-type virus to infectious mononucleosis. *Proceedings of the National Academy of Science,* **59**, 94–101.

137. Hippocrates. Translated in F Adams, *The Genuine Works of Hippocrates*. William Wood and Company, New York, 1891.
138. Holmes WE. (1944). *Bacillary and rickettsial infections*. The Macmillan Co, New York, 676 pp.
139. Holzel A. (1974). Bray's discovery of pathogenic *Esch. coli* as a cause of infantile gastroenteritis. *Archives of Diseases of Childhood*, **49**, 668.
140. Hooke R. (1665). *Micrographia or some physiological descriptions of minute bodies*. J. Martyn and J. Allestry, London, with 38 plates.
141. Hope-Simpson RE. (1965). The nature of herpes zoster: a long-term study and a new hypothesis. *Proceedings of the Royal Society of Medicine*, **58**, 9–20.
142. Hornick RB, Woodward TE. (1967). Appraisal of typhoid vaccine in experimentally infected human subjects. *Transactions of the American Clinical and Climatological Association*, **78**, 70–8.
143. Hube X. *De morbo scrofulose*, cited and translated in (285).
144. Hughes ML. (1893). The natural history of certain fevers occurring in the Mediterranean. *Mediterranean Naturalist*, **2**, 299–300, 325–7, 332–4.
145. Hughes ML. (1897). *Mediterranean, Malta or undulant fever*. MacMillan, London, 232 pp.
146. Hunter J. (1786). *A treatise on venereal disease*. Sherwood, Nealy and Jones, London, second edition with an introduction and commentary by Joseph Adams, 449 pp., 1918.
147. Hutchinson J. (1858). Report on the effects of infantile syphilis in marring the development of the teeth. *Transactions of the Pathological Society of London*, **9**, 449–55.
148. Hutchinson J. (1879). An address on syphilis as an imitator. *British Medical Journal*, **i**, 499–501.
149. Ibn Khaldun. (*c*.1400). Cited in BI Williams, African trypanosomiasis. In: FEG Cox, editor. *The Wellcome Trust illustrated history of tropical diseases*. The Wellcome Trust, London, pp. 178–91, 1996.
150. Ishigami T. (1905). *Textbook on plague*, revised by Shibasaburo Kitasato, translated, enlarged and illustrated with pathogenic horticulture by Donald MacDonald. Adelaide, 180 pp.
151. Ivanosvski D. (1892). [Concerning the mosaic disease of the tobacco plant]. [*Bulletin of the Imperial Academy of St Petersburg*], **35**, 67–70. In Russian, translated in (179).
152. Jackson C, Wunner H, editors (2007). *Rabies*, second edition, Elsevier Inc., New York, 2007.
153. Jenner E. (1796). Reprinted in J Baron, *The life of Edward Jenner*. Henry Coulbourn, London, volume 1, 624 pp., 1838.
154. Jenner E. (1801). *The origin of vaccine inoculation*. Printed by DN Shury, London, 8 pp.
155. Jobim JM da Cruz. (1835). Discurso sobre as moléstias que mais affligem a clase pobre do Rio de Janeiro, Sociedade de Medicina, Rio de Janeiro, Brazil, 1835. Cited and translated in (218).

156. Katsurada S. (1904). [The etiology of a parasitic disease]. *Iji Shimbun*, No. 669, 1325–32. In Japanese, translated in (157).
157. Kean BH, Mott JE, Russell AJ. (1978). *Tropical medicine and parasitology: classic investigations*. Cornell University Press, 677 pp.
158. Kitasato S. (1889). Ueber den Tetanusbacillus. *Zentralblatt für Hygien und Infektionskrankheiten*, **7**, 225–33, cited and translated in (179).
159. Kitasato S. (1894), The bacillus of bubonic plague. *Lancet*, **ii**, 428–30.
160. Knowles R, Napier LE, Smith RO. (1924). On a *Herpetomonas* found in the gut of the sandfly, *Phlebotomus argentipes*, fed on kala-azar patients. *Indian Medical Gazette*, **59**, 593–7.
161. Koch R. (1874). In his notebook, cited and translated in (55).
162. Koch R. (1876). Die Aetiologie der Milzbrand-Krankheit, begründet auf die Entwicklungsgeschichte der Bacillus anthracis. *Beiträge zur Biologie der Pflanzen*, **2**, 277–310, cited and translated in (55).
163. Koch R. (1882). Die Äetiologie der Tuberculose. *Berliner klinische Wochenschrift*, **15**, 221–30, translated in (54).
164. Koch R. (1884). Report to the Minister in Berlin, cited and translated in (55).
165. Koch R. (1903). Die Bekampfung des Typhus. *Veröffentlichengun aus dem Gebiete des Militär-Sanitättswesens*, **21**, cited and translated in (55).
166. Koen JS. (1919). A practical method for field diagnosis of swine diseases. *American Journal of Veterinary Medicine*, **14**, 469.
167. Koplik H. (1896). The diagnosis of the invasion of measles from a study of the exanthema as it appears on the buccal mucous membrane. *Archives of Pediatrics*, **13**, 918–22.
168. Krugman S, Giles JP, Hammond JP. (1967). Infectious hepatitis: evidence for two distinctive clinical, epidemiological and immunological types of infection. *Journal of the American Medical Association*, **200**, 365–73.
169. Krugman S. (1986).The Willowbrook hepatitis studies revisited: ethical aspects. *Reviews of Infectious Diseases*, **8**, 157–62.
170. Küchenmeister F. (*c*.1853). Cited in R Leuckart, *die Parasiten des Menschen*. C.F. Winter'sche Verlagshandlung 1879, translated by WE Hoyle, *The parasites of man*, Young J Pentland, Edinburgh, 771 pp., 1886.
171. Küchenmeister F. (1855). Experimenteller Nachweiss, dass *Cysticercus cellulosae* innerhald des menschlichen Darmkanales sich in *Taenia solium* umwandelt. *Wiener medizinische Wochenschrift*, **5**, 1–4, translated in (157).
172. Küchenmeister F. (1855) *Die in und an dem Körper des lebenden Menschen vorkommenden Parasiten*, 846 pp., translated by E Lankester, *On animal and vegetable parasites of the human body*. The Sydenham Society, London, 452 pp., 1857.
173. Laennec RT. (1819). *De l'auscultation médiate ou traité du diagnostic de maladies des poumons et du coeur, fondé principalement sur ce nouveau moyen d'exploration*. Brosson et Chaudé Libraires, Paris, translated in (205).
174. Lambl V. (1859). Mikroskopische untersuchungen der darm-excrete. *Vierteljahrschrift für die Praktische Helkunde, Prag*, **i**, 1–58, translated in (157).

175. Landsteiner K, Popper E. (1909). Übertragung der Poliomyelitis acuta auf Affen. *Zeitschrift für Immunotätsforschung und Experimentelle Therapie*, **ii**, 337–90, translated in S Flexner, PA Lewis, Experimental epidemic poliomyelitis in monkeys. *Journal of Experimental Medicine*, **12**, 227–7, 1910.
176. Langenbeck B. (1839). Auffindung von Pilzen auf der Schleimhaut der Speiseröhre einer Typhusleiche. *Neue Notizen aus dem Gebiete der Natur- und Heilkunde*, **12**, 146–7, translated in M Knoke and H Bernhardt, The first description of an oesophageal candidosis by Bernhard von Langenbeck in 1839, *Mycoses* **49**, 283–7.
177. Laveran A. (1880). Un nouveau parasite trouvé dans le sang des malades atteints de fièvre palustre. *Bulletins et Mémoires de Société des Hôpitaux de Paris*, **17**, 158–64, translated in (157).
178. Laveran A, Mesnil. (1912). *Trypanosomes et trypanosomiasis*. Masson et Cie, Paris, second edition, 999 pp.
179. Lechevalier HA, Solotorovsky, M. (1974). *Three centuries of microbiology*. Dover Publications, New York, 536 pp.
180. Leeuwenhoek A van. (1681). The collected letters of Antoni Van Leeuwenhoek, Letter No. 66, 4 November, translated in (157).
181. Leeuwenhoek A van. (1681). The collected letters of Antoni Van Leeuwenhoek, translated in (64).
182. Leeuwenhoek A van. (1684). Microscopical observations about animals in the scurf of the teeth, cited and translated in (54).
183. Leishman WB. (1903). On the possibility of the occurrence of trypanosomiasis in India. *British Medical Journal*, **i**, 1251–4.
184. Lesh (Lösch) FA. (1875). Massenhafte entwickelung von amöben in dickdarm. *Archiv für Pathologische Anatomie und Physiologie und Klinische Medicin* (Virchow), **65**, 196–211, translated in (157).
185. Lever AM, Berkhout B. (2008). 2008 Nobel prize in medicine for discoverers of HIV. *Retrovirology*, **5**, 91–2.
186. Lewis T. On a haematozoon inhabiting blood, its relation to chyluria and other diseases. Abstracted in *Lancet*, **ii**, 889–90, 1872 and **i**, 56–67, 1873.
187. Lister J. (1867). On the antiseptic principle in the treatment of surgery. *Lancet*, **ii**, 353–6, 668–9.
188. Lister J. (1867). On the use of carbolic acid. *Lancet*, **ii**, 595.
189. Lister J. (1874). Cited in JT Wrench, *Lord Lister: his life and work*. Fisher and Unwin, London, 1913, 396 pp.
190. Lister J. (1881). Relation of microorganisms to inflammation. *Lancet*, **ii**, 695.
191. Loeffler F. (1884). Untersuchung über die Bedeutung der Mikroorganismen für die Entstehung der Diphtherie beim Menschen, bei der Taube und beim Kalb. *Mittheilungen aus dem Kaiserlichen Gesundheitsamt*, **2**, 421–99 (article dated December 1883), translated in (179).
192. Loeffler F, Frosch P. (1898). Berichte der Kommission zur Erfuschung der Maul- und Klauenseuche bei dem Institut für Infektions-Krankheiten im

Berlin. *Zentralblatt für Bakteriologie und Parasitenkunde*, **23**, 371–91, translated in (179).

193. Looss A. (1898). Zur Lebensgeschichte des *Ankylostoma duodenale*. *Centralblatt für Bakteriologie und Parasitenkunde*, **24**, 441–88, translated in (157).
194. Looss A. (1901). Über das Eindrigen der Ankylostomalarven in die menschliche Haut. *Centralblatt für Bakteriologie und Parasitenkunde*, **29**, 733–9, translated in (157).
195. Louis PCA (1828). *Anatomical, pathological and therapeutic researches upon the disease known under the name of gastro-enterite*, translated from the original French by HI Bowditch, Boston, 1836.
196. Low GC. (1900). A recent observation on *Filaria nocturna* in *Culex*. *British Medical Journal*, **i**, 1456–7.
197. Lowson JA. (1894). The plague at Hong Kong. *Lancet*, **ii**, 325.
198. Lumsden WH. (1974). Some episodes in the history of African trypanosomiasis. *Proceedings of the Royal Society of Medicine*, **67**, 789–96.
199. Lürman A. (1995). Eine Ikterusepidemie. *Berliner klinische Wochenschrift*, **22**, 20, translated in R Schmid, History of viral hepatitis: a tale of dogmas and misinterpretations. *Journal of Gastroenterology and Hepatology*, **16**, 719–22, 2001.
200. MacArthur JR. (1953). Cited in Zigas V, Gajdusek DC. Kuru: clinical study of a new syndrome resembling paralysis agitans in natives of the Eastern Highlands of Australian New Guinea. *Medical Journal of Australia*, **44**, 745–54, 1957.
201. Macarthur WP. (1927). 'Old time typhus in Britain'. *Transactions of the Royal Society of Tropical Medicine and Hygiene*, **20**, 487.
202. MacCallum FO. (1944). Transmission experiments in animals and man with material from infective hepatitis and homologous serum jaundice. *Proceedings of the Royal Society of Medicine*, **37**, 449–53.
203. MacCallum FO. (1946). Homologous serum hepatitis. *Proceedings of the Royal Society of Medicine*, **39**, 655–7.
204. Macdonald H, MacDonald EJ. (1933). Experimental pertussis. *Journal of Infectious Diseases*, **53**, 328–30.
205. Major RH. (1948). *Classic descriptions of disease*. Blackwell, Oxford, 679 pp., 1948.
206. Manson P. (*c*.1877). Cited in DR Linicombe (editor), Patrick Manson as a parasitologist: a critical review. *International Review of Tropical Medicine*. Academic Press, New York, pp. 77–129, 1961.
207. Manson P. (*c*.1877). Cited in PH Manson-Bahr and A Alcock, *The life and works of Sir Patrick Manson*. Cassell and Co, London, 283 pp., 1927.
208. Manson P. (1894). On the nature and significance of the crescentic and flagellated bodies in malarial blood. *British Medical Journal*, **ii**, 1306–8.
209. Manwaring WH (1932). Experimental scarlet fever in children. *California and Western Medicine*, **36**, 200.
210. Marchiafava E, Bignami A. (1892). *Sulla febbri malariche estivo-autumnali*. Rome, translated in (157).
211. Marshall B. (1983). Unidentified curved bacillus on gastric epithelium. *Lancet*, **i**, 1273.

212. Marshall BJ *et al.* (1985). Attempt to fulfil Koch's postulates for pyloric Campylobacter. *Medical Journal of Australia*, **142**, 436–9.
213. Marshall BJ. (1985). The pathogenesis of non-ulcer dyspepsia. *Medical Journal of Australia*, **143**, 379.
214. Marshall BJ, Goodwin CS. (1987). Revised nomenclature of *Campylobacter pyloridis*. *International Journal of Systematic Bacteriology*, **37**, 8.
215. Marshall BJ. (1988). The *Campylobacter pylori* story. *Scandinavian Journal of Gastroenterology*, Supplement 146, 58–66.
216. Mead R. (1767). *The medical works of Richard Mead*. Printed for Thomas Ewing, Dublin, 511 pp.
217. Mézeray F. (1673). *Histoire Chronologique -de France*. Amsterdam, three volumes, translated in *Medical and Physical Journal*, **34**, 137, 1815.
218. Miles MA. (2004). The discovery of Chagas disease: progress and prejudice. *Medical Clinics of North America*, **18**, 247–60.
219. Miyairi K, Suzuki M. (1913). [On the development of *Schistosoma japonicum*]. *Tokyo Iji Shinshi* No. 1836, pp. 1–5. In Japanese, abstracted in KS Warren, *Schistosomiasis: the evolution of a medical literature*. MIT Press, Cambridge, 1307 pp., 1973.
220. Molinet. J (*c.*1500). *Traité de la maladie vénerienne par le Sieur de la Martinière*. Paris, 1664, cited and translated in (260).
221. Morris RJ. (1976). *Cholera 1832: the social response to an epidemic*. Croom Helm, London, 228 pp.
222. Mott FW. (1906). Histological observations on sleeping sickness and other trypanosome infections. *Reports of Sleeping Sickness Commission*, No. VII, 5–46.
223. Naumann M. (1871). Cited in (84).
224. Needham JT. (1749). *Observations upon the generation, composition and decomposition of animal and vegetable substances*. London, 52 pp.
225. Needham JT. (1769). A summary on some observations on the generation, composition and decomposition of animal and vegetable substance. Cited in (84).
226. Neisser A. (1879). Ueber eine der Gonorrhoe eigentumliche Micrococcus-form. *Zentralblatt für die Medizinischen Wissenschaften*, **17**, 497–500, translated in JD Oriel, Eminent venerologists. 1. Albert Neisser, *Genitourinary Medicine*, **65**, 229–34.
227. New South Wales coroner report. (2010) <http://www.abc.net.au/news/stories/2010/05/31/2913760.htm>. Accessed on 2 January 2011.
228. Nicolle C. (1908). Nouvelles acquisitions sur la kala-azar: cultures, inoculation au chien, étiologie. *Comptes Rendus Hebdomadaires des Sciences*, **146**, 489–99, translated in (157).
229. Nicolle C, Comte C, Conseil E. (1910). Experimental transmission of exanthematous typhus by body lice (*Pediculus vestimenti*). *Annales de l'Institut Pasteur*, **24**, 261–7. Also cited in (179).
230. Nicolle C. (1928). Nobel Lecture. <http://nobelprize.org/nobel_prizes/medicine/laureates/1928/nicolle-lecture.html>. Accessed 3 February 2011.

231. Nobel Prize Committee. <http://nobelprize.org/nobel_prizes/medicine/laureates/1997/>. Accessed on 16 July 2011.
232. Ogston A. (*c.*1878). Cited in SD Elek, *Staphylococcus pyogenes and its relation to disease,* E&S Livingstone, Edinburgh, 767 pp., 1959.
233. Ogston A. (1880). Ueber Abscesse. *Archiv für klinische Chirugie,* **2**, translated in *Reviews of Infectious Diseases,* **6**, 122–8, 1984.
234. Ogston A. (1881). Report on microorganisms in surgical diseases. *British Medical Journal,* **i**, 369.
235. Ogston A. (1882). Micrococcus poisoning. *Journal of Anatomy and Physiology,* **16**, 526–67.
236. Oldstone, MBA. (1998). *Viruses, plagues and history*. Oxford University Press, New York, 211 pp.
237. Osler, W. (1890). On the *Amoeba coli* in dysentery and in dysenteric liver abscess. *Bulletin of the Johns Hopkins Hospital.* **1**, 53–4.
238. Pappworth MH. (1971). The Willowbrook experiments. *Lancet,* **i**, 1181.
239. Paracelsus. (1573) *De natura rerum,* cited and translated in (126).
240. Paré A. (1979 [(*c.*1560]. Cited and translated in RH Kampmeier, Syphilis as a cause of aneursym. *Sexually Transmitted Diseases,* **6**, 270–2.
241. Pasteur L. (1860). De l'origine des ferments: nouvelle experiences relative au générations dites spontanées. *Comptes Rendus de l'Académie des Sciences,* **50**, 849–852, translated in (84).
242. Pasteur L. (1860b). Nouvelles expériences relatives aux générations dites spontanées. *Comptes Rendus de l'Académie des Sciences,* **51**, 348–52, translated in (84).
243. Pasteur L. (1861). *Sur les corpuscles organisés qui existent dans l'atmosphère. Examen de la doctrine des generations spontanées.* Société Chimique, translated in (54).
244. Pasteur L. (1871). Cited in (84).
245. Pasteur L, Joubert J, Chamberland C. (1878). La théorie des germes et ses applications à la médicine et à la chirugie. *Comptes Rendus de l' Académie des Sciences,* **86**, 1037–43. Cited in (84).
246. Pasteur L. (1884). Letter to Isidore Straus, translated in (55).
247. Pepys S. (1659–1669). *The diary of Samuel Pepys from 1659 to 1669,* edited by HB Wheatley <http://digital.library.upenn.edu/webbin/gutbook/lookup?num=4200>. Accessed on 21 July 2011.
248. Pettenkofer M von. (*c.*1892). Cited and translated in LK Altman, *Who goes first; the story of self-experimentation in medicine.* Random House, New York, 430 pp., 1987.
249. Pfeiffer E. (1889). Drüsenfieber *Jahrbuch für Kinderheilkunde,* **29**, 257–64, translated in (205).
250. Piazza M de. (*c.*1347). *Cronaca,* edited by A Giuffiedi. ILA Palma, Palermo, 1980, translated in J Aberth, *The Black Death: the great mortality of 1348–1350, a brief history with documents.* Palgrave MacMillan, 199 pp., 2005.
251. Piper DW. (1985). Bacteria, gastritis, acid hyposecretion and peptic ulcer. *Medical Journal of Australia,* **142**, 31.

252. Pires T. Cited in A Cortesão, *The Suma Oriental of Tomé Pires*. Hakluyt Society, London, 578 pp., 1944.
253. Planck M. (1948). *Wissenschaftliche Selbstbiographie. Mit einem Bildnis und der von Max von Laue gehaltenen Traueransprache*. Johann Ambrosius Barth Verlag, Leipzig, translated in F. Gaynor, *Scientific autobiography and other papers*. New York, 1949.
254. Pope A. (1751). *Works of Alexander Pope esq.: in nine volumes complete*. London, nine volumes.
255. Pouchet FA. (1859). *Hétérogonie*. Paris, 1859, translated in W Bulloch, *The history of bacteriology*. Oxford University Press, Oxford, 442 pp., 1938.
256. Procopius of Caesarea. (c.560). *History of the wars I*, edited by HB Dewing. MacMillan, New York, 1914; translated in IW Sherman, *Twelve diseases that changed the world*. ASM Press, Washington DC, 219 pp., 2007.
257. Prusiner SB (1982). Novel proteinaceous infectious particles cause scrapie. *Science*, **216**, 136–44.
258. Prüss J (1536). Cited and translated in (64).
259. Pusey WA. (1933). *The history and epidemiology of syphilis*. Charles C Thomas, Springfield, Illinois, 113 pp.
260. Quétel C. (1990). *History of syphilis*, translated by J Braddock and B Pike, Polity Press, Cambridge, 342 pp.
261. Quincke HI, Roos E. (1893). Ueber amöben enteritis. *Berliner klinische Wochenschrift*, **30**, 1089–94, translated in (157).
262. Randolph, Lord. (1562). Cited in WI Beveridge, *Influenza: the last great plague*. Heinemann, London, 124 pp., 1977.
263. Redi F 1909 [1688]. *Esperienze intorno alla generazione degl'insetti*, tr. in M Bigelow, *Experiments on the Generation of Insects*, Chicago, Opencourt.
264. Reed W, Carroll J, Agramonte A, Lazear JW. (1900). The etiology of yellow fever: a preliminary note. *Philadelphia Medical Journal*, **6** Reprinted in *Reviews of Infectious Diseases*, **5**, 1103–11, 1983.
265. Reed W. (1902). Recent researches concerning the etiology, propagation, and prevention of yellow fever by the United States Army Commission. *Journal of Hygiene*, **2**,101–19.
266. Renoult A. (1803). Sur l'hématurie que les Européens éprouvent en Égypte. *Journal Générale de Médecine de Chiriugie et de Pharmacie*, **17**, 366–70, cited and translated in (126).
267. Rentdorff R. (1954). The experimental transmission of human intestinal protozoan parasites. *American Journal of Hygiene*, **59**, 209–20.
268. Rentdorff R. (1978). In: W Jakubowski, JC Hoff (editors): Waterborne transmission of giardiasis. Proceedings of a symposium, Sept. 18–20. US Environmental Protection Agency, Cincinnati, Ohio, June 1979 (EPA-600/9-79-001), pp. 174–91.
269. Rhodes R. (1997). *Deadly feasts: science and the hunt for answers in the CJD crisis*. Touchstone Books, London, 278 pp.

270. Ricketts HT. (1909). A micro-organism which apparently has a specific relationship to Rocky Mountain Spotted Fever. *Journal of the American Medical Association*, **52**, 379–80.
271. Ricketts HT, Wilder RM. (1910). The etiology of the typhus fever (tabardillo) of Mexico City: a further preliminary report. *Journal of the American Medical Association*, **54**, 1373–5.
272. Ricord P. (*c*.1843). Cited and translated in JD Oriel, Eminent venereologists. 3. Philippe Ricord, *Genitourinary Medicine*, **65**, 388–93. 1989.
273. Rocha Lima H de. (1916). Zur Aetiologie des Fleckfiebers. *Berliner klinische Wochenschrift*, **53**, 567–72, translated in N Hahon, *Selected papers on the Rickettsiae*. Harvard University Press, Cambridge, 382 pp., 1974.
274. Rogers L. (1904). Preliminary note on the development of *Trypanosoma* in cultures of the Cunningham-Leishman-Donovan bodies of cachexial fever and kala-azar. *Lancet*, **ii**, 215–16.
275. Romanowsky DL. (1891). Zur Frage der Parasitologie und Therapie der Malaria. *St. Petersburger medicinische Wochenschrift*, **8**, 297–302, translated in (157).
276. Ross R. (1897). On some peculiar pigmented cells found in two mosquitos fed on malarial blood. *British Medical Journal*, **ii**, 1786–8.
277. Ross R. (1923). *Memoirs with a full account of the great malaria problem and its solution*. J Murray, London, 547 pp.
278. Rothman D, Rothman S. (1984). *The Willowbrook wars*. HarperCollins, Cambridge, 405 pp.
279. Rush B. (1818). An account of the bilious remitting yellow fever as it appeared in Philadelphia in the year 1793. *Medical inquiries and observations*. Benj. and Thos. Kite, Philadelphia, two volumes, reprinted in (205).
280. Russell A. (1856). Of the Mal d'Aleppo. *Natural history of Aleppo and parts adjacent*. A Millar, London, 266 pp.
281. Sabin AB, Ward R. (1941). The natural history of human poliomyelitis. 1 Distribution of virus in nervous and non-nervous tissues. *Journal of Experimental Medicine*, **73**, 771–93.
282. Schaudinn F. (1903). Untersuchungen über die fortpflanzung einiger Rhizopen. *Arbeiten aus dem Kaiserlichen Gesundheitsamte*, **19**, 547–76, translated in (157).
283. Schaudinn F, Hoffmann E. (1905). Vorläufiger Berichte über das Vorkommen von Spirochaeten in syphilitischen Krankheitsproducten und bei Papillomen. *Arbeiten aus dem kaiserlichen Gesundheitsamte*, **22**, 527–34, translated in LD Longo, *American Journal of Obstetrics and Gynecology*, **132**, 571–2, 1978.
284. Schoenlein JL. (1839). Zur Pathogenieder impetigenes. *Archiv für Anatomie, Physiologie und wissenschaftliche Medizin*, p. 82, cited and translated in (285).
285. Seeliger HP. (1985). The discovery of *Achorion schoenleinii*. *Mycosen*, **28**, 161–82.
286. Semmelweis I. (1861). *Aetiologie, der Begriff und die Prophylaxis des Kindbettfiebers*, 543 pp., translated by K. Codell Carter. University of Wisconsin Press, Madison, 263 pp., 1983.

287. Shiga K. (1898). Über den Dysenteriebacillus (Bacillus dysentericus). *Zentralblatt für Bakteriologie, Parasitenkunde und Infektionskrankheiten*, **23**, 599–600, **24**, 817–28, translated in (179).
288. Shulman ST, Friedman HC, Sims RH. (2007). Theodor Escherich: The first pediatric infectious diseases physician? *Clinical Infectious Diseases*, **45**, 1025–9.
289. Simond PL. (1898). Translated in M Simond, ML Godley, PD Mouriquant. Paul-Louis Simond and his discovery of plague transmission by rat fleas: a centenary. *Journal of the Royal Society of Medicine*, **91**, 101–4, 1998.
290. Smart WR. (1877). On dengue or dandy fever. *British Medical Journal*, **i**, 382–3.
291. Smith W, Andrewes CH, Laidlaw PP. (1933). A virus obtained from influenza patients. *Lancet*, **ii**, 66–8.
292. Snow J. (1849 and 1855). *Snow on cholera, being a reprint of two papers by John Snow, MD, together with a biographical memoir by B.W. Richardson, MD*. London. Humphrey Milford, Oxford University Press, 191 pp., 1936.
293. Solomon HC. (1911). The etiology of trachoma. *Transactions of the American Microscopical Society*, **30**, 41–55.
294. Stanley WM. (1935). Isolation of a crystalline protein possessing the properties of tobacco mosaic virus. *Science*, **81**, 644–54.
295. Stark W. (*c*.1770). *The works of the late Dr William Stark*, compiled by J Carmichael Smith. J Johnson, London, 1788, cited in SL Cummins, Some early British phthisologists, *Proceedings of the Royal Society of Medicine*, **37**, 517–24, 1944 27–5.
296. Steenstrup JJ. (1842). *Om Fortplantning og Udviling geinnem vexlende Generations*. CA Reitzel, Kjøbenhaven, 76 pp., translated in G Busk, *On the alternation of generations*. The Ray Society, London, 132 pp., 1845.
297. Stow J (1598). *A survey of London*, edited by H Morley. George Routledge and Sons, Ludgate Hill, 448 pp., 1890.
298. Sutnick A. (1966). Cited in BS Blumberg, Australia antigen and the biology of hepatitis B. *Science*, **197**, 17–25.
299. Swaminath CS, Shortt HE, Anderson LA. (1942). Transmission of Indian kala-azar to man by the bites of *Phlebotomus argentipes*, Ann. and Brun. *Indian Journal of Medical Research*, **30**, 473–7 and an editorial note.
300. Sydenham T. (*c*.1680). *The works of Thomas Sydenham*, translated from the Latin by RG Latham. The Sydenham Society, London, two volumes, 1848–50.
301. Sylvius F. (1679). *Opera Medica*. Cited and translated in (205).
302. Talley NJ. (1985). The pathogenesis of non-ulcer dyspepsia. *Medical Journal of Australia*, **143**, 173.
303. Tidy HL. (1923). Glandular fever and infective mononucleosis. *Proceedings of the Royal Society of Medicine*, **16**, 70–2.
304. Tyson E. (1663). *Lumbricus latus. Philosophical Transactions of the Royal Society*, **13**, 113–44.
305. Underwood M. (1789). *A treatise on diseases of children*. J Matthews, second edition, two volumes.
306. Vallisnieri A. (1726). *Esperienze ed osservazione intorno all'Origini. Sviluppi e Costumi di varij Insetti*. Padua, cited and translated in (64).

307. Van Arsdale WW. (1886). On the present state of knowledge in bacterial science in its surgical relations. *Annals of Surgery*, **3**, 321–33.
308. Veale H. (1866). History of an epidemic of rotheln, with observations on its pathology. *Edinburgh Medical Journal*, **12**, 404–14.
309. Vianna G de O. (1911). Contribuções para o estudo da anatomia patologica da 'molestia de Carlos Chagas'. *Memórias do Instituto Oswaldo Cruz*, **3**, 276–94, cited and translated in (218).
310. Vieusseux G. (1806). Mémoire sur le maladie qui a régné à Genève au printemps de 1805. *Journal de Médecine, Chirurgie, Pharmacie*, **11**, 163–82, translated in (205).
311. Vigo J de. (1534). *Opera Omnia domini Joannis de Vigo in Chyrugi*. J Moilin, Lyon, translated in (205).
312. Villemin J. (1868). *Études sur la tuberculose*. J-B Baillière et fils, Paris, 640 pp.
313. von Siebold CT. (1844) Parasiten. In: *Handwörterbuch der Physiologie mit Rücksit auf physiologische Pathologie*, edited by R Wagner. Braun Schweig, 2:650–76, translated in (172).
314. von Siebold CT. (1854). *Über die Band- und Blasenwürmer*. W Englemann, Leipzig, 115 pp., translated by TH Huxley. The Sydenham Society, London, 88 pp., 1857.
315. Walker EL, Sellards AW. (1913). Experimental entamebic dysentery. *Philippine Journal of Science*, **8**, 253–330.
316. Warren JR. (1983). Unidentified curved bacillus on gastric epithelium. *Lancet*, **i**, 1273.
317. Webster LT. (1942). *Rabies*. The MacMillan Co, New York, 168 pp.
318. Weller TH. (1996). Varicella: historical perspective and clinical overview. *Journal of Infectious Diseases*, **17** (Supplement 3), 306–9.
319. Wells WC. (1812). Observations on the dropsy, which succeeds scarlet fever. *Transactions of a Society for the Improvement of Medical and Chirurgical Knowledge*, **3**, 167–86.
320. Wichmann JE. (1764). *Aetiologie der Krätze*. Hanover, 1764, cited and translated in (64).
321. Willis T. (1684). *Practice of Physick*. London, cited in RY Keers, *Pulmonary tuberculosis—a journey down the centuries*. Baillière Tindall, London, 265 pp., 1978.
322. Willis T. (1685). *The London practice of physick*. Thomas Basset, London, 672 pp.
323. Wilson E. (1843). Course of lectures on diseases of the skin: their history, pathology and treatment. *Lancet*, **i**, 697–700.
324. Winterbottom TM. (1803). An account of the native Africans in the neighbourhood of Sierra Leone. J Jatchard & J Mawman, London, volume 2, 283 pp.
325. Wohlbach SB. (1925). The rickettsiae and their relationship to disease. *Journal of the American Medical Association*, **84**, 723–8.
326. Wollstein M. (1905). A study of the bacteriology of pertussis, with special reference to the agglutination of the patient's blood. *Journal of Experimental Medicine*, **7**, 335–42.
327. Woodward TE. (1970). The unmasking of typhoid fever. *South African Medical Journal*, **44**, 99–106.

328. Wordsworth, William. Cited in Anonymous, *British Medical Journal*, **i**, 46, 1861.
329. Wucherer OE. (1868). Noticia preliminary sobre vermes de un espécie ainda não descripta. *Gazeta Médica da Bahia*, **3**, 97–9, translated in (157).
330. Wunderlich C. (1868). *Das Verhalten der Eigenwärme in Krankheiten*. Wigande, Leipzig.
331. Yersin A. (1894). La peste bubonique à Hong-Kong. *Annales de l'Institut Pasteur*, **8**, 662–7, translated in (179).
332. Yersin A. (1894). Cited and translated in (179).

GLOSSARY

abortion expulsion from the uterus of an incompletely developed offspring
abscess collection of pus
aetiology (etiology) cause
agar a gelatinous substance extracted from seaweed used for the production of media on which to grow bacteria
agglutination sticking or clumping together
allergy increased sensitivity to a foreign substance
alveolus air sacs in the lungs at the ends of branching breathing tubes
amastigote stage in the development of a trypanosome
amniotic membrane thin tissue that surrounds the amniotic fluid in an embryo
aneurysm swelling in the aorta or an artery
anthelmintic anti-worm drug
antibody a protein produced by the immune system to react with an antigen
antigen a foreign substance introduced into the body which stimulates an immune response
arachnoiditis inflammation of the arachnoid, the middle of three membranes that surround the brain and spinal cord
arthritis inflammation of a joint
arthropod invertebrate animal which has a chitinous exoskeleton and a segmented body to which jointed appendages are articulated in pairs.
ascites fluid in the abdominal (peritoneal) cavity
autoinoculation inoculation of a person with material from a lesion occurring elsewhere in that person
autonomic nerves nerves which control unconscious, involuntary activity of organs like the heart and gut
axon the filamentous process of a nerve cell, often very long, which carries impulses to its target tissue
bacillus rod-shaped bacterium
bacteriophage a virus that infects bacteria
bacterium unicellular organism lacking cytoplasmic organelles and nuclear membrane, usually with a cell wall containing peptidoglycan
balanitis inflammation of the foreskin of the penis
Berkefeld filter a filter made from diatomaceous earth through which viruses but not bacteria can pass
biopsy surgical removal of tissue for examination
bovine of cattle
bronchus branching airway below the trachea
bubo swollen, inflamed lymph node
cadaver dead body
capsomere the outer covering of protein that protects the DNA or RNA of a virus
caseation cheese-like material in a tubercle
cataract clouding of the lens

CD4 lymphocytes particular lymphocytes carrying an antigen on their surface called CD4 which are targeted by the human immunodeficiency virus

cDNA library a collection of DNA sequences generated from (and therefore complementary to) messenger RNA sequences

cell culture complex process by which cells are grown from tissues under controlled conditions and in which viruses will grow

cellulitis inflammation of the fatty tissue beneath the skin

cercaria a larval stage in the development of a schistosome

cerebellum the part of the brain near the base and back of the skull which coordinates muscle activity

cerebrospinal fluid the fluid that bathes the brain and spinal cord

cervicitis inflammation of the cervix

cervix the neck-like narrow passage at the lower end of the uterus

Chamberland filter a filter made from porcelain through which viruses but not bacteria can pass

chancre an ulcer, especially on the genitals

chorea irregular, involuntary movements

chorioallantoic membrane a vascular membrane in the egg of birds and reptiles formed by the apposition of the chorion and allantois and which lines the inner surface of the eggshell

chyluria milky urine

coccus spherical bacterium

colon large intestine not including the distal rectum

colony of bacteria a conglomeration of bacteria formed by the repeated multiplication of a single bacterium on an agar plate

commensal a parasite that lives off its host but does not harm it

conjunctiva the membrane on the surface of the sclera, the white of the eye

consumption tuberculosis

contagion infection by touch

cornea the clear part at the front of the eye

crithidium a stage in the development of a trypanosome (now called an epimastigote) because of their resemblance to protozoa of the genus *Crithidia*

culture of bacteria the growth of bacteria under controlled conditions in the laboratory

cyst a thin-walled body containing fluid

cysticercus a cystic stage in the development of tapeworms

cytopathic effect degenerative changes of cells in cell culture caused by viruses

cytoplasm the part of a cell between the nucleus and the cell membrane

dermatophyte a fungus infecting the skin

diarrhoea frequent, fluid bowel actions

diatom a single-celled alga with a cell wall of silicon

diploid a cell with sets of paired chromosomes, one from each parent

DNA deoxyribonucleic acid, a sequence of nucleotides that contains the genetic instructions used in the development and function of living organisms

DNA polymerase an enzyme that promotes the replication of DNA

duodenum the first part of the small intestine

dura mater the outermost of the three membranes that surround the brain and spinal cord

dysentery diarrhoea with blood and mucus in the stools

electron microscope a microscope that uses electrons to magnify objects beyond that which can be seen by an ordinary light microscope

embryonated an egg that contains an embryo, i.e. a fertilized egg which has begun cell division

encephalitis inflammation of the brain

endarteritis inflammation of the internal lining of arteries (blood vessels)

endocarditis inflammation of the valves of the heart

envelope of a virus an outer structure of many viruses that surrounds the nucleocapsid

enzyme a protein that catalyses or accelerates a biochemical reaction

eosinophil a white blood cell that has pink-staining granules in its cytoplasm

epimastigote a stage in the development of trypanosomes

epithelium the layer of cells that lines the mucous membranes of the mouth, respiratory, gastrointestinal, and urinary tracts

erysipelas inflammation of the dermis, one of the layers of the skin

eschar a scab

eukaryote an organism with cells with a distinct nucleus and organelles in the cytoplasm

excystation the process in protozoa by which a trophozoite emerges from a cyst

facultative anaerobe a bacterium which will grow in the presence or absence of oxygen

fermentation the chemical breakdown of sugars

fibrinous containing fibrin, an elastic, insoluble, whitish protein produced from the protein fibrinogen by the coagulation of the blood

filterable virus an infectious agent small enough to pass through a Berkefeld or Chamberland filter, indicating that it is smaller than a bacterium

flagellum a whip-like appendage to a cell

flea a small, wingless, jumping insect

fomite an inanimate object capable of transferring an infection

fungus a eukaryotic organism of the kingdom Fungi which lacks chlorophyll and vascular tissue and which includes yeasts and moulds

gamete a haploid male or female germ cell that is able to unite with another of the opposite sex to form a zygote.

gametocyte a cell from which gametes develop by meiosis

ganglion a group of nerve cells forming a nerve centre outside the brain and spinal cord

gangrene dead tissue

gastric of the stomach

gastritis inflammation of the stomach

gastroscopy a procedure to view the stomach with a flexible tube

gel a solid jelly-like material used in biochemical and immunological analyses

gelatine a medium for growing colonies of bacteria

general paresis of the insane a disorder of the brain produced by syphilis

genus a grouping of similar species of organisms

gingivitis inflammation of the gums

Gram stain the procedure which colours bacteria either purple (Gram-positive) or pink (Gram-negative)

granuloma an aggregation of white cells

gumma a rubbery tumour produced by syphilis

haemagglutinin an antibody or an antigen (e.g. on a flu virus) that causes red blood cells to clump together
haematuria passage of blood in the urine
haemolysis breakdown of red blood cells
haemophilia an inherited bleeding disorder
haploid a cell with a single set of unpaired chromosomes
helminth worm
hepatitis inflammation of the liver
hermaphrodite an organism which has both male and female sexual organs
heterophile antibodies antibodies that react with antigens of cells from a different species
histological section thin slice of tissue which can be viewed under the microscope
hydrocoele an accumulation of fluid around the testis
hypha a threadlike filament of a fungus
icosahedral twenty sides
ileum the third and last part of the small intestine
immunization the process of making a person or animal resistant to another infectious agent
inclusion body an abnormal structure in the cytoplasm or nucleus of a cell
inflammation a. clinical—an area of swelling, heat, redness, pain, and tenderness. b. microscopical—infiltration of an area with white blood cells and fluid
infusoria an old-fashioned term for a microbe or unicellular organism
insect an arthropod with six legs and 0, 1, or 2 pairs of wings
instar a phase between two moults of an arthropod or crustacean
jaundice yellowing of the eyes and skin
jejunum the second or middle part of the small intestine
kinetoplast a mass of mitochondrial DNA adjacent to the basal body of a flagellum in a trypanosome or similar protozoon
Koch's postulates the criteria necessary for determining whether a particular infectious agent causes a particular disease. See p.193 for details
larva a stage in the development of an invertebrate animal
laryngitis inflammation of the larynx = voice-box
latency the process by which an infectious organism remains dormant but alive and potentially able to cause disease
leucocyte white blood cell
louse a small, wingless, flattened insect
lymphadenopathy enlargement of the lymph nodes = glands
lymphatic vessel which carries lymph fluid to the lymph nodes thence the bloodstream
lymphoblast an immature lymphocyte
lymphocyte a white cell with a single round nucleus which is involved in the generation of antibodies or other components of the immune response
macule a flat, red lesion in the skin
media material for growing infectious agents
meningitis inflammation of the meninges = membranes around the brain and spinal cord
mesentery a double-layered membrane which attaches the small intestine to the back of the abdominal cavity
methylene blue a synthetic dye used as a bacterial stain
miasma an unpleasant or unhealthy smell, vapour, or atmosphere
microfilaria a larval stage of a filarial worm

microfilaraemia microfilariae in the blood
micrometre one-millionth of a metre
microvilli microscopic folds on the surface of an epithelial cell which increase the absorptive area
miracidium a larval stage in the life cycle of a schistosome
mite a minute arthropod with four pairs of legs
mitochondrion a structure found in the cytoplasm of a cell involved in the production of energy, also known as cellular respiration
molecular biology the branch of biology that studies the structure and function of the proteins and nucleic acids essential to life
monocyte a large white blood cell with an ovoid or kidney-shaped nucleus
mould a filamentous fungus
mucous membrane a membrane composed of overlying epithelial cells and underlying connective tissue which lines the mouth, respiratory, gastrointestinal, and urinary tracts
muscardine a fungal disease of silkworms
nagana a trypanosomal infection of livestock in Africa
nanometre one billionth of a metre
nephritis inflammation of the kidneys
neuralgia nerve pain
neuraminidase an enzyme that breaks down substances containing neuraminic acid
neutrophil a white blood cell with several lobes to the nucleus
nit a louse egg, usually attached to a hair or other object
nodule a small lump
nucleocapsid the basic structure of a virus, consisting of a core of nucleic acid enclosed in a protein coat
nucleus the central part of a cell, surrounded by a nuclear membrane, which contains the chromosomes
oedema fluid in the tissues
oesophagus gullet
oocyst a thick-walled zygote of a malarial-type parasite found in mosquitoes
ophthalmia inflammation of the eye
orchitis inflammation of the testis
osteomyelitis inflammation of the bone
otitis inflammation of the ear
ovum egg
pannus (in this book) infiltration of the cornea by blood vessels and inflammatory cells
papule a small, raised, solid swelling in the skin
paralysis inability to move
parasite an organism that lives at the expense of another
parotid gland salivary gland below the ear
pathogen an organism that causes disease
pathogenesis the mechanism by which disease develops
pathogenic disease-causing
PCR polymerase chain reaction—a technique for replicating strands of nucleic acids an enormous number of times
pelvic inflammatory disease inflammation in the female pelvis especially involving the ovaries and Fallopian tubes
peptic ulcer a stomach (gastric) or duodenal ulcer
peptidoglycan a complex chemical that gives firmness to the bacterial cell wall
percutaneous through the skin
peritonitis inflammation of the peritoneal = abdominal cavity

petechia small haemorrhage in the skin
Peyer's patches collections of lymphoid cells in the submucosa of the distal ileum
pharyngitis inflammation of the pharynx at the back of the throat
phthisis tuberculosis
plaque assay a technique for counting the numbers of viruses
plasma blood from which red and white blood cells and platelets have been removed
platelet a small cell fragment without a nucleus in the blood
pleura the lining of the cavity between the lungs and chest wall
pleural effusion fluid in the pleural cavity
pneumonia inflammation of the lungs
polymerase chain reaction see PCR
polymorphonuclear leucocyte see neutrophil
porcelain a hard, white, translucent ceramic made by firing pure kaolin clay
portal vein the vein that carries blood from the intestines to the liver
prion when pathogenic, a protein of abnormal shape that causes degeneration in the brain
proctitis inflammation of the rectum
proglottid an individual bit of a tapeworm
prokaryote a unicellular organism without a nuclear membrane or organelles in the cytoplasm
prostration marked weakness
protozoon a single-celled animal
pseudomembrane a conglomeration of inflammatory exudate and dead tissue that sticks together to look like a membrane
pseudopod an extension from an amoeba that permits it to move
puerperal at childbirth
pulmonary of the lungs
pupa the stage in the life cycle of an insect between a larva and an adult
pus a thick yellowish or greenish liquid produced in infected tissue
pustule a small blister or pimple in the skin containing pus
QED quod erat demonstrandum, Latin for thus it has been shown
reverse transcriptase an enzyme that generates a strand of DNA from RNA
ribosome a minute structure in the cytoplasm where proteins are synthesized
RNA ribonucleic acid—a sequence of nucleotides that is coded for by DNA
saprophyte an organism that lives on dead or decaying matter
schizont a multinucleate stage in the development of malarial parasites
Schwann cell a cell that surrounds a nerve fibre
sciatic nerve the large nerves that pass on each side from the back to supply the legs
sclera the structure that forms the whites of the eyes
scolex head of a tapeworm
septicaemia bacteria in the bloodstream that causes fever and other symptoms
septum a partition—in this book, in fungal hyphae
serum the fluid left after blood has clotted
species a group of similar organisms capable of interbreeding
spirochaete a spiral bacterium
spleen an organ in the upper left part of the abdominal cavity involved in the protection against bacterial infections but one can live without it
spore a one-celled, reproductive unit capable of giving rise to a new

individual, sometimes with a wall around it to protect it from the environment

sporocyst a stage in the life cycle of schistosomes

sporozoite a motile, spore-like stage in the life cycle of malarial parasites

stomatitis inflammation of the mouth

submucosa a layer of loose connective tissue beneath the mucous membrane

T cells a subgroup of lymphocytes involved in the cell-mediated immune reaction

tabes dorsalis disease of the spinal cord produced by syphilis

thallus tangled mass of fungal filaments

thrush infection with *Candida*

tinea infection of the skin with a fungus

tonsillitis inflammation of the tonsils

toxin a poison

trachea the windpipe

tracheotomy an operation to make an opening in the trachea

triatomine a family of biting bugs involved in the transmission of Chagas disease

trismus tightening of the muscles around the jaw; lockjaw

trophozoite a stage in the life cycle of protozoa

trypanosome a protozoon with a flagellum

trypomastigote a stage in the life cycle of a trypanosome

tubercle a collection of white cells and other cells, often with central caseation, seen in tuberculosis

ulcer a break in the skin

urethritis inflammation of the urethra

vaccination an old-fashioned term for immunization

vacuole a cavity or inclusion in the cytoplasm filled with fluid, food, or waste

variolation deliberate infection of the skin with a mild form of the smallpox virus

virulence disease-causing capacity

virus an infective agent with only DNA or RNA which must multiply inside cells

Wasserman reaction a blood test for the diagnosis of syphilis

worm an invertebrate animal with a long, slender, soft body and no limbs

yeast a unicellular fungus

yolk sac A membranous sac containing yolk attached to the embryos of reptiles and birds

zoonosis infection of a human acquired from an animal

zymodeme a variant of an enzyme with a different amino acid sequence that can be distinguished by various biochemical techniques

NOTES ON PRONUNCIATION

Many of the words in this book will look foreign to you and you may be uncertain how to pronounce them. There is no right way and no wrong way to pronounce many of these words. Some words are pronounced differently by speakers in different parts of the world. Some words are pronounced differently by different speakers in the same part of the world. The same speaker may even change his or her pronunciation with the passage of time.

Take *Wuchereria bancrofti* and *Schistosoma mansoni* as examples. When I first learnt about these organisms at a parasitology course in Sydney in 1968 my teacher pronounced the species names as bancroft-ee and manson-ee. This made eminent sense to me. The Latin ending 'i' means 'of' when attached to the end of a person's name or the root of a word. For the etymologically minded it is the second declension, singular, genitive. When I learnt Latin as a schoolboy, the example we used was amici—pronounced amic-ee—'of a friend'. So, I followed my teacher's example. When I went to Cleveland, Ohio, in the United States, to work in a parasitology laboratory, the main subject of investigation was *Schistosoma mansoni*. Everyone pronounced it manson-eye, so to make myself understood, I followed suit. Then I introduced *Wuchereria bancrofti* into that same laboratory. What was I to call it? In the interests of consistency I pronounced it bancroft-eye. But now, 35 years later and back in Australia, I have reverted to my roots and say bancroft-ee.

What about the genus names? If you think of schizophrenia (which is universally pronounced 'skitzophrenia'), you might reasonably assume that we are dealing with 'skistosoma'. But I have never heard anybody pronounce it other than 'shistosoma'. *Wuchereria* is even worse. I first learnt it as 'Wook (as in look) -er-ear-ia'. But years later I discovered that it was named after a Brazilian doctor who had a German father named Wucherer. Perhaps it should be pronounced Vuch (ch as in loch) –ereria.

So much for technical names. What about common ones? Have a look at p.79 for the pronunciation of 'fungi'. Should antennae be pronounced antenn-eye, antenn-ee, or anntenay (as in hay)? Reverting to Latin again, it is the first declension, nominative, plural which I learnt as 'eye' as in agricolae (the farmers). So perhaps it should be antenn-eye. But almost everyone says antenn-ee. And what about 'antibiotic'? Those of us raised in the British tradition will say antee-by-ot (as in hot) -ic whereas an American may well utter an-tie-bee-oh-tic!

So it is all a matter of opinion and usage varies enormously. Look at a word, think how it sounds to you, and say whatever rolls off your tongue most easily. Who can say you are wrong?

FURTHER READING

You may be wondering what the diseases look like in people suffering from these various infections. Unfortunately, space has precluded including them in this book. Fear not, help is at hand. All you need do is go to www.google.com or its subsidiary in your country, click on 'Images' then type in the name of the disease or the causative organism and you will be immediately presented with a plethora of photographs.

If you would rather have a book with photographs, there are many to choose from including:

Cox FE. (1996). *The Wellcome Trust illustrated history of tropical diseases,* The Wellcome Trust, London, 452 pp.

Dobson M. (2008). *The extraordinary stories behind history's deadliest killers,* Quercus Publishing Plc, London, 255 pp.

Edmond RT, Welsby PD, Rowland HA. (2003). *Color atlas of infectious diseases,* 4th edition, Mosby, Edinburgh.

Lambert HP, Farrar W. (1982). *Infectious diseases illustrated: an integrated text and color atlas,* WB Saunders Co, Philadelphia, 297 pp.

Peters W, Pasvol G. (2007). *Atlas of tropical medicine and parasitology,* 6th edition, Elsevier-Mosby, Edinburgh, 429 pp.

If you are looking for definitive, authoritative information on a specific infection, then I recommend the following:

Mandell GJ, Bennet JE, Dolin R. (2010). *Mandell, Douglas, and Bennett's principles and practice of infectious diseases,* Churchill Livingstone/Elsevier, Philadelphia, 2 volumes, 4320 pp.

Warrell DA, Cox TM, Firth JD, Török E (editors) (2012). *Oxford textbook of medicine: infection,* Oxford University Press, Oxford, 864 pp.

The following books review the history of either groups of infections or specific infections:

Beveridge WI. (1977). *Influenza: the last great plague.* Heinemann, London, 124 pp.

Brock TD. (1961). *Milestones in microbiology.* Prentice Hall, Englewood Cliffs, New Jersey, 276 pp.

Brock TD. (1988). *Robert Koch, a life in medicine and bacteriology.* Science Tech Publishers, Madison, Wisconsin, 364 pp.

Busvine JR. (1976). *Insects, hygiene and history.* Athlone Press, 262 pp.

Descours L. (1928). *Pasteur and his work*, translated by AF and BH Wedd. Fisher and Unwin, London, 256 pp.

Duclaux E. (1896). *Pasteur: the history of a mind*, translated by Erwin F. Smith and Florence Hedges. Scarecrow Reprint Corp, Methuen, New Jersey, 363 pp., reprinted 1973.

Grove DI. (1990). *A History of human helminthology*. CAB International, Wallingford, pp. 848. Free download from <http://www.users.on.net/~david.grove/BOOK.PDF>.

Kean BH, Mott JE, Russell AJ. (1978). *Tropical medicine and parasitology: classic investigations*. Cornell University Press, Ithaca, 677 pp.

Keers RY (1978). *Pulmonary tuberculosis—a journey down the centuries*, Baillière Tindall, London, 265 pp.

Kiple KF (editor). (1993). *The Cambridge world history of human disease*, Cambridge University Press, Cambridge, 1176 pp.

Kohn GC. (2007). *The encyclopaedia of plague and pestilence: from ancient times to the present*, Infobase Publishing, New York, 529 pp.

Lechevalier HA, Solotorovsky, M. (1974). *Three centuries of microbiology*. Dover Publications, New York, 536 pp.

Major RH. (1948). *Classic descriptions of disease*. Blackwell, Oxford, 679 pp.

McNeill WH. (1976). *Plagues and people*, Doubleday, Garden City, New York, 340 pp.

Oldstone, MBA. (2009). *Viruses, plagues and history: past, present and future*. Oxford University Press, New York, 400 pp.

Quétel C. (1990). *History of syphilis*, translated by J Braddock and B Pike, Polity Press, Cambridge, 342 pp.

Rhodes R. (1997). *Deadly feasts: science and the hunt for answers in the CJD crisis*. Touchstone Books, London, 278 pp.

Rothman D, Rothman S. (1984). *The Willowbrook wars*. HarperCollins, Cambridge, 405 pp.

FIGURE CREDITS

1. Author's portfolio
2. From F.H. Garrison, *An Introduction to the History of Medicine*, W.B. Saunders Company, Philadelphia, pp.996, 1292. Figure on p.256
3. Modified from K.D. Chatterjee, *Parasitology (protozoology and helminthology)*, sixth edition, Calcutta, pp.218, 1967
4. Reproduced with permission from Professor Marshall Lightowlers, Victoria, Australia and Dr. David Jenkins, NSW, Australia
5. Modified from R. Leuckart: *Die Parasiten des Menschen*, CD Winter'sche Verlagshandlung, Leipzig, volume 1, 1886
6. Modified from K.D. Chatterjee, *Parasitology (protozoology and helminthology)*, sixth edition, Calcutta, pp.218, 1967
7. Modified from K.D. Chatterjee, *Parasitology (protozoology and helminthology)*, sixth edition, Calcutta, pp.218, 1967
8. Author's portfolio. From D.I. Grove, *A history of human helminthology*, CABI, Wallingford, pp.848 + plates, 1990
9. Author's portfolio. From D.I. Grove, *A history of human helminthology*, CABI, Wallingford, pp.848 + plates, 1990
10. Modified from K.D. Chatterjee, *Parasitology (protozoology and helminthology)*, sixth edition, Calcutta, pp.218, 1967
11. Author's portfolio. From D.I. Grove, *A history of human helminthology*, CABI, Wallingford, pp.848 + plates, 1990
12. Author's portfolio
13. Author's portfolio. From D.I. Grove, *A history of human helminthology*, CABI, Wallingford, pp.848 + plates, 1990
14. Author's portfolio. From D.I. Grove, *A history of human helminthology*, CABI, Wallingford, pp.848 + plates, 1990
15. Author's portfolio
16. Modified from K.D. Chatterjee, *Parasitology (protozoology and helminthology)*, sixth edition, Calcutta, pp.218, 1967
17. Author's portfolio. From D.I. Grove, *A history of human helminthology*, CABI, Wallingford, pp.848 + plates, 1990
18. Author's portfolio
19. Photo: Centres for Disease Control and Prevention (CDC) Public Health Image Library
20. Author's portfolio. From D.I. Grove, *A history of human helminthology*, CABI, Wallingford, pp.848 + plates, 1990
21. Author's portfolio. From D.I. Grove, *A history of human helminthology*, CABI, Wallingford, pp.848 + plates, 1990
22. Author's portfolio. From D.I. Grove, *A history of human helminthology*, CABI, Wallingford, pp.848 + plates, 1990
23. Modified from K.D. Chatterjee, *Parasitology (protozoology and helminthology)*, sixth edition, Calcutta, pp.218, 1967
24. Author's portfolio. From D.I. Grove, *A history of human helminthology*, CABI, Wallingford, pp.848 + plates, 1990
25. Photo: Centres for Disease Control and Prevention (CDC) Public Health Image Library

26. Author's portfolio. From D.I. Grove, *A history of human helminthology*, CABI, Wallingford, pp.848 + plates, 1990
27. Photo: Centres for Disease Control and Prevention (CDC) Public Health Image Library/Dr. Mae Melvin
28. Author's portfolio. From D.I. Grove, *A history of human helminthology*, CABI, Wallingford, pp.848 + plates, 1990
29. Author's portfolio
30. Author's portfolio. From D.I. Grove, *A history of human helminthology*, CABI, Wallingford, pp.848 + plates, 1990
31. Modified from K.D. Chatterjee, *Parasitology (protozoology and helminthology)*, sixth edition, Calcutta, pp.218, 1967
32. Photo: Centres for Disease Control and Prevention (CDC) Public Health Image Library/ Reed & Carnrick Pharmaceuticals
33. From: Hooke, Robert (1635–1703). *Micrographia: or some physiological descriptions of minute bodies made by magnifying glasses: with observations and inquiries thereupon.* (MDCLXVII [1667])/ Wikimedia
34. Author's portfolio
35. US Centers for Disease Control, Public Health Image Library Image Number 4077, CDC/W.H.O. (World Health Organisation)
36. Photo: Centres for Disease Control and Prevention (CDC) Public Health Image Library/Donated by the World Health Organisation, Geneva, Switzerland
37. From R.M. Diaz Diaz, *History of acarus scabiei*, Piel 18:471–473, 2003
38. From K. Mellanby, *Scabies*, Oxford University Press, Oxford, pp. 81, 1943, Fig.9
39. The Bodleian Library, University of Oxford, Shelfmark 1544 d. 42, pp. 8, fig. 3, K. Mellanby, *Scabies*, Oxford University Press, Oxford, pp. 81, 1943, Fig.3
40. Wellcome Library, London
41. From H.P. Schmeidebach, N Kampe, Robert Remak (1815–1865), *A Case Study in Jewish Emancipation in the Mid-19th Century German Community*, The Leo Baeck Institute Yearbook 34:95–129, 1989
42. From R.H. Major, *Classic descriptions of disease*, Blackwell, Oxford, p.226, 1948
43. From H.P. Seeliger, *The discovery of Achorion shenleinii*, Mykosen 28:161–182, 1985
44. From G.C. Ainsworth, *An Introduction to the History of Mycology*, Cambridge University Press, Cambridge, pp.359, 1976, Fig.72 Unattributed
45. From M. Knoke and H. Bernhardt, *The First Description of An Oesophageal Candidosis* by Bernhard von Langenbeck,1839, Mycoses 49:283–287, 2006
46. From J.A. Barnett, A History of Research on Yeasts 12, *Yeast* 25:385–417, 2008
47. From J.A. Barnett, A History of Research on Yeasts 12, *Yeast* 25:385–417, 2008
48. The Bodleian Library, University of Oxford, Shelfmark 19113 d. 105, plt. 1 fig. 3, Robin C. 1853. *Histoire Naturelle des Végétaux Parasites qui Croissent sur l'Homme et sur les Animaux Vivants*. Atlas de 15 Planches. Baillière: Paris
49. From J.A. Barnett, A History of Research on Yeasts 12, *Yeast* 25:385–417, 2008
50. The Wellcome Library, London
51. Portrait of Antony van Leeuwenhoek, 1686, by Jan Verkolje/ http://ihm.nlm.nih.gov/images/B16786/Wikimedia
52. Public Domain/Wikimedia Commons http://en.wikipedia.org/wiki/Vilém_Dušan_Lambl
53. From B.H. Kean, K.E. Mott, A.J. Russell, *Tropical Medicine & Parasitology: Classic Investigations*, Cornell University Press, pp. 677, 1978, Part of Plate 22
54. Public Domain/Wikipedia Commons http://en.wikipedia.org/wiki/File:Alfred_Giard_1846-1908.jpg
55. From A. Martínez-Palomo (editor), *Amebiasis*, Elsevier, Amsterdam, pp.269, 1986, Fig.1. Chapter by M Martínez-Báez
56. From R. Leuckart: *Die Parasiten des Menschen*, CD Winter'sche Verlagshandlung, Leipzig, volume 1, 1886, Fig.94
57. From B.H. Kean, K.E. Mott, A.J. Russell, *Tropical Medicine & Parasitology: Classic Investigations*, Cornell University Press, pp. 677, 1978, Plate 14
58. Reprinted from B.H. Kean, K.E. Mott, and A.J. Russell, *Tropical Medicine & Parasitology: Classic Investigations*. Copyright © 1978 by Cornell University Press. Used by permission of the publisher, Cornell University Press

59. From A. Martínez-Palomo (editor), *Amebiasis*, Elsevier, Amsterdam, pp.269, 1986, Fig.2. Chapter by M Martínez-Báez
60. From B.H. Kean, K.E. Mott, A.J. Russell, *Tropical Medicine & Parasitology: Classic Investigations*, Cornell University Press, pp. 677, 1978, Plate 19
61. Reprinted from J. Felsen, Bacillary Dysentery, Colitis and Enteritis, W.B. Sunders Co, Philadelphia, pp. 618, 1945, Fig.16
62. From The National Library of Medicine/Public Domain http://wwwihm.nlm.nih.gov/ihm/images/B/14/555.jpg
63. Reprinted from W.H. Wernsdorfer, I. MacGregor, *Principles and Practice of Malariology*, Churchill Livingstone, Edinburgh, volume 1, 1988. Composite of Figures 25.2 and 25.4; Chapter by T. Harinasuta and D. Bunnag
64. From W. Schreiber, *Infectio—Infectious Diseases in the History of Medicine*, Editiones Roche, Basle, pp.232, 1987. Figure on p.218. Unattributed
65. From B.H. Kean, K.E. Mott, A.J. Russell, *Tropical Medicine & Parasitology: Classic Investigations*, Cornell University Press, pp. 677, 1978, Plate 5
66. Public Domain/Wikimedia Commons http://en.wikipedia.org/wiki/File:Ettore_Marchiafava.jpg
67. Public Domain/Wikimedia Commons http://commons.wikimedia.org/wiki/File:Camillo_Golgi.jpg
68. Reprinted from W.H. Wernsdorfer, I. MacGregor, *Principles and Practice of Malariology*, Churchill Livingstone, Edinburgh, volume 1, 1988. Figure 25.6; Chapter by T. Harinasuta and D. Bunnag
69. From B.H. Kean, K.E. Mott, A.J. Russell, *Tropical Medicine & Parasitology: Classic Investigations*, Cornell University Press, pp. 677, 1978, Plate 7
70. Reprinted from W.H. Wernsdorfer, I. MacGregor, *Principles and Practice of Malariology*, Churchill Livingstone, Edinburgh, volume 1, 1988. Figures 1.2; Chapter by L.J. Bruce-Chwatt
71. From B.H. Kean, K.E. Mott, A.J. Russell, *Tropical Medicine & Parasitology: Classic Investigations*, Cornell University Press, pp. 677, 1978, Plate 6
72. From W. Schreiber, *Infectio—Infectious Diseases in the History of Medicine*, Editiones Roche, Basle, pp.232, 1987. Figure on p.219. Unattributed
73. From B.H. Kean, K.E. Mott, A.J. Russell, *Tropical Medicine & Parasitology: Classic Investigations*, Cornell University Press, pp. 677, 1978, Plate 9
74. From F. Rocco, *The Miraculous Fever Tree*, HarperCollins, New York, pp.348, 2003. Attributed National Library of Medicine (U.S.)
75. Photo: Centres for Disease Control and Prevention (CDC) Public Health Image Library/James Gathany
76. From P.C.C. Garnham et al, *The Pre-Erythrocytic Stage of Plasmodium Ovale*, Transactions of the Royal Society of Tropical Medicine and Hygiene, 49:158–167, 1955. Fig.61955, Elsevier
77. American Society of Tropical Medicine & Hygiene
78. Photo: Centres for Disease Control and Prevention (CDC) Public Health Image Library/Alexander J. da Silva, PhD/Melanie Moser
79. Author's portfolio
80. Author's portfolio
81. From R.M. Forde, *Journal of Tropical Medicine* 5: 261–263, 1903
82. Wellcome Library, London
83. From B.H. Kean, K.E. Mott, A.J. Russell, *Tropical Medicine & Parasitology: Classic Investigations*, Cornell University Press, pp. 677, 1978, Plate 28
84. From Wikimedia Commons http://en.wikipedia.org/wiki/File:Aldo_Castellani's_portrait_by_Spiro_Judor_(1936)
85. From A. Laveran, F. Mesnil, *Trypanosomes et Trypanosomiasis*, Masson et Cie, Paris, second edition, pp.999, 1912, Fig.16
86. Photo: Centres for Disease Control and Prevention (CDC) Public Health Image Library/Alexander J. da Silva, PhD/Melanie Moser
87. Wellcome Library, London
88. From B.H. Kean, K.E. Mott, A.J. Russell, *Tropical Medicine & Parasitology: Classic Investigations*, Cornell University Press, pp. 677, 1978, Plate 34 Fig.5
89. From C.A. Hoare, Early Discoveries Regarding the Parasite of Oriental Sore, *Transactions of the Royal Society of Tropical Medicine and Hygiene* 32:67–92, 1938, Plate 1

90. Public Domain/The National Library of Medicine/Wikimedia Commons http://commons.wikimedia.org/wiki/File:William_Boog_Leishman.jpg
91. Public Domain
92. Public Domain/Wikimedia Commons http://en.wikipedia.org/wiki/File:Charles_Nicolle_at_microscope.jpg
93. From A. Laveran, F. Mesnil, *Trypanosomes et Trypanosomiasis*, Masson et Cie, Paris, second edition, pp.999, 1912, Figure 20
94. Photo: Centres for Disease Control and Prevention (CDC) Public Health Image Library/ Frank Collins
95. Photo: Centres for Disease Control and Prevention (CDC) Public Health Image Library/ Alexander J. da Silva, PhD/Blaine Mathison
96. Photo: Centres for Disease Control and Prevention (CDC) Public Health Image Library/World Health Organization
97. From M.A. Miles, *The Discovery of Chagas Disease*, Medical Clinics of North America 18:247–260, 2004, Fig. 1. Attributed to the Archive of the Casa Oswaldo Cruz
98. Public Domain/Wikimedia Commons http://en.wikipedia.org/wiki/File: Oswaldo_cruz.jpg
99. Photo: Centres for Disease Control and Prevention (CDC) Public Health Image Library/Dr. Mae Melvin
100. Reprinted from B.H. Kean, K.E. Mott, and A.J. Russell, *Tropical Medicine & Parasitology*: Classic Investigations. Copyright © 1978 by Cornell University Press. Used by permission of the publisher, Cornell University Press
101. Photo: Centres for Disease Control and Prevention (CDC) Public Health Image Library/ Alexander J. da Silva, PhD/Melanie Moser
102. From W. Schreiber, *Infectio—Infectious Diseases in the History of Medicine*, Editiones Roche, Basle, pp.232, 1987. Figure on bottom of p.61. Unattributed
103. From W. Bulloch, *The History of Bacteriology*, Oxford University Press, Oxford, pp.422, 1938 and 1960
104. Wellcome Library, London
105. From W.J. Sinclair, *Semmelweis: His Life and Doctrine*, University Press, Manchester, pp.369, 1909. Frontispiece
106. Author's portfolio
107. From G.L. Geison, *The Private Science of Louis Pasteur*, Princeton University Press, Princeton NJ, pp.378, 1995 Fig 5, Musėe Pasteur Paris
108. Wellcome Library, London
109. From W. Bulloch: *The History of Bacteriology*, Oxford University Press, Oxford, pp.422, 1938 and 1960, Fig.14
110. From K. Walker, *Joseph Lister*, Hutchinson, London, pp.195, 1956. Frontispiece
111. From W.D. Foster, *A History of Medical Bacteriology and Immunology*, William Heinemann Medical Books, pp.227, 1970, Fig.7 attributed: Crookshank, 1886
112. From T.D. Brock, *Robert Koch: A Life in Medicine and Bacteriology*, Science Tech Publishers, Madison, Wisconsin, Fig.11.3, Not attributed
113. From P. Baldry, *The Battle Against Bacteria*, Cambridge University Press, second edition, pp.179, 1976, Fig.17
114. Author's portfolio. From D.I. Grove, *A History of Human Helminthology*, pp.848, 1990
115. From W. Bulloch: The History of Bacteriology, Oxford University Press, Oxford, pp.422, 1938 and 1960, Fig 26
116. From R.H. Major, *Classic Descriptions of Disease*, Blackwell, Oxford, p.59, 1948
117. Author's portfolio
118. Author's portfolio
119. From S.L. Cummins, *Tuberculosis in History*, Baillière, Tindall & Cox, pp.205, 1949, Plate facing 112
120. Author's portfolio
121. From S.L. Cummins, *Tuberculosis in History*, Baillière, Tindall & Cox, pp.205, 1949, Plate facing 136
122. From S.L. Cummins, *Tuberculosis in History*, Baillière, Tindall & Cox, pp.205, 1949, Plate facing 176

123. From T.D. Brock, *Robert Koch: A Life in Medicine and Bacteriology*, Science Tech Publishers, Madison, Wisconsin, Fig.141c, Not attributed
124. Public Domain/Wikimedia Commons http://en.wikipedia.org/wiki/File:Daniel_Cornelius_Danielssen.jpg
125. Public Domain/Wikimedia Commons http://commons.wikimedia.org/wiki/File:Gerhard_Henrik_Armauer_Hansen.jpg
126. From G.A. Hansen, C. Looft, *Leprosy in its Clinical and Pathological Effects*, translated by N. Walker, John Wright & Co, Bristol, pp.162, 1895, Plate 6, Fig.1
127. From G.A. Hansen, C. Looft, *Leprosy in its Clinical and Pathological Effects*, translated by N. Walker, John Wright & Co, Bristol, pp.162, 1895, Plate 11, Fig. 2
128. Wellcome Library, London
129. Author's portfolio
130. From http://clendening.kumc.edu/dc/pc/garre02.jpg
131. Author's portfolio
132. Wikimedia/Google Books
133. The Bodleian Library, University of Oxford, Reprinted from G Payling Wright, *An Introduction to Pathology*, Longmans Green & Co, London, pp.660, 1958 Fig.16.1d
134. Author's portfolio
135. Authors' portfolio
136. From R.H. Major, *Classic Descriptions of Disease*, Blackwell, Oxford, p.43, 1948
137. Authors' portfolio
138. From W. Schreiber, *Infectio—Infectious Diseases in the History of Medicine*, Editiones Roche, Basle, pp.232, 1987. Figure on p.76. Unattributed
139. From W.A. Pusey, *The History and Epidemiology of Syphilis*, Charles C. Thomas, Springfield, Illinois, pp.113, 1933, p.54
140. Wellcome Library, London
141. From W.A. Pusey, *The History and Epidemiology of Syphilis*, Charles C. Thomas, Springfield, Illinois, pp.113, 1933, p.64
142. Mary Evans Picture Library
143. Public Domain/The National Library of Medicine/Wikimedia Commons http://en.wikipedia.org/wiki/File:Anton_Weichselbaum.jpg
144. From W. Schreiber, *Infectio—Infectious Diseases in the History of Medicine*, Editiones Roche, Basle, pp.232, 1987. Figure on p.178. Unattributed
145. From W.A. Pusey, *The History and Epidemiology of Syphilis*, Charles C. Thomas, Springfield, Illinois, pp.113, 1933, p.71
146. Photo: Centres for Disease Control and Prevention (CDC) Public Health Image Library
147. From *Robert Koch: A Life in Medicine and Bacteriology*, Science Tech Publishers, Madison, Wisconsin, fig.10.1, Not attributed
148. From http://images.nobelprize.org/nobel_prizes/medicine/laureates/1919/bordet_postcard.jpg
149. Wellcome Library, London
150. From N. Longmate: *King Cholera*, Hamish Hamilton, London, pp 271, 1966, lower image facing p.21
151. From N. Longmate: *King Cholera*, Hamish Hamilton, London, pp 271, 1966 facing p.117
152. Authors' portfolio
153. From J. Snow, *On the Mode of Communication of Cholera*, John Churchill, London, second edition, 1855
154. From J. Snow, *On the Mode of Communication of Cholera*, John Churchill, London, second edition, 1855
155. From *Robert Koch: A Life in Medicine and Bacteriology*, Science Tech Publishers, Madison, Wisconsin, fig.15.2, Not attributed
156. Public Domain/Wikimedia Commons http://es.wikipedia.org/wiki/Archivo: Filippo_Pacini.jpg
157. The Bodleian Library, University of Oxford, reprinted from G Payling Wright, *An Introduction to Pathology*, Longmans Green & Co, London, pp.660, 1958 Fig.16.1a
158. Public Domain/Wikimedia Commons http://en.wikipedia.org/wiki/File:Pierre-Charles_Alexandre_Louis.jpg
159. Public Domain/Photograph Collection, Archives and Special Collections, Dickinson College, Carlisle, PA http://hd.housedivided.dickinson.edu/node/2370

160. From The National Library of Medicine/Public Domain http://www.nlm.nih.gov/exhibition/cholera/images/b03865.jpg
161. Roger McFadden
162. From S.T. Shulman et al, Theodor Escherich, The First Pediatric Infectious Diseases Physician? *Clinical Infectious Diseases* 45:1025–1029, 2007
163. Wellcome Library, London
164. Photo: Centres for Disease Control and Prevention (CDC) Public Health Image Library
165. The Bodleian Library, University of Oxford, shelfmark 1561 d. 56, p. frontispiece, reprinted from J Felsen, *Bacillary dysentery, colitis and enteritis*, WB Sunders Co, Philadelphia, pp. 618, 1945 Frontispiece
166. Public Domain/Wikimedia Commons http://upload.wikimedia.org/wikipedia/commons/d/d8/Opisthotonus_in_a_patient_suffering_from_tetanus_-_Painting_by_Sir_Charles_Bell_-_1809.jpg
167. Public Domain
168. Photo: Centres for Disease Control and Prevention (CDC) Public Health Image Library
169. From W. Schreiber, *Infectio—Infectious Diseases in the History of Medicine*, Editiones Roche, Basle, pp.232, 1987. Figure on p.10. Unattributed
170. From Anonymous, The Plague at Hong Kong, *Lancet* ii:325, 1895
171. Wellcome Library, London
172. The Bodleian Library, University of Oxford, reprinted from TH Davey, WPH Lightbody, *Control of disease in the tropics*, HK Lewis & Co, London, pp.408, 1956. Figure 37c
173. From Wikimedia Commons http://upload.wikimedia.org/wikipedia/commons/thumb/e/e8/Davidbruce.JPG/461px-Davidbruce.JPG
174. From M.L. Hughes, *Mediterranean, Malta or Undulant Fever*, MacMillan and Cop, London, pp.231, 1897. Frontispiece
175. The Bodleian Library, University of Oxford, shelfmark 1563 e. 25, p. 143, frontispiece, reprinted from ML Hughes, Mediterranean, Malta or undulant fever, MacMillan and Cop, London, pp.231, 1897. Frontispiece
176. From M.L. Hughes, *Mediterranean, Malta or Undulant Fever*, MacMillan and Cop, London, pp.231, 1897. Chart VI, upper panel
177. Public Domain/From Wikimedia Commons http://sl.wikipedia.org/wiki/Slika:Bernhard_Bang_2.jpg
178. Photo: Centres for Disease Control and Prevention (CDC) Public Health Image Library
179. Photo: Centres for Disease Control and Prevention (CDC) Public Health Image Library/Dr. Joseph McDade
180. By courtesy and permission of Trevor Steele
181. Author's portfolio
182. C. Northcott/ http://images.nobelprize.org/nobel_prizes/medicine/laureates/2005/marshall_postcard.jpg
183. Author's portfolio
184. Author's portfolio
185. From S.B. Wohlbach, J.L. Todd, F.W. Palfrey, *The Etiology and Pathology of Typhus Being the Main Report of the Typhus Research Commission of the League or Red Cross Societies to Poland*, Harvard University Press, Cambridge, pp.222+plates, 1922,. Chart 2, upper panel
186. From D. Gross, G. Schaefer, 100th anniversary of the death of Howard Taylor Ricketts, *Microbes and* Infection, 13(1):10–3, 2010
187. From S. Zia, The Cultivation of Mexican and European Typhus Rickettsiae in the Chorio-Allantoic Membrane of the Chick Embryo, American Journal of Pathology, 10:211–218, 1934. Free full text article in Pubmed Central
188. From L. Jaenicke, Stanislaus von Prowazek (1875–1915)—Prodigy Between Working Bench and Coffee house, Protist 152: 157–166, 2001
189. From L. Halberstedter, S. von Prowazek, Zur Aetiologie des Trachoms Deutsche Klinische Wochenschrift 33:1285, 1907
190. Taylor & Francis Group LLC Books
191. Modified from GL Mandell, JE Bennett, R Dolin, Principles and Practice of Infectious Diseases, fifth edition, pp.3661 , 2000. Fig 177-2 Chapter, Introduction to Chlamydial diseases by WE Stamm, RB Jones, BE Batteiger

192. Author's portfolio
193. Author's portfolio
194. From A.P. Waterson and L. Wilkinson, An Introduction to the History of Virology, Cambridge University Press, 1978. Not attributed
195. From F.R. Lillie, *The Development of the Chick*, Henry Holt & Co., New York, and George Bell & Sons, London, 1908
196. Public Domain
197. From J. Baron, *The Life of Edward Jenner*, frontispiece to volume 1, Henry Colburn Publisher, London, 1838
198. Photo: Centres for Disease Control and Prevention (CDC) Public Health Image Library/ Fred Murphy
199. Photo: Centres for Disease Control and Prevention (CDC) Public Health Image Library/ Dr. Daniel P. Perl
200. Courtesy of Dr F A Murphy
201. From H.A. Kelly, *Walter Reed and Yellow Fever*, The Medical Standard Book Company, Baltimore, second edition, pp.310, 1906. Figure facing p.114
202. From H.A. Kelly, *Walter Reed and Yellow Fever*, The Medical Standard Book Company, Baltimore, second edition, pp.310, 1906. Frontispiece
203. Photo: Centres for Disease Control and Prevention (CDC) Public Health Image Library/ Erskine Palmer, Ph.D.
204. Photo: Centres for Disease Control and Prevention (CDC) Public Health Image Library/ Prof. Frank Hadley Collins, Dir., Cntr. for Global Health and Infectious Diseases, Univ. of Notre Dame
205. Photo: Centres for Disease Control and Prevention (CDC) Public Health Image Library/ Frederick Murphy, Cynthia Goldsmith
206. From T.M. Daniel, F.C. Robbins, *Polio*; University of Rochester Press, Rochester, pp. 202, 1997. Figure 1
207. Author's portfolio
208. Photo: Centres for Disease Control and Prevention (CDC) Public Health Image Library/ Dr. Joseph J. Esposito; F. A. Murphy
209. Photo: Centres for Disease Control and Prevention (CDC) Public Health Image Library/ Dr. Edwin P. Ewing, Jr.
210. Photo: Centres for Disease Control and Prevention (CDC) Public Health Image Library/ Cynthia S. Goldsmith; William Bellini, Ph.D.
211. RPAH Museum & Archives
212. Photo: Centres for Disease Control and Prevention (CDC) Public Health Image Library/ Dr. Fred Murphy; Sylvia Whitfield
213. Photo: Centres for Disease Control and Prevention (CDC) Public Health Image Library/ Courtesy of A. Harrison and F. A. Murphy
214. From *British Medical Journal* 5 July 1902, p.47
215. Photo: Centres for Disease Control and Prevention (CDC) Public Health Image Library/ Dr. Erskine Palmer; B.G. Partin
216. Author's portfolio
217. Public Domain/Wikimedia Commons/Ed Uthman http://en.wikipedia.org/wiki/File:Infectious_Mononucleosis_3.jpg
218. Photo: Centres for Disease Control and Prevention (CDC) Public Health Image Library/ F. A. Murphy
219. Public Domain/Wikimedia Commons/NASA http://upload.wikimedia.org/wikipedia/commons/c/c1/Baruch_Samuel_Blumberg_by_Tom_Trower_%28NASA%29.jpg
220. Photo: Centres for Disease Control and Prevention (CDC) Public Health Image Library
221. Photo: Centres for Disease Control and Prevention (CDC) Public Health Image Library/Betty Partin
222. Author's portfolio
223. With permission from Prof. Barre Barré-Sinoussi © The Institut Pasteur
224. Photo: Centres for Disease Control and Prevention (CDC) Public Health Image Library/ A. Harrison, P. Feorino; E. L. Palmer
225. Photo by Keystone/Hulton Archive/Getty Images

226. Photo: Centres for Disease Control and Prevention (CDC) Public Health Image Library/ Teresa Hammett
227. Courtesy of Simon & Schuster, reprinted from Richard Rhodes, *Deadly feasts: science and the hunt for answers in the CJD crisis*, Touchstone Books, London, 1997 and 1998. Figure on p.185
228. Photo by Raphael Gaillarde/Gamma-Rapho via Getty Images

The Publishers and author apologize for any errors or omissions in the above list. If contacted they will be happy to rectify these at the earliest opportunity.

PERSON INDEX

SUBJECT INDEX